Linear Algebra

Linear Algebra offers a unified treatment of both matrix-oriented and theoretical approaches to the course, which will be useful for classes with a mix of mathematics, physics, engineering, and computer science students. Major topics include singular value decomposition, the spectral theorem, linear systems of equations, vector spaces, linear maps, matrices, eigenvalues and eigenvectors, linear independence, bases, coordinates, dimension, matrix factorizations, inner products, norms, and determinants.

CAMBRIDGE MATHEMATICAL TEXTBOOKS

Cambridge Mathematical Textbooks is a program of undergraduate and beginning graduate level textbooks for core courses, new courses, and interdisciplinary courses in pure and applied mathematics. These texts provide motivation with plenty of exercises of varying difficulty, interesting examples, modern applications, and unique approaches to the material.

ADVISORY BOARD
John B. Conway, *George Washington University*
Gregory F. Lawler, *University of Chicago*
John M. Lee, *University of Washington*
John Meier, *Lafayette College*
Lawrence C. Washington, *University of Maryland, College Park*

A complete list of books in the series can be found at www.cambridge.org/mathematics

Recent titles include the following:

Chance, Strategy, and Choice: An Introduction to the Mathematics of Games and Elections, S. B. Smith
Set Theory: A First Course, D. W. Cunningham
Chaotic Dynamics: Fractals, Tilings, and Substitutions, G. R. Goodson
Introduction to Experimental Mathematics, S. Eilers & R. Johansen
A Second Course in Linear Algebra, S. R. Garcia & R. A. Horn
Exploring Mathematics: An Engaging Introduction to Proof, J. Meier & D. Smith
A First Course in Analysis, J. B. Conway
Introduction to Probability, D. F. Anderson, T. Seppäläinen & B. Valkó
Linear Algebra, E. S. Meckes & M. W. Meckes

Linear Algebra

ELIZABETH S. MECKES
Case Western Reserve University, Cleveland, OH, USA

MARK W. MECKES
Case Western Reserve University, Cleveland, OH, USA

CAMBRIDGE
UNIVERSITY PRESS

CAMBRIDGE
UNIVERSITY PRESS

University Printing House, Cambridge CB2 8BS, United Kingdom

One Liberty Plaza, 20th Floor, New York, NY 10006, USA

477 Williamstown Road, Port Melbourne, VIC 3207, Australia

314–321, 3rd Floor, Plot 3, Splendor Forum, Jasola District Centre, New Delhi – 110025, India

79 Anson Road, #06–04/06, Singapore 079906

Cambridge University Press is part of the University of Cambridge.

It furthers the University's mission by disseminating knowledge in the pursuit of education, learning, and research at the highest international levels of excellence.

www.cambridge.org
Information on this title: www.cambridge.org/9781107177901
DOI: 10.1017/9781316823200

First published 2018

A catalog record for this publication is available from the British Library.

Library of Congress Cataloging-in-Publication Data
Names: Meckes, Elizabeth S., author. | Meckes, Mark W., author.
Title: Linear algebra / Elizabeth S. Meckes (Case Western Reserve University, Cleveland, OH, USA), Mark W. Meckes (Case Western Reserve University, Cleveland, OH, USA).
Description: Cambridge : Cambridge University Press, [2018] | Includes bibliographical references and index.
Identifiers: LCCN 2017053812 | ISBN 9781107177901 (alk. paper)
Subjects: LCSH: Algebras, Linear–Textbooks.
Classification: LCC QA184.2 .M43 2018 | DDC 512/.5–dc23
LC record available at https://lccn.loc.gov/2017053812

ISBN 978-1-107-17790-1 Hardback

To Juliette and Peter

Contents

Preface

It takes some chutzpah to write a linear algebra book. With so many choices already available, one must ask (and our friends and colleagues did): what is new here?

The most important context for the answer to that question is the intended audience. We wrote the book with our own students in mind; our linear algebra course has a rather mixed audience, including majors in mathematics, applied mathematics, and our joint degree in mathematics and physics, as well as students in computer science, physics, and various fields of engineering. Linear algebra will be fundamental to most if not all of them, but they will meet it in different guises; this course is furthermore the only linear algebra course most of them will take.

Most introductory linear algebra books fall into one of two categories: books written in the style of a freshman calculus text and aimed at teaching students to do computations with matrices and column vectors, or full-fledged "theorem–proof" style rigorous math texts, focusing on abstract vector spaces and linear maps, with little or no matrix computation. This book is different. We offer a unified treatment, building both the basics of computation and the abstract theory from the ground up, emphasizing the connections between the matrix-oriented viewpoint and abstract linear algebraic concepts whenever possible. The result serves students better, whether they are heading into theoretical mathematics or towards applications in science and engineering. Applied math students will learn Gaussian elimination and the matrix form of singular value decomposition (SVD), but they will also learn how abstract inner product space theory can tell them about expanding periodic functions in the Fourier basis. Students in theoretical mathematics will learn foundational results about vector spaces and linear maps, but they will also learn that Gaussian elimination can be a useful and elegant theoretical tool.

Key features of this book include:

- **Early introduction of linear maps:** Our perspective is that mathematicians invented vector spaces so that they could talk about linear maps; for this reason, we introduce linear maps as early as possible, immediately after the introduction of vector spaces.

- **Key concepts referred to early and often:** In general, we have introduced topics we see as central (most notably eigenvalues and eigenvectors) as early as we could, coming back to them again and again as we introduce new concepts which connect to these central ideas. At the end of the course, rather than having just learned the definition of an eigenvector a few weeks ago, students will have worked with the concept extensively throughout the term.
- **Eases the transition from calculus to rigorous mathematics:** Moving beyond the more problem-oriented calculus courses is a challenging transition; the book was written with this transition in mind. It is written in an accessible style, and we have given careful thought to the motivation of new ideas and to parsing difficult definitions and results after stating them formally.
- **Builds mathematical maturity:** Over the course of the book, the style evolves from extremely approachable and example-oriented to something more akin to the style of texts for real analysis and abstract algebra, paving the way for future courses in which a basic comfort with mathematical language and rigor is expected.
- **Fully rigorous, but connects to computation and applications:** This book was written for a proof-based linear algebra course, and contains the necessary theoretical foundation of linear algebra. It also connects that theory to matrix computation and geometry as often as possible; for example, SVD is considered abstractly as the existence of special orthonormal bases for a map; from a geometric point of view emphasizing rotations, reflections, and distortions; and from a more computational point of view, as a matrix factorization. Orthogonal projection in inner product spaces is similarly discussed in theoretical, computational, and geometric ways, and is connected with applied minimization problems such as linear least squares for curve-fitting and approximation of smooth functions on intervals by polynomials.
- **Pedagogical features:** There are various special features aimed at helping students learn to read a mathematics text: frequent "Quick Exercises" serve as checkpoints, with answers upside down at the bottom of the page. Each section ends with a list of "Key Ideas," summarizing the main points of the section. Features called "Perspectives" at the end of some chapters collect the various viewpoints on important concepts which have been developed throughout the text.
- **Exercises:** The large selection of problems is a mix of the computational and the theoretical, the straightforward and the challenging. There are answers or hints to selected problems in the back of the book.

The book begins with linear systems of equations over $\mathbb{R}$, solution by Gaussian elimination, and the introduction of the ideas of pivot variables and free variables. Section 1.3 discusses the geometry of $\mathbb{R}^n$ and geometric viewpoints on linear systems. We then move into definitions and examples of abstract fields and vector spaces.

Chapter 2 is on linear maps. They are introduced with many examples; the usual cohort of rotations, reflections, projections, and multiplication by matrices in $\mathbb{R}^n$, and more abstract examples like differential and integral operators on function spaces. Eigenvalues are first introduced in Section 2.1; the representation of arbitrary linear maps on $\mathbb{F}^n$ by matrices is proved in Section 2.2. Section 2.3 introduces matrix multiplication as the matrix representation of composition, with an immediate derivation of the usual formula. In Section 2.5, the range, kernel, and eigenspaces of a linear map are introduced. Finally, Section 2.6 introduces the Hamming code as an application of linear algebra over the field of two elements.

Chapter 3 introduces linear dependence and independence, bases, dimension, and the Rank–Nullity Theorem. Section 3.5 introduces coordinates with respect to arbitrary bases and the representation of maps between abstract vector spaces as matrices; Section 3.6 covers change of basis and introduces the idea of diagonalization and its connection to eigenvalues and eigenvectors. Chapter 3 concludes by showing that all matrices over algebraically closed fields can be triangularized.

Chapter 4 introduces general inner product spaces. It covers orthonormal bases and the Gram–Schmidt algorithm, orthogonal projection with applications to least squares and function approximation, normed spaces in general and the operator norm of linear maps and matrices in particular, isometries, and the QR decomposition.

Chapter 5 covers the singular value decomposition and the spectral theorem. We begin by proving the main theorem on the existence of SVD and the uniqueness of singular values for linear maps, then specialize to the matrix factorization. There is a general introduction to adjoint maps and their properties, followed by the Spectral Theorem in the Hermitian and normal cases. Geometric interpretation of SVD and truncations of SVD as low-rank approximation are discussed in Section 5.2. The four fundamental subspaces associated to a linear map, orthogonality, and the connection to the Rank–Nullity Theorem are discussed in Section 5.3.

Finally, Chapter 6 is on determinants. We have taken the viewpoint that the determinant is best characterized as the unique alternating multilinear form on matrices taking value 1 at the identity; we derive many of its properties from that characterization. We introduce the Laplace expansion, give an algorithm for computing determinants via row operations, and prove the sum over permutations formula. The last is presented as a nice example of the power of linear algebra: there is no long digression on combinatorics, but instead permutations are quickly identified with permutation matrices, and concepts like the sign of a permutation arise naturally as familiar linear algebraic constructions. Section 6.3 introduces the characteristic polynomial and the Cayley–Hamilton Theorem, and Section 6.4 concludes the chapter with applications of the determinant to volume and Cramer's rule.

In terms of student prerequisites, one year of calculus is sufficient. While calculus is not needed for any of the main results, we do rely on it for some examples and exercises (which could nevertheless be omitted). We do not expect students to have taken a rigorous mathematics course before. The book is written assuming some basic background on sets, functions, and the concept of a proof; there is an appendix containing what is needed for the student's reference (or crash course).

Finally, some thanks are in order. To write a textbook that works in the classroom, it helps to have a classroom to try it out in. We are grateful to the CWRU Math 307 students from Fall 2014, Spring and Fall 2015, and Spring 2016 for their roles as cheerful guinea pigs.

A spectacular feature of the internet age is the ability to get help typesetting a book from someone half-way around the world (where it may in fact be 2 in the morning). We thank the users of tex.stackexchange.com for generously and knowledgeably answering every question we came up with.

We began the project of writing this book while on sabbatical at the Institut de Mathématiques de Toulouse at the University of Toulouse, France. We thank the Institut for its warm hospitality and the Simons Foundation for providing sabbatical support. We also thank the National Science Foundation and the Simons Foundation for additional support.

And lastly, many thanks to Sarah Jarosz, whose album *Build Me Up From Bones* provided the soundtrack for the writing of this book.

Elizabeth Meckes
Mark Meckes

Cleveland, Ohio, USA

To the Student

This will be one of the most important classes you ever take. Linear algebra and calculus are the foundations of modern mathematics and its applications; the language and viewpoint of linear algebra is so thoroughly woven into the fabric of mathematical reasoning that experienced mathematicians, scientists, and engineers can forget it is there, in the same way that native speakers of a language seldom think consciously about its formal structure. Achieving this fluency is a big part of that nebulous goal of "mathematical maturity."

In the context of your mathematical education, this book marks an important transition. In it, you will move away from a largely algorithmic, problem-centered viewpoint toward a perspective more consciously grounded in rigorous theoretical mathematics. Making this transition is not easy or immediate, but the rewards of learning to think like a mathematician run deep, no matter what your ultimate career goals are. With that in mind, we wrote this book to be *read* – by you, the student. Reading and learning from an advanced mathematics text book is a skill, and one that we hope this book will help you develop.

There are some specific features of this book aimed at helping you get the most out of it. Throughout the book, you will find "Quick Exercises," whose answers are usually found (upside down) at the bottom of the page. These are exercises which you should be able to do fairly easily, but for which you may need to write a few lines on the back of an envelope. They are meant to serve as checkpoints; do them! The end of each section lists "Key Ideas," summarizing (sometimes slightly informally) the big picture of the section. Certain especially important concepts on which there are many important perspectives are summarized in features called "Perspectives" at the end of some chapters. There is an appendix covering the basics of sets, functions, and complex number arithmetic, together with some formal logic and proof techniques. And of course, there are many exercises. Mathematics isn't something to know, it's something to do; it is through the exercises that you really learn how.

1

Linear Systems and Vector Spaces

1.1 Linear Systems of Equations

Bread, Beer, and Barley

We begin with a very simple example. Suppose you have 20 pounds of raw barley and you plan to turn some of it into bread and some of it into beer. It takes one pound of barley to make a loaf of bread and a quarter pound of barley to make a pint of beer. You could use up all the barley on 20 loaves of bread, or alternatively, on 80 pints of beer (although that's probably not advisable). What are your other options? Before rolling up our sleeves and figuring that out, we make the following obvious-seeming observation, without which rolling up our sleeves won't help much.

> It's very difficult to talk about something that has no name.

That is, before we can *do math* we have to have something concrete to *do math to*. Therefore: let x be the number of loaves of bread you plan to bake and y be the number of pints of beer you want to wash it down with. Then the information above can be expressed as

$$x + \frac{1}{4}y = 20, \tag{1.1}$$

and now we have something real (to a mathematician, anyway) to work with.

This object is called a **linear equation in two variables**. Here are some things to notice about it:

- There are infinitely many solutions to equation (1.1) (assuming you're okay with fractional loaves of bread and fractional pints of beer).
- We only care about the positive solutions, but even so, there are still infinitely many choices. (It's important to notice that our interest in positive solutions is a feature of the real-world situation being modeled, but it's not built into the model itself. We just have to remember what we're doing when interpreting solutions. This caveat may or may not be true of other models.)

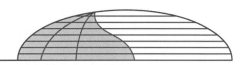

- We could specify how much bread we want and solve for how much beer we can have, or vice versa. Or we could specify a fixed ratio of bread to beer (i.e., fix the value of $c = x/y$) and solve.
- For the graphically inclined, we can draw a picture of all the solutions of this equation in the x–y plane, as follows:

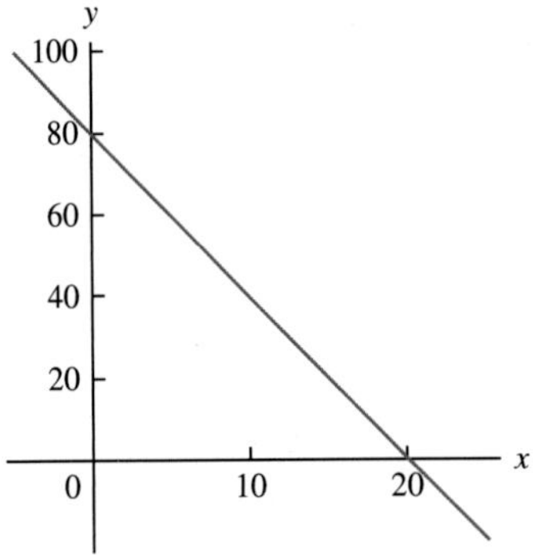

Figure 1.1 Graph of the solutions of equation (1.1).

Each point in the x–y plane corresponds to a quantity of bread and of beer, and if a point lies on the line above, it means that it is possible to exactly use up all of the barley by making those quantities of bread and beer.

We have, of course, drastically oversimplified the situation. For starters, you also need yeast to make both bread and beer. It takes 2 teaspoons yeast to make a loaf of bread and a quarter teaspoon to make a pint of beer. Suppose you have meticulously measured that you have exactly 36 teaspoons of yeast available for your fermentation processes. We now have what's called a **linear system of equations**, as follows:

$$\begin{aligned} x + \frac{1}{4}y &= 20 \\ 2x + \frac{1}{4}y &= 36. \end{aligned} \tag{1.2}$$

You could probably come up with a couple of ways to solve this system. Just to give one, if we subtract the first equation from the second, we get

$$x = 16.$$

If we then plug $x = 16$ into the first equation and solve for y, we get

$$16 + \frac{1}{4}y = 20 \qquad \Longleftrightarrow \qquad y = 16.$$

(The symbol $\Longleftrightarrow$ above is read aloud as "if and only if," and it means that the two equations are **equivalent**; i.e., that any given value of y makes the equation on the left true *if and only if* it makes the equation on the right true. Please learn to use this symbol when appropriate; it is not correct to use an equals sign instead.)

We now say that the **solution** to the linear system (1.2) is $x = 16, y = 16$. In particular, we've discovered that there's exactly one way to use up all the barley and all the yeast to make bread and beer.

Here are some things to notice:

- We can represent the situation modeled by system (1.2) graphically:

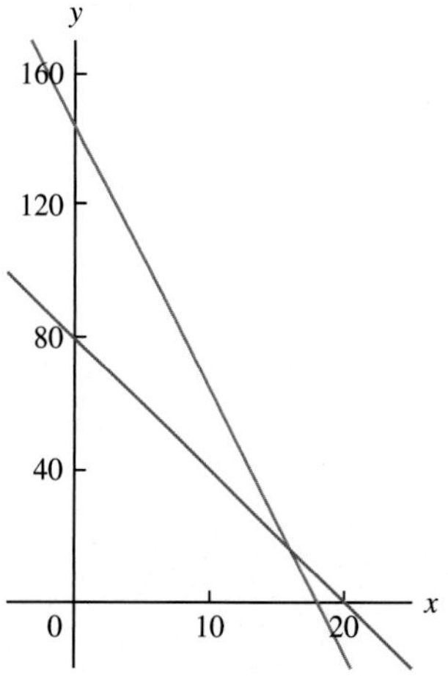

Figure 1.2 Graph of the solutions of the equations in system (1.2): blue for the first equation and red for the second.

In Figure 1.2, each line represents all the solutions to one of the equations in system (1.2). The intersection of these two lines (at the point (16, 16)) represents the unique way of using up all of both your ingredients.

- If the amount of yeast had been different, we might have ended up with a solution that did not have both $x, y > 0$.

Quick Exercise #1. Give a quantity of yeast for which the corresponding system has a solution, but with either $x < 0$ or $y < 0$.

- If you switch to a different bread recipe that only requires 1 teaspoon (tsp) of yeast per loaf, we might instead have infinitely many solutions (as we did before considering the yeast), or none. (How?)

Suppose now that your menu gets more interesting: someone comes along with milk and dried rosemary. Your bread would taste better with a little of each put into the dough; also, you can use the milk to make a simple cheese, which would also be nice if you flavored it with rosemary. The beer might be good flavored with rosemary, too.* Suppose you use 1 cup of milk per loaf of bread, and 8 cups

*What great philosopher of the modern era said "A man can live on packaged food from here 'til Judgment Day if he's got enough rosemary."?

QA #1: Anything less than 20 teaspoons of yeast will result in $x < 0$.

of milk per round of cheese. You put 2 tsps of rosemary in each loaf of bread, 1 tsp in each round of cheese and 1/4 tsp in each pint of beer. Then we have a new variable z – the number of rounds of cheese – and two new equations, one for the milk and one for the rosemary. Suppose you have 11 gallons (i.e., 176 cups) of milk and 56 tsps of rosemary. Our linear system now becomes:

$$\begin{aligned} x + \frac{1}{4}y &= 20 \\ 2x + \frac{1}{4}y &= 36 \\ x + 8z &= 176 \\ 2x + \frac{1}{4}y + z &= 56. \end{aligned} \tag{1.3}$$

If we go ahead and solve the system, we will find that $x = 16, y = 16, z = 20$. Since we've been given the solution, though, it's quicker just to check that it's correct.

Quick Exercise #2. Check that this solution is correct.

In any case, it seems rather lucky that we could solve the system at all; i.e., with four ingredients being divided among three products, we were able to exactly use up everything. A moment's reflection confirms that this was a lucky accident: since it worked out so perfectly, we can see that if you'd had more rosemary but the same amount of everything else, you wouldn't be able to use up everything exactly. This is a first example of a phenomenon we will meet often: *redundancy*. The system of equations (1.3) is redundant: if we call the equations E_1, E_2, E_3, E_4, then

$$E_4 = \left(\frac{1}{8}\right) E_1 + \left(\frac{7}{8}\right) E_2 + \left(\frac{1}{8}\right) E_3.$$

In particular, if (x, y, z) is a solution to all three equations E_1, E_2, and E_3, then it satisfies E_4 automatically. Thus E_4 tells us nothing about the values of (x, y, z) that we couldn't already tell from just the first three equations. The solution $x = 16, y = 16, z = 20$ is in fact the unique solution to the first three equations, and it satisfies the redundant fourth equation for free.

Linear Systems and Solutions

Now we'll start looking at more general systems of equations that resemble the ones that came up above. Later in this book, we will think about various types of numbers, some of which you may never have met. But for now, we will restrict

QA #2: If you did anything more complicated than just plug in the numbers 16, 16, and 20, then you're making it too hard.

our attention to equations involving real numbers and real variables, i.e., unknown real numbers.

Definition The set of all real numbers is denoted by $\mathbb{R}$. The notation $t \in \mathbb{R}$ is read "t is in $\mathbb{R}$" or "t is an element of $\mathbb{R}$" or simply "t is a real number."

Definition A **linear system of m equations in n variables over $\mathbb{R}$** is a set of equations of the form

$$\begin{aligned} a_{11}x_1 + \cdots + a_{1n}x_n &= b_1 \\ &\vdots \\ a_{m1}x_1 + \cdots + a_{mn}x_n &= b_m. \end{aligned} \tag{1.4}$$

Here, $a_{ij} \in \mathbb{R}$ for each pair (i, j) with $1 \leq i \leq m$ and $1 \leq j \leq n$; $b_i \in \mathbb{R}$ for each $1 \leq i \leq m$; and $x_1, \ldots, x_n$ are the n variables of the system. Such a system of equations is also sometimes called an **$m \times n$ linear system.**

A **solution** of the linear system (1.4) is a set of real numbers $c_1, \ldots, c_n$ such that

$$\begin{aligned} a_{11}c_1 + \cdots + a_{1n}c_n &= b_1 \\ &\vdots \\ a_{m1}c_1 + \cdots + a_{mn}c_n &= b_m. \end{aligned}$$

That is, a solution is a list of values that you can plug in for $x_1, \ldots, x_n$ so that the equations in system (1.4) are satisfied.

What makes a system *linear* is that only a couple simple things can be done to the variables: they are multiplied by constants, and added to other variables multiplied by constants. The variables aren't raised to powers, multiplied together, or plugged into complicated functions like logarithms or cosines.

A crucial thing to notice about the definition above is that a solution is defined to be "something that works," and *not* "something you find by 'solving.'" That is, if it acts like a solution, it's a solution – it doesn't matter where you got it. For example, if you are asked to verify that $x = 1, y = -1$ is a solution to the linear system

$$\begin{aligned} 3x + 2y &= 1 \\ x - y &= 2, \end{aligned}$$

all you need to do is plug in $x = 1, y = -1$ and see that the equations are both true. You *do not* need to solve the system and then observe that what you found is that $x = 1, y = -1$. (This point was mentioned earlier in the answer to Quick

Exercise #2.) Doing that in this case won't have resulted in any great harm, just wasted a few minutes of your time. We could have cooked up ways to waste a lot more of your time, though.

This is an instance of the following central principle of mathematics.

The Rat Poison Principle

Q: How do you find out if something is rat poison?

A: You feed it to a rat.

The point is that if someone hands you something and asks if it is rat poison, the answer doesn't depend on where or how she got it. The crucial thing is to see whether it does the job. The Rat Poison Principle is important because it tells you what to do in abstract situations when solving isn't an option.

Example Suppose that we know that $x_1 = c_1, \ldots, x_n = c_n$ is a solution to the linear system

$$\begin{aligned} a_{11}x_1 + \cdots + a_{1n}x_n &= 0 \\ &\vdots \\ a_{m1}x_1 + \cdots + a_{mn}x_n &= 0. \end{aligned} \tag{1.5}$$

(A system like this, in which all the constants on the right-hand side are 0, is called **homogeneous**.) This simply means that

$$\begin{aligned} a_{11}c_1 + \cdots + a_{1n}c_n &= 0 \\ &\vdots \\ a_{m1}c_1 + \cdots + a_{mn}c_n &= 0. \end{aligned}$$

Now if $r \in \mathbb{R}$ is any constant, then

$$\begin{aligned} a_{11}(rc_1) + \cdots + a_{1n}(rc_n) &= r(a_{11}c_1 + \cdots + a_{1n}c_n) = r0 = 0 \\ &\vdots \\ a_{m1}(rc_1) + \cdots + a_{mn}(rc_n) &= r(a_{m1}c_1 + \cdots + a_{mn}c_n) = r0 = 0. \end{aligned}$$

This tells us that $x_1 = rc_1, \ldots, x_n = rc_n$ is another solution of the system (1.5). So, without solving anything, we've found a way to take one known solution of this linear system and come up with infinitely many different solutions. ▲

We end this section with some more terminology that will be important in talking about linear systems and their solutions.

Definition A linear system is called **consistent** if a solution exists and is called **inconsistent** if it has no solution.

A solution $c_1, \ldots, c_n$ of an $m \times n$ linear system is called **unique** if it is the *only* solution; i.e., if whenever $c'_1, \ldots, c'_n$ is a solution to the same system, $c_i = c'_i$ for $1 \leq i \leq n$.

Quick Exercise #3. Which of the three systems (1.1) (a 1×2 system), (1.2), (1.3) are consistent? Which have unique solutions?

KEY IDEAS

- A linear system of equations over $\mathbb{R}$ is a set of equations of the form

$$\begin{aligned} a_{11}x_1 + \cdots + a_{1n}x_n &= b_1 \\ &\vdots \\ a_{m1}x_1 + \cdots + a_{mn}x_n &= b_m. \end{aligned}$$

  The list of numbers $c_1, \ldots, c_n$ is a solution to the system if you can plug each c_i in for x_i in the equations above and the resulting equations are all true.
- The Rat Poison Principle: if you feed it to a rat and the rat dies, it's rat poison. Things in math are often defined by what they do.
- A linear system is consistent if it has at least one solution; otherwise it is inconsistent.
- A solution to a linear system is unique if it is the only solution.

EXERCISES

1.1.1 Suppose that you're making bread, beer, and cheese using the recipes in this section. As before, you have 176 cups of milk and 56 tsps of rosemary, but you now have only 35 tsps of yeast and as much barley as you need.

(a) How much bread, beer, and cheese can you make to exactly use up your yeast, milk, and rosemary?

(b) How much barley will you need?

1.1.2 For each of the following linear systems, graph the set of solutions of each equation. Is the system consistent or inconsistent? If it is consistent, does it have a unique solution?

QA #3: All three are consistent, as we've seen. We've also seen that the system (1.2) has a unique solution but the equation (1.1) does not. We simply told you that the solution of (1.3) is unique; we'll see why that's true in the next section.

(a) $2x - y = 7$, $x + y = 2$

(b) $3x + 2y = 1$, $x - y = -3$, $2x + y = 0$

(c) $x + y = 2$, $2x - y = 1$, $x - y = -1$

(d) $4x - 2y = -6$, $-2x + y = 2$

(e) $x + 2y - z = 0$, $-x + y + z = 0$, $x - y - z = 0$

(f) $x + y + z = 0$, $x - y + z = 0$, $x + y - z = 0$

1.1.3 For each of the following linear systems, graph the set of solutions of each equation. Is the system consistent or inconsistent? If it is consistent, does it have a unique solution?

(a) $x + 2y = 5$, $2x + y = 0$

(b) $x - y = -1$, $2x - y = 4$, $x + y = 3$

(c) $-2x + y = 0$, $x + y = -3$, $x - y = 1$

(d) $2x - 4y = 2$, $-x + 2y = -1$

(e) $0x + 0y + z = 0$, $0x + y + 0z = 0$, $x + 0y + 0z = 0$

(f) $0x + 0y + z = 0$, $x + 0y + z = 1$, $x + 0y - z = -1$

1.1.4 Modify the redundant linear system (1.3) to make it inconsistent. Interpret your new system in terms of the effort to use up all of the ingredients when making the bread, beer, and cheese.

1.1.5 Give geometric descriptions of the sets of all solutions for each of the following linear systems.

(a) $0x + 0y + z = 0$

(b) $0x + 0y + z = 0$, $0x + y + 0z = 0$

(c) $0x + 0y + z = 0$, $0x + y + 0z = 0$, $x + 0y + 0z = 0$

(d) $0x + 0y + z = 0$, $0x + y + 0z = 0$, $x + 0y + 0z = 0$, $x + y + z = 0$

(e) $0x + 0y + z = 0$, $0x + y + 0z = 0$, $x + 0y + 0z = 0$, $x + y + z = 1$

1.1.6 Give geometric descriptions of the sets of all solutions for each of the following linear systems.

(a) $0x + y + 0z = 1$

(b) $x + 0y + 0z = 0$, $0x + y + 0z = 0$

(c) $0x + y + z = 0$, $x + 0y + z = 0$, $x + y + 0z = 0$

(d) $0x + y + z = 0$, $x + 0y + z = 0$, $x + y + 0z = 0$, $x + y + z = 0$

(e) $0x + y + z = 0$, $x + 0y + z = 0$, $x + y + 0z = 0$, $x + y + z = 1$

1.1.7 Suppose that $f(x) = ax^2 + bx + c$ is a quadratic polynomial whose graph passes through the points $(-1, 1)$, $(0, 0)$, and $(1, 2)$.

(a) Find a linear system satisfied by a, b, and c.

(b) Solve the linear system to determine what the function f is.

1.1.8 Let $f(x) = a + b\cos x + c\sin x$ for some $a, b, c \in \mathbb{R}$. (This is a simple example of a **trigonometric polynomial**.) Suppose you know that

$$f(0) = -1, \qquad f(\pi/2) = 2, \qquad \text{and} \qquad f(\pi) = 3.$$

(a) Find a linear system satisfied by a, b, and c.

(b) Solve the linear system to determine what the function f is.

1.1.9 Show that $\left(\frac{de - bf}{ad - bc}, \frac{af - ec}{ad - bc}\right)$ is a solution of the 2×2 linear system

$$\begin{aligned} ax + by &= e \\ cx + dy &= f, \end{aligned}$$

as long as $ad \neq bc$.

1.1.10 Suppose that $x_1 = c_1, \ldots, x_n = c_n$ and $x_1 = d_1, \ldots, x_n = d_n$ are both solutions of the linear system (1.4). Under what circumstances is $x_1 = c_1 + d_1, \ldots, x_n = c_n + d_n$ also a solution?

1.1.11 Suppose that $x_1 = c_1, \ldots, x_n = c_n$ and $x_1 = d_1, \ldots, x_n = d_n$ are both solutions of the linear system (1.4), and t is a real number. Show that $x_1 = tc_1 + (1 - t)d_1, \ldots, x_n = tc_n + (1 - t)d_n$ is also a solution to system (1.4).

1.2 Gaussian Elimination

The Augmented Matrix of a Linear System

In the last section we solved some small linear systems in *ad hoc* ways; i.e., we weren't particularly systematic about it. You can easily imagine that if you're going to try to solve a much larger system by hand, or (more realistically) if you're going to program a computer to do the algebra for you, you need a more systematic approach.

The first thing we'll need is a standardized way to write down a linear system. For a start, we should treat all the variables the same way in every equation. That means writing them in the same order, and including all of the variables in every equation. If one of the equations as originally written doesn't contain one of the variables, we can make it appear by including it with a coefficient of 0. In the examples in Section 1.1 we already kept the variables in a consistent order, but didn't always have all the variables written in every equation. So, for example, we should rewrite the system (1.3) as

$$\begin{aligned} x + \frac{1}{4}y + 0z &= 20 \\ 2x + \frac{1}{4}y + 0z &= 36 \\ x + 0y + 8z &= 176 \\ 2x + \frac{1}{4}y + z &= 56. \end{aligned} \tag{1.6}$$

The next thing to notice is that there's a lot of notation which we don't actually need, as long as we remember that what we're working with is a linear system. The coefficients on the left, and the constants on the right, contain all the information about the system, and we can completely do away with the variable names and the $+$ and $=$ signs. We *do*, however, need to keep track of the relative positions of those numbers. We do this by writing them down in what's called a matrix.

Definition Let m and n be natural numbers. An $m \times n$ **matrix** A over $\mathbb{R}$ is a doubly indexed collection of real numbers

$$A = [a_{ij}]_{\substack{1 \le i \le m \\ 1 \le j \le n}}.$$

That is, for each ordered pair of integers (i, j) with $i \in \{1, 2, \ldots, m\}$ and $j \in \{1, 2, \ldots, n\}$, there is a corresponding real number a_{ij}. The number a_{ij} is called the (i, j) **entry** of A. We sometimes denote the (i, j) entry of A by $[A]_{ij}$.

We denote the set of $m \times n$ matrices over $\mathbb{R}$ by $M_{m,n}(\mathbb{R})$.

One way a (small) matrix can be specified is by just writing down all of its entries, like this:

$$A = \begin{bmatrix} 1 & \frac{1}{4} & 0 \\ 2 & \frac{1}{4} & 0 \\ 1 & 0 & 8 \\ 2 & \frac{1}{4} & 1 \end{bmatrix}. \tag{1.7}$$

When we do this, an $m \times n$ matrix has m rows and n columns, and the (i, j) entry a_{ij} appears in the ith row (from the top) and the jth column (from the left). For example, the matrix above is a 4×3 matrix, and $a_{21} = 2$. To remember which is m and which is n, and which is i and which is j, keep in mind:

Rows, then columns.

The matrix in equation (1.7) is called the **coefficient matrix** of the linear system (1.6). In order to record all of the information in a linear system, we need to have not just the coefficients but also the numbers on the right-hand sides of the equations. We organize this in the so-called augmented matrix of the system.

Definition Given an $m \times n$ linear system

$$\begin{aligned} a_{11}x_1 + \cdots + a_{1n}x_n &= b_1 \\ &\vdots \\ a_{m1}x_1 + \cdots + a_{mn}x_n &= b_m, \end{aligned}$$

its **augmented matrix** is the $m \times (n+1)$ matrix

$$\left[\begin{array}{ccc|c} a_{11} & \cdots & a_{1n} & b_1 \\ \vdots & \ddots & \vdots & \vdots \\ a_{m1} & \cdots & a_{mn} & b_m \end{array}\right].$$

The bar between the last two columns of the augmented matrix isn't really necessary, of course, but it is customary to separate the coefficients in the left-hand sides of the equations from the constants on the right-hand sides in this way.

Quick Exercise #4. Write down the coefficient matrix and the augmented matrix of the system

$$\begin{aligned} x - z &= 2 \\ y + z &= 0 \\ y - x &= 3. \end{aligned}$$

Row Operations

To systematize our *ad hoc* approach to solving linear systems, we start by listing the things we did in the last section, and figuring out what they do to the augmented matrix of the system. The main things we generally do that don't change the solutions to the system are:

1. add a multiple of one equation to another,
2. multiply both sides of an equation by a nonzero constant.

There's another thing we can do, which we probably wouldn't bother with if the equations are all written out, but can be useful for the augmented matrix, which is:

3. switch the order that two equations are written in.

QA #4: $\begin{bmatrix} 1 & 0 & -1 \\ 0 & 1 & 1 \\ -1 & 1 & 0 \end{bmatrix}$ and $\left[\begin{array}{ccc|c} 1 & 0 & -1 & 2 \\ 0 & 1 & 1 & 0 \\ -1 & 1 & 0 & 3 \end{array}\right]$. Watch the order of the variables!

This means we can do any of the following operations to the augmented matrix without changing the solutions to the corresponding system.

Row Operations

R1: Add a multiple of one row to another row.
R2: Multiply a row by a nonzero constant.
R3: Switch any two rows.

Note that row operation **R1** also lets us subtract a multiple of one row from another (by adding a negative multiple) and that row operation **R2** lets us divide a row by any nonzero constant.

The process by which we solve a linear system once it's written as an augmented matrix is called **Gaussian elimination**. The basic idea is to try to get the part of the augmented matrix to the left of the bar to have all 1s on the diagonal (i.e., the (i, i) entries) and zeroes everywhere else, because an augmented matrix like that leads to a system whose solutions you can just read off.

Quick Exercise #5. Translate the following augmented matrix back into a linear system of equations and solve it.

$$\left[\begin{array}{ccc|c} 1 & 0 & 0 & 2 \\ 0 & 1 & 0 & -5 \\ 0 & 0 & 1 & 3 \end{array}\right]$$

Before describing the algorithm in the abstract, we'll first use it in an example.

The linear system describing how much bread, beer, and cheese we can make with our ingredients was first written in (1.3) and rewritten to include all the coefficients (even the zeroes) in (1.6). The augmented matrix for the system is

$$\left[\begin{array}{ccc|c} 1 & \frac{1}{4} & 0 & 20 \\ 2 & \frac{1}{4} & 0 & 36 \\ 1 & 0 & 8 & 176 \\ 2 & \frac{1}{4} & 1 & 56 \end{array}\right].$$

The top-left entry is already a 1, so we start by using the top row to clear out all the other numbers in the first column. For example, we subtract $2\times$ the first row from the second (using row operation **R1**), to get the new system

QA #5: $x = 2$, $y = -5$, $z = 3$.

$$\left[\begin{array}{ccc|c} 1 & \frac{1}{4} & 0 & 20 \\ 0 & -\frac{1}{4} & 0 & -4 \\ 1 & 0 & 8 & 176 \\ 2 & \frac{1}{4} & 1 & 56 \end{array}\right].$$

Similarly, we subtract the first row from the third row and $2\times$ the first row from the fourth row to get

$$\left[\begin{array}{ccc|c} 1 & \frac{1}{4} & 0 & 20 \\ 0 & -\frac{1}{4} & 0 & -4 \\ 0 & -\frac{1}{4} & 8 & 156 \\ 0 & -\frac{1}{4} & 1 & 16 \end{array}\right].$$

Next we use row operation R2 to multiply the second row by -4, and get

$$\left[\begin{array}{ccc|c} 1 & \frac{1}{4} & 0 & 20 \\ 0 & 1 & 0 & 16 \\ 0 & -\frac{1}{4} & 8 & 156 \\ 0 & -\frac{1}{4} & 1 & 16 \end{array}\right].$$

Next we use the second row to clear out all the other numbers in lower rows from the second column, to get

$$\left[\begin{array}{ccc|c} 1 & \frac{1}{4} & 0 & 20 \\ 0 & 1 & 0 & 16 \\ 0 & 0 & 8 & 160 \\ 0 & 0 & 1 & 20 \end{array}\right].$$

Next multiply the third row by $\frac{1}{8}$:

$$\left[\begin{array}{ccc|c} 1 & \frac{1}{4} & 0 & 20 \\ 0 & 1 & 0 & 16 \\ 0 & 0 & 1 & 20 \\ 0 & 0 & 1 & 20 \end{array}\right]$$

and subtract the third row from the fourth:

$$\left[\begin{array}{ccc|c} 1 & \frac{1}{4} & 0 & 20 \\ 0 & 1 & 0 & 16 \\ 0 & 0 & 1 & 20 \\ 0 & 0 & 0 & 0 \end{array}\right]. \tag{1.8}$$

At this point there are two slightly different routes we can take. We can do one more row operation, subtracting $\frac{1}{4}$ times the second row from the first row, to get

$$\left[\begin{array}{ccc|c} 1 & 0 & 0 & 16 \\ 0 & 1 & 0 & 16 \\ 0 & 0 & 1 & 20 \\ 0 & 0 & 0 & 0 \end{array}\right].$$

If we turn this back into a linear system, we get

$$\begin{aligned} x + 0y + 0z &= 16 \\ 0x + y + 0z &= 16 \\ 0x + 0y + z &= 20 \\ 0x + 0y + 0z &= 0, \end{aligned} \qquad \Longleftrightarrow \qquad \begin{aligned} x &= 16 \\ y &= 16 \\ z &= 20 \\ 0 &= 0. \end{aligned}$$

It is crucially important that the last equation is $0 = 0$, i.e., something that's trivially true, as opposed to, say, $0 = 1$, which would mean that there were no solutions to the system (why?).

Alternatively, if we skip the last row operation and turn the augmented matrix (1.8) into a linear system, we get

$$\begin{aligned} x + \frac{1}{4}y + 0z &= 20 \\ 0x + y + 0z &= 16 \\ 0x + 0y + z &= 20 \\ 0x + 0y + 0z &= 0, \end{aligned} \qquad \Longleftrightarrow \qquad \begin{aligned} x + \frac{1}{4}y &= 20 \\ y &= 16 \\ z &= 20 \\ 0 &= 0. \end{aligned}$$

Here we are given y and z explicitly, and knowing y, we can substitute it into the first equation to get

$$x + 4 = 20 \qquad \Longrightarrow \qquad x = 16.$$

Thus we get the same unique solution $x = 16, y = 16, z = 20$. This process, where we substitute variable values from the later equations into the earlier equations, is called **back-substitution**.

Quick Exercise #6. Identify which row operations are being performed in each step below.

$$\left[\begin{array}{cc|c} 3 & 2 & 1 \\ 1 & -1 & 2 \end{array}\right] \xrightarrow{(a)} \left[\begin{array}{cc|c} 1 & -1 & 2 \\ 3 & 2 & 1 \end{array}\right] \xrightarrow{(b)} \left[\begin{array}{cc|c} 1 & -1 & 2 \\ 0 & 5 & -5 \end{array}\right]$$

$$\xrightarrow{(c)} \left[\begin{array}{cc|c} 1 & -1 & 2 \\ 0 & 1 & -1 \end{array}\right] \xrightarrow{(d)} \left[\begin{array}{cc|c} 1 & 0 & 1 \\ 0 & 1 & -1 \end{array}\right]$$

Does it Always Work?

To formalize the approach we took above, we first need to carefully define our goal. We said we wanted to try to have all 1s on the diagonal and zeroes everywhere else, but this isn't necessarily possible. The following definition describes a general form that turns out always to be possible to achieve and to allow us to easily solve any consistent system.

QA #6: (a) **R3**: switch the rows. (b) **R1**: add $-3\times$ the first row to the second row. (c) **R2**: multiply the second row by $\frac{1}{5}$. (d) **R1**: add the second row to the first row.

Definition A matrix A is in **row-echelon form** (or REF for short) if:

1. every row containing any nonzero entries is above every row which contains only zeroes,
2. the first nonzero entry in each row is 1,
3. if $i < j$ then the first nonzero entry in the ith row is strictly farther to the left than the first nonzero entry in the jth row.

If A is in row-echelon form, then each entry which is the first nonzero entry in its row is called a **pivot**. We say that A is in **reduced row-echelon form** (or RREF for short) if:

1. A is in row-echelon form,
2. any column which contains a pivot contains no other nonzero entries.

Quick Exercise #7. Determine whether each of the matrices below is in row-echelon or reduced row-echelon forms.

(a) $\left[\begin{array}{cc|c} 1 & 1 & 0 \\ 0 & 2 & 1 \\ 0 & 0 & 1 \end{array}\right]$ (b) $\left[\begin{array}{cc|c} 1 & -3 & 0 \\ 0 & 0 & 1 \\ 0 & 0 & 0 \end{array}\right]$ (c) $\left[\begin{array}{cc|c} 1 & 1 & 0 \\ 0 & 1 & 2 \\ 0 & 0 & 0 \end{array}\right]$

(d) $\left[\begin{array}{cc|c} 1 & 0 & 0 \\ 0 & 1 & 0 \\ 0 & 0 & 1 \end{array}\right]$ (e) $\left[\begin{array}{cc|c} 1 & 0 & 0 \\ 0 & 0 & 0 \\ 0 & 1 & 0 \end{array}\right]$

Our reason for introducing these somewhat complicated conditions is that it is always possible to put a matrix into RREF via row operations. Just as importantly, when the augmented matrix of a linear system is in RREF, it is easy to tell at a glance whether the system has any solutions, and only a little bit more work to describe them all.

Theorem 1.1 *Every linear system can be put into reduced row-echelon form using row operations.*

Proof This proof is an example of a "proof by algorithm," in which we prove that it's always possible to do something by explicitly describing a procedure for doing it. As mentioned before, the process we will use is called **Gaussian elimination** after C. F. Gauss.*

*When the process is used, as here, specifically to put a matrix into RREF, it is sometimes called *Gauss–Jordan elimination.*

QA #7: (a) Neither. (b) RREF. (c) REF, but not RREF. (d) RREF. (e) Neither.

Find the first column from the left that has any nonzero entries in it. Using row operation **R3** if necessary, get a nonzero entry at the top of the column, and then use row operation **R2** to make that top entry 1. This 1 is now a pivot in the first row. Then use row operation **R1** to make all the other entries of the column containing this pivot 0, by adding the appropriate multiples of the top row to all the other rows. The matrix now might have some columns of all zeroes on the left, and the first nonzero column has a 1 at the top and zeroes below.

Now find the next column to the right with nonzero entries *below* the first row. By using row operation **R3** to move one of the lower rows up if necessary, get a nonzero entry in the second row of this column; use **R2** to make that entry a 1. This 1 is a pivot in the second row. Notice that it must be to the right of the pivot in the first row, since all the entries below and to the left of that first pivot are 0. Now use the second row of the matrix and row operation **R1** to make all the lower entries (i.e., below the second row) in the same column as this pivot 0.

Repeat the process in the last paragraph to create a pivot in each successive row of the matrix, until this is no longer possible, which can only happen when the remaining rows only contain zeroes. At this point the resulting matrix is in REF, but not necessarily RREF.

Finally, use row operation **R1** to make all the entries in the same column as any pivot 0. The resulting matrix will now be in RREF. ▲

Quick Exercise #8. (a) Go back and check that our solution of the system (1.6) starting on page 12 followed the Gaussian elimination algorithm described in the proof of Theorem 1.1.

(b) Explain in what way the row operations in Quick Exercise #6 don't precisely follow the algorithm as described above.*

Recall that each column of the augmented matrix of a linear system corresponds to one of the variables in the original system.

Definition A variable in a linear system is called a **pivot variable** if the corresponding column in the RREF of the augmented matrix of the system contains a pivot; otherwise, it is called a **free variable**.

Each row contains at most one pivot (because the pivot is the *first* nonzero entry in the row), so a row containing a pivot gives a formula for the corresponding pivot variable in terms of free variables. For example, let's go back to thinking about the supply of barley, yeast, and milk. Suppose you make bread, beer, and cheese according to your earlier recipes, but you decide to add in the option of

*That doesn't mean that there's anything *wrong* with what we did there; we just found the RREF in a slightly different way from what the algorithm says.

QA #8: (a) ✓ (b) The algorithm as described would start by multiplying the first row by $\frac{1}{5}$, not by switching the rows.

muesli for breakfast. Say it takes half a pound of barley and a cup of milk to make a bowl of muesli. If w is the number of bowls of muesli you plan to make, we have a new system of three equations (one for each ingredient) in four variables (remember, x is the number of loaves of bread, y is the number of pints of beer, and z is the number of rounds of cheese):

$$\begin{aligned} x + \frac{1}{4}y + \frac{1}{2}w &= 20 \\ 2x + \frac{1}{4}y &= 36 \\ x + 8z + w &= 176. \end{aligned} \tag{1.9}$$

The augmented matrix of the system (if we order the variables x, y, z, w) is

$$\left[\begin{array}{cccc|c} 1 & \frac{1}{4} & 0 & \frac{1}{2} & 20 \\ 2 & \frac{1}{4} & 0 & 0 & 36 \\ 1 & 0 & 8 & 1 & 176 \end{array}\right]$$

and in RREF it is

$$\left[\begin{array}{cccc|c} 1 & 0 & 0 & -\frac{1}{2} & 16 \\ 0 & 1 & 0 & 4 & 16 \\ 0 & 0 & 1 & \frac{3}{16} & 20 \end{array}\right] \tag{1.10}$$

(see Exercise 1.2.5). From here, we see that x, y, z are all pivot variables and w is a free variable. (Note: it doesn't always happen that the free variables come last!) The formulas we get for x, y, z in terms of w from the augmented matrix are

$$\begin{aligned} x &= 16 + \frac{1}{2}w \\ y &= 16 - 4w \\ z &= 20 - \frac{3}{16}w. \end{aligned} \tag{1.11}$$

So you can decide how much muesli you want (w), and that determines how much of everything else you can make. For instance, if you decide that you want to make 2 bowls of muesli, you should make 17 loaves of bread, 8 pints of beer, and $19\frac{5}{8}$ rounds of cheese.

Quick Exercise #9. Check that $(x, y, z, w) = (17, 8, 19\frac{5}{8}, 2)$ really is a solution of the system (1.9). Remember the Rat Poison Principle!

There's something surprising that we can see from the formulas in the system (1.11): increasing the amount of muesli you want actually *increases* the amount of bread you end up making, even though making muesli uses up some of the ingredients you need for bread. You can see from the system (1.11) how this works: making muesli leads you to make less beer and cheese, which makes it more plausible that you end up with more bread. Trying to reason this through without actually solving the system is a bad idea.

Alternatively, instead of putting a system all the way into RREF, we can just put it into REF (just skip the last paragraph of the proof of Theorem 1.1). We can then solve for the pivot variables in terms of the free variables using back-substitution, as discussed in an earlier example beginning on page 14. In principle this takes essentially the same amount of work as the method we discussed above. In practice, the back-substitution method can be better for solving linear systems numerically (i.e., with a computer which represents numbers as terminating decimals). This is basically because going all the way to RREF has more of a tendency to introduce round-off errors.

Issues like minimizing round-off errors are addressed by the field of numerical linear algebra and are beyond the scope of this book. We will mostly work with RREF because it is simpler to use it to understand existence and uniqueness of solutions, which we will turn to next. However, examining the proof of Theorem 1.1 reveals an important fact about the relationship between REF and RREF: if a matrix is put into REF, then it already has the same pivots as its RREF. This will save a little time in what follows.

Pivots and Existence and Uniqueness of Solutions

This process for solving for the pivot variables in terms of the free variables always works the same way, with one exception. If the last nonzero row of the augmented matrix has a pivot in the last column (the one to the right of the bar), we get the equation $0 = 1$, and so the system is inconsistent. As long as we're not in that situation, any choice of values of the free variables yields a solution of the system.

Theorem 1.2 *Suppose the augmented matrix* A *of a linear system is in REF. Then the system is consistent if and only if there is no pivot in the last column of* A.

Proof First note that since the pivots of a matrix A in REF are the same as if it were put into RREF, we can assume that in fact A is in RREF.

If there is no pivot in the last column, as explained above we get formulas for each of the pivot variables in terms of the free variables. If we set each of the free variables (if there are any) equal to 0, then we get a solution to the system.*

If there is a pivot in the last column, then the system includes the equation $0 = 1$, and therefore there is no solution. ▲

*There's nothing special about the choice of 0 here. Any other choice of values of the free variables would work, so for convenience we just made the simplest choice possible.

Example An $n \times n$ linear system is called **upper triangular** if the coefficient $a_{ij} = 0$ whenever $i > j$:

$$\begin{aligned} a_{11}x_1 + a_{12}x_2 + a_{13}x_3 + \cdots + a_{1n}x_n &= b_1 \\ a_{22}x_2 + a_{23}x_3 + \cdots + a_{2n}x_n &= b_2 \\ a_{33}x_3 + \cdots + a_{3n}x_n &= b_3 \\ &\vdots \\ a_{nn}x_n &= b_n. \end{aligned} \tag{1.12}$$

If $a_{ii} \neq 0$ for each i, then the augmented matrix of this system can be put into REF simply by dividing the ith row by a_{ii} for each i. Then each of the first n columns contains a pivot, so the last column does not. By Theorem 1.2, we conclude that in this case the system (1.12) must be consistent. ▲

Theorem 1.3 *Suppose the augmented matrix* A *of a linear system in n variables is in REF and the system is consistent. Then the system has a unique solution if and only if there is a pivot in each of the first n columns of* A.

Proof As in the proof of Theorem 1.2, we can assume that A is in RREF. As explained above, a system in RREF gives us formulas for each of the pivot variables in terms of the free variables, and we can set the values of the free variables as we like. Suppose there is at least one free variable, say x_i. Then there is a solution of the system in which we set all of the free variables to be 0, and there is another solution of the system in which we set $x_i = 1$ and the rest of the free variables (if there are any) to be 0. (Notice that this may or may not affect the values of the pivot variables, but the two solutions we've just described are definitely different because the value of x_i is different.)

Therefore the system has multiple solutions whenever there are any free variables; i.e., whenever any of the first n columns of A doesn't contain a pivot.

On the other hand, if each of the first n columns of A contains a pivot and the system is consistent, then A is of the form

$$\left[\begin{array}{cccc|c} 1 & 0 & \cdots & 0 & b_1 \\ 0 & 1 & \cdots & 0 & b_2 \\ \vdots & \vdots & \ddots & \vdots & \vdots \\ 0 & 0 & \cdots & 1 & b_n \\ 0 & 0 & \cdots & 0 & 0 \\ \vdots & \vdots & \ddots & \vdots & \vdots \\ 0 & 0 & \cdots & 0 & 0 \end{array}\right].$$

(The final rows of all zeroes may or may not be there, depending on whether the original system had some redundant equations or not.) That is, the system consists of n equations of the form $x_i = b_i$ for $i = 1, \ldots, n$, and so there is exactly one solution. ▲

Example We again consider the upper triangular system (1.12). As we saw, if $a_{ii} \neq 0$ for each i, then the augmented matrix of the system can be put into an REF with a pivot in each of the first n columns. By Theorem 1.3, the system (1.12) has a unique solution.

On the other hand, if $a_{ii} = 0$ for some i, then for the smallest such i, when the augmented matrix is put into REF, the ith column won't contain a pivot. So in that case Theorem 1.3 implies that the linear system (1.12) is either inconsistent, or is consistent with multiple solutions. ▲

Corollary 1.4 *If $m < n$ then it is impossible for an $m \times n$ linear system to have a unique solution.*

Proof By Theorem 1.1, a given linear system can be put into RREF. If m is less than n, then it's impossible for each of the first n columns of the augmented matrix to contain a pivot, since there are at most m pivots. By Theorem 1.3, this means that the system cannot have a unique solution. ▲

Definition An $m \times n$ linear system is called **underdetermined** if $m < n$.

This term is motivated by Corollary 1.4, which says that an underdetermined linear system – which has fewer equations than variables – doesn't contain enough information to determine a unique solution. However, it is possible for a system to be both underdetermined and inconsistent.

Quick Exercise #10. Give an example of a linear system which is both underdetermined and inconsistent.

Extending this intuition, we also have the following terminology.

Definition An $m \times n$ linear system is called **overdetermined** if $m > n$.

The idea here is that an overdetermined system – which has more equations than variables – typically contains too many restrictions to have any solution at all. We will return to this idea later.

QA #10: For example, $x + y + z = 0$ and $x + y + z = 1$.

KEY IDEAS

- A matrix is a doubly indexed collection of numbers: $A = [a_{ij}]_{\substack{1 \le i \le m \\ 1 \le j \le n}}$.
- Linear systems can be written as augmented matrices, with the matrix of coefficients on the left and the numbers b_i on the right.
- Gaussian elimination is a systematic way of using row operations on an augmented matrix to solve a linear system. The allowed operations are: adding a multiple of one row to another (R1), multiplying a row by a nonzero number (R2), or swapping two rows (R3).
- A matrix is in row-echelon form (REF) if zero rows are at the bottom, the first nonzero entry in every row is a 1, and the first nonzero entry in any row is to the right of the first nonzero entry in the rows above.
- A pivot is the first nonzero entry in its row in a matrix in REF. The matrix is in reduced row-echelon form (RREF) if every pivot is alone in its column.
- Columns (except the last one) in the augmented matrix of a linear system correspond to variables. Pivot variables are those whose columns are pivot columns after row-reducing the matrix; the other variables are called free variables.
- A system is consistent if and only if there is no pivot in the last column of the RREF of its augmented matrix. A consistent system has a unique solution if and only if there is a pivot in every column but the last, i.e., if every variable is a pivot variable.

EXERCISES

1.2.1 Identify which row operations are being performed in each step below. (Not just the general type, but which multiple of which row is added to which other row, etc.)

Hint: There isn't necessarily a single correct answer for every step.

$$\begin{bmatrix} 1 & 2 & 3 \\ 4 & 5 & 6 \\ 7 & 8 & 9 \\ 10 & 11 & 12 \end{bmatrix} \xrightarrow{\text{(a)}} \begin{bmatrix} 1 & 2 & 3 \\ 4 & 5 & 6 \\ -3 & -3 & -3 \\ 10 & 11 & 12 \end{bmatrix} \xrightarrow{\text{(b)}} \begin{bmatrix} 4 & 5 & 6 \\ 1 & 2 & 3 \\ -3 & -3 & -3 \\ 10 & 11 & 12 \end{bmatrix}$$

$$\xrightarrow{\text{(c)}} \begin{bmatrix} 1 & 2 & 3 \\ 1 & 2 & 3 \\ -3 & -3 & -3 \\ 10 & 11 & 12 \end{bmatrix} \xrightarrow{\text{(d)}} \begin{bmatrix} 1 & 2 & 3 \\ 1 & 2 & 3 \\ -3 & -3 & -3 \\ 0 & -9 & -18 \end{bmatrix} \xrightarrow{\text{(e)}} \begin{bmatrix} 1 & 2 & 3 \\ 1 & 2 & 3 \\ -3 & -3 & -3 \\ 0 & 1 & 2 \end{bmatrix}$$

$$\xrightarrow{\text{(f)}} \begin{bmatrix} 1 & 2 & 3 \\ 1 & 2 & 3 \\ 1 & 1 & 1 \\ 0 & 1 & 2 \end{bmatrix} \xrightarrow{\text{(g)}} \begin{bmatrix} 0 & 1 & 2 \\ 1 & 2 & 3 \\ 1 & 1 & 1 \\ 0 & 1 & 2 \end{bmatrix} \xrightarrow{\text{(h)}} \begin{bmatrix} 0 & 1 & 2 \\ 0 & 1 & 2 \\ 1 & 1 & 1 \\ 0 & 1 & 2 \end{bmatrix}.$$

1.2.2 Identify which row operations are being performed in each step below. (Not just the general type, but which multiple of which row is added to which other row, etc.)
Hint: There isn't necessarily a single correct answer for every step.

$$\begin{bmatrix} 1 & 2 & 3 \\ 4 & -5 & 6 \\ 7 & 9 & 8 \\ 6 & 7 & 5 \end{bmatrix} \xrightarrow{(a)} \begin{bmatrix} 1 & 2 & 3 \\ 4 & -5 & 6 \\ 6 & 7 & 5 \\ 7 & 9 & 8 \end{bmatrix} \xrightarrow{(b)} \begin{bmatrix} 1 & 2 & 3 \\ 0 & -13 & -6 \\ 6 & 7 & 5 \\ 7 & 9 & 8 \end{bmatrix} \xrightarrow{(c)} \begin{bmatrix} 1 & 2 & 3 \\ 0 & 13 & 6 \\ 6 & 7 & 5 \\ 7 & 9 & 8 \end{bmatrix}$$

$$\xrightarrow{(d)} \begin{bmatrix} 1 & -11 & -3 \\ 0 & 13 & 6 \\ 6 & 7 & 5 \\ 7 & 9 & 8 \end{bmatrix} \xrightarrow{(e)} \begin{bmatrix} 1 & -11 & -3 \\ 0 & 13 & 6 \\ 7 & -4 & 2 \\ 7 & 9 & 8 \end{bmatrix} \xrightarrow{(f)} \begin{bmatrix} 1 & -11 & -3 \\ 0 & 13 & 6 \\ 7 & -4 & 2 \\ 0 & 13 & 6 \end{bmatrix}$$

$$\xrightarrow{(g)} \begin{bmatrix} 1 & -11 & -3 \\ 0 & 0 & 0 \\ 7 & -4 & 2 \\ 0 & 13 & 6 \end{bmatrix} \xrightarrow{(h)} \begin{bmatrix} 1 & -11 & -3 \\ 0 & 13 & 6 \\ 7 & -4 & 2 \\ 0 & 0 & 0 \end{bmatrix}.$$

1.2.3 Find the RREF of each of the following matrices.

(a) $\begin{bmatrix} -2 & 0 & 1 \\ 1 & 3 & -4 \end{bmatrix}$ (b) $\left[\begin{array}{cccc|c} 2 & 3 & 4 & -1 & 1 \\ 1 & 0 & -1 & 0 & 3 \\ 2 & 2 & 2 & -1 & 2 \end{array}\right]$ (c) $\begin{bmatrix} -2 & 1 \\ 0 & 3 \\ 1 & -4 \end{bmatrix}$

(d) $\begin{bmatrix} 1 & 2 & -3 \\ 4 & -5 & 6 \\ 7 & -8 & 9 \end{bmatrix}$ (e) $\begin{bmatrix} 1 & 1 & 1 & 1 \\ 1 & 2 & 2 & 2 \\ 1 & 2 & 3 & 3 \\ 1 & 2 & 3 & 4 \end{bmatrix}$ (f) $\left[\begin{array}{cccc|c} -1 & 2 & 1 & 0 & 1 \\ 2 & 3 & 1 & -1 & 1 \\ -1 & -2 & 0 & 2 & -1 \\ 2 & 2 & 1 & 0 & 1 \end{array}\right]$

1.2.4 Find the RREF of each of the following matrices.

(a) $\begin{bmatrix} 3 & 0 & 2 \\ 1 & -4 & 1 \end{bmatrix}$ (b) $\left[\begin{array}{cccc|c} 2 & 1 & 3 & 0 & 0 \\ -1 & 1 & -3 & 2 & -3 \\ 1 & 0 & 2 & -1 & 1 \end{array}\right]$ (c) $\begin{bmatrix} -3 & 0 \\ 1 & 2 \\ 4 & -1 \end{bmatrix}$

(d) $\begin{bmatrix} 1 & 2 & 3 \\ 4 & 5 & 6 \\ 7 & 8 & 9 \end{bmatrix}$ (e) $\begin{bmatrix} 1 & 2 & 3 & 4 \\ 2 & 2 & 3 & 4 \\ 3 & 3 & 3 & 4 \\ 4 & 4 & 4 & 4 \end{bmatrix}$ (f) $\left[\begin{array}{cccc|c} 2 & 1 & 0 & 3 & 9 \\ -1 & 1 & 3 & 1 & -1 \\ 1 & 2 & 3 & -2 & -4 \\ 0 & 2 & 4 & 1 & 0 \end{array}\right]$

1.2.5 Show that the matrix (1.10) is the RREF of the linear system (1.9).

1.2.6 Find all the solutions of each of the following linear systems.

(a) $x + 3y + 5z = 7$
$2x + 4y + 6z = 8$

(b) $x - 2y + z + 3w = 3$
$-x + y - 3z - w = -1$

(c) $x + 0y + z = 2$
$x - y + 2z = -1$
$x + y = 5$

(d) $3x + y - 2z = -3$
$x + 0y + 2z = 4$
$-x + 2y + 3z = 1$
$2x - y + z = -6$

(e) $3x + y - 2z = -3$
$x + 0y + 2z = -4$
$-x + 2y + 3z = 1$
$2x - y + z = -6$

1.2.7 Find all the solutions of each of the following linear systems.

(a) $x + 2y + 3z = 4$
$5x + 6y + 7z = 5$

(b) $x - 3y + 2z - w = 1$
$2x + y - z + 5w = 2$

(c) $2x - y + z = 3$
$-x + 3y + 2z = 1$
$x + 2y + 3z = 4$

(d) $x - y + 2z = 3$
$2x + y - z = 1$
$-x + 2y - z = -2$
$3x - 2y + z = 4$

(e) $x - y + 2z = 1$
$2x + y - z = 3$
$-x + 2y - z = -4$
$3x - 2y + z = 2$

1.2.8 Suppose that $f(x) = ax^3 + bx^2 + cx + d$ is a cubic polynomial whose graph passes through the points $(-1, 2)$, $(0, 3)$, $(1, 4)$, and $(2, 15)$. Determine the values of a, b, c, and d.

1.2.9 Suppose that $f(x) = ax^3 + bx^2 + cx + d$ is a cubic polynomial whose graph passes through the points $(-2, 2)$, $(-1, -1)$, $(0, 2)$, and $(1, 5)$. Determine the values of a, b, c, and d.

1.2.10 Show that if $ad - bc \neq 0$, then the linear system

$$ax + by = e$$
$$cx + dy = f$$

has a unique solution.
Hint: Consider separately the cases $a \neq 0$ and $a = 0$.

1.2.11 Suppose you're making bread, beer, and cheese using the recipes of Section 1.1 (summarized in the left-hand side of the equations (1.3)). Is it possible to exactly use up 1 pound of barley, 2 tsps of yeast, 5 cups of milk, and 2 tsps of rosemary?

1.2.12 (a) Explain why every linear system over $\mathbb{R}$ has either zero solutions, one unique solution, or infinitely many solutions.
(b) Why doesn't this contradict the fact that the system of equations

$$\begin{aligned} x^2 - y &= 1 \\ x + 2y &= 3 \end{aligned}$$

has exactly two solutions?

1.2.13 Give examples of linear systems of each of the following types, if possible. Explain how you know they have the claimed properties, or else explain why there is no such system.
(a) Underdetermined and inconsistent.
(b) Underdetermined with a unique solution.
(c) Underdetermined with more than one solution.
(d) Overdetermined and inconsistent.
(e) Overdetermined with more than one solution.
(f) Overdetermined with a unique solution.
(g) Square (i.e., $n \times n$ for some n) and inconsistent.
(h) Square with a unique solution.
(i) Square with more than one solution.

1.2.14 Under what conditions is the following linear system consistent? Give your answer in terms of an equation or equations that must be satisfied by the b_i.

$$\begin{aligned} 2x + y &= b_1 \\ x - y + 3z &= b_2 \\ x - 2y + 5z &= b_3 \end{aligned}$$

1.2.15 Under what conditions is the following linear system consistent? Give your answer in terms of an equation or equations that must be satisfied by the b_i.

$$\begin{aligned} x_1 + 2x_2 - x_3 + 7x_4 &= b_1 \\ -x_2 + x_3 - 3x_4 &= b_2 \\ 2x_1 - 2x_3 + 2x_4 &= b_3 \\ -3x_1 + x_2 + 4x_3 &= b_4 \end{aligned}$$

1.3 Vectors and the Geometry of Linear Systems

Vectors and Linear Combinations

In this section we will introduce another way of writing linear systems which will let us take advantage of our understanding of the geometry of vectors.

Definition An n-**dimensional vector over** $\mathbb{R}$ is an ordered list of n real numbers. In this book, we use the convention that all vectors are **column vectors** and are written in the form

$$\mathbf{v} = \begin{bmatrix} v_1 \\ v_2 \\ \vdots \\ v_n \end{bmatrix},$$

where $v_i \in \mathbb{R}$ for each i. The set of all n-dimensional vectors over $\mathbb{R}$ is denoted $\mathbb{R}^n$.

It follows from this definition that $\begin{bmatrix} v_1 \\ \vdots \\ v_n \end{bmatrix} = \begin{bmatrix} w_1 \\ \vdots \\ w_n \end{bmatrix}$ if and only if $v_i = w_i$ for $i = 1, 2, \ldots, n$. This means that a single equation between n-dimensional vectors contains the same information as n equations between numbers.

You may have seen notation like $\vec{\mathbf{v}}$ in a vector-calculus class, as a way of clearly distinguishing vectors from real numbers. As you will see throughout this course, part of the power of linear algebra is its universality, and in particular, we will come to see "vectors" as much more general objects than just little arrows in space. For this reason, we avoid such notation.

The most basic things we can do with vectors are multiply them by real numbers and add them together, as follows.

Definition

1. Let $\mathbf{v} \in \mathbb{R}^n$ and $c \in \mathbb{R}$. The **scalar multiple** $c\mathbf{v}$ of $\mathbf{v}$ by c is defined by multiplying each of the components of $\mathbf{v}$ by c; i.e.,

$$\text{if} \quad \mathbf{v} = \begin{bmatrix} v_1 \\ v_2 \\ \vdots \\ v_n \end{bmatrix}, \quad \text{then} \quad c\mathbf{v} := \begin{bmatrix} cv_1 \\ cv_2 \\ \vdots \\ cv_n \end{bmatrix}.^*$$

   When working in the context of vectors in $\mathbb{R}^n$, we refer to real numbers as **scalars**.

*The symbol ":=" means "is defined to be." The colon goes on the side of the thing being defined; you can write $b =: a$, and it means that you know what b is already and you're defining a.

2. Let $\mathbf{v}, \mathbf{w} \in \mathbb{R}^n$. The **sum** $\mathbf{v} + \mathbf{w}$ of $\mathbf{v}$ and $\mathbf{w}$ is defined by adding the vectors component-wise; i.e.,

$$\text{if} \quad \mathbf{v} = \begin{bmatrix} v_1 \\ v_2 \\ \vdots \\ v_n \end{bmatrix} \quad \text{and} \quad \mathbf{w} = \begin{bmatrix} w_1 \\ w_2 \\ \vdots \\ w_n \end{bmatrix}, \quad \text{then} \quad \mathbf{v} + \mathbf{w} := \begin{bmatrix} v_1 + w_1 \\ v_2 + w_2 \\ \vdots \\ v_n + w_n \end{bmatrix}.$$

3. A **linear combination** of $\mathbf{v}_1, \mathbf{v}_2, \ldots, \mathbf{v}_k \in \mathbb{R}^n$ is a vector of the form

$$\sum_{i=1}^{k} c_i \mathbf{v}_i = c_1 \mathbf{v}_1 + c_2 \mathbf{v}_2 + \cdots + c_k \mathbf{v}_k,$$

where $c_1, c_2, \ldots, c_k \in \mathbb{R}$.

4. The **span** of $\mathbf{v}_1, \mathbf{v}_2, \ldots, \mathbf{v}_k \in \mathbb{R}^n$ is the set $\langle \mathbf{v}_1, \mathbf{v}_2, \ldots, \mathbf{v}_k \rangle$ of all linear combinations of $\mathbf{v}_1, \mathbf{v}_2, \ldots, \mathbf{v}_k$. That is,

$$\langle \mathbf{v}_1, \mathbf{v}_2, \ldots, \mathbf{v}_k \rangle := \left\{ \sum_{i=1}^{k} c_i \mathbf{v}_i \,\middle|\, c_1, c_2, \ldots, c_k \in \mathbb{R} \right\}.$$

Examples

1. For $i = 1, \ldots, n$, let $\mathbf{e}_i \in \mathbb{R}^n$ be the vector with ith entry 1 and all other entries 0.* The vectors $\mathbf{e}_1, \ldots, \mathbf{e}_n$ are called the **standard basis vectors of** $\mathbb{R}^n$. Given any vector $\mathbf{x} = \begin{bmatrix} x_1 \\ \vdots \\ x_n \end{bmatrix} \in \mathbb{R}^n$, we can write $\mathbf{x} = x_1 \mathbf{e}_1 + \cdots + x_n \mathbf{e}_n$. Therefore $\langle \mathbf{e}_1, \ldots, \mathbf{e}_n \rangle = \mathbb{R}^n$.

2. Let $\mathbf{v}_1 = \begin{bmatrix} 1 \\ 0 \end{bmatrix}$ and $\mathbf{v}_2 = \begin{bmatrix} 1 \\ 2 \end{bmatrix}$ in $\mathbb{R}^2$. Given any $\mathbf{x} = \begin{bmatrix} x \\ y \end{bmatrix} \in \mathbb{R}^2$,

$$\begin{bmatrix} x \\ y \end{bmatrix} = \left(x - \frac{y}{2} \right) \mathbf{v}_1 + \frac{y}{2} \mathbf{v}_2.$$

Therefore $\langle \mathbf{v}_1, \mathbf{v}_2 \rangle = \mathbb{R}^2$.

3. Let

$$\mathbf{v}_1 = \begin{bmatrix} 1 \\ 0 \end{bmatrix}, \qquad \mathbf{v}_2 = \begin{bmatrix} 0 \\ 1 \end{bmatrix}, \qquad \text{and} \qquad \mathbf{v}_3 = \begin{bmatrix} 1 \\ 2 \end{bmatrix}.$$

*Notice that exactly what $\mathbf{e}_i$ means depends on the value of n. For example, when $n = 2$, $\mathbf{e}_1 = \begin{bmatrix} 1 \\ 0 \end{bmatrix}$, but when $n = 3$, $\mathbf{e}_1 = \begin{bmatrix} 1 \\ 0 \\ 0 \end{bmatrix}$. Nevertheless, to keep the notation from getting too complicated, we don't explicitly specify n when writing $\mathbf{e}_i$.

Combining the above two examples, we can see, for example, that

$$\begin{aligned}\begin{bmatrix}1\\1\end{bmatrix} &= 1\mathbf{v}_1 + 1\mathbf{v}_2 + 0\mathbf{v}_3\\ &= \frac{1}{2}\mathbf{v}_1 + 0\mathbf{v}_2 + \frac{1}{2}\mathbf{v}_3.\end{aligned}$$

This shows us that, in general, a linear combination of vectors $\mathbf{v}_1, \ldots, \mathbf{v}_k$ can be written in the form $\sum_{i=1}^{k} c_k \mathbf{v}_k$ in more than one way. ▲

The Vector Form of a Linear System

Recall our very first linear system:

$$\begin{aligned} x + \frac{1}{4}y &= 20\\ 2x + \frac{1}{4}y &= 36. \end{aligned} \tag{1.13}$$

Making use of the vector notation we've just introduced, we can rewrite this system of two equations over $\mathbb{R}$ as a single equation involving vectors in $\mathbb{R}^2$:

$$x\begin{bmatrix}1\\2\end{bmatrix} + y\begin{bmatrix}\frac{1}{4}\\\frac{1}{4}\end{bmatrix} = \begin{bmatrix}20\\36\end{bmatrix}. \tag{1.14}$$

Quick Exercise #11. Verify that equation (1.14) really is the same as the system (1.13).

Although the problem is exactly the same as before, this way of writing the system changes our perspective. From this formulation, we can see that the system (1.14) has a solution if and only if the vector $\begin{bmatrix}20\\36\end{bmatrix}$ lies in the span

$$\left\langle \begin{bmatrix}1\\2\end{bmatrix}, \begin{bmatrix}\frac{1}{4}\\\frac{1}{4}\end{bmatrix} \right\rangle.$$

The coefficients x and y which let us write the vector $\begin{bmatrix}20\\36\end{bmatrix}$ as a linear combination of $\begin{bmatrix}1\\2\end{bmatrix}$ and $\begin{bmatrix}\frac{1}{4}\\\frac{1}{4}\end{bmatrix}$ are exactly the amounts of bread and beer that we can make with our barley and yeast.

In general, an $m \times n$ linear system

$$\begin{aligned} a_{11}x_1 + a_{12}x_2 + \cdots + a_{1n}x_n &= b_1 \\ a_{21}x_1 + a_{22}x_2 + \cdots + a_{2n}x_n &= b_2 \\ &\vdots \\ a_{m1}x_1 + a_{m2}x_2 + \cdots + a_{mn}x_n &= b_m \end{aligned} \tag{1.15}$$

can be rewritten as a single equation stating that one m-dimensional vector is a linear combination of n other m-dimensional vectors with unknown coefficients:

$$x_1 \begin{bmatrix} a_{11} \\ a_{21} \\ \vdots \\ a_{m1} \end{bmatrix} + x_2 \begin{bmatrix} a_{12} \\ a_{22} \\ \vdots \\ a_{m2} \end{bmatrix} + \cdots + x_n \begin{bmatrix} a_{1n} \\ a_{2n} \\ \vdots \\ a_{mn} \end{bmatrix} = \begin{bmatrix} b_1 \\ b_2 \\ \vdots \\ b_m \end{bmatrix}. \tag{1.16}$$

Note that the augmented matrix for the system in equation (1.15) is what you get by taking the column vectors from equation (1.16) and making them the columns of the matrix:

$$\left[\begin{array}{cccc|c} a_{11} & a_{12} & \cdots & a_{1n} & b_1 \\ a_{21} & a_{22} & \cdots & a_{2n} & b_2 \\ \vdots & \vdots & \ddots & \vdots & \vdots \\ a_{m1} & a_{m2} & \cdots & a_{mn} & b_m \end{array}\right].$$

Example Let $\mathbf{v}_1 = \begin{bmatrix} 1 \\ 1 \\ -1 \end{bmatrix}$ and $\mathbf{v}_2 = \begin{bmatrix} 2 \\ -1 \\ 0 \end{bmatrix}$. The linear system

$$\begin{aligned} x + 2y &= 1 \\ x - y &= 1 \\ -x &= 1 \end{aligned}$$

is inconsistent (check that!), and so

$$\begin{bmatrix} 1 \\ 1 \\ 1 \end{bmatrix} \notin \langle \mathbf{v}_1, \mathbf{v}_2 \rangle .$$

▲

Quick Exercise #12. Show that every vector $\begin{bmatrix} x \\ y \end{bmatrix} \in \mathbb{R}^2$ can be written as a linear combination of $\begin{bmatrix} 1 \\ 0 \end{bmatrix}$, $\begin{bmatrix} 0 \\ 1 \end{bmatrix}$, and $\begin{bmatrix} 1 \\ 2 \end{bmatrix}$ in infinitely many ways.

QA #12: The matrix $\left[\begin{array}{ccc|c} 1 & 0 & 1 & x \\ 0 & 1 & 2 & y \end{array}\right]$ is already in RREF with no pivot in the last column, and the corresponding system has a free variable.

The Geometry of Linear Combinations

Recall from vector calculus how we visualize a vector in $\mathbb{R}^2$ or $\mathbb{R}^3$: if we have a vector, say $\begin{bmatrix} 1 \\ 2 \end{bmatrix} \in \mathbb{R}^2$, we draw an arrow in the x–y plane from the origin to the point with coordinates $(1, 2)$ like this:

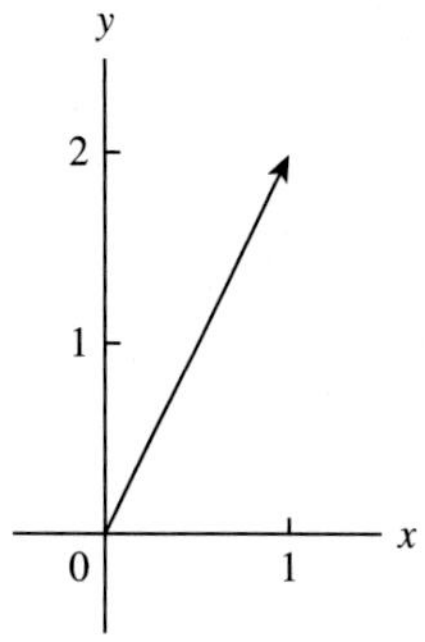

Figure 1.3 The arrow representing the vector $\begin{bmatrix} 1 \\ 2 \end{bmatrix}$ in $\mathbb{R}^2$.

We can calculate the length of the arrow using the Pythagorean Theorem: if $\mathbf{v} = \begin{bmatrix} x \\ y \end{bmatrix}$, then the **length** or **norm** or **magnitude** of the vector $\mathbf{v}$ is written $\|\mathbf{v}\|$ and is given by

$$\|\mathbf{v}\| = \sqrt{x^2 + y^2}.$$

In $\mathbb{R}^3$, we try to do basically the same thing, although drawing space on paper is always a challenge. Here is a picture of the vector $\begin{bmatrix} 1 \\ 2 \\ 3 \end{bmatrix}$:

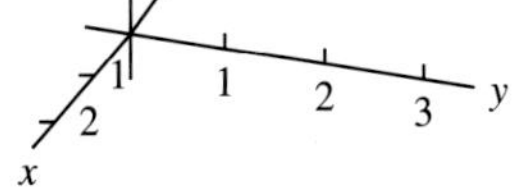

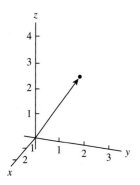

Figure 1.4 The arrow representing the vector $\begin{bmatrix} 1 \\ 2 \\ 3 \end{bmatrix}$ in $\mathbb{R}^3$.

The length of a vector $\mathbf{v} = \begin{bmatrix} x \\ y \\ z \end{bmatrix}$ is given by $\|\mathbf{v}\| = \sqrt{x^2 + y^2 + z^2}$.

The operations of scalar multiplication and addition that we defined above then have geometric interpretations. If we multiply a vector $\mathbf{v}$ by a positive scalar c, then the vector $c\mathbf{v}$ points in the same direction as $\mathbf{v}$ but has length $\|c\mathbf{v}\| = c\,\|\mathbf{v}\|$; if c is negative, then $c\mathbf{v}$ points in the opposite direction from $\mathbf{v}$ and its length is $|c|\,\|\mathbf{v}\|$.

Quick Exercise #13. If v is the vector

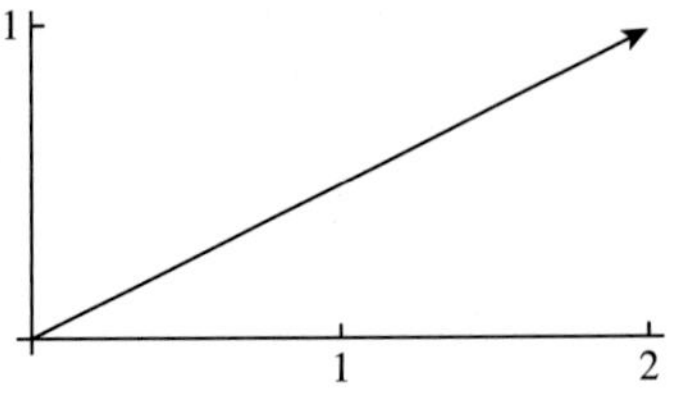

then identify which of the following are pictures of $-\mathbf{v}$, $0\mathbf{v}$, $\frac{1}{2}\mathbf{v}$, $2\mathbf{v}$, and $\pi\mathbf{v}$:

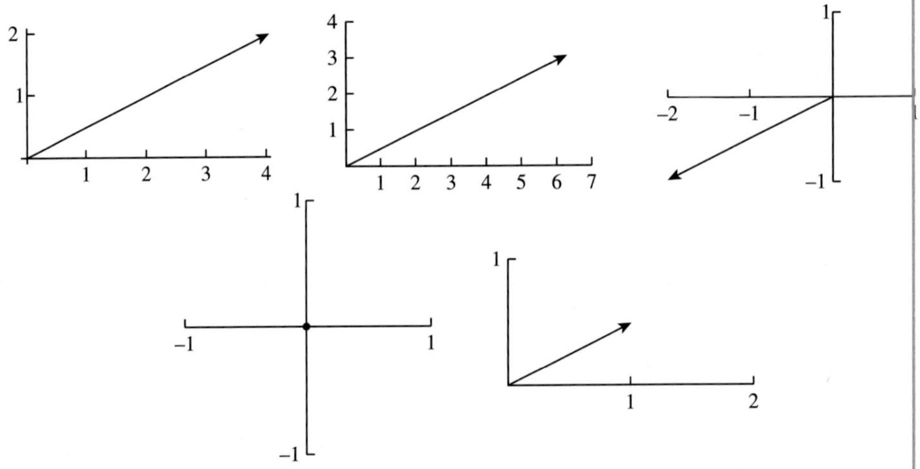

We visualize addition of $\mathbf{v}$ and $\mathbf{w}$ by moving one of the vectors, say $\mathbf{v}$, so that it is based at the tip of the other; the vector $\mathbf{v}+\mathbf{w}$ is then the one from the origin to the tip of the translated $\mathbf{v}$:

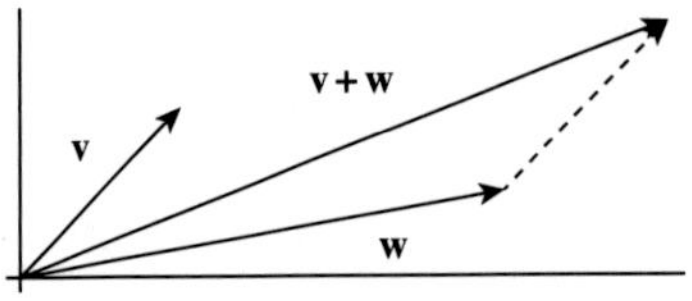

Figure 1.5 Geometric meaning of vector addition.

Examples

1. Returning to the example from page 26 of $\mathbf{v}_1 = \begin{bmatrix} 1 \\ 0 \end{bmatrix}$ and $\mathbf{v}_2 = \begin{bmatrix} 1 \\ 2 \end{bmatrix}$ in $\mathbb{R}^2$, it is geometrically clear that $\langle \mathbf{v}_1, \mathbf{v}_2 \rangle = \mathbb{R}^2$ (as we've already verified algebraically), since it's possible to get to any point in the plane using steps in the $\mathbf{v}_1$ direction together with steps in the $\mathbf{v}_2$ direction:

QA #13: In order, they are $2\mathbf{v}$, $\pi\mathbf{v}$, $-\mathbf{v}$, $0\mathbf{v}$, and $\frac{1}{2}\mathbf{v}$.

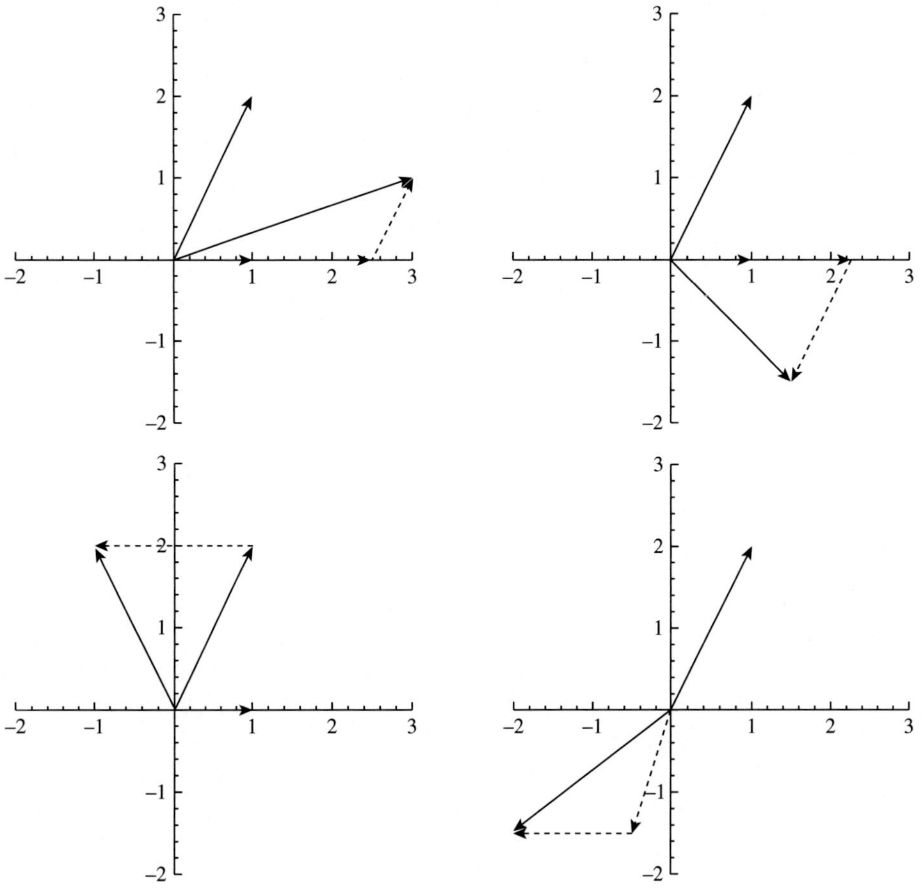

Figure 1.6 Geometrically representing different vectors in $\mathbb{R}^2$ as linear combinations of $\begin{bmatrix} 1 \\ 0 \end{bmatrix}$ and $\begin{bmatrix} 1 \\ 2 \end{bmatrix}$.

2. Returning next to the example from page 28 of $\mathbf{v}_1 = \begin{bmatrix} 1 \\ 1 \\ -1 \end{bmatrix}$ and $\mathbf{v}_2 = \begin{bmatrix} 2 \\ -1 \\ 0 \end{bmatrix}$, since $\mathbf{v}_1$ and $\mathbf{v}_2$ do not point in the same direction, the span $\langle \mathbf{v}_1, \mathbf{v}_2 \rangle$ is a plane in $\mathbb{R}^3$. We found algebraically that the vector $\begin{bmatrix} 1 \\ 1 \\ 1 \end{bmatrix}$ does not lie in this plane.

▲

With the more geometric viewpoint on vectors in mind, let's return to the 2×2 linear system written in vector form in equation (1.14):

$$x \begin{bmatrix} 1 \\ 2 \end{bmatrix} + y \begin{bmatrix} \frac{1}{4} \\ \frac{1}{4} \end{bmatrix} = \begin{bmatrix} 20 \\ 36 \end{bmatrix}.$$

Geometrically, asking if the system has a solution is the same as asking whether it is possible to get from the origin in the x–y plane to the point $(20, 36)$ by first going some distance in the direction of $\begin{bmatrix} 1 \\ 2 \end{bmatrix}$ (or the opposite direction), and from there going some distance in the direction of $\begin{bmatrix} \frac{1}{4} \\ \frac{1}{4} \end{bmatrix}$ (or the opposite direction). The answer, as we've already seen, is yes:

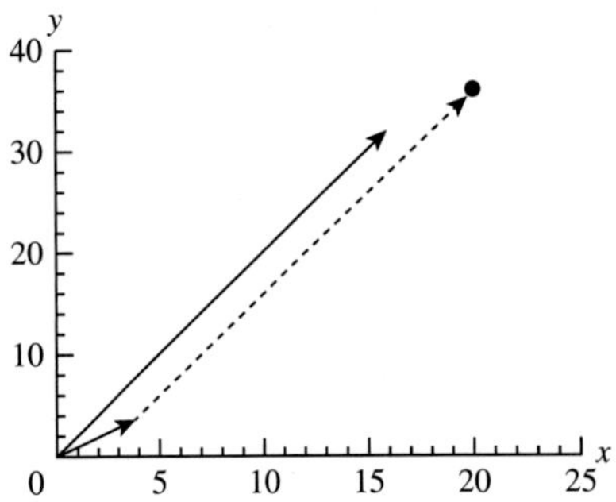

Figure 1.7 Geometric demonstration that the system (1.14) is consistent.

Moreover, this can be done in one and only one way: it's easy to see in the picture that if we change the value of x, we either overshoot or undershoot, and there's no distance we could then move in the direction of $\begin{bmatrix} \frac{1}{4} \\ \frac{1}{4} \end{bmatrix}$ to hit the point $(20, 36)$.

The situation can change a lot if we make a small change to the system. For example, if you actually had 40 teaspoons of yeast, and you suspect that your beer brewing would work better if you used half a teaspoon of yeast per pint rather than only a quarter, you would have the system

$$\begin{aligned} x + \frac{1}{4}y &= 20 \\ 2x + \frac{1}{2}y &= 40, \end{aligned} \tag{1.17}$$

or in vector form:

$$x\begin{bmatrix} 1 \\ 2 \end{bmatrix} + y\begin{bmatrix} \frac{1}{4} \\ \frac{1}{2} \end{bmatrix} = \begin{bmatrix} 20 \\ 40 \end{bmatrix}. \tag{1.18}$$

In this case, it is possible to get from the origin in the x–y plane to the point $(20, 40)$ by first going some distance in the direction of $\begin{bmatrix} 1 \\ 2 \end{bmatrix}$, and from there going some distance in the direction of $\begin{bmatrix} \frac{1}{4} \\ \frac{1}{2} \end{bmatrix}$, and there are in fact infinitely many ways to do it: the vectors $\begin{bmatrix} 1 \\ 2 \end{bmatrix}$ and $\begin{bmatrix} \frac{1}{4} \\ \frac{1}{2} \end{bmatrix}$ both point directly from the origin toward the point $(20, 40)$:

What's happening here is that there is redundancy in the geometry; we don't need both of the vectors $\begin{bmatrix} 1 \\ 2 \end{bmatrix}, \begin{bmatrix} \frac{1}{4} \\ \frac{1}{2} \end{bmatrix}$ to get to the point $(20, 40)$. We can also see the

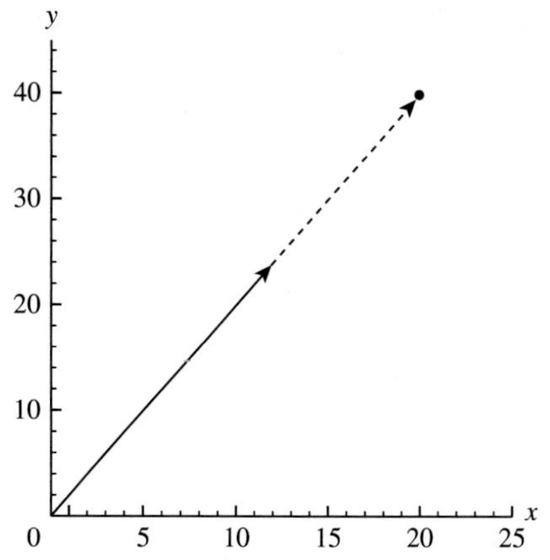

Figure 1.8 Geometric illustration of one solution of the system (1.18).

redundancy in the system from an algebraic point of view: in the system (1.17), the second equation is just the first equation multiplied by 2, so it contains no new information.

The situation would again have been entirely different if you had wanted to use more yeast in your brewing process, but you didn't happen to have any extra lying around. Then the new system would have been

$$\begin{aligned} x + \frac{1}{4}y &= 20 \\ 2x + \frac{1}{2}y &= 36, \end{aligned} \tag{1.19}$$

or in vector form:

$$x \begin{bmatrix} 1 \\ 2 \end{bmatrix} + y \begin{bmatrix} \frac{1}{4} \\ \frac{1}{2} \end{bmatrix} = \begin{bmatrix} 20 \\ 36 \end{bmatrix}. \tag{1.20}$$

The system (1.20) is inconsistent.

Quick Exercise #14. (a) Draw a picture that shows that the system (1.20) is inconsistent.
(b) Show algebraically that the system (1.20) is inconsistent.

The Geometry of Solutions

So far in this section, the vectors whose geometry we've exploited have been the columns of the augmented matrix of a linear system. The vectors in question came

QA #14: (a) 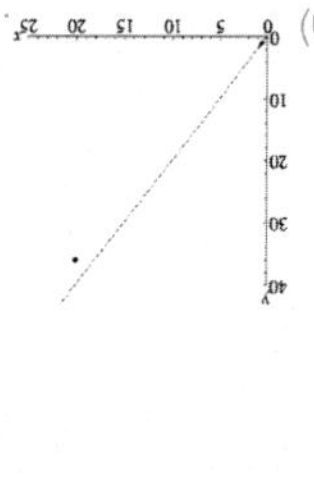 (b) $36 \neq 40$. Alternatively, the RREF is $\left[\begin{array}{cc|c} 1 & \frac{1}{4} & 20 \\ 0 & 0 & 1 \end{array}\right]$.

directly from the coefficients of the system. Another way that geometry can shed light on linear systems is by viewing the solutions of a system as vectors. Recall the system

$$\begin{aligned} x + \frac{1}{4}y + \frac{1}{2}w &= 20 \\ 2x + \frac{1}{4}y &= 36 \\ x + 8z + w &= 176 \end{aligned} \tag{1.21}$$

describing how much bread, beer, cheese, and muesli can be made from our supply of barley, yeast, and milk. A solution of this system consists of values for x, y, z, w, which we can write as a vector $\begin{bmatrix} x \\ y \\ z \\ w \end{bmatrix} \in \mathbb{R}^4$. In Section 1.2, we determined that x, y, and z are the pivot variables, and we found the following formulas for them in terms of the free variable w:

$$\begin{aligned} x &= 16 + \frac{1}{2}w \\ y &= 16 - 4w \\ z &= 20 - \frac{3}{16}w. \end{aligned} \tag{1.22}$$

If we add to these the trivially true formula $w = w$, then the system (1.22) is equivalent to the vector equation

$$\begin{bmatrix} x \\ y \\ z \\ w \end{bmatrix} = \begin{bmatrix} 16 + \frac{1}{2}w \\ 16 - 4w \\ 20 - \frac{3}{16}w \\ w \end{bmatrix} = \begin{bmatrix} 16 \\ 16 \\ 20 \\ 0 \end{bmatrix} + w \begin{bmatrix} \frac{1}{2} \\ -4 \\ -\frac{3}{16} \\ 1 \end{bmatrix}.$$

Geometrically, this says that the set of solutions of the system (1.21) is the line in $\mathbb{R}^4$ through the point $(16, 16, 20, 0)$ with direction vector $\begin{bmatrix} \frac{1}{2} \\ -4 \\ -\frac{3}{16} \\ 1 \end{bmatrix}$.

The same basic approach works with multiple free variables. Consider the 3×5 linear system

$$\begin{aligned} x_1 + 2x_2 + 5x_4 &= -3 \\ -x_1 - 2x_2 + x_3 - 6x_4 + x_5 &= 2 \\ -2x_1 - 4x_2 - 10x_4 + x_5 &= 8. \end{aligned} \tag{1.23}$$

Quick Exercise #15. Show that the RREF of the augmented matrix of the system (1.23) is

$$\left[\begin{array}{ccccc|c} 1 & 2 & 0 & 5 & 0 & -3 \\ 0 & 0 & 1 & -1 & 0 & -3 \\ 0 & 0 & 0 & 0 & 1 & 2 \end{array}\right].$$

This gives us the formulas

$$\begin{aligned} x_1 &= -3 - 2x_2 - 5x_4 \\ x_3 &= 3 + x_4 \\ x_5 &= 2 \end{aligned}$$

for the pivot variables x_1, x_3, x_5 in terms of the free variables x_2 and x_4. In vector form, this becomes

$$\begin{bmatrix} x_1 \\ x_2 \\ x_3 \\ x_4 \\ x_5 \end{bmatrix} = \begin{bmatrix} -3 - 2x_2 - 5x_4 \\ x_2 \\ -3 + x_4 \\ x_4 \\ 2 \end{bmatrix} = \begin{bmatrix} -3 \\ 0 \\ -3 \\ 0 \\ 2 \end{bmatrix} + x_2 \begin{bmatrix} -2 \\ 1 \\ 0 \\ 0 \\ 0 \end{bmatrix} + x_4 \begin{bmatrix} -5 \\ 0 \\ 1 \\ 1 \\ 0 \end{bmatrix}.$$

In general, we can write the solutions of a consistent linear system with k free variables as a fixed vector plus an arbitrary linear combination of k other fixed vectors.

KEY IDEAS

- $\mathbb{R}^n$ is the set of column vectors of length n.
- A linear combination of the vectors $\mathbf{v}_1, \ldots, \mathbf{v}_n$ is a vector of the form $c_1\mathbf{v}_1 + \cdots + c_n\mathbf{v}_n$ for real numbers $c_1, \ldots, c_n$.
- The span of a list of vectors is the set of all linear combinations of those vectors.

EXERCISES

1.3.1 For each pair of vectors below, draw sketches illustrating $\mathbf{v}$, $\mathbf{w}$, $\mathbf{v} + \mathbf{w}$, $\mathbf{v} - \mathbf{w}$, and $2\mathbf{v} - 3\mathbf{w}$.

(a) $\mathbf{v} = \begin{bmatrix} 1 \\ 2 \end{bmatrix}$, $\mathbf{w} = \begin{bmatrix} 2 \\ 1 \end{bmatrix}$ (b) $\mathbf{v} = \begin{bmatrix} -1 \\ 1 \end{bmatrix}$, $\mathbf{w} = \begin{bmatrix} 2 \\ 1 \end{bmatrix}$ (c) $\mathbf{v} = \begin{bmatrix} -2 \\ -1 \end{bmatrix}$, $\mathbf{w} = \begin{bmatrix} 3 \\ -1 \end{bmatrix}$ (d) $\mathbf{v} = \begin{bmatrix} -1 \\ 1 \end{bmatrix}$, $\mathbf{w} = \begin{bmatrix} -1 \\ 1 \end{bmatrix}$ (e) $\mathbf{v} = \begin{bmatrix} 3 \\ -3 \end{bmatrix}$, $\mathbf{w} = \begin{bmatrix} 2 \\ -2 \end{bmatrix}$

1.3.2 For each pair of vectors below, draw sketches illustrating $\mathbf{v}$, $\mathbf{w}$, $\mathbf{v}+\mathbf{w}$, $\mathbf{v}-\mathbf{w}$, and $3\mathbf{v}-2\mathbf{w}$.

(a) $\mathbf{v}=\begin{bmatrix}3\\1\end{bmatrix}$, $\mathbf{w}=\begin{bmatrix}0\\1\end{bmatrix}$ (b) $\mathbf{v}=\begin{bmatrix}1\\-1\end{bmatrix}$, $\mathbf{w}=\begin{bmatrix}1\\2\end{bmatrix}$ (c) $\mathbf{v}=\begin{bmatrix}-1\\-2\end{bmatrix}$, $\mathbf{w}=\begin{bmatrix}3\\1\end{bmatrix}$ (d) $\mathbf{v}=\begin{bmatrix}-1\\2\end{bmatrix}$, $\mathbf{w}=\begin{bmatrix}1\\-2\end{bmatrix}$ (e) $\mathbf{v}=\begin{bmatrix}1\\-1\end{bmatrix}$, $\mathbf{w}=\begin{bmatrix}2\\-2\end{bmatrix}$

1.3.3 Describe, in vector form, the set of all solutions of each of the following linear systems.

(a) $\begin{aligned} 2x-y+z&=3\\ -x+y&=-2\\ x+2y+3z&=-1\end{aligned}$ (b) $\begin{aligned} 2x-y+z&=0\\ -x+y&=1\\ x+2y+3z&=5\end{aligned}$

(c) $\begin{aligned} x+y+z&=1\\ x-2y+3z&=4\end{aligned}$ (d) $\begin{aligned} x_1+2x_2+3x_3+4x_4&=5\\ 6x_1+7x_2+8x_3+9x_4&=10\\ 11x_1+12x_2+13x_3+14x_4&=15\end{aligned}$

(e) $\begin{aligned} x-2y&=0\\ -3x+6y&=0\\ 2x-4y&=0\end{aligned}$ (f) $\begin{aligned} x-2y&=1\\ -3x+6y&=-3\\ 2x-4y&=2\end{aligned}$

1.3.4 Describe, in vector form, the set of all solutions of each of the following linear systems.

(a) $\begin{aligned} x+2y+3z&=2\\ 2x-y+z&=4\\ -x+y&=-2\end{aligned}$ (b) $\begin{aligned} x+2y+3z&=4\\ 2x-y+z&=3\\ -x+y&=-1\end{aligned}$

(c) $\begin{aligned} 2x-y+3z&=0\\ x+2y-3z&=1\end{aligned}$ (d) $\begin{aligned} x_1-2x_2+3x_3-4x_4&=5\\ -6x_1+7x_2-8x_3+9x_4&=-10\\ 11x_1-12x_2+13x_3-14x_4&=15\end{aligned}$

(e) $\begin{aligned} -x+y&=0\\ 2x-2y&=0\\ -3x+3y&=0\end{aligned}$ (f) $\begin{aligned} -x+y&=3\\ 2x-2y&=2\\ -3x+3y&=1\end{aligned}$

1.3.5 (a) Under what conditions is the vector $\begin{bmatrix} b_1 \\ b_2 \\ b_3 \\ b_4 \end{bmatrix}$ in

$\left\langle \begin{bmatrix} 1 \\ 0 \\ 2 \\ -3 \end{bmatrix}, \begin{bmatrix} 2 \\ -1 \\ 0 \\ 1 \end{bmatrix}, \begin{bmatrix} -1 \\ 1 \\ -2 \\ 4 \end{bmatrix} \right\rangle$? Give your answer in terms of an equation or equations satisfied by the entries $b_1, \ldots, b_4$.

(b) Which of the following vectors are in the span

$$\left\langle \begin{bmatrix} 1 \\ 0 \\ 2 \\ -3 \end{bmatrix}, \begin{bmatrix} 2 \\ -1 \\ 0 \\ 1 \end{bmatrix}, \begin{bmatrix} -1 \\ 1 \\ -2 \\ 4 \end{bmatrix} \right\rangle?$$

(i) $\begin{bmatrix} 0 \\ 3 \\ -1 \\ 1 \end{bmatrix}$ (ii) $\begin{bmatrix} 1 \\ -1 \\ 3 \\ 0 \end{bmatrix}$ (iii) $\begin{bmatrix} 4 \\ 2 \\ 0 \\ -1 \end{bmatrix}$ (iv) $\begin{bmatrix} 1 \\ 6 \\ 2 \\ 3 \end{bmatrix}$

1.3.6 (a) Under what conditions is the vector $\begin{bmatrix} b_1 \\ b_2 \\ b_3 \\ b_4 \end{bmatrix}$ in

$\left\langle \begin{bmatrix} 1 \\ 0 \\ -1 \\ 0 \end{bmatrix}, \begin{bmatrix} 0 \\ 2 \\ 1 \\ 1 \end{bmatrix}, \begin{bmatrix} 2 \\ -1 \\ 0 \\ 0 \end{bmatrix} \right\rangle$? Give your answer in terms of an equation or equations satisfied by the entries $b_1, \ldots, b_4$.

(b) Which of the following vectors are in the span

$$\left\langle \begin{bmatrix} 1 \\ 0 \\ -1 \\ 0 \end{bmatrix}, \begin{bmatrix} 0 \\ 2 \\ 1 \\ 1 \end{bmatrix}, \begin{bmatrix} 2 \\ -1 \\ 0 \\ 0 \end{bmatrix} \right\rangle?$$

(i) $\begin{bmatrix} 0 \\ 3 \\ -1 \\ 1 \end{bmatrix}$ (ii) $\begin{bmatrix} 1 \\ -1 \\ 3 \\ 0 \end{bmatrix}$ (iii) $\begin{bmatrix} 4 \\ 2 \\ 0 \\ -1 \end{bmatrix}$ (iv) $\begin{bmatrix} 1 \\ 6 \\ 2 \\ 3 \end{bmatrix}$

1.3.7 Describe all possible ways of writing $\begin{bmatrix} -1 \\ 1 \end{bmatrix}$ as a linear combination of the vectors $\begin{bmatrix} 1 \\ 2 \end{bmatrix}$, $\begin{bmatrix} 3 \\ 4 \end{bmatrix}$, and $\begin{bmatrix} 5 \\ 6 \end{bmatrix}$.

1.3.8 Describe all possible ways of writing $\begin{bmatrix} 2 \\ 1 \end{bmatrix}$ as a linear combination of the vectors $\begin{bmatrix} 3 \\ 1 \end{bmatrix}$, $\begin{bmatrix} 4 \\ 1 \end{bmatrix}$, and $\begin{bmatrix} 5 \\ 1 \end{bmatrix}$.

1.3.9 Describe all possible ways of writing $\begin{bmatrix} 1 \\ 2 \\ 3 \\ 4 \end{bmatrix}$ as a linear combination of the vectors

$$\begin{bmatrix} 1 \\ 1 \\ 0 \\ 0 \end{bmatrix}, \begin{bmatrix} 1 \\ 0 \\ 1 \\ 0 \end{bmatrix}, \begin{bmatrix} 1 \\ 0 \\ 0 \\ 1 \end{bmatrix}, \begin{bmatrix} 0 \\ 1 \\ 1 \\ 0 \end{bmatrix}, \begin{bmatrix} 0 \\ 1 \\ 0 \\ 1 \end{bmatrix}, \begin{bmatrix} 0 \\ 0 \\ 1 \\ 1 \end{bmatrix}.$$

1.3.10 Consider a linear system in vector form:

$$x_1\mathbf{v}_1 + \cdots + x_n\mathbf{v}_n = \mathbf{b},$$

where $\mathbf{v}_1, \ldots, \mathbf{v}_n, \mathbf{b} \in \mathbb{R}^m$. Show that the system is consistent if and only if

$$\mathbf{b} \in \langle \mathbf{v}_1, \ldots, \mathbf{v}_n \rangle .$$

1.3.11 Use the Pythagorean Theorem twice to derive the formula on page 29 for the length of a vector in $\mathbb{R}^3$. Include a picture in your solution.

1.3.12 Let $\mathbf{v}_1, \mathbf{v}_2, \mathbf{v}_3 \in \mathbb{R}^3$, and suppose that the linear system

$$x\mathbf{v}_1 + y\mathbf{v}_2 + z\mathbf{v}_3 = \mathbf{0}$$

has infinitely many solutions. Show that $\mathbf{v}_1, \mathbf{v}_2, \mathbf{v}_3$ lie in a plane containing the origin in $\mathbb{R}^3$.

1.3.13 Two vectors $\mathbf{v}, \mathbf{w} \in \mathbb{R}^n$ are called **collinear** if $\mathbf{v} = a\mathbf{w}$ or $\mathbf{w} = a\mathbf{v}$ for some $a \in \mathbb{R}$. Show that the span of any two nonzero vectors in $\mathbb{R}^2$ which are not collinear is all of $\mathbb{R}^2$.

1.3.14 Give an example of three vectors $\mathbf{u}, \mathbf{v}, \mathbf{w} \in \mathbb{R}^3$, such that no two of them are collinear (see Exercise 1.3.13), but $\langle \mathbf{u}, \mathbf{v}, \mathbf{w} \rangle \neq \mathbb{R}^3$.

1.3.15 Give a geometric description of the set of solutions of a consistent linear system over $\mathbb{R}$ in three variables if it has k free variables, for $k = 0, 1, 2, 3$.

1.4 Fields*

General Fields

So far, we've assumed that all the numbers we've worked with – as coefficients or as possible values of variables – have been real numbers. The set $\mathbb{R}$ of real numbers is part of a hierarchy of different types of numbers: in one direction, there is the more general class of complex numbers, and in the other direction, there are special classes of real numbers, like integers or rational numbers. We may be interested in solving linear systems with complex coefficients, or we may have a system with integer coefficients and be looking for integer solutions. The first question we want to address is how much of what we've done in the previous sections carries over to other number systems.

Our algorithm for solving linear systems was Gaussian elimination, described in the proof of Theorem 1.1, so we begin by examining that algorithm to see what algebraic operations we used.

Quick Exercise #16. Reread the proof of Theorem 1.1, and keep track of what types of algebraic operation are used.

As you've presumably confirmed, there are only four operations involved: addition, subtraction, multiplication, and division, which suggests the following definition.

Definition A **field** consists of a set $\mathbb{F}$ with operations $+$ and $\cdot$, and distinct elements $0, 1 \in \mathbb{F}$, such that all of the following properties hold:

$\mathbb{F}$ is closed under addition: For each $a, b \in \mathbb{F}$, $a + b \in \mathbb{F}$.

$\mathbb{F}$ is closed under multiplication: For each $a, b \in \mathbb{F}$, $ab = a \cdot b \in \mathbb{F}$.

Addition is commutative: For each $a, b \in \mathbb{F}$, $a + b = b + a$.

Addition is associative: For each $a, b, c \in \mathbb{F}$, $a + (b + c) = (a + b) + c$.

0 is an additive identity: For each $a \in \mathbb{F}$, $a + 0 = 0 + a = a$.

Every element has an additive inverse: For each $a \in \mathbb{F}$, there is an element $b \in \mathbb{F}$ such that $a + b = b + a = 0$.

Multiplication is commutative: For each $a, b \in \mathbb{F}$, $ab = ba$.

Multiplication is associative: For each $a, b, c \in \mathbb{F}$, $a(bc) = (ab)c$.

*Most of the rest of this book works with scalars from an arbitrary field (except Chapters 4 and 5, in which scalars are always real or complex), but this is only essential in Section 2.6. Readers who are interested only in real and complex scalars can skip to Section 1.5, mentally replacing each occurence of $\mathbb{F}$ with $\mathbb{R}$ or $\mathbb{C}$.

QA #16: It's on page 15.

> **1 is a multiplicative identity:** For each $a \in \mathbb{F}$, $a1 = 1a = a$.
>
> **Every nonzero element has a multiplicative inverse:** For each $a \in \mathbb{F}$ such that $a \neq 0$, there is an element $c \in \mathbb{F}$ such that $ac = ca = 1$.
>
> **Distributive law:** For each $a, b, c \in \mathbb{F}$, $a(b + c) = ab + ac$.

For notational convenience, we will denote the additive inverse of a by $-a$, so that

$$a + (-a) = (-a) + a = 0,$$

and the multiplicative inverse (for $a \neq 0$) by a^{-1}, so that

$$aa^{-1} = a^{-1}a = 1.$$

We will also call the addition of an additive inverse **subtraction** and denote it as usual:

$$a - b := a + (-b).$$

Similarly, we call multiplication by a multiplicative inverse **division** and denote it by

$$\frac{a}{b} := ab^{-1}.$$

> **Bottom Line:** a field consists of a bunch of things that you can add, subtract, multiply, and divide with, and those operations behave basically as you expect.

Examples

1. **The real numbers:** the set $\mathbb{R}$ of real numbers, together with ordinary addition and multiplication, satisfies all the properties above, so $\mathbb{R}$ is a field.
2. **The complex numbers:** the set

   $$\mathbb{C} := \{a + ib \mid a, b \in \mathbb{R}\}$$

   of complex numbers, with addition and multiplication defined by

   $$(a + ib) + (c + id) := a + c + i(b + d)$$

   and

   $$(a + ib) \cdot (c + id) = ac - bd + i(ad + bc)$$

   is a field. In particular, recall the following formula for the multiplicative inverse of a complex number.

Quick Exercise #17. Show that if $a, b \in \mathbb{R}$ and a and b are not both 0, then $(a+ib)\left(\frac{a}{a^2+b^2} - i\frac{b}{a^2+b^2}\right) = 1$.

Notice that this exercise doesn't just show that $(a+ib)^{-1} = \frac{a}{a^2+b^2} - i\frac{b}{a^2+b^2}$; more fundamentally, it shows that a complex number which is a multiplicative inverse of $(a+ib)$ exists at all, which is not *a priori** obvious.

3. **The integers:** Let $\mathbb{Z}$ denote the set of integers.† Then $\mathbb{Z}$ is *not* a field: nonzero elements other than 1 and -1 do not have multiplicative inverses in $\mathbb{Z}$.
4. **The rational numbers:** Let

$$\mathbb{Q} := \left\{ \frac{m}{n} \;\middle|\; m, n \in \mathbb{Z},\; n \neq 0 \right\}.$$

Then $\mathbb{Q}$, with the usual addition and multiplication, is a field.‡

5. **The field of two elements:** Define $\mathbb{F}_2 := \{0, 1\}$ with the following addition and multiplication tables:

+	0	1
0	0	1
1	1	0

·	0	1
0	0	0
1	0	1

One can simply check that the properties of the field are satisfied. Note in particular that in $\mathbb{F}_2$, $1 = -1$; this makes $\mathbb{F}_2$ a very different place from $\mathbb{R}$, where the only number equal to its own additive inverse is 0.

Working with the field $\mathbb{F}_2$ is often useful in computer science, since computers internally represent data as strings of 0s and 1s.

Quick Exercise #18. Show that if $a, b \in \mathbb{F}_2$, then $a + b = 1$ if and only if:

$a = 1$ or $b = 1$, but *not* both.

This means that addition in $\mathbb{F}_2$ models the logical operation called "exclusive or." ▲

Technically, a field consists of the set $\mathbb{F}$ together with the operations $+$ and $\cdot$, but we will normally just use $\mathbb{F}$ to refer to both, with the operations understood. We do need to be careful, however, because the same symbols may mean different

*This is a Latin phrase that mathematicians use a lot. It refers to what you know before doing any deep thinking.

† The letter $\mathbb{Z}$ is traditionally used because the German word for integers is "Zahlen."

‡ $\mathbb{Q}$ stands for "quotient."

QA #17: $(a+ib)\left(\frac{a-ib}{a^2+b^2}\right) = \frac{a^2+b^2}{a^2+b^2} = 1.$

things in different contexts: for example, we saw above that $1 + 1 = 0$ is true in $\mathbb{F}_2$, but not in $\mathbb{R}$.

Arithmetic in Fields

The following theorem collects a lot of "obvious" facts about arithmetic in fields. In the field $\mathbb{R}$, these are just the familiar rules of arithmetic, but it's useful to know that they continue to be true in other fields. When defining a field, we required as few things as we could (so that checking that something is a field is as quick as possible), but all of the following properties follow quickly from the ones in the definition.

Theorem 1.5 *All of the following hold for any field* $\mathbb{F}$.

1. *The additive identity is unique.*
2. *The multiplicative identity is unique.*
3. *Additive inverses are unique.*
4. *Multiplicative inverses are unique.*
5. *For every* $a \in \mathbb{F}$, $-(-a) = a$.
6. *For every nonzero* $a \in \mathbb{F}$, $(a^{-1})^{-1} = a$.
7. *For every* $a \in \mathbb{F}$, $0a = 0$.
8. *For every* $a \in \mathbb{F}$, $(-1)a = -a$.
9. *If* $a, b \in \mathbb{F}$ *and* $ab = 0$, *then either* $a = 0$ *or* $b = 0$.

In interpreting the final statement of the theorem, recall the following.

In mathematical English, the statement "A or B" always means "A or B or both."

Proof 1. What it means to say that the additive identity is unique is that 0 is the only element of $\mathbb{F}$ that functions as an additive identity; i.e., if there is an element $a \in \mathbb{F}$ with $a + b = b$ for each $b \in \mathbb{F}$, then in fact $a = 0$.
Suppose then that $a \in \mathbb{F}$ satisfies $a + b = b$ for every $b \in \mathbb{F}$. Taking $b = 0$ implies that $a + 0 = 0$. By definition, 0 is an additive identity, so $a + 0 = a$; i.e. $0 = a + 0 = a$.
2. Exercise (see Exercise 1.4.16).
3. Exercise (see Exercise 1.4.17).
4. Let $a \in \mathbb{F}$ be nonzero, and let $b \in \mathbb{F}$ be such that $ab = 1$. Then

$$a^{-1} = a^{-1}1 = a^{-1}(ab) = (a^{-1}a)b = 1b = b.$$

5. For any $b \in \mathbb{F}$, the notation $-b$ means the unique additive inverse of b in $\mathbb{F}$ (its uniqueness is part 3 of this theorem). That means that we need to show that a

acts as the additive inverse of $-a$, which means $a + (-a) = 0$. But this is true by the definition of $-a$.

6. Exercise (see Exercise 1.4.18).
7. Since $0 = 0 + 0$,

$$0a = (0 + 0)a = 0a + 0a,$$

and so

$$0 = 0a - 0a = (0a + 0a) - 0a = 0a + (0a - 0a) = 0a + 0 = 0a.$$

(Note that all we did above was write out, in excruciating detail, "subtract $0a$ from both sides of $0a = 0a + 0a$.")

8. Exercise (see Exercise 1.4.19).
9. Suppose that $a \neq 0$. Then by part 7,

$$b = 1b = (a^{-1}a)b = a^{-1}(ab) = a^{-1}0 = 0.$$ ▲

Remark To prove the statement $a = 0$ or $b = 0$ in part 9, we employed the strategy of assuming that $a \neq 0$, and showing that in that case we must have $b = 0$. This is a common strategy when proving that (at least) one of two things must happen: show that if one of them doesn't happen, then the other one has to. ▲

> **Bottom Line, again:** a field consists of a bunch of things that you can add, subtract, multiply, and divide with, and those operations behave basically as you expect.

Example To illustrate that the properties in the definition of a field are all that are needed for lots of familiar algebra, suppose that x is an element of some field $\mathbb{F}$ and $x^2 = 1$ (where, as usual, $x^2 := x \cdot x$). Then we know

$$0 = 1 - 1 = x^2 - 1.$$

By Theorem 1.5 part 8,

$$0 = x^2 - 1 = x^2 + x - x - 1 = x \cdot x + x \cdot 1 + (-1)x + (-1),$$

and then using the distributive law twice,

$$0 = x(x + 1) + (-1)(x + 1) = (x - 1)(x + 1).$$

By Theorem 1.5 part 9, this means that either $x - 1 = 0$ or $x + 1 = 0$. In other words, we've proved that $x = 1$ and $x = -1$ are the only solutions of $x^2 = 1$ in any field.* ▲

*One unexpected thing can happen here, though: in $\mathbb{F}_2$, $-1 = 1$ and so the equation $x^2 = 1$ actually has only one solution!

Linear Systems over a Field

Our motivation for the definition of a field was that it provided the context in which we could solve linear systems of equations. For completeness, we repeat the definitions that we gave over $\mathbb{R}$ for general fields.

Definition A **linear system over $\mathbb{F}$ of m equations in n variables** is a set of equations of the form

$$\begin{aligned} a_{11}x_1 + \cdots + a_{1n}x_n &= b_1, \\ &\vdots \\ a_{m1}x_1 + \cdots + a_{mn}x_n &= b_m. \end{aligned} \tag{1.24}$$

Here, $a_{i,j} \in \mathbb{F}$ for each pair (i,j) with $1 \le i \le m$ and $1 \le j \le n$, $b_i \in \mathbb{F}$ for each $1 \le i \le m$, and $x_1, \ldots, x_n$ are the n variables of the system.

A **solution** of the linear system (1.24) is a set of elements $c_1, \ldots, c_n \in \mathbb{F}$ such that

$$\begin{aligned} a_{11}c_1 + \cdots + a_{1n}c_n &= b_1, \\ &\vdots \\ a_{m1}c_1 + \cdots + a_{mn}c_n &= b_m. \end{aligned}$$

Definition Let m and n be natural numbers. An $m \times n$ **matrix** A over $\mathbb{F}$ is a doubly indexed collection of elements of $\mathbb{F}$:

$$\mathrm{A} = \left[a_{ij}\right]_{\substack{1 \le i \le m \\ 1 \le j \le n}}.$$

That is, for each ordered pair of integers (i,j) with $i \in \{1, 2, \ldots, m\}$ and $j \in \{1, 2, \ldots, n\}$, there is a corresponding $a_{ij} \in \mathbb{F}$, called the (i,j) **entry** of A. We sometimes denote the (i,j) entry of A by $[\mathrm{A}]_{ij}$.

We denote the set of $m \times n$ matrices over $\mathbb{F}$ by $\mathrm{M}_{m,n}(\mathbb{F})$.

Definition Given an $m \times n$ linear system

$$\begin{aligned} a_{11}x_1 + \cdots + a_{1n}x_n &= b_1, \\ &\vdots \\ a_{m1}x_1 + \cdots + a_{mn}x_n &= b_m, \end{aligned}$$

over $\mathbb{F}$, its **augmented matrix** is the $m \times (n+1)$ matrix

$$\left[\begin{array}{ccc|c} a_{11} & \cdots & a_{1n} & b_1 \\ \vdots & \ddots & \vdots & \vdots \\ a_{m1} & \cdots & a_{mn} & b_m \end{array}\right].$$

Recall the definition of row-echelon form and reduced row-echelon form on page 15; since the only numbers appearing in the definition are 1 and 0, we can re-interpret the definition as having been made in an arbitrary field. We then have the following result exactly as in the real case.

Theorem 1.6 *Let $\mathbb{F}$ be a field. Every linear system over $\mathbb{F}$ can be put into reduced row-echelon form using row operations R1, R2, and R3.*

Proof The proof is identical to that of Theorem 1.1; it is a good exercise to go back through that proof and identify which field properties are used in each step (and to confirm that the properties from the definition of a field and Theorem 1.5 are all that's needed to make the proof work). ▲

Quick Exercise #19. Go back and do that.

We can use RREF to compute solutions, or to analyze their existence and uniqueness just as before. The next two results are, similarly, proved just like Theorems 1.2 and 1.3.

Theorem 1.7 *Suppose the augmented matrix* A *of a linear system is in REF. Then the system is consistent if and only if there is no pivot in the last column of* A.

Theorem 1.8 *Suppose the augmented matrix* A *of a linear system in n variables is in REF and the system is consistent. Then the system has a unique solution if and only if there is a pivot in each of the first n columns of* A.

Example Consider the linear system

$$\begin{aligned} x + iz &= -1 \\ -ix - y + (-1+i)z &= 3i \\ iy + (1+2i)z &= 2 \end{aligned}$$

over $\mathbb{C}$. We perform Gaussian elimination to find the RREF of the augmented matrix:

$$\left[\begin{array}{ccc|c} 1 & 0 & i & -1 \\ -i & -1 & -1+i & 3i \\ 0 & i & 1+2i & 2 \end{array}\right] \to \left[\begin{array}{ccc|c} 1 & 0 & i & -1 \\ 0 & -1 & -2+i & 2i \\ 0 & i & 1+2i & 2 \end{array}\right]$$
$$\to \left[\begin{array}{ccc|c} 1 & 0 & i & -1 \\ 0 & 1 & 2-i & -2i \\ 0 & i & 1+2i & 2 \end{array}\right] \to \left[\begin{array}{ccc|c} 1 & 0 & i & -1 \\ 0 & 1 & 2-i & -2i \\ 0 & 0 & 0 & 0 \end{array}\right]. \tag{1.25}$$

Quick Exercise #20. Identify which row operations were performed in (1.25).

Since the pivots are only in the first two columns, the system is consistent, but does not have a unique solution. We can solve for the pivot variables x and y in terms of the free variable z, which can be any complex number:

$$x = -1 - iz$$
$$y = -2i - (2-i)z.$$

In other words, the set of solutions of the system is

$$\left\{ \begin{bmatrix} -1 - iz \\ -2i - (2-i)z \\ z \end{bmatrix} \middle| \, z \in \mathbb{C} \right\}.$$

KEY IDEAS

- A field is a set of things you can add and multiply, with arithmetic which works the same way as the arithmetic of real numbers.
- The most important fields are $\mathbb{R}$ (the real numbers), $\mathbb{C}$ (the complex numbers), $\mathbb{Q}$ (the rational numbers), and $\mathbb{F}_2$ (the field of two elements – 0 and 1).
- Linear systems with coefficients in a general field can be treated in exactly the same way as linear systems over $\mathbb{R}$.

EXERCISES

1.4.1 Let $\mathbb{F} = \{a + b\sqrt{5} : a, b \in \mathbb{Q}\}$. Show that $\mathbb{F}$ is a field.
Hint: Since $\mathbb{F} \subseteq \mathbb{R}$, you can take things like associativity, commutativity, and the distributive law as known. What you need to check is that $0, 1 \in \mathbb{F}$, that the sum and product of two numbers in $\mathbb{F}$ is actually in $\mathbb{F}$, and that the additive and multiplicative inverses of a number in $\mathbb{F}$ are in $\mathbb{F}$.

QA #20: Add i times row 1 to row 2; multiply row 2 by -1; add $-i$ times row 2 to row 3.

1.4.2 Let $\mathbb{F} = \{a + bi : a, b \in \mathbb{Q}\}$, where $i^2 = -1$. Show that $\mathbb{F}$ is a field.

1.4.3 Explain why each of the following is not a field.

(a) $\{x \in \mathbb{R} \mid x \neq 0\}$

(b) $\{x \in \mathbb{R} \mid x \geq 0\}$

(c) $\{x + iy \mid x, y \geq 0\} \cup \{x + iy \mid x, y \leq 0\}$

(d) $\left\{\frac{m}{2^n} \mid m, n \in \mathbb{Z}\right\}$

1.4.4 Solve the following 2×2 system, where a, b, c, d, f are all nonzero elements of a field $\mathbb{F}$. For each step, carefully explain exactly which properties from the definition of a field and Theorem 1.5 you are using.

$$\begin{aligned} ax + by &= c \\ dx &= f. \end{aligned}$$

1.4.5 Find all solutions of each of the following linear systems over $\mathbb{C}$.

(a) $\begin{aligned} x + (2 - i)y &= 3 \\ -ix + y &= 1 + i \end{aligned}$ (b) $\begin{aligned} x + iz &= 1 \\ ix + y &= 2i \\ 2x + iy + 3iz &= 1 \end{aligned}$ (c) $\begin{aligned} x + 2iy &= 3 \\ 4ix + 5y &= 6i \end{aligned}$

1.4.6 Find all solutions of each of the following linear systems over $\mathbb{F}_2$.

(a) $\begin{aligned} x + y + z &= 0 \\ x + z &= 1 \end{aligned}$ (b) $\begin{aligned} x + z &= 1 \\ x + y + z &= 0 \\ y + z &= 1 \end{aligned}$

1.4.7 Show that if $ad - bc \neq 0$, then the linear system

$$\begin{aligned} ax + by &= e \\ cx + dy &= f \end{aligned}$$

over a field $\mathbb{F}$ has a unique solution.

Hint: Consider separately the cases $a \neq 0$ and $a = 0$.

1.4.8 Consider a linear system

$$\begin{aligned} a_{11}x_1 + \cdots + a_{1n}x_n &= b_1 \\ &\vdots \\ a_{m1}x_1 + \cdots + a_{mn}x_n &= b_m, \end{aligned}$$

with $a_{ij} \in \mathbb{R}$ and $b_i \in \mathbb{R}$ for each i and j.

(a) Suppose that the system has a solution $\mathbf{x} \in \mathbb{C}^n$. Show that the system has a solution $\mathbf{y} \in \mathbb{R}^n$.

(b) If the system has a *unique* solution $\mathbf{x} \in \mathbb{R}^n$, then $\mathbf{x}$ is also the *unique* solution in $\mathbb{C}^n$ (i.e., you cannot find more solutions by going to $\mathbb{C}^n$).

(c) Give an example of a (necessarily non-linear) equation over $\mathbb{R}$ that has a solution in $\mathbb{C}$ but no solution in $\mathbb{R}$.

1.4.9 Consider a linear system

$$\begin{aligned} a_{11}x_1 + \cdots + a_{1n}x_n &= b_1 \\ &\vdots \\ a_{m1}x_1 + \cdots + a_{mn}x_n &= b_m, \end{aligned}$$

with $a_{ij} \in \mathbb{Z}$ and $b_i \in \mathbb{Z}$ for each i and j.

(a) Suppose that the system has a solution $\mathbf{x} \in \mathbb{R}^n$. Show that the system has a solution $\mathbf{y} \in \mathbb{Q}^n$.

(b) If the system has a *unique* solution $\mathbf{x} \in \mathbb{Q}^n$, then $\mathbf{x}$ is also the *unique* solution in $\mathbb{R}^n$ (i.e., you cannot find more solutions by going to $\mathbb{R}^n$).

(c) Give an example of a linear equation with coefficients in $\mathbb{Z}$ that has a solution in $\mathbb{Q}$ but no solution in $\mathbb{Z}$.

1.4.10 An $n \times n$ linear system over a field $\mathbb{F}$ is called **upper triangular** if the coefficient $a_{ij} = 0$ whenever $i > j$:

$$\begin{aligned} a_{11}x_1 + a_{12}x_2 + a_{13}x_3 + \cdots + a_{1n}x_n &= b_1 \\ a_{22}x_2 + a_{23}x_3 + \cdots + a_{2n}x_n &= b_2 \\ a_{33}x_3 + \cdots + a_{3n}x_n &= b_3 \\ &\vdots \\ a_{nn}x_n &= b_n. \end{aligned} \tag{1.26}$$

Show that if $a_{ii} \neq 0$ for each i, then the system is consistent with a unique solution.

1.4.11 Find the RREF of $\begin{bmatrix} 0 & 1 & 1 \\ 1 & 0 & 1 \\ 1 & 1 & 0 \end{bmatrix}$,

(a) over $\mathbb{R}$,

(b) over $\mathbb{F}_2$.

1.4.12 Give examples of linear systems of each of the following types, if possible. Explain how you know they have the claimed properties, or else explain why there is no such system. *Note*: in the first part, you are free to choose the base field however you like.

(a) Consistent with exactly two solutions.

(b) Consistent over $\mathbb{R}$ with exactly two solutions.

1.4.13 Suppose that $\mathbb{F}, \mathbb{K}$ are fields* with $\mathbb{F} \subsetneq \mathbb{K}$ (and that the operations of addition and multiplication are the same). Suppose that a given linear

*The letter $\mathbb{K}$ (or K or k) is often used to denote a field. It stands for the German word *Körper*, which ordinarily translates as "body," but in mathematics means (this kind of) "field."

system over $\mathbb{F}$ has a unique solution. Show that you cannot find more solutions by thinking of the system as being over $\mathbb{K}$ instead.

1.4.14 Let p be a prime number, and let $\mathbb{F}_p := \{0, 1, \dots, p-1\}$. Define addition and multiplication in $\mathbb{F}_p$ as follows.

Let $a, b \in \mathbb{F}_p$. Interpret a and b as natural numbers, and write their (ordinary) sum as

$$a + b = kp + r,$$

where $0 \le r \le p - 1$ (i.e., r is the remainder when you divide $a + b$ by p). Define addition in $\mathbb{F}_p$ by

$$a + b := r \qquad (\text{in } \mathbb{F}_p).$$

Similarly, write the (usual) product of a and b as

$$ab = k'p + r',$$

with $0 \le r' \le p - 1$, and define the product of a and b in $\mathbb{F}_p$ by

$$ab := r' \qquad (\text{in } \mathbb{F}_p).$$

Show that, with these definitions, $\mathbb{F}_p$ is a field.

The operations defined here are called **addition mod** p and **multiplication mod** p ("mod" is short for "modulo"), and it is customary to use the same notation as for ordinary addition and multiplication, so watch out!

Hint: The trickiest part is checking that each nonzero element has a multiplicative inverse. To do that, first show that if $a, b, c \in \{1, \dots, p-1\}$ and $ab = ac$ (in $\mathbb{F}_p$), then $b = c$. Use this to show that for a fixed $a \in \{1, \dots, p-1\}$, the products $a1, a2, \dots, a(p-1)$ in $\mathbb{F}_p$ are all distinct numbers in $\{1, \dots, p-1\}$, and therefore one of them must be 1.

1.4.15 Show that $\{0, 1, 2, 3\}$, with addition and multiplication mod 4 (defined as in Exercise 1.4.14), is not a field.

1.4.16 Prove part 2 of Theorem 1.5.

1.4.17 Prove part 3 of Theorem 1.5.

1.4.18 Prove part 6 of Theorem 1.5.

1.4.19 Prove part 8 of Theorem 1.5.

1.5 Vector Spaces

General Vector Spaces

In this section we will introduce the abstract concept of a vector which generalizes the familiar arrows in space; we begin with some motivating examples.

Examples

1. Consider the following linear system over $\mathbb{R}$:

$$\begin{aligned} x_1 + 2x_2 + 5x_4 &= 0 \\ -x_1 - 2x_2 + x_3 - 6x_4 + x_5 &= 0 \\ -2x_1 - 4x_2 - 10x_4 + x_5 &= 0. \end{aligned}$$

In vector form, the solutions to the system are

$$\begin{bmatrix} x_1 \\ x_2 \\ x_3 \\ x_4 \\ x_5 \end{bmatrix} = x_2 \begin{bmatrix} -2 \\ 1 \\ 0 \\ 0 \\ 0 \end{bmatrix} + x_4 \begin{bmatrix} -5 \\ 0 \\ 1 \\ 1 \\ 0 \end{bmatrix}.$$

That is, we can think of the solutions of the system as the subset

$$S := \left\langle \begin{bmatrix} -2 \\ 1 \\ 0 \\ 0 \\ 0 \end{bmatrix}, \begin{bmatrix} -5 \\ 0 \\ 1 \\ 1 \\ 0 \end{bmatrix} \right\rangle \subseteq \mathbb{R}^5.$$

Note that, while this is a set of the familiar sort of vectors, it is a very special subset of $\mathbb{R}^5$; because it is the span of a list of vectors, it is closed under vector addition and scalar multiplication. That is, S has the property that if we add two vectors in S we get a third vector in S, and that if we multiply any vector in S by a scalar we get a new vector in S. Of course, elements of S are vectors in $\mathbb{R}^5$, so it's not news that we can add them and multiply by scalars, but what is very important is that those operations *preserve solutions of the linear system.*

2. Let $C[a, b]$ denote the set of continuous real-valued functions defined on $[a, b]$. The elements of $C[a, b]$ don't immediately look much like the vectors we worked with in Section 1.3, but they do share some important features. In particular, if $f, g \in C[a, b]$, then we can define a new function $f + g$ by adding pointwise: for each $x \in [a, b]$, we define

$$(f + g)(x) := f(x) + g(x).$$

It is a calculus fact that the new function $f + g$ is continuous because f and g are. Also, given $c \in \mathbb{R}$ and $f \in C[a, b]$, we can define a function $cf \in C[a, b]$ by

$$(cf)(x) := cf(x)$$

for each $x \in [a, b]$.

So even though functions don't look a lot like elements of $\mathbb{R}^n$, the set $C[a, b]$ does possess operations of addition and scalar multiplication like $\mathbb{R}^n$ does. ▲

With these examples in mind, we make the following abstract definition. From now on, $\mathbb{F}$ is always understood to be a field.

Definition A **vector space** over $\mathbb{F}$ consists of a set V together with operations $+$ (called **(vector) addition**) and $\cdot$ (called **scalar multiplication**) and an element $0 \in V$ such that all of the following properties hold:

The sum of two vectors is a vector: For each $v, w \in V$, $v + w \in V$.

A scalar multiple of a vector is a vector: For each $c \in \mathbb{F}$ and $v \in V$, $c \cdot v \in V$.

Vector addition is commutative: For each $v, w \in V$, $v + w = w + v$.

Vector addition is associative: For each $u, v, w \in V$, $u + (v + w) = (u + v) + w$.

The zero vector is an identity for vector addition: For each $v \in V$, $v + 0 = 0 + v = v$.

Every vector has an additive inverse: For each $v \in V$, there is an element $w \in V$ such that $v + w = w + v = 0$.

Multiplication by 1: For each $v \in V$, $1 \cdot v = v$.

Scalar multiplication is "associative": For each $a, b \in \mathbb{F}$ and $v \in V$, $a \cdot (b \cdot v) = (ab) \cdot v$.

Distributive law #1: For each $a \in \mathbb{F}$ and $v, w \in V$, $a \cdot (v + w) = a \cdot v + a \cdot w$.

Distributive law #2: For each $a, b \in \mathbb{F}$ and $v \in V$, $(a + b) \cdot v = a \cdot v + b \cdot v$.

The field $\mathbb{F}$ is called the **base field** for V. Elements of V are called **vectors** and elements of $\mathbb{F}$ are called **scalars**.

Note that this definition of "vector" is a *generalization* of what we had before: before, "vector" referred to a column vector in $\mathbb{F}^n$ but, from now on, "vector" will simply mean an element of a vector space; it might be an element of $\mathbb{F}^n$, but it might be something else entirely.

Bottom Line: a vector space over $\mathbb{F}$ consists of a bunch of things (called vectors) that you can add to each other and multiply by elements in $\mathbb{F}$, and those operations behave basically as you expect.

When $\mathbb{F}$ is the field of real numbers $\mathbb{R}$ or the field of complex numbers $\mathbb{C}$, we refer to V as a **real vector space** or a **complex vector space**, respectively.

Some notational issues: In practice, we often omit the $\cdot$ in scalar multiplication and simply write cv. We usually just write V with the base field $\mathbb{F}$ and the operations $+$ and $\cdot$ understood, but we sometimes write $(V, +, \cdot)$, or even $(V, \mathbb{F}, +, \cdot)$ for emphasis. As in the case of fields, for $v \in V$ we will use the notation $-v$ to denote the additive inverse of v, so that $v+(-v) = 0$. Similarly, we write $v+(-w) = v-w$.

Note that there is potential ambiguity in the notation; because we've chosen not to decorate our vectors with arrows or otherwise distinguish them typographically from scalars, the same notation is used for the element $0 \in \mathbb{F}$ and the element $0 \in V$; moreover, it is possible in our notation to write down nonsensical expressions (for example, the sum of a scalar and a vector, or a vector times vector) without it being immediately obvious that they are nonsense.

Quick Exercise #21. Suppose that V and W are vector spaces over $\mathbb{F}$ (with no vectors in common), and that

$$u, v \in V, \; u \neq 0, \qquad w \in W, \qquad \text{and} \qquad a, b \in \mathbb{F}, \; a \neq 0.$$

For each of the following expressions, determine what it represents (i.e., scalar, vector in V, or vector in W) or if it's nonsense.
(a) $au + bv$ (b) $v + w$ (c) u^{-1} (d) $a^{-1}bw$ (e) $b + a^{-1}u$ (f) $(b + a^{-1})u$ (g) uv

While we will take advantage of the generality of the definition of a vector space in many ways, it is important to keep in mind our prototype example of a vector space:

Proposition 1.9 *Let $\mathbb{F}$ be a field and $n \in \mathbb{N}$. Then*

$$\mathbb{F}^n := \left\{ \begin{bmatrix} a_1 \\ \vdots \\ a_n \end{bmatrix} \middle| \; a_1, \ldots, a_n \in \mathbb{F} \right\}$$

is a vector space over $\mathbb{F}$.

Sketch of proof In $\mathbb{F}^n$ we define vector addition and scalar multiplication in $\mathbb{F}^n$ by

$$\begin{bmatrix} a_1 \\ \vdots \\ a_n \end{bmatrix} + \begin{bmatrix} b_1 \\ \vdots \\ b_n \end{bmatrix} := \begin{bmatrix} a_1 + b_1 \\ \vdots \\ a_n + b_n \end{bmatrix} \qquad \text{and} \qquad c \begin{bmatrix} a_1 \\ \vdots \\ a_n \end{bmatrix} := \begin{bmatrix} ca_1 \\ \vdots \\ ca_n \end{bmatrix},$$

and define the zero vector to be the vector $\mathbf{0}$ all of whose entries are 0. All of the properties from the definition of a vector space hold essentially trivially, but in principle need to be checked. For example, the following manipulations verify the first distributive law:

QA #21: (a) vector in V (b) nonsense (c) nonsense (d) vector in W (e) nonsense (f) vector in V (g) nonsense

$$c\left(\begin{bmatrix}a_1\\ \vdots\\ a_n\end{bmatrix}+\begin{bmatrix}b_1\\ \vdots\\ b_n\end{bmatrix}\right)=c\begin{bmatrix}a_1+b_1\\ \vdots\\ a_n+b_n\end{bmatrix}=\begin{bmatrix}c(a_1+b_1)\\ \vdots\\ c(a_n+b_n)\end{bmatrix}$$
$$=\begin{bmatrix}ca_1+cb_1\\ \vdots\\ ca_n+cb_n\end{bmatrix}=\begin{bmatrix}ca_1\\ \vdots\\ ca_n\end{bmatrix}+\begin{bmatrix}cb_1\\ \vdots\\ cb_n\end{bmatrix}=c\begin{bmatrix}a_1\\ \vdots\\ a_n\end{bmatrix}+c\begin{bmatrix}b_1\\ \vdots\\ b_n\end{bmatrix}.$$

Checking the other properties is similarly trivial and tedious. ▲

Notation: If $\mathbf{v} \in \mathbb{F}^n$, we write $[\mathbf{v}]_i$ for the ith entry of $\mathbf{v}$.

The general definition of a vector space above is tailor-made so that the following operations are possible.

Definition Let V be a vector space over $\mathbb{F}$.

1. A **linear combination** of $v_1, v_2, \ldots, v_k \in V$ is a vector of the form
$$\sum_{i=1}^{k} c_i v_i = c_1 v_1 + c_2 v_2 + \cdots + c_k v_k,$$
where $c_1, c_2, \ldots, c_k \in \mathbb{F}$.
2. The **span** of $v_1, v_2, \ldots, v_k \in V$ is the set $\langle v_1, v_2, \ldots, v_k\rangle$ of all linear combinations of $v_1, v_2, \ldots, v_k$. That is,
$$\langle v_1, v_2, \ldots, v_k\rangle := \left\{ \sum_{i=1}^{k} c_i v_i \;\middle|\; c_1, c_2, \ldots, c_k \in \mathbb{F} \right\}.$$

Quick Exercise #22. Show that if $v \in \langle v_1, \ldots, v_k\rangle$, then $\langle v_1, \ldots, v_k\rangle = \langle v_1, \ldots, v_k, v\rangle$.

Examples of Vector Spaces

Example From Proposition 1.9 above, $\mathbb{F}^n$ is a vector space over $\mathbb{F}$, with addition and scalar multiplication defined entry-wise. ▲

Example Consider the following system of equations over $\mathbb{F}$:

$$\begin{aligned} a_{11}x_1 + \cdots + a_{1n}x_n &= 0 \\ &\vdots \\ a_{m1}x_1 + \cdots + a_{mn}x_n &= 0. \end{aligned} \tag{1.27}$$

QA #22: Clearly, every linear combination of $v_1, \ldots, v_k$ is also a linear combination of $v_1, \ldots, v_k, v$. If $v = \sum_{i=1}^{k} d_i v_i$, then $c_1 v_1 + \cdots + c_k v_k + dv = (c_1 + dd_1)v_1 + \cdots + (c_k + dd_k)v_k$.

A system like this, in which all the constants on the right hand side are 0, is called **homogeneous.**

Let $S \subseteq \mathbb{F}^n$ be the set of solutions of the system. Then S is a vector space (with addition and scalar multiplication defined as in $\mathbb{F}^n$).

In fact, we don't need to check that all of the properties in the definition of a vector space hold, because vector addition and scalar multiplication in $\mathbb{F}^n$ are already known to satisfy things like commutativity, associativity, distributive laws, etc. What we do have to check is that S has a zero vector (which is obvious: plugging in $x_j = 0$ for each j clearly gives a solution), and that if we just use the definitions of addition and scalar multiplication from $\mathbb{F}^n$, it's still true that

- the sum of two vectors is a vector,
- a scalar multiple of a vector is a vector,
- each vector has an additive inverse.

The point here is that in the present context "vector" means an element of S. So we have to show, for example, that if $\mathbf{v}, \mathbf{w} \in S$, then $\mathbf{v} + \mathbf{w} \in S$; *a priori*, all we know is that $\mathbf{v} + \mathbf{w} \in \mathbb{F}^n$.

To do this, let

$$\mathbf{v} = \begin{bmatrix} v_1 \\ \vdots \\ v_n \end{bmatrix} \quad \text{and} \quad \mathbf{w} = \begin{bmatrix} w_1 \\ \vdots \\ w_n \end{bmatrix}$$

and suppose that $\mathbf{v}, \mathbf{w} \in S$; i.e.,

$$\begin{aligned} a_{11}v_1 + \cdots + a_{1n}v_n = 0 &= a_{11}w_1 + \cdots + a_{1n}w_n, \\ &\vdots \\ a_{m1}v_1 + \cdots + a_{mn}v_n = 0 &= a_{m1}w_1 + \cdots + a_{mn}w_n. \end{aligned}$$

Then

$$\begin{aligned} a_{11}(v_1 + w_1) + \cdots + a_{1n}(v_n + w_n) &= (a_{11}v_1 + \cdots + a_{1n}v_n) + (a_{11}w_1 \\ &\quad + \cdots + a_{1n}w_n) = 0 \\ &\vdots \\ a_{m1}(v_1 + w_1) + \cdots + a_{mn}(v_n + w_n) &= (a_{m1}v_1 + \cdots + a_{mn}v_n) + (a_{m1}w_1 \\ &\quad + \cdots + a_{mn}w_n) = 0, \end{aligned}$$

and so $\mathbf{v} + \mathbf{w} \in S$.

Similarly, if $c \in \mathbb{F}$ and $\mathbf{v} \in S$ as above, then

$$\begin{aligned} a_{11}(cv_1) + \cdots + a_{1n}(cv_n) &= c(a_{11}v_1 + \cdots + a_{1n}v_n) = 0 \\ &\vdots \\ a_{m1}(cv_1) + \cdots + a_{mn}(cv_n) &= c(a_{m1}v_1 + \cdots + a_{mn}v_n) = 0, \end{aligned}$$

so $c\mathbf{v} \in S$. It is easy to check that

$$-\mathbf{v} = \begin{bmatrix} -v_1 \\ \vdots \\ -v_n \end{bmatrix} = (-1)\mathbf{v},$$

so taking $c = -1$ above shows that $-\mathbf{v} \in S$. ▲

This example illustrates the following situation, in which one vector space sits inside of a larger space and inherits the operations of addition and scalar multiplication from it.

Definition Let $(V, +, \cdot)$ be a vector space over $\mathbb{F}$. A subset $U \subseteq V$ is called a **subspace** of V if $(U, +, \cdot)$ is itself a vector space.

As we discussed in the example of the set of solutions to a homogeneous linear system, to check whether a subset of a vector space is a subspace, the properties from the definition of a vector space which need to be verified are exactly those that require "a vector" to exist; in this context, "vector" means "element of U." This observation is summarized in the following theorem.

Theorem 1.10 *Let V be a vector space over $\mathbb{F}$. A subset $U \subseteq V$ is a subspace of V if and only if*

- $0 \in U$;
- *if $u_1, u_2 \in U$, then $u_1 + u_2 \in U$;*
- *if $u \in U$ and $c \in \mathbb{F}$, then $cu \in U$;*
- *if $u \in U$, then $-u \in U$.*

If the second condition in Theorem 1.10 holds, we say that U is **closed under addition**, and if the third conditions holds, we say that U is **closed under scalar multiplication.***

Examples

1. Using Theorem 1.10, it is trivial to check that if $v_1, \ldots, v_k \in V$, then $\langle v_1, \ldots, v_k \rangle$ is a subspace of V.
2. Let $\mathbb{F}$ be a field, $m, n \in \mathbb{N}$, and let $\mathrm{M}_{m,n}(\mathbb{F})$ denote the set of $m \times n$ matrices over $\mathbb{F}$. When $m = n$ we just write $\mathrm{M}_n(\mathbb{F})$.
   The zero matrix in $\mathrm{M}_{m,n}(\mathbb{F})$ is the matrix with all entries equal to 0.

*The fourth condition doesn't usually get a name. We'll see why at the end of this section.

We define the sum of two matrices entry-wise: if $A = \left[a_{ij}\right]_{\substack{1 \le i \le m \\ 1 \le j \le n}}$ and $B = \left[b_{ij}\right]_{\substack{1 \le i \le m \\ 1 \le j \le n}}$, then the matrix $A + B$ is defined by

$$A + B := \left[a_{ij} + b_{ij}\right]_{\substack{1 \le i \le m \\ 1 \le j \le n}}.$$

Similarly, if $c \in \mathbb{F}$, the matrix cA is defined by

$$cA := \left[ca_{ij}\right]_{\substack{1 \le i \le m \\ 1 \le j \le n}}.$$

Checking that $M_{m,n}(\mathbb{F})$ is a vector space over $\mathbb{F}$ is identical to the proof that $\mathbb{F}^n$ is a vector space.

3. Consider the set $\mathbb{F}^\infty$ of infinite sequences with entries in $\mathbb{F}$:

$$\mathbb{F}^\infty := \left\{(a_i)_{i \ge 1} \;\middle|\; a_i \in \mathbb{F}, 1 \le i < \infty\right\}.$$

The zero sequence is the sequence with each entry equal to 0.
Addition and scalar multiplication are again defined entry-wise: for $(a_i)_{i\ge1}, (b_i)_{i\ge1} \in \mathbb{F}^\infty$ and $c \in \mathbb{F}$,

$$(a_i)_{i\ge1} + (b_i)_{i\ge1} := (a_i + b_i)_{i\ge1} \qquad \text{and} \qquad c(a_i)_{i\ge1} := (ca_i)_{i\ge1}.$$

Then $\mathbb{F}^\infty$ is a vector space over $\mathbb{F}$.

4. Let $c_0 \subseteq \mathbb{R}^\infty$ be the sequences in $\mathbb{R}$ with entries tending to zero; i.e., for $(a_i)_{i\ge1} \in \mathbb{R}^\infty$, $(a_i)_{i\ge1} \in c_0$ if and only if

$$\lim_{i\to\infty} a_i = 0.$$

Then c_0 is a subspace of $\mathbb{R}^\infty$:

- if $a_i = 0$ for all i, then $\lim_{i\to\infty} a_i = 0$;
- if $\lim_{i\to\infty} a_i = \lim_{i\to\infty} b_i = 0$, then $\lim_{i\to\infty}(a_i + b_i) = \lim_{i\to\infty} a_i + \lim_{i\to\infty} b_i = 0$;
- if $\lim_{i\to\infty} a_i = 0$, and $c \in \mathbb{R}$, then $\lim_{i\to\infty}(ca_i) = c \lim_{i\to\infty} a_i = 0$;
- if $\lim_{i\to\infty} a_i = 0$, then $\lim_{i\to\infty}(-a_i) = 0$.

It was important in this example that the limit of the sequence entries was required to be 0; if c_s is the set of sequences in $\mathbb{R}$ with entries tending to $s \neq 0$, then c_s is not a subspace of $\mathbb{R}^\infty$.

Quick Exercise #23. Why not?

5. We saw at the beginning of the section that if we let $C[a, b]$ be the set of all $f : [a, b] \to \mathbb{R}$ which are continuous, then $C[a, b]$ is a vector space over $\mathbb{R}$. ▲

QA #23: The quickest reason: the zero sequence is not in c_s.

There are many other classes of functions which are vector spaces; any time the property (or properties) defining a set of functions is (or are) possessed by the zero function and preserved under addition and scalar multiplication, the set in question is a vector space. A vector space whose elements are functions is called a **function space**. Some examples are the following:

- If $D[a,b]$ denotes the set of $f : [a,b] \to \mathbb{R}$ which are differentiable, then $D[a,b]$ is a vector space: the zero function is differentiable, the sum of differentiable functions is differentiable, and any multiple of a differentiable function is differentiable.
- More generally, if $k \in \mathbb{N}$ and $D^k[a,b]$ is the set of $f : [a,b] \to \mathbb{R}$ which are k-times differentiable, then $D^k[a,b]$ is a vector space.
- The set

$$\mathcal{P}_n(\mathbb{F}) := \left\{ a_0 + a_1x + \cdots + a_nx^n \;\middle|\; a_0, a_1, \ldots, a_n \in \mathbb{F} \right\}$$

of formal polynomials in x over $\mathbb{F}$ of degree n or less is a vector space: $0 \in \mathcal{P}_n(\mathbb{F})$ (take $a_i = 0$ for each i), the sum of two polynomials is a polynomial, its degree is not larger than either of the summands, and any scalar multiple of a polynomial is a polynomial.

Here, we are assuming the usual rules for adding polynomials and multiplying them by scalars, which over the familiar fields of $\mathbb{R}$ and $\mathbb{C}$ are just special cases of the rules for adding functions and multiplying them by scalars. There is a bit of sleight-of-hand here, though; as we said above, the elements of $\mathcal{P}_n(\mathbb{F})$ are *formal* polynomials, which means that they are just symbols that look like what is inside the brackets. They are not, and in general cannot be treated as, functions into which arguments can be plugged in. Exercise 1.5.12 gives one illustration of the reason. In the case of formal polynomials, we *define* addition of polynomials by addition of corresponding coefficients as the *definition* of addition, and similarly multiplication by a scalar by multiplying each of the corresponding coefficients is the definition of scalar multiplication in $\mathcal{P}_n(\mathbb{F})$.

Arithmetic in Vector Spaces

As with fields, there are a lot of basic facts that are true for any vector space but aren't included in the definition. The following theorem collects a lot of these "obvious" facts about arithmetic in vector spaces.

Theorem 1.11 (The "obvious" facts about arithmetic in vector spaces) *The following hold for any vector space V over $\mathbb{F}$.*

1. *The additive identity in V is unique.*
2. *Additive inverses of vectors are unique.*
3. *For every $v \in V$, $-(-v) = v$.*

4. *For every* $v \in V$, $0v = 0$.
5. *For every* $a \in \mathbb{F}$, $a0 = 0$.
6. *For every* $v \in V$, $(-1)v = -v$.
7. *If* $a \in \mathbb{F}$, $v \in V$, *and* $av = 0$, *then either* $a = 0$ *or* $v = 0$.

Notice that in the equation $0v = 0$ in part 4, the two 0s refer to different things; on the left, it is the scalar $0 \in \mathbb{F}$ and on the right, it is the zero vector $0 \in V$. Similarly, in part 7, the conclusion is that either $a = 0 \in \mathbb{F}$ or $v = 0 \in V$.

All of these properties are essentially identical to those we proved about arithmetic in fields, and moreover the proofs are similar. Still, they are formally different statements and do need to be proved. There are minor differences: for example, parts 4 and 5 are different from each other here, whereas by commutativity of multiplication, the analogous statements are the same in a field. By way of illustration, we will prove two parts of the theorem above, and leave the rest as exercises.

Proof of Theorem 1.11, parts 5 and 7 To prove part 5, since $0 + 0 = 0$, if $a \in \mathbb{F}$,

$$a0 = a0 + a0;$$

adding $-a0$ to both sides gives that

$$0 = a0.$$

To prove part 7, let $a \in \mathbb{F}$ and $v \in V$, and suppose that $av = 0$. If $a \neq 0$, then a has a multiplicative inverse $a^{-1} \in \mathbb{F}$, so

$$v = (1)v = (a^{-1}a)v = a^{-1}(av) = a^{-1}0 = 0.$$ ▲

Note that it was convenient in proving part 7 to start with the assumption that $a \neq 0$ rather than $v \neq 0$, because it gave us something concrete (namely a^{-1}) to work with; the assumption $v \neq 0$ is a little less useful because it doesn't immediately hand us something to work with.

Bottom Line, again: a vector space over $\mathbb{F}$ consists of a bunch of things that you can add to each other and multiply by elements in $\mathbb{F}$, and those operations behave basically as you expect.

A simple but useful consequence of Theorem 1.11 is that with it the criteria for $U \subseteq V$ to be a subspace (Theorem 1.10) can be slightly simplified: by part 6, the additive inverse $-v$ is actually the scalar multiple $(-1)v$, and so if U is closed under scalar multiplication, $u \in U$ implies that $-u \in U$. We therefore have the following slightly shorter list of properties to check:

> $U \subseteq V$ is a subspace if and only if
> - $0 \in U$;
> - if $u_1, u_2 \in U$, then $u_1 + u_2 \in U$ (i.e., U is closed under addition);
> - if $u \in U$ and $c \in \mathbb{F}$, then $cu \in U$ (i.e., U is closed under scalar multiplication).

You may be wondering why we don't drop the first condition, using part 4 of Theorem 1.11 to say that since $0 = 0v$ for any v, the first part also follows if U is closed under scalar multiplication. In fact, this works as long as we can find a $v \in U$ so that we can write $0 = 0v$. That is, if we remove the first condition, we have to add the condition that U has to be nonempty, and so we don't really gain anything.

Quick Exercise #24. Show that if $U \subseteq V$ is closed under addition and scalar multiplication, and U contains any vector u, then U is a subspace of V.

KEY IDEAS

- A vector space is a set of objects (called vectors) that you can add together and multiply by scalars.
- The prototype example of a vector space over $\mathbb{F}$ is $\mathbb{F}^n$, the set of column vectors with n entries.
- A subspace of a vector space is a subset which is also a vector space. It needs to contain the zero vector and be closed under addition and scalar multiplication.
- Some examples of vector spaces include: solutions of a homogeneous linear system, spaces of matrices, spaces of infinite sequences, and spaces of functions.

EXERCISES

1.5.1 Determine which of the following are subspaces of the given vector space, and justify your answers.

(a) The x-axis in $\mathbb{R}^3$.

(b) The set $\left\{ \begin{bmatrix} x \\ y \end{bmatrix} \middle| \, x, y \geq 0 \right\}$ in $\mathbb{R}^2$ (i.e., the first quadrant of the plane).

(c) The set $\left\{ \begin{bmatrix} x \\ y \end{bmatrix} \middle| \, x, y \geq 0 \right\} \cup \left\{ \begin{bmatrix} x \\ y \end{bmatrix} \middle| \, x, y \leq 0 \right\}$ in $\mathbb{R}^2$ (i.e., the first and third quadrants of the plane).

QA #24: By part 4 of Theorem 1.11, $0 = 0u$, and so $0 \in U$ since U is closed under scalar multiplication.

(d) The set of solutions of the linear system

$$\begin{aligned} x - y + 2z &= 4 \\ 2x - 5z &= -1. \end{aligned}$$

1.5.2 Show that the set of solutions of a *nonhomogeneous* $m \times n$ linear system is never a subspace of $\mathbb{F}^n$.

1.5.3 The **trace** of an $n \times n$ matrix $\mathrm{A} = \left[a_{ij}\right]_{\substack{1 \le i \le n \\ 1 \le j \le n}}$ is

$$\operatorname{tr} \mathrm{A} = \sum_{i=1}^{n} a_{ii}.$$

Show that $S = \{\mathrm{A} \in \mathrm{M}_n(\mathbb{F}) \mid \operatorname{tr} \mathrm{A} = 0\}$ is a subspace of $\mathrm{M}_n(\mathbb{F})$.

1.5.4 Determine which of the following are subspaces of $C[a, b]$. Here $a, b, c \in \mathbb{R}$ are fixed and $a < c < b$.

(a) $V = \left\{f \in C[a, b] \mid f(c) = 0\right\}$.
(b) $V = \left\{f \in C[a, b] \mid f(c) = 1\right\}$.
(c) $V = \left\{f \in D[a, b] \mid f'(c) = 0\right\}$.
(d) $V = \left\{f \in D[a, b] \mid f' \text{ is constant}\right\}$.
(e) $V = \left\{f \in D[a, b] \mid f'(c) = 1\right\}$.

1.5.5 (a) Show that $\mathbb{C}$ is a vector space over $\mathbb{R}$.
(b) Show that $\mathbb{Q}$ is not a vector space over $\mathbb{R}$.

1.5.6 Show that every complex vector space is also a real vector space.

1.5.7 Suppose that $\mathbb{K}$ is a field, $\mathbb{F} \subseteq \mathbb{K}$, and $\mathbb{F}$ is a field with the operations inherited from $\mathbb{K}$ (i.e., $\mathbb{F}$ is a **subfield** of $\mathbb{K}$). Show that every vector space V over $\mathbb{K}$ is also a vector space over $\mathbb{F}$.

1.5.8 Show that if $\mathbb{F}$ is a subfield of $\mathbb{K}$ (see Exercise 1.5.7), then $\mathbb{K}$ is a vector space over $\mathbb{F}$.

1.5.9 Let $\mathbb{F}$ be any field and let A be any nonempty set. Prove that the set V of all functions $f : A \to \mathbb{F}$ is a vector space over $\mathbb{F}$, equipped with pointwise addition and scalar multiplication.

1.5.10 Let V be a vector space, and suppose that U and W are both subspaces of V. Show that

$$U \cap W := \{v \mid v \in U \text{ and } v \in W\}$$

is a subspace of V.

1.5.11 Show that if U_1, U_2 are subspaces of a vector space V, then

$$U_1 + U_2 := \{u_1 + u_2 \mid u_1 \in U_1,\ u_2 \in U_2\}$$

is also a subspace of V.

1.5.12 Show that $x^2 + x = 0$ for every $x \in \mathbb{F}_2$. Therefore if the polynomial $p(x) = x^2 + x$ over $\mathbb{F}_2$ is interpreted as a function $p : \mathbb{F}_2 \to \mathbb{F}_2$, as opposed to a formal expression, it is the same as the function $p(x) = 0$.

1.5.13 Prove that $\{\mathbf{v} \in \mathbb{R}^n \mid v_i > 0, i = 1, \ldots, n\}$ is a real vector space, with "addition" given by component-wise multiplication, "scalar multiplication" given by

$$\lambda \begin{bmatrix} v_1 \\ \vdots \\ v_n \end{bmatrix} = \begin{bmatrix} v_1^\lambda \\ \vdots \\ v_n^\lambda \end{bmatrix},$$

and the "zero" vector given by $\mathbf{0} = \begin{bmatrix} 1 \\ \vdots \\ 1 \end{bmatrix}$.

1.5.14 Let

$$V = \left\{ (p_1, \ldots, p_n) \,\middle|\, p_i > 0,\ i = 1, \ldots, n;\ \sum_{i=1}^{n} p_i = 1 \right\}.$$

(Geometrically, V is a simplex.) Define an addition operation on V by

$$(p_1, \ldots, p_n) + (q_1, \ldots, q_n) := \frac{(p_1 q_1, \ldots, p_n q_n)}{p_1 q_1 + \cdots + p_n q_n},$$

where the division by $p_1 q_1 + \cdots + p_n q_n$ is ordinary division of a vector by a scalar. Define a scalar multiplication operation by

$$\lambda(p_1, \ldots, p_n) := \frac{(p_1^\lambda, \ldots, p_n^\lambda)}{p_1^\lambda + \cdots + p_n^\lambda}.$$

Show that V is a real vector space.

1.5.15 Let S be any set. For subsets $A, B \subseteq S$, **symmetric difference** is the set $A \Delta B$ defined by the property that s is in $A \Delta B$ if and only if s is in either A or B, but not in both of them. That is,

$$A \Delta B = (A \cup B) \setminus (A \cap B) = (A \setminus B) \cup (B \setminus A)$$

in set-theoretic notation.

Let V be the collection of all subsets of S. For $A, B \in V$ (so $A, B \subseteq S$), define $A + B = A \Delta B$. Furthermore, define $0A = \emptyset$ and $1A = A$. Prove that V is a vector space over the field $\mathbb{F}_2$, with zero vector the empty set $\emptyset$.

Hint: Check the conditions in the definition of vector space in order, to be sure you don't miss any. In proving the properties of scalar multiplication, since $\mathbb{F}_2$ has so few elements you can check them all case-by-case. It will be helpful for proving a couple of the properties to think first about what $A \Delta A$ is.

1.5.16 Prove part 1 of Theorem 1.11.

1.5.17 Prove part 2 of Theorem 1.11.

1.5.18 Prove part 3 of Theorem 1.11.

1.5.19 Prove part 4 of Theorem 1.11.

1.5.20 Prove part 6 of Theorem 1.11.

2

Linear Maps and Matrices

Throughout this chapter (and, in fact, for much of the rest of the book) we will be working with multiple vector spaces at the same time; it will always be understood that all of our vector spaces have the same base field $\mathbb{F}$.

2.1 Linear Maps

Recognizing Sameness

One of the most fundamental problems of mathematics is to be able to determine when two things which appear to be different are actually the same in some meaningful way. We have already come across some examples of this phenomenon, without explicitly noting them as such. For example, consider the following vector spaces over $\mathbb{R}$:

$$\mathbb{R}^n = \left\{ \begin{bmatrix} a_1 \\ \vdots \\ a_n \end{bmatrix} \middle| \; a_1, \ldots, a_n \in \mathbb{R} \right\},$$

and

$$\mathcal{P}_d(\mathbb{R}) = \left\{ a_0 + a_1 x + \cdots + a_d x^d \;\middle|\; a_0, a_1, \ldots, a_d \in \mathbb{R} \right\}.$$

On the face of it, these two sets appear to be completely different; we have vectors in n-dimensional space on the one hand, and polynomials on the other. Even so, the two spaces feel kind of the same: really, they consist of ordered lists of numbers, and the vector space operations work the same way, that is, componentwise.

Seeing and being able to formally describe this kind of identical structure in different guises is part of what mathematics is for, and motivates the following definition.

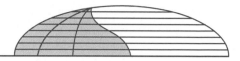

Definition Let V and W be vector spaces. A function $T : V \to W$ is called a **linear map** (or **linear transformation**, or **linear operator**) if it has both of the following properties.

Additivity: For each $u, v \in V$, $T(u + v) = T(u) + T(v)$.
Homogeneity: For each $v \in V$ and $a \in \mathbb{F}$, $T(av) = aT(v)$.

We refer to V as the **domain** of T and W as the **codomain** of T. The set of all linear maps from V to W is denoted $\mathcal{L}(V, W)$. When V and W are the same space, we write $\mathcal{L}(V) := \mathcal{L}(V, V)$.

That is, a linear map between vector spaces is exactly one that *respects the vector space structure*: performing the vector space operations of addition and scalar multiplication works the same way whether you do it before or after applying the linear map. The map thus serves as a kind of translation between the two settings which lets us formalize the idea that the operations work the same way in both spaces.

Sometimes there is no need for translation: if V is a vector space, the **identity operator*** on V is the function $I : V \to V$ such that for each $v \in V$,

$$I(v) = v.$$

Quick Exercise #1. Show that the map $I : V \to V$ is linear.

For a more substantial example, consider the following map $T : \mathbb{R}^n \to \mathcal{P}_{n-1}(\mathbb{R})$:

$$T\left(\begin{bmatrix} a_1 \\ \vdots \\ a_n \end{bmatrix}\right) := a_1 + a_2 x + \cdots + a_n x^{n-1}. \tag{2.1}$$

Then T is linear:

$$T\left(\begin{bmatrix} a_1 \\ \vdots \\ a_n \end{bmatrix} + \begin{bmatrix} b_1 \\ \vdots \\ b_n \end{bmatrix}\right) = T\left(\begin{bmatrix} a_1 + b_1 \\ \vdots \\ a_n + b_n \end{bmatrix}\right)$$
$$= (a_1 + b_1) + (a_2 + b_2)x + \cdots + (a_n + b_n)x^{n-1}$$

*This is another place where we will embrace ambiguity in our notation: we use the same symbol for the identity operator on every vector space.

QA #1: $I(v + w) = v + w = I(v) + I(w)$ and $I(av) = av = aI(v)$.

$$= \left(a_1 + a_2x + \cdots + a_nx^{n-1}\right) + \left(b_1 + b_2x + \cdots + b_nx^{n-1}\right)$$
$$= T\left(\begin{bmatrix} a_1 \\ \vdots \\ a_n \end{bmatrix}\right) + T\left(\begin{bmatrix} b_1 \\ \vdots \\ b_n \end{bmatrix}\right),$$

and

$$T\left(c\begin{bmatrix} a_1 \\ \vdots \\ a_n \end{bmatrix}\right) = T\left(\begin{bmatrix} ca_1 \\ \vdots \\ ca_n \end{bmatrix}\right) = ca_1 + ca_2x + \cdots + ca_nx^{n-1}$$
$$= c\left(a_1 + a_2x + \cdots + a_nx^{n-1}\right) = cT\left(\begin{bmatrix} a_1 \\ \vdots \\ a_n \end{bmatrix}\right),$$

thus formalizing our observation that the vector space operations on vectors and polynomials are fundamentally the same.

Linear Maps in Geometry

Even though we motivated linear maps as a way of connecting the vector space operations on two different spaces, linear maps play many other fundamental roles within mathematics. For example, it turns out that many familiar examples of geometric and algebraic transformations define linear maps; we will in particular see below that reflections and rotations are linear. The common thread between transforming vectors into polynomials and transforming arrows in space through rotations and reflections is the fact that all of these transformations preserve the vector space structure.

Examples

1. Let T be the map of $\mathbb{R}^3$ given by reflection across the x–y plane:

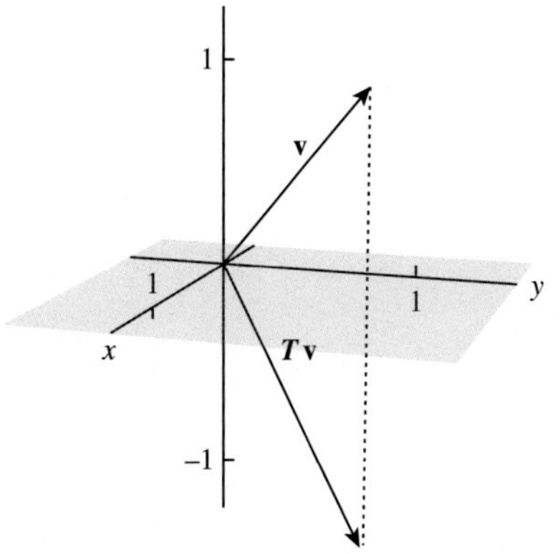

Figure 2.1 Reflection across the x–y plane.

The following picture demonstrates that T is additive:

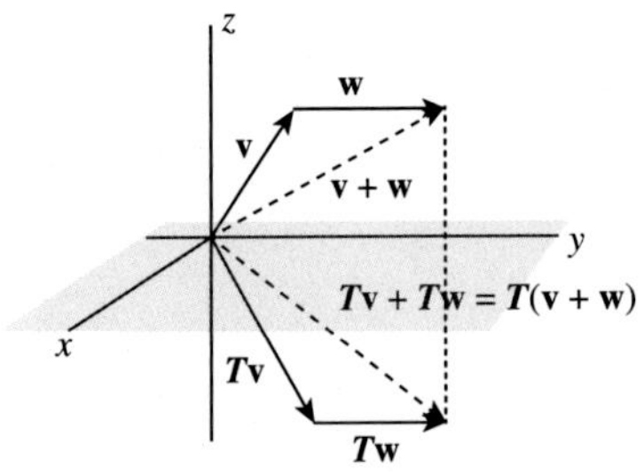

Figure 2.2 Additivity of reflection.

We leave the (simpler) corresponding picture demonstrating homogeneity as Exercise 2.1.1.

2. Let R_θ be the map on $\mathbb{R}^2$ given by rotation counterclockwise by θ radians:

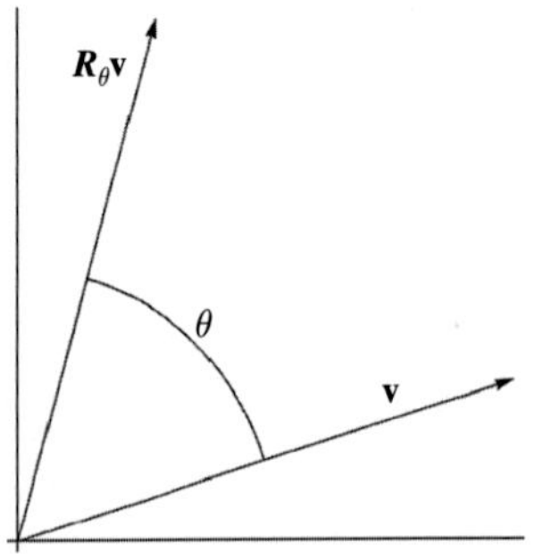

Figure 2.3 Rotation by θ.

Again, one can see the linearity of R_θ in pictures:

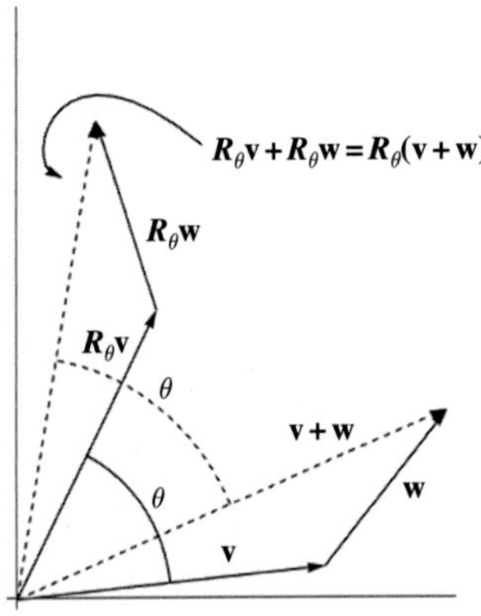

Figure 2.4 Additivity of rotation.

▲

Matrices as Linear Maps

Let $A = \left[a_{ij}\right]_{\substack{1\le i\le m\\ 1\le j\le n}}$ be an $m \times n$ matrix over $\mathbb{F}$. Then A acts as a linear operator from $\mathbb{F}^n$ to $\mathbb{F}^m$ via **matrix–vector multiplication,*** as follows: for $\mathbf{v} = \begin{bmatrix} v_1 \\ \vdots \\ v_n \end{bmatrix} \in \mathbb{F}^n$, define $A\mathbf{v}$ by

$$A\mathbf{v} := \begin{bmatrix} \sum_{j=1}^{n} a_{1j}v_j \\ \vdots \\ \sum_{j=1}^{n} a_{mj}v_j \end{bmatrix}. \tag{2.2}$$

Quick Exercise #2. Show that

$$\begin{bmatrix} | & & | \\ \mathbf{a}_1 & \cdots & \mathbf{a}_n \\ | & & | \end{bmatrix} \begin{bmatrix} v_1 \\ \vdots \\ v_n \end{bmatrix} = \sum_{j=1}^{n} v_j\mathbf{a}_j. \tag{2.3}$$

That is, $A\mathbf{v}$ *is a linear combination of the columns of* A, *with coefficients given by the entries of* $\mathbf{v}$.

The function from $\mathbb{F}^n$ to $\mathbb{F}^m$ defined this way is indeed linear: Given $\mathbf{v} = \begin{bmatrix} v_1 \\ \vdots \\ v_n \end{bmatrix}$ and $\mathbf{w} = \begin{bmatrix} w_1 \\ \vdots \\ w_n \end{bmatrix}$, by (2.3),

$$A(\mathbf{v}+\mathbf{w}) = \sum_{j=1}^{n}(v_j + w_j)\mathbf{a}_j = \sum_{j=1}^{n} v_j\mathbf{a}_j + \sum_{j=1}^{n} w_j\mathbf{a}_j = A\mathbf{v} + A\mathbf{w}.$$

Checking homogeneity is similar.

Formula (2.3) makes it easy to see that if $\mathbf{e}_j \in \mathbb{F}^n$ is the vector with a 1 in the jth position and zeroes everywhere else, then $A\mathbf{e}_j = \mathbf{a}_j$; that is,

*Watch the order of m and n carefully: an $m \times n$ matrix goes from $\mathbb{F}^n$ to $\mathbb{F}^m$!

$\mathbf{A}\mathbf{e}_j$ is exactly the jth column of the matrix $\mathbf{A}$.

As in $\mathbb{R}^n$, the vectors $\mathbf{e}_1, \ldots, \mathbf{e}_n \in \mathbb{F}^n$ are called the **standard basis vectors in** $\mathbb{F}^n$.

Examples

1. The $n \times n$ **identity matrix** $\mathbf{I}_n \in \mathrm{M}_n(\mathbb{F})$ is the $n \times n$ matrix with 1s on the diagonal and zeroes elsewhere. In terms of columns,

$$\mathbf{I}_n = \begin{bmatrix} | & & | \\ \mathbf{e}_1 & \cdots & \mathbf{e}_n \\ | & & | \end{bmatrix}.$$

By (2.3), it follows that

$$\mathbf{I}_n\mathbf{v} = \sum_{j=1}^{n} v_j \mathbf{e}_j = \mathbf{v}$$

for each $\mathbf{v} \in \mathbb{F}^n$. That is, the identity matrix $\mathbf{I}_n$ acts as the identity operator $I \in \mathcal{L}(\mathbb{F}^n)$, hence its name.

2. An $n \times n$ matrix $\mathbf{A}$ is called **diagonal** if $a_{ij} = 0$ whenever $i \neq j$. That is,

$$\mathbf{A} = \begin{bmatrix} a_{11} & & 0 \\ & \ddots & \\ 0 & & a_{nn} \end{bmatrix}.$$

We sometimes denote the diagonal matrix with diagonal entries $d_1, \ldots, d_n \in \mathbb{F}$ as

$$\mathrm{diag}(d_1, \ldots, d_n) := \begin{bmatrix} d_1 & & 0 \\ & \ddots & \\ 0 & & d_n \end{bmatrix}.$$

Matrix–vector products are particularly simple with diagonal matrices, as the following quick exercise shows.

Quick Exercise #3. Let $\mathbf{D} = \mathrm{diag}(d_1, \ldots, d_n)$, and let $\mathbf{x} \in \mathbb{F}^n$. Show that

$$\mathbf{D}\mathbf{x} = \begin{bmatrix} d_1 x_1 \\ \vdots \\ d_n x_n \end{bmatrix}.$$

3. Consider the 3×3 matrix $\mathrm{A} = \begin{bmatrix} 0 & 1 & 0 \\ 0 & 0 & 1 \\ 1 & 0 & 0 \end{bmatrix}$. Then for $\begin{bmatrix} x \\ y \\ z \end{bmatrix} \in \mathbb{R}^3$,

$$\mathrm{A}\begin{bmatrix} x \\ y \\ z \end{bmatrix} = \begin{bmatrix} y \\ z \\ x \end{bmatrix}.$$

That is, reordering the coordinates of a vector in this way is a linear transformation of $\mathbb{R}^3$. More generally, any reordering of coordinates is a linear map on $\mathbb{F}^n$. ▲

Eigenvalues and Eigenvectors

For some linear maps from a vector space to itself, there are special vectors called eigenvectors, on which the map acts in a very simple way. When they exist, these vectors play an important role in understanding and working with the maps in question.

Definition Let V be a vector space over $\mathbb{F}$ and $T \in \mathcal{L}(V)$. The vector $v \in V$ is an **eigenvector** for T with **eigenvalue*** $\lambda \in \mathbb{F}$ if $v \neq 0$ and

$$Tv = \lambda v.$$

If $\mathrm{A} \in \mathrm{M}_n(\mathbb{F})$, the eigenvalues and eigenvectors of A are the eigenvalues and eigenvectors of the map in $\mathcal{L}(\mathbb{F}^n)$ defined by $\mathrm{v} \mapsto \mathrm{Av}$.

That is, eigenvectors are nonzero vectors on which T acts by scalar multiplication. Geometrically, one can think of this as saying that if v is an eigenvector of T, then applying T may change its length but not its direction.† More algebraically, if $v \in V$ is an eigenvector of a linear map T, then the set $\langle v \rangle$ of all scalar multiples of v is **invariant** under T; that is, if $w \in \langle v \rangle$, then $T(w) \in \langle v \rangle$ as well.

*These terms are half-way translated from the German words *Eigenvektor* and *Eigenwert*. The German adjective "eigen" can be translated as "own" or "proper," so an eigenvector of T is something like "T's very own vector" or "a right proper vector for T." The fully English phrases "characteristic vector/value" and "proper vector/value" are sometimes used, but they're considered quite old-fashioned these days.

† The exact opposite direction counts as the same.

> **Quick Exercise #4.** Prove the preceding statement: if $v \in V$ is an eigenvector of a linear map T and $w \in \langle v \rangle$, then $T(w) \in \langle v \rangle$ as well.

Examples

1. Consider the linear map $T : \mathbb{R}^2 \to \mathbb{R}^2$ given by

$$T\begin{bmatrix} x \\ y \end{bmatrix} := \begin{bmatrix} 1 & 0 \\ 0 & 2 \end{bmatrix}\begin{bmatrix} x \\ y \end{bmatrix} = \begin{bmatrix} x \\ 2y \end{bmatrix}.$$

It is easy to visualize what this map does to $\mathbb{R}^2$:

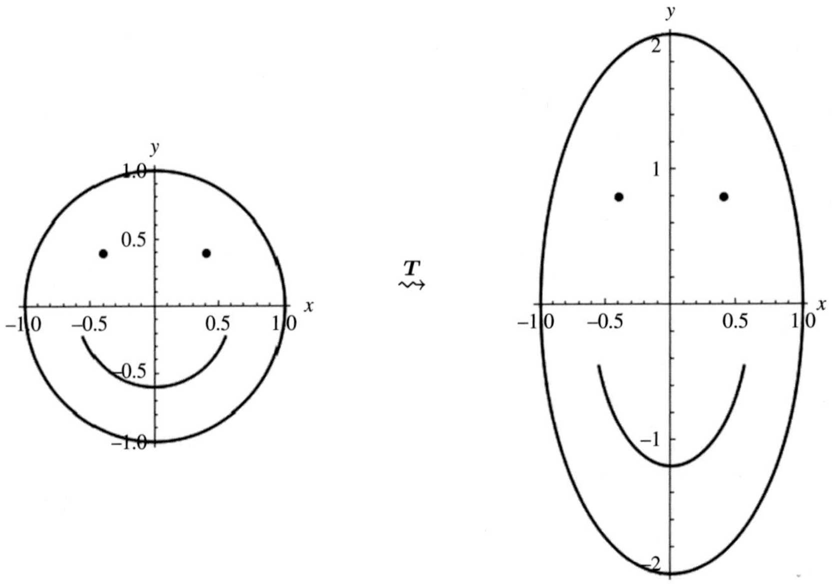

Figure 2.5 Result of multiplication by $\begin{bmatrix} 1 & 0 \\ 0 & 2 \end{bmatrix}$.

The vector $\mathbf{e}_1 := \begin{bmatrix} 1 \\ 0 \end{bmatrix}$ is an eigenvector with eigenvalue 1: T does not change it at all. The vector $\mathbf{e}_2 := \begin{bmatrix} 0 \\ 1 \end{bmatrix}$ is an eigenvector with eigenvalue 2: T stretches $\mathbf{e}_2$ by a factor of 2 but does not change its direction.

2. Recall the map $T : \mathbb{R}^3 \to \mathbb{R}^3$ given by reflection across the x–y plane:

QA #4: If v is an eigenvector, then for some $\lambda \in \mathbb{F}$, $T(v) = \lambda v$. Then $w \in \langle v \rangle \Rightarrow w = cv \Rightarrow T(w) = T(cv) = cT(v) = c\lambda v \in \langle v \rangle$.

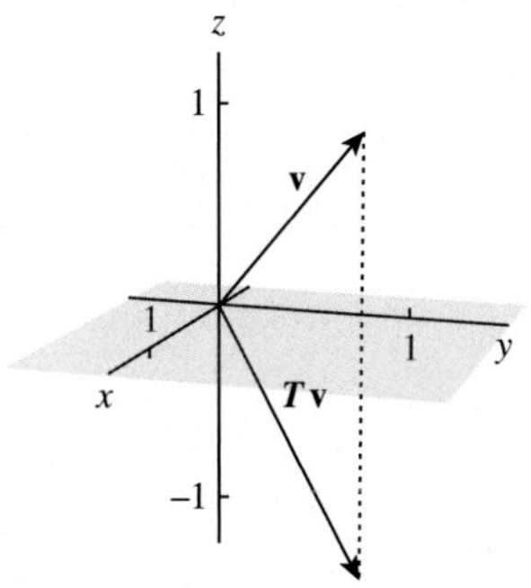

Figure 2.6 Reflection across the x–y plane.

It's easy to recognize some eigenvalues and eigenvectors of T: any vector that lies in the x–y plane is unchanged by T and is thus an eigenvector of T with eigenvalue 1. On the other hand, if $\mathbf{v}$ lies along the z-axis, then its reflection across the x–y plane is exactly $-\mathbf{v}$, and so any such $\mathbf{v}$ is an eigenvector of T with eigenvalue -1.

3. Recall the map $R_\theta : \mathbb{R}^2 \to \mathbb{R}^2$ given by rotation by θ:

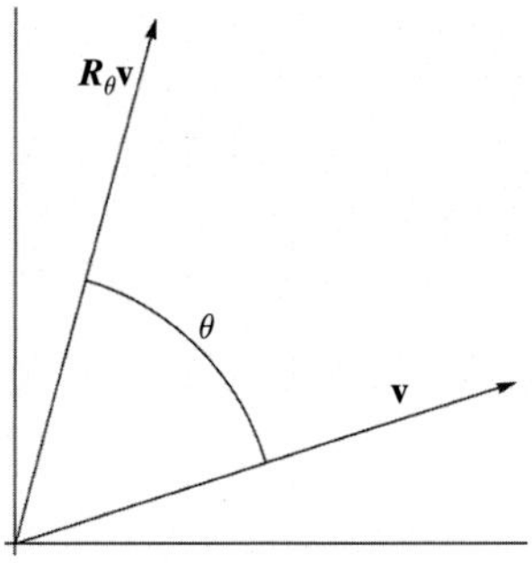

Figure 2.7 Rotation by θ.

It is not too hard to visualize that (unless $\theta = 0$ or $\theta = \pi$), R_θ has no eigenvectors.

Quick Exercise #5. What are the eigenvectors (and corresponding eigenvalues) of R_π? What about R_0?

4. Recall the diagonal matrix

$$\mathbf{D} = \mathbf{diag}(d_1, \ldots, d_n) = \begin{bmatrix} d_1 & & 0 \\ & \ddots & \\ 0 & & d_n \end{bmatrix}.$$

Notice that the jth column of $\mathbf{D} = \mathbf{diag}(d_1, \ldots, d_n)$ is exactly $d_j \mathbf{e}_j$. Thus

$$\mathbf{D}\mathbf{e}_j = d_j \mathbf{e}_j,$$

and so each $\mathbf{e}_j$ is an eigenvector of $\mathbf{D}$ with eigenvalue d_j.

QA #5: (b) Every $\mathbf{v} \neq 0$ is an eigenvector of R_π with eigenvalue -1, and an eigenvector of R_0 with eigenvalue 1 (R_0 is just the identity).

5. Consider the matrix

$$A = \begin{bmatrix} 0 & 1 \\ -1 & 0 \end{bmatrix}.$$

We want to find all possible x, y, and λ such that x and y are not both 0 and

$$\begin{bmatrix} 0 & 1 \\ -1 & 0 \end{bmatrix} \begin{bmatrix} x \\ y \end{bmatrix} = \lambda \begin{bmatrix} x \\ y \end{bmatrix}.$$

This reduces to the very simple linear system

$$\begin{aligned} y &= \lambda x \\ -x &= \lambda y. \end{aligned} \tag{2.4}$$

Substituting the first equation into the second, we get that $y = -\lambda^2 y$, so either $y = 0$ or $\lambda^2 = -1$. But if $y = 0$ then the second equation tells us $x = 0$ as well, which is not allowed. Therefore any eigenvalue λ of A satisfies $\lambda^2 = -1$.

At this point we should notice that we've been vague about the field $\mathbb{F}$. If we think of A as a matrix in $M_2(\mathbb{R})$, then we've found that A has no real eigenvalues. But if we let ourselves work with *complex* scalars, then $\lambda = \pm i$ are both possibilities. Once we pick one of these values of λ, the linear system (2.4) shows us that every vector $\begin{bmatrix} x \\ ix \end{bmatrix}$ for $x \in \mathbb{C}$ is an eigenvector of A with eigenvalue i, and every vector $\begin{bmatrix} x \\ -ix \end{bmatrix}$ for $x \in \mathbb{C}$ is an eigenvector of A with eigenvalue $-i$.

6. Here is an example of finding the eigenvalues and eigenvectors of a map without the aid of geometric intuition or unusually simple algebra. Consider the matrix

$$A = \begin{bmatrix} 3 & -1 \\ 1 & 1 \end{bmatrix} \in M_2(\mathbb{R})$$

acting as a linear map on $\mathbb{R}^2$. We want to find all possible real numbers x, y, and λ such that

$$\begin{bmatrix} 3 & -1 \\ 1 & 1 \end{bmatrix} \begin{bmatrix} x \\ y \end{bmatrix} = \lambda \begin{bmatrix} x \\ y \end{bmatrix}$$

and such that $\begin{bmatrix} x \\ y \end{bmatrix} \neq \begin{bmatrix} 0 \\ 0 \end{bmatrix}$. Subtracting* $\lambda \begin{bmatrix} x \\ y \end{bmatrix}$ from both sides shows that the matrix equation above is equivalent to the linear system given by

*Subtraction is a surprisingly good trick!

$$\left[\begin{array}{cc|c} 3-\lambda & -1 & 0 \\ 1 & 1-\lambda & 0 \end{array}\right] \xrightarrow{\text{R3}} \left[\begin{array}{cc|c} 1 & 1-\lambda & 0 \\ 3-\lambda & -1 & 0 \end{array}\right]$$

$$\xrightarrow{\text{R1}} \left[\begin{array}{cc|c} 1 & 1-\lambda & 0 \\ 0 & -(1-\lambda)(3-\lambda)-1 & 0 \end{array}\right]. \tag{2.5}$$

We can see from the second row that for $\begin{bmatrix} x \\ y \end{bmatrix}$ to be a solution, either $y = 0$ or

$$-(1-\lambda)(3-\lambda) - 1 = -(\lambda-2)^2 = 0.$$

If $y = 0$, then from the first row we get that $x = 0$ as well, but $\begin{bmatrix} 0 \\ 0 \end{bmatrix}$ doesn't count as an eigenvector. We thus need $\lambda = 2$ for the equation to have a solution; in that case, the system in (2.5) becomes

$$\left[\begin{array}{cc|c} 1 & -1 & 0 \\ 0 & 0 & 0 \end{array}\right],$$

so that any $\begin{bmatrix} x \\ y \end{bmatrix}$ with $x = y$ is an eigenvector. That is, we have shown that $\lambda = 2$ is the only eigenvalue of the map given by multiplication by A, and the corresponding eigenvectors are those vectors $\begin{bmatrix} x \\ x \end{bmatrix}$ with $x \neq 0$. ▲

The Matrix–Vector Form of a Linear System

The notation of matrix–vector multiplication introduced earlier in this section allows us to connect linear systems of equations and linear maps, as follows. Let

$$\begin{aligned} a_{11}x_1 + \cdots + a_{1n}x_n &= b_1 \\ &\vdots \\ a_{m1}x_1 + \cdots + a_{mn}x_n &= b_m \end{aligned} \tag{2.6}$$

be a linear system over $\mathbb{F}$. If $A := \left[a_{ij}\right]_{\substack{1\le i\le m \\ 1\le j\le n}}$ is the matrix of coefficients, $\mathbf{x} := \begin{bmatrix} x_1 \\ \vdots \\ x_n \end{bmatrix}$ is the column of variables, and $\mathbf{b} := \begin{bmatrix} b_1 \\ \vdots \\ b_m \end{bmatrix}$ is the vector made of the scalars on the right-hand sides of the equations, then we can rewrite (2.6) very concisely as

$$Ax = b. \tag{2.7}$$

So, for example, the linear system (1.3), summarizing how we can use up all our ingredients when making bread, beer, and cheese, can be written in matrix form as

$$\begin{bmatrix} 1 & \frac{1}{4} & 0 \\ 2 & \frac{1}{4} & 0 \\ 1 & 0 & 8 \\ 2 & \frac{1}{4} & 1 \end{bmatrix} \begin{bmatrix} x \\ y \\ z \end{bmatrix} = \begin{bmatrix} 20 \\ 36 \\ 176 \\ 56 \end{bmatrix}.$$

Example Let us return to our consideration of bread, beer, and barley, with a more personal twist. We've effectively described a small economy consisting of a farmer who grows barley, a baker who turns barley into bread, and a brewer who turns barley into beer.

More particularly, index the industries involved as: 1 – farming (barley); 2 – baking (bread); 3 – brewing (beer), and define the quantities a_{ij} to be the number of units produced by industry i which are consumed to produce each unit of industry j. So, for example, a_{12} is the number of pounds of barley needed to produce one loaf of bread, while a_{21} is the number of loaves of bread the farmers eat per pound of barley produced.

Suppose that the vector $x = \begin{bmatrix} x_1 \\ x_2 \\ x_3 \end{bmatrix}$ describes the quantities of barley, bread, and beer produced. If everything that is produced is also consumed within our little economy, then x is called a **feasible production plan**, or FPP. The total amount of barley consumed in order to produce x_1 further pounds of barley, x_2 loaves of bread, and x_3 pints of beer is

$$a_{11}x_1 + a_{12}x_2 + a_{13}x_3,$$

and if x is an FPP, then this must be equal to x_1, the number of pounds of barley produced. By similar reasoning, saying that x is an FPP is exactly requiring that the following system of equations be satisfied:

$$\begin{aligned} a_{11}x_1 + a_{12}x_2 + a_{13}x_3 &= x_1 \\ a_{21}x_1 + a_{22}x_2 + a_{23}x_3 &= x_2 \\ a_{31}x_1 + a_{32}x_2 + a_{33}x_3 &= x_3. \end{aligned} \tag{2.8}$$

In matrix form, this can be written

$$Ax = x,$$

for A the matrix with entries a_{ij}; that is, the requirement that $\mathbf{x}$ is an FPP means exactly that the vector $\mathbf{x}$ is an eigenvector of A with eigenvalue 1. Since 1 may or may not be an eigenvalue of A, this means that there may or may not be any feasible production plan for our economy.

More concretely, suppose that our economy is described by the matrix

$$A = \begin{bmatrix} 0 & 1 & 1/4 \\ 1/6 & 1/2 & 1/8 \\ 1/3 & 1 & 1/4 \end{bmatrix}.$$

(The first row comes from the barley usage we discussed way back in equation (1.1), together with the assumption that the farmer does not use raw barley for anything.) Then $\mathbf{x} = \begin{bmatrix} 3/4 \\ 1/2 \\ 1 \end{bmatrix}$ is an eigenvector of A with eigenvalue 1.

Quick Exercise #6. Verify the last statement. (Remember the Rat Poison Principle!)

This tells us that if, in one given unit of time, the farmer produces 3/4 lb of barley, the baker produces 1/2 loaf of bread, and the brewer produces 1 pint of beer, then it will be possible for all their products to be exactly used up. ▲

KEY IDEAS

- Linear maps are functions between vector spaces which preserve the vector space operations.
- Rotations and reflections of space are linear.
- Multiplication by a matrix is linear.
- Eigenvectors are nonzero vectors such that $Tv = \lambda v$ for some λ. The scalar λ is called the eigenvalue.
- Linear systems can be written in the form $A\mathbf{x} = \mathbf{b}$, where A is a matrix, $\mathbf{x}$ is a vector of unknowns, and $\mathbf{b}$ is a vector.

EXERCISES

2.1.1 (a) Draw a picture illustrating that the reflection across the x–y plane in $\mathbb{R}^3$ is homogeneous.

(b) Draw a picture illustrating that the rotation of $\mathbb{R}^2$ counterclockwise by θ radians is homogeneous.

2.1.2 If L is a line through the origin in $\mathbb{R}^2$, the **orthogonal projection** onto L is the map $P : \mathbb{R}^2 \to \mathbb{R}^2$ defined as follows. Given $\mathbf{x} \in \mathbb{R}^2$, draw a right triangle whose hypotenuse is given by $\mathbf{x}$, with one leg on the line L:

QA #6: Simply compute $A\mathbf{x}$, and see that it turns out to be $\mathbf{x}$.

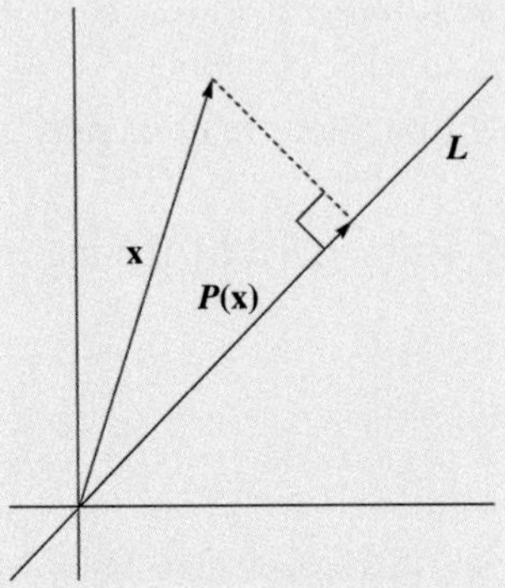

The vector which lies along the leg on L is $P(\mathbf{x})$. Draw pictures illustrating that P is a linear map.

2.1.3 Compute each of the following matrix–vector products, or state that it is not possible.

(a) $\begin{bmatrix} 1 & 2 & -3 \\ -4 & 5 & -6 \end{bmatrix}\begin{bmatrix} -2 \\ 7 \\ 1 \end{bmatrix}$ (b) $\begin{bmatrix} 1 & 2 & -3 \\ -4 & 5 & -6 \end{bmatrix}\begin{bmatrix} -2 \\ 7 \end{bmatrix}$

(c) $\begin{bmatrix} -2 & \frac{1}{3} & \pi & \sqrt{2} \\ \frac{2}{7} & e & 0 & 10 \end{bmatrix}\begin{bmatrix} -1 \\ 0 \\ \frac{2}{3} \\ 2 \end{bmatrix}$ (d) $\begin{bmatrix} 1 & 2 & 3 & 4 \end{bmatrix}\begin{bmatrix} 5 \\ 6 \\ 7 \\ 8 \end{bmatrix}$

(e) $\begin{bmatrix} 0.1 & 2.3 \\ -1.4 & 3.2 \\ -1.5 & 0.6 \end{bmatrix}\begin{bmatrix} 2.1 \\ 3.4 \end{bmatrix}$ (f) $\begin{bmatrix} i & 2-i \\ 3+2i & 1 \end{bmatrix}\begin{bmatrix} 1+3i \\ -2 \end{bmatrix}$

2.1.4 Compute each of the following matrix–vector products, or state that it is not possible.

(a) $\begin{bmatrix} 1 & 2 \\ -3 & -4 \\ 5 & -6 \end{bmatrix}\begin{bmatrix} -2 \\ 7 \\ 1 \end{bmatrix}$ (b) $\begin{bmatrix} 1 & 2 \\ -3 & -4 \\ 5 & -6 \end{bmatrix}\begin{bmatrix} -2 \\ 7 \end{bmatrix}$

(c) $\begin{bmatrix} 4 & 5 & 6 \end{bmatrix}\begin{bmatrix} 1 \\ 2 \\ 3 \end{bmatrix}$ (d) $\begin{bmatrix} 1 & 0 & 1 & 1 \\ 0 & 1 & 1 & 1 \\ 1 & 1 & 1 & 0 \end{bmatrix}\begin{bmatrix} -1 \\ 2 \\ -3 \\ 4 \end{bmatrix}$

(e) $\begin{bmatrix} 1.2 & -0.7 \\ 2.5 & -1.6 \end{bmatrix}\begin{bmatrix} 2.4 \\ -0.8 \end{bmatrix}$ (f) $\begin{bmatrix} 1 & 1+i & 4 \\ -i & -2 & 3-2i \end{bmatrix}\begin{bmatrix} 2+i \\ -i \\ 1 \end{bmatrix}$

2.1.5 Let $C(\mathbb{R})$ be the vector space (over $\mathbb{R}$) of continuous functions $f : \mathbb{R} \to \mathbb{R}$. Define $T : C(\mathbb{R}) \to C(\mathbb{R})$ by

$$[Tf](x) = f(x)\cos(x).$$

Show that T is a linear map.

2.1.6 Let $C[0, 1]$ be the real vector space of continuous functions $f : [0, 1] \to \mathbb{R}$. Define $T : C[0, 1] \to \mathbb{R}$ by

$$Tf = \int_0^1 f(x)\, dx.$$

Show that T is a linear map.

2.1.7 Verify that $\begin{bmatrix} 1 \\ -1 \\ 1 \\ -1 \end{bmatrix}$ is an eigenvector of $\begin{bmatrix} 1 & 2 & 3 & 4 \\ 2 & 3 & 4 & 1 \\ 3 & 4 & 1 & 2 \\ 4 & 1 & 2 & 3 \end{bmatrix}$. What is the corresponding eigenvalue?

2.1.8 Verify that $\begin{bmatrix} 1 \\ 1 \\ 0 \\ -1 \end{bmatrix}$ is an eigenvector of $\begin{bmatrix} 1 & 1 & 1 & 1 \\ 1 & 2 & 2 & 2 \\ 1 & 2 & 3 & 3 \\ 1 & 2 & 3 & 4 \end{bmatrix}$. What is the corresponding eigenvalue?

2.1.9 Find all the eigenvalues and eigenvectors of each of the following matrices over $\mathbb{C}$.

(a) $\begin{bmatrix} 1 & 1 \\ 1 & 1 \end{bmatrix}$ (b) $\begin{bmatrix} i & 2+i \\ 2-i & -i \end{bmatrix}$ (c) $\begin{bmatrix} 2 & -7 \\ 0 & 2 \end{bmatrix}$ (d) $\begin{bmatrix} 0 & 0 & 1 \\ 0 & -2 & 0 \\ 1 & 0 & 0 \end{bmatrix}$

2.1.10 Find all the eigenvalues and eigenvectors of each of the following matrices over $\mathbb{C}$.

(a) $\begin{bmatrix} 1 & 2 \\ 3 & 4 \end{bmatrix}$ (b) $\begin{bmatrix} i & 2+i \\ 0 & 1 \end{bmatrix}$ (c) $\begin{bmatrix} 1 & 2 \\ -1 & 2 \end{bmatrix}$ (d) $\begin{bmatrix} 0 & 1 & 1 \\ 1 & 0 & 1 \\ 1 & 1 & 0 \end{bmatrix}$

2.1.11 Find all the eigenvalues and eigenvectors of the orthogonal projection $P \in \mathcal{L}(\mathbb{R}^2)$ described in Exercise 2.1.2.

2.1.12 Find all the eigenvalues and eigenvectors of the matrix $\mathrm{A} \in \mathrm{M}_n(\mathbb{R})$ with $a_{ij} = 1$ for each i and j.

2.1.13 Consider a linear system in matrix form:

$$\mathrm{A}\mathbf{x} = \mathbf{b},$$

where $\mathrm{A} \in \mathrm{M}_{m,n}(\mathbb{F})$ and $\mathbf{b} \in \mathbb{F}^m$ are given. Suppose that $\mathbf{x}, \mathbf{y} \in \mathbb{F}^n$ are both solutions. Under what conditions is $\mathbf{x} + \mathbf{y}$ also a solution?

Remark: You might recognize this as essentially the same as Exercise 1.1.10. Don't just refer to your answer to that exercise; use what you've now learned about the matrix form of a linear system to answer the question more easily.

2.1.14 Suppose that $A \in M_n(\mathbb{F})$ and that for each $j = 1, \ldots, n$, e_j is an eigenvector of A. Prove that A is a diagonal matrix.

2.1.15 Suppose that $A \in M_n(\mathbb{F})$ and that every $v \in \mathbb{F}^n$ is an eigenvector of A. Prove that $A = \lambda I_n$ for some $\lambda \in \mathbb{F}$.

2.1.16 Suppose that $S, T \in \mathcal{L}(V)$ and that $\lambda \neq 0$ is an eigenvalue of ST. Prove that λ is also an eigenvalue of TS.
Hint: You need to be careful about the fact that an eigenvalue must correspond to a *nonzero* eigenvector. For the case when $\lambda = 0$, see Exercise 3.4.14.

2.1.17 Define the function $S : \mathbb{F}^\infty \to \mathbb{F}^\infty$ on the space of infinite sequences of elements of $\mathbb{F}$ by $S((a_1, a_2, \ldots)) = (0, a_1, a_2, \ldots)$.
(a) Show that S is a linear map.
(b) Show that S has no eigenvalues.
Hint: Consider separately the cases $\lambda = 0$ and $\lambda \neq 0$.

2.2 More on Linear Maps

Isomorphism

We motivated the idea of linear maps as a way to formalize the sameness of vector space operations in two spaces, using the spaces $\mathbb{R}^n$ and $\mathcal{P}_{n-1}(\mathbb{R})$ as our first example. But in addition to the vector space operations working the same way, there is another, perhaps more basic, way in which $\mathbb{R}^n$ and $\mathcal{P}_{n-1}(\mathbb{R})$ are the same, namely that the lists of numbers in both cases are the same length, so that a vector in $\mathbb{R}^n$ contains no more and no less information than a polynomial in $\mathcal{P}_{n-1}(\mathbb{R})$. That is, not just the operations but the actual spaces seem to be the same.

Definition An **isomorphism** between vector spaces V and W is a bijective* linear map $T : V \to W$. If there exists an isomorphism from $V \to W$, then V is said to be **isomorphic** to W.

Quick Exercise #7. Show that the identity operator on V is an isomorphism.

*For the definitions of bijective. injective, and surjective, see Appendix Section A.1.

QA #7: We saw in Quick Exercise #1 of Section 2.1 that I is linear. We also need to check that I is bijective, which is easy.

Our map $T : \mathbb{R}^n \to \mathcal{P}_{n-1}(\mathbb{R})$ defined in (2.1) is an isomorphism: if

$$a_1 + a_2x + \cdots + a_nx^{n-1} = b_1 + b_2x + \cdots + b_nx^{n-1},$$

then $a_i = b_i$ for each $i \in \{1, \dots, n\}$; that is,

$$\begin{bmatrix} a_1 \\ \vdots \\ a_n \end{bmatrix} = \begin{bmatrix} b_1 \\ \vdots \\ b_n \end{bmatrix},$$

and so T is injective.

Given a polynomial $a_0 + a_1x^n + \cdots + a_{n-1}x^{n-1}$,

$$a_0 + a_1x^n + \cdots + a_{n-1}x^{n-1} = T\left(\begin{bmatrix} a_0 \\ \vdots \\ a_{n-1} \end{bmatrix}\right),$$

and so T is surjective.

That is, the sense in which $\mathbb{R}^n$ and $\mathcal{P}_{n-1}(\mathbb{R})$ are the same is exactly that they are isomorphic vector spaces over $\mathbb{R}$.

Proposition A.2 in the Appendix says that a function $f : X \to Y$ is bijective if and only if it is invertible, that is, if and only if it has an inverse function $f^{-1} : Y \to X$. In the context of linear maps, we restate this as follows.

Proposition 2.1 *Let $T \in \mathcal{L}(V, W)$. Then T is an isomorphism if and only if T is invertible.*

A priori, a linear map $T \in \mathcal{L}(V, W)$ could turn out to have an inverse function which is not linear. The following result shows that this cannot actually happen.

Proposition 2.2 *If $T : V \to W$ is linear and invertible, then $T^{-1} : W \to V$ is also linear.*

Proof To check the additivity of T^{-1}, first observe that by additivity of T and the fact that $T \circ T^{-1} = I \in \mathcal{L}(W)$,

$$T\big(T^{-1}(w_1) + T^{-1}(w_2)\big) = T\big(T^{-1}(w_1)\big) + T\big(T^{-1}(w_2)\big) = w_1 + w_2.$$

Applying T^{-1} to both sides of the equation above then gives

$$T^{-1}(w_1) + T^{-1}(w_2) = T^{-1}(w_1 + w_2).$$

Similarly, if $c \in \mathbb{F}$ and $w \in W$, then

$$T(cT^{-1}(w)) = cT(T^{-1}(w)) = cw,$$

and so

$$cT^{-1}(w) = T^{-1}(cw).$$

▲

Example Consider the subspace $U = \left\langle \begin{bmatrix} 1 \\ 1 \\ -1 \end{bmatrix}, \begin{bmatrix} 2 \\ -1 \\ 0 \end{bmatrix} \right\rangle \subseteq \mathbb{R}^3$. Geometrically, it is clear that U is a plane, and so it should be isomorphic to $\mathbb{R}^2$. If we define $S : \mathbb{R}^2 \to U$ by

$$S \begin{bmatrix} x \\ y \end{bmatrix} := x \begin{bmatrix} 1 \\ 1 \\ -1 \end{bmatrix} + y \begin{bmatrix} 2 \\ -1 \\ 0 \end{bmatrix},$$

then it is easy to check that S is linear. It is also clear that S is surjective, since every vector in U can be written as $x \begin{bmatrix} 1 \\ 1 \\ -1 \end{bmatrix} + y \begin{bmatrix} 2 \\ -1 \\ 0 \end{bmatrix}$ for some x and y. To see that S is injective, suppose that

$$x_1 \begin{bmatrix} 1 \\ 1 \\ -1 \end{bmatrix} + y_1 \begin{bmatrix} 2 \\ -1 \\ 0 \end{bmatrix} = x_2 \begin{bmatrix} 1 \\ 1 \\ -1 \end{bmatrix} + y_2 \begin{bmatrix} 2 \\ -1 \\ 0 \end{bmatrix},$$

or equivalently,

$$u \begin{bmatrix} 1 \\ 1 \\ -1 \end{bmatrix} + v \begin{bmatrix} 2 \\ -1 \\ 0 \end{bmatrix} = \begin{bmatrix} 0 \\ 0 \\ 0 \end{bmatrix},$$

where $u = x_1 - x_2$ and $v = y_1 - y_2$. Solving this linear system for u and v shows that the only solution is $u = v = 0$, so $x_1 = x_2$ and $y_1 = y_2$. That is, S is injective and thus an isomorphism.

From this, we can also conclude that there is a well-defined bijective linear map $T : U \to \mathbb{R}^2$ with

$$T \left(x \begin{bmatrix} 1 \\ 1 \\ -1 \end{bmatrix} + y \begin{bmatrix} 2 \\ -1 \\ 0 \end{bmatrix} \right) = \begin{bmatrix} x \\ y \end{bmatrix};$$

T is exactly S^{-1}. ▲

Properties of Linear Maps

Knowing that a map is linear puts quite a lot of restrictions on what the images of sets can look like. As a first result, for linear maps the image of the zero vector is completely determined.

Theorem 2.3 *If* $T \in \mathcal{L}(V, W)$, *then* $T(0) = 0$.

Proof By linearity,

$$T(0) = T(0 + 0) = T(0) + T(0);$$

subtracting $T(0)$ from both sides gives

$$0 = T(0).$$

▲

The result above says that if, for example, a function $T : V \to V$ is known to be linear, then 0 is necessarily a **fixed point** for T; that is, $T(0) = 0$.

For further restrictions on what linear maps can do to sets, see Exercises 2.2.10 and 2.2.11.

If S and T are linear maps such that the composition $S \circ T$ makes sense, we usually write it simply as ST. If $T \in \mathcal{L}(V)$, then we denote $T \circ T$ by T^2, and more generally, T^k denotes the k-fold composition of T with itself.

Proposition 2.4 *Let* U, V, *and* W *be vector spaces. If* $T : U \to V$ *and* $S : V \to W$ *are both linear, then* ST *is linear.*

Proof See Exercise 2.2.19. ▲

Recall that addition and scalar multiplication of functions is defined pointwise, as follows.

Definition Let V and W be vector spaces over $\mathbb{F}$ and let $S, T : V \to W$ be linear maps. Then

$$(S + T)(v) := S(v) + T(v)$$

and for $c \in \mathbb{F}$,

$$(cT)(v) := c(T(v)).$$

These definitions give the collection of linear maps between two vector spaces its own vector space structure:

Theorem 2.5 *Let* V *and* W *be vector spaces over* $\mathbb{F}$, *let* $S, T : V \to W$ *be linear maps, and let* $c \in \mathbb{F}$. *Then* $S + T$ *and* cT *are linear maps from* V *to* W. *Moreover,* $\mathcal{L}(V, W)$ *is itself a vector space over* $\mathbb{F}$ *with these operations.*

Proof See Exercise 2.2.20. ▲

The operations interact nicely with composition of linear maps.

Theorem 2.6 (Distributive Laws for Linear Maps) *Let U, V, and W be vector spaces and let $T, T_1, T_2 \in \mathcal{L}(U, V)$ and $S, S_1, S_2 \in \mathcal{L}(V, W)$. Then*

1. $S(T_1 + T_2) = ST_1 + ST_2$,
2. $(S_1 + S_2)T = S_1T + S_2T$.

Proof We will give the proof of part 1; part 2 is trivial.

Let $u \in U$. By the linearity of S,

$$S(T_1 + T_2)(u) = S\big[T_1(u) + T_2(u)\big] = S\big[T_1(u)\big] + S\big[T_2(u)\big] = (ST_1 + ST_2)(u).$$

That is,

$$S(T_1 + T_2) = ST_1 + ST_2.$$

▲

Composition of linear maps satisfies an associative law (part 1 of Lemma A.1) and distributive laws, just like multiplication of real numbers, or elements of a field in general. For this reason, we often think of composition of linear maps as a kind of product. There's one very important difference, though: composition does *not* satisfy a commutative law. That is, ST (which means "do T first, then do S") is in general different from TS (which means "do S first, then do T"). For this reason, parts 1 and 2 of Theorem 2.6 have to be stated separately, unlike the single corresponding distributive law in fields.

Quick Exercise #8. What are the domains and codomains of ST and TS in each of the following situations?

(a) $S \in \mathcal{L}(V, W)$ and $T \in \mathcal{L}(W, V)$.
(b) $S \in \mathcal{L}(U, V)$ and $T \in \mathcal{L}(V, W)$.

Theorem 2.7 *A function $T : V \to W$ is a linear map if and only if, for every list $v_1, \ldots, v_k \in V$ and $a_1, \ldots, a_k \in \mathbb{F}$,*

$$T\left(\sum_{i=1}^{k} a_i v_i\right) = \sum_{i=1}^{k} a_i T(v_i).$$

QA #8: (a) $ST \in \mathcal{L}(W)$, but $TS \in \mathcal{L}(V)$. (b) $TS \in \mathcal{L}(U, W)$, but ST is not even defined (unless W happens to be a subspace of U).

Proof Suppose that $T : V \to W$ is linear, and let $v_1, \ldots, v_k \in V$ and $a_1, \ldots, a_k \in \mathbb{F}$. Then by repeated applications of the additivity of T, followed by an application of homogeneity to each term,

$$\begin{aligned} T\left(\sum_{i=1}^{k} a_i v_i\right) &= T\left(\sum_{i=1}^{k-1} a_i v_i\right) + T(a_k v_k) \\ &= T\left(\sum_{i=1}^{k-2} a_i v_i\right) + T(a_{k-1} v_{k-1}) + T(a_k v_k) \\ &\;\vdots \\ &= T(a_1 v_1) + \cdots + T(a_k v_k) \\ &= a_1 T(v_1) + \cdots + a_k T(v_k). \end{aligned}$$

Conversely, if $T : V \to W$ has the property that for every list $v_1, \ldots, v_k \in V$ and $a_1, \ldots, a_k \in \mathbb{F}$,

$$T\left(\sum_{i=1}^{k} a_i v_i\right) = \sum_{i=1}^{k} a_i T(v_i),$$

then taking $k = 2$ and $a_1 = a_2 = 1$ gives that, for every v_1, v_2,

$$T(v_1 + v_2) = T(v_1) + T(v_2),$$

and taking $k = 1$ gives that for every $a \in \mathbb{F}$ and $v \in V$,

$$T(av) = aT(v).$$ ▲

The Matrix of a Linear Map

We saw in Section 2.1 that multiplication of a vector by a matrix $\mathrm{A} \in \mathrm{M}_{m,n}(\mathbb{F})$ gives a linear map $\mathbb{F}^n \to \mathbb{F}^m$. In fact, it turns out that *every* linear map from $\mathbb{F}^n$ to $\mathbb{F}^m$ is of this type.

Theorem 2.8 *Let $T \in \mathcal{L}(\mathbb{F}^n, \mathbb{F}^m)$. Then there is a unique matrix $\mathrm{A} \in \mathrm{M}_{m,n}(\mathbb{F})$ such that for every $\mathrm{v} \in \mathbb{F}^n$,*

$$T(\mathrm{v}) = \mathrm{Av}.$$

We call A *the* **matrix of** T, *or say that* T *is* **represented by** A.

Proof Recall that we saw when defining matrix–vector multiplication that if e_j is the vector with 1 in the jth entry and zeroes elsewhere, then Ae_j is exactly the jth column of A. This means that, given T, if there is a matrix so that $T(\mathrm{v}) = \mathrm{Av}$ for every $\mathrm{v} \in \mathbb{F}^n$, then its jth column must be given by $T\mathrm{e}_j$. That is, there's only

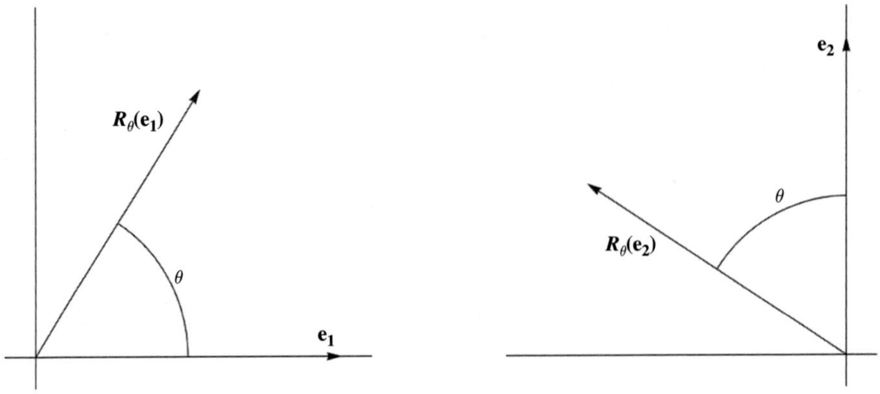

Figure 2.8 The effect of R_θ on $\mathbf{e}_1$ (left) and on $\mathbf{e}_2$ (right).

one way this could possibly work,* and if it does work, then the uniqueness of the matrix comes for free.

To confirm that it really does work, for each $j = 1, \ldots, n$, define $\mathbf{a}_j = T(\mathbf{e}_j) \in \mathbb{F}^m$, and then define

$$\mathrm{A} := \begin{bmatrix} | & & | \\ \mathbf{a}_1 & \cdots & \mathbf{a}_n \\ | & & | \end{bmatrix}.$$

Then given any vector $\mathbf{v} = \begin{bmatrix} v_1 \\ \vdots \\ v_n \end{bmatrix} \in \mathbb{F}^n$,

$$T(\mathbf{v}) = T\left(\sum_{j=1}^{n} v_j \mathbf{e}_j\right) = \sum_{j=1}^{n} v_j T(\mathbf{e}_j) = \sum_{j=1}^{n} v_j \mathbf{a}_j = \mathrm{A}\mathbf{v}$$

by the linearity of T and equation (2.3). ▲

It's worth emphasizing what we saw in the first paragraph of the proof above:

> The jth column of the matrix of T is $T\mathbf{e}_j$.

Example Let $\theta \in (0, 2\pi)$, and recall the linear map $R_\theta : \mathbb{R}^2 \to \mathbb{R}^2$ given by (counterclockwise) rotation by θ. To determine which matrix A_θ is the matrix of R_θ, we need to compute $R_\theta(\mathbf{e}_j)$ for $j = 1, 2$. This is trigonometry:

*Which makes this what the mathematician Tim Gowers calls a "just do it" proof.

$$R_\theta(\mathbf{e}_1) = \begin{bmatrix} \cos(\theta) \\ \sin(\theta) \end{bmatrix} \quad \text{and} \quad R_\theta(\mathbf{e}_2) = \begin{bmatrix} -\sin(\theta) \\ \cos(\theta) \end{bmatrix}.$$

The matrix A_θ of R_θ is therefore

$$\mathrm{A}_\theta = \begin{bmatrix} \cos(\theta) & -\sin(\theta) \\ \sin(\theta) & \cos(\theta) \end{bmatrix}.$$

From this we can tell that for each $\begin{bmatrix} x \\ y \end{bmatrix} \in \mathbb{R}^2$,

$$R_\theta\left(\begin{bmatrix} x \\ y \end{bmatrix}\right) = \begin{bmatrix} \cos(\theta) & -\sin(\theta) \\ \sin(\theta) & \cos(\theta) \end{bmatrix}\begin{bmatrix} x \\ y \end{bmatrix} = \begin{bmatrix} x\cos(\theta) - y\sin(\theta) \\ x\sin(\theta) + y\cos(\theta) \end{bmatrix}.$$

Notice what linearity has bought us: doing two very simple trigonometry calculations gives us the answers to infinitely many more complicated trigonometry problems. This fact is extremely useful, for example in computer graphics, where you may want to quickly rotate a large number of different points. ▲

Quick Exercise #9. Let $R : \mathbb{R}^2 \to \mathbb{R}^2$ be the reflection across the line $y = x$:

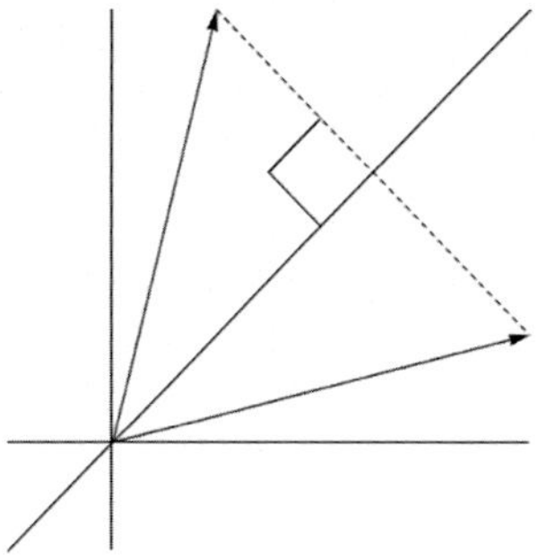

Find the matrix of R.

Theorem 2.8 gives a function from the space $\mathcal{L}(\mathbb{F}^n, \mathbb{F}^m)$ of linear maps to the space $\mathrm{M}_{m,n}(\mathbb{F})$ of matrices. The following theorem shows that this function is actually an isomorphism; $\mathcal{L}(\mathbb{F}^n, \mathbb{F}^m)$ and $\mathrm{M}_{m,n}(\mathbb{F})$ are the same space in different guises. In particular, addition and scalar multiplication of linear maps correspond to addition and scalar multiplication of matrices.

Theorem 2.9 *Let $C : \mathcal{L}(\mathbb{F}^n, \mathbb{F}^m) \to \mathrm{M}_{m,n}(\mathbb{F})$ be the map which associates a linear map to its matrix. Then C is an isomorphism of vector spaces.*

QA #9: $R\begin{bmatrix} 0 \\ 1 \end{bmatrix} = \begin{bmatrix} 1 \\ 0 \end{bmatrix}$ and $R\begin{bmatrix} 1 \\ 0 \end{bmatrix} = \begin{bmatrix} 0 \\ 1 \end{bmatrix}$, so the matrix of R is $\begin{bmatrix} 0 & 1 \\ 1 & 0 \end{bmatrix}$.

Proof The easiest way to show that C is invertible is to show that the map $D : \mathrm{M}_{m,n}(\mathbb{F}) \to \mathcal{L}(\mathbb{F}^m, \mathbb{F}^n)$ defined by $[D(\mathrm{A})](\mathrm{v}) = \mathrm{Av}$ is an inverse of C. This is done in Exercise 2.2.18. It remains to show that C is linear.

Let $T_1, T_2 \in \mathcal{L}(\mathbb{F}^n, \mathbb{F}^m)$. Then the jth column of the matrix of $T_1 + T_2$ is given by

$$(T_1 + T_2)(\mathrm{e}_j) = T_1\mathrm{e}_j + T_2\mathrm{e}_j,$$

which is the sum of the jth columns of $C(T_1)$ and $C(T_2)$, so

$$C(T_1 + T_2) = C(T_1) + C(T_2).$$

Similarly, if $a \in \mathbb{F}$ and $T \in \mathcal{L}(\mathbb{F}^n, \mathbb{F}^m)$, then the jth column of aT is given by

$$(aT)(\mathrm{e}_j) = a\big(T(\mathrm{e}_j)\big),$$

so that

$$C(aT) = aC(T).$$ ▲

Some Linear Maps on Function and Sequence Spaces

We saw in Section 2.1 that many familiar geometric operations define linear maps. It is also the case that familiar operations from calculus are linear operations on function spaces.

Examples

1. Let $C^1[a, b]$ be the set of continuously differentiable* functions $f : [a, b] \to \mathbb{R}$. We have already seen (see the examples on page 57) that $C^1[a, b]$ is a vector space. The **differentiation operator** $D : C^1[a, b] \to C[a, b]$ defined by

   $$Df(x) := f'(x)$$

   is a linear map: if $f, g \in C^1[a, b]$, then

   $$D(f + g)(x) = (f + g)'(x) = f'(x) + g'(x) = Df(x) + Dg(x),$$

   and if $c \in \mathbb{R}$, then

   $$D(cf)(x) = (cf)'(x) = cf'(x) = cDf(x).$$

2. Let $h \in C[a, b]$ be fixed. Then we can define a multiplication operator $M_h : C[a, b] \to C[a, b]$ by

   $$M_hf(x) := h(x)f(x).$$

   It is straightforward to check that M_h is linear. (Note that for our notation $M_h : C[a, b] \to C[a, b]$ to be legitimate, one also has to recall from calculus that if f and h are continuous, then so is hf.)

*That is, having one continuous derivative.

3. Let $k : \mathbb{R}^2 \to \mathbb{R}$ be continuous, and define an operator (called an **integral operator**) $T_k : C[a,b] \to C[a,b]$ by

$$T_k f(x) := \int_a^b k(x,y) f(y)\, dy. \tag{2.9}$$

The operator T_k is linear: if $f, g \in C[a,b]$, then by the linearity of the integral,

$$\begin{aligned} T_k(f+g)(x) &= \int_a^b k(x,y)\big(f(y)+g(y)\big)\, dy \\ &= \int_a^b k(x,y) f(y)\, dy + \int_a^b k(x,y) g(y)\, dy = T_k f(x) + T_k g(x). \end{aligned}$$

Similarly, if $c \in \mathbb{R}$,

$$T_k(cf)(x) = \int_a^b k(x,y) c f(y)\, dy = c \int_a^b k(x,y) f(y)\, dy = c T_k f(x).$$

The function k is called the **(integral) kernel** of the operator T_k.

Although it may appear more exotic, defining a linear map in terms of an integral kernel in this way is exactly analogous to defining a linear map on $\mathbb{R}^n$ using a matrix. Begin by thinking of a vector $\mathrm{v} = \begin{bmatrix} v_1 \\ \vdots \\ v_n \end{bmatrix} \in \mathbb{R}^n$ as a function $\mathrm{v} : \{1, \dots, n\} \to \mathbb{R}$ given by

$$\mathrm{v}(i) = v_i,$$

and think of an $m \times n$ matrix $\mathrm{A} = \big[a_{ij}\big]_{\substack{1 \le i \le m \\ 1 \le j \le n}}$ as a function $\mathrm{A} : \{(i,j) \mid 1 \le i \le m, 1 \le j \le n\} \to \mathbb{R}$ given by

$$\mathrm{A}(i,j) = a_{ij}.$$

If A acts on $\mathbb{R}^n$ by matrix–vector multiplication, then the ith coordinate of Av is

$$\big[\mathrm{Av}\big](i) = \sum_{j=1}^{n} a_{ij} v_j = \sum_{j=1}^{n} \mathrm{A}(i,j)\mathrm{v}(j).$$

The formula (2.9) is exactly what we get if we replace the set $\{1, \dots, n\}$ with the interval $[a,b]$, turn sums into integrals, and change the names of A and v to k and f.

4. Let $\mathbb{F}^\infty$ be the vector space of infinite sequences with entries in $\mathbb{F}$. For each $k \in \mathbb{N}$, define the shift operator $S_k : \mathbb{F}^\infty \to \mathbb{F}^\infty$ by

$$S_k\big((c_n)_{n \ge 1}\big) := (c_{n+k})_{n \ge 1};$$

that is, S_k shifts the sequence to begin at the $(k+1)$th entry (and throws away the first k entries). It is easy to check that S_k is a linear operator on $\mathbb{F}^\infty$. Moreover, S_k is also a linear operator from c_0 to itself (see Example 4 on page 56). The linearity is not in question; the point is that if $(c_n)_{n\geq 1} \in c_0$, then so is $S_k\big((c_n)_{n\geq 1}\big)$, since for any k, $\lim_{n\to\infty} a_n = 0$ if and only if $\lim_{n\to\infty} a_{n+k} = 0$. ▲

Quick Exercise #10. What is $S_k S_\ell$?

KEY IDEAS

- Isomorphisms are invertible linear maps between spaces.
- The composition of linear maps, sum of linear maps, and scalar multiple of a linear map, and inverse of a linear map are all linear.
- Any linear map from $\mathbb{F}^n$ to $\mathbb{F}^m$ can be represented by multiplication by a matrix.
- Differentiation, multiplication by a fixed function, and integration against an integral kernel are all linear maps on function spaces.

EXERCISES

2.2.1 Define $T : \mathbb{R}^3 \to \mathbb{R}^3$ by

$$T\left(\begin{bmatrix} x \\ y \\ z \end{bmatrix}\right) = \begin{bmatrix} y \\ z \\ 0 \end{bmatrix}.$$

(a) Is T linear?
(b) Is T injective?
(c) Is T surjective?
Justify all your answers.

2.2.2 Give an explicit isomorphism between $\mathbb{R}^2$ and the set of solutions of the linear system (over $\mathbb{R}$)

$$\begin{aligned} w - x + 3z &= 0 \\ w - x + y + 5z &= 0 \\ 2w - 2x - y + 4z &= 0. \end{aligned}$$

2.2.3 Give an explicit bijective function between $\mathbb{C}^2$ and the set of solutions of the linear system (over $\mathbb{C}$)

$$\begin{aligned} w + ix - 2y &= 1 + i \\ -w + x + 3iz &= 2. \end{aligned}$$

QA #10: $S_{k+\ell}$.

2.2.4 Find the matrix which represents reflection across the x–y plane in $\mathbb{R}^3$.

2.2.5 Find the matrix which represents the linear map T in Exercise 2.2.1.

2.2.6 Find the matrix of the orthogonal projection onto the line $y = x$ in $\mathbb{R}^2$ (see Exercise 2.1.2).

2.2.7 Find the matrix of the linear map $T : \mathbb{R}^2 \to \mathbb{R}^2$ which first reflects across the y-axis, then rotates counterclockwise by $\pi/4$ radians, then stretches by a factor of 2 in the y-direction.

2.2.8 Show that if 0 is an eigenvalue of $T \in \mathcal{L}(V)$, then T is not injective, and therefore not invertible.

2.2.9 Suppose that $T \in \mathcal{L}(V)$ is invertible and $v \in V$ is an eigenvector of T with eigenvalue $\lambda \in \mathbb{F}$. Show that v is also an eigenvector of T^{-1}. What is the corresponding eigenvalue?

2.2.10 If $\mathrm{x}, \mathrm{y} \in \mathbb{R}^n$, the **line segment** between x and y is the set

$$L := \{(1-t)\mathrm{x} + t\mathrm{y} \mid 0 \le t \le 1\}.$$

Show that if $T : \mathbb{R}^n \to \mathbb{R}^n$ is linear, then $T(L)$ is also a line segment.

2.2.11 Show that the image of the unit square $\{(x, y) \mid 0 \le x, y \le 1\}$ in $\mathbb{R}^2$ under a linear map is a parallelogram.

2.2.12 Suppose that $T \in \mathcal{L}(U, V)$, $S \in \mathcal{L}(V, W)$, and $c \in \mathbb{F}$. Show that $S \circ (cT) = c(ST)$.

2.2.13 Consider the linear map $T : \mathbb{R}^2 \to \mathbb{R}^2$ given by the matrix

$$\mathrm{A} = \begin{bmatrix} -1 & \frac{3}{2} \\ 0 & 2 \end{bmatrix}.$$

(a) Show that $\begin{bmatrix} 1 \\ 0 \end{bmatrix}$ and $\begin{bmatrix} 1 \\ 2 \end{bmatrix}$ are eigenvectors, and determine the corresponding eigenvalues.

(b) Draw the image of the unit square $\{(x, y) \mid 0 \le x, y \le 1\}$ under T.

2.2.14 Consider the linear map $T : \mathbb{R}^2 \to \mathbb{R}^2$ given by the matrix

$$\mathrm{A} = \begin{bmatrix} 2 & 1 \\ 1 & 2 \end{bmatrix}.$$

(a) Show that $\begin{bmatrix} 1 \\ 1 \end{bmatrix}$ and $\begin{bmatrix} 1 \\ -1 \end{bmatrix}$ are eigenvectors, and determine the corresponding eigenvalues.

(b) Draw the image of the unit square $\{(x, y) \mid 0 \le x, y \le 1\}$ under T.

2.2.15 Let $C^\infty(\mathbb{R})$ be the vector space of all infinitely differentiable functions on $\mathbb{R}$ (i.e., functions which can be differentiated infinitely many times), and let $D : C^\infty(\mathbb{R}) \to C^\infty(\mathbb{R})$ be the differentiation operator $Df = f'$.

Show that every $\lambda \in \mathbb{R}$ is an eigenvalue of D, and give a corresponding eigenvector.

2.2.16 Define $T : C[0,\infty) \to C[0,\infty)$ by

$$Tf(x) = \int_0^x f(y)\, dy.$$

(Note that, by the Fundamental Theorem of Calculus, Tf is an antiderivative of f with $Tf(0) = 0$.)

(a) Show that T is linear.

(b) Show that T is an integral operator (as in the example beginning on page 87), although with a discontinuous kernel $k(x,y)$.

2.2.17 Suppose that $T \in \mathcal{L}(U,V)$ and $S \in \mathcal{L}(V,W)$.

(a) Show that if ST is injective, then T is injective.

(b) Show that if ST is surjective, then S is surjective.

2.2.18 Here is an outline of the proof of the invertibility of C in Theorem 2.9.

(a) Define $D : \mathrm{M}_{m,n}(\mathbb{F}) \to \mathcal{L}(\mathbb{F}^n, \mathbb{F}^m)$ by $D(\mathrm{A})\mathrm{v} = \mathrm{Av}$ (i.e., $D(\mathrm{A})$ is just the linear map "multiply the vector by A"), and show that D is linear.

(b) Prove that CD is the identity map on $\mathrm{M}_{m,n}(\mathbb{F})$. That is, given A, if we define T by $T(\mathrm{v}) = \mathrm{Av}$, then the matrix of T is A.

(c) Prove that DC is the identity map on $\mathcal{L}(\mathbb{F}^n, \mathbb{F}^m)$. That is, if T is a linear map with matrix A, and if we define S by $S(\mathrm{v}) = \mathrm{Av}$, then $S = T$.

Fill in the details.

2.2.19 Prove Proposition 2.4.

2.2.20 Prove Theorem 2.5.

2.3 Matrix Multiplication

Definition of Matrix Multiplication

Suppose that we have an $m \times n$ matrix A and an $n \times p$ matrix B over $\mathbb{F}$. As we saw in Section 2.1, we can think of A as a linear map from $\mathbb{F}^n$ to $\mathbb{F}^m$ and B as a linear map from $\mathbb{F}^p$ to $\mathbb{F}^n$, so we can form the composition of the two:

$$\mathrm{v} \mapsto \mathrm{A}\,(\mathrm{Bv})\,. \tag{2.10}$$

By Proposition 2.4, this is a linear map from $\mathbb{F}^p$ to $\mathbb{F}^m$, so Theorem 2.8 tells us that it is represented by some $m \times p$ matrix. The question then arises: what is the matrix of this map in terms of A and B? The answer is that we *define* the product of two matrices A and B so that the matrix of the composition in formula (2.10) is the matrix product of A and B.*

Definition Let A be an $m \times n$ matrix over $\mathbb{F}$ and B be an $n \times p$ matrix over $\mathbb{F}$. The product AB is the unique $m \times p$ matrix over $\mathbb{F}$, such that for all $\mathbf{v} \in \mathbb{F}^p$,

$$\mathrm{A}(\mathrm{B}\mathbf{v}) = (\mathrm{AB})\mathbf{v}.$$

This is a fine definition, since we know that each linear map is associated with a unique matrix, but of course we'd also like a formula for the entries of AB in terms of the entries of A and B. We can get this from the definition and the formula we already have for matrix–vector multiplication:

$$[\mathrm{AB}\mathbf{v}]_i = [\mathrm{A}(\mathrm{B}\mathbf{v})]_i = \sum_{k=1}^{n} a_{ik}(\mathrm{B}\mathbf{v})_k = \sum_{k=1}^{n} a_{ik}\left(\sum_{j=1}^{p} b_{kj}v_j\right) = \sum_{j=1}^{p}\left(\sum_{k=1}^{n} a_{ik}b_{kj}\right)v_j.$$

Comparing this to the formula for the ith entry of a matrix–vector product in equation (2.2), we see that for $1 \le i \le m$ and $1 \le j \le n$,

$$[\mathrm{AB}]_{ij} = \sum_{k=1}^{n} a_{ik}b_{kj}. \tag{2.11}$$

It's important to note that, just as it only makes sense to form the composition $S \circ T$ of two linear maps S and T if T takes values in the domain of S, it only makes sense to multiply two matrices A and B if the product $\mathrm{B}\mathbf{v}$ is the right length to be multiplied by A. If B is an $m \times n$ matrix, then $\mathrm{B}\mathbf{v}$ makes sense for $\mathbf{v} \in \mathbb{F}^n$ and $\mathrm{B}\mathbf{v} \in \mathbb{F}^m$; if A is a $p \times q$ matrix, then $\mathrm{A}\mathbf{w}$ makes sense if $\mathbf{w} \in \mathbb{F}^q$, so for $\mathrm{A}(\mathrm{B}\mathbf{v})$ to make sense, we must have $q = m$. That is:

For $\mathrm{AB} = \mathrm{C}$ to be defined, the number of *columns* of A must be equal to the number of *rows* of B:

$$\underbrace{\mathrm{A}}_{m\times n}\underbrace{\mathrm{B}}_{n\times p} = \underbrace{\mathrm{C}}_{m\times p}$$

In that case, AB has the same number of rows as A and the same number of columns as B. (Rows, then columns again!)

*In case when you first saw the formula (2.11), you thought it looked weird, now you know why we do it this way.

Quick Exercise #11. Let

$$A = \begin{bmatrix} 1 & -1 \\ 0 & 2 \end{bmatrix}, \qquad B = \begin{bmatrix} 0 & 2 & 0 \\ -1 & 0 & 1 \end{bmatrix}, \qquad \text{and} \qquad C = \begin{bmatrix} 2 & 1 \\ 0 & -2 \\ 1 & 0 \end{bmatrix}.$$

Determine whether each of the following matrix expressions is defined, and the size of each matrix that is defined:
(a) AB (b) BA (c) AC (d) CA (e) BC (f) CB

Examples

1. Let $Z_{m\times n}$ denote the $m \times n$ matrix whose entries are all 0. Since it is clear that composing with the zero map always gives the zero map, and that composing with the identity map does nothing, it must be true that for $A \in M_{m,n}(\mathbb{F})$

$$AZ_{n\times p} = Z_{m\times p}, \qquad Z_{p\times m}A = Z_{p\times n},$$

and

$$AI_n = A, \qquad I_mA = A.$$

Exercise 2.3.6 asks you to confirm these formulas from the multiplication rule in equation (2.11).

2. Recall the map $R_\theta : \mathbb{R}^2 \to \mathbb{R}^2$ which rotates the plane counterclockwise by an angle of θ radians. We computed in the last section (see the example on page 84) that the matrix of R_θ is

$$\begin{bmatrix} \cos(\theta) & -\sin(\theta) \\ \sin(\theta) & \cos(\theta) \end{bmatrix}.$$

Consider the map $T : \mathbb{R}^2 \to \mathbb{R}^2$ given by first rotating counterclockwise by θ radians and then reflecting in the x-axis. Rather than working out what happens to e_1 and e_2 after performing these two operations, we note that the matrix of reflection across the x-axis is

$$\begin{bmatrix} 1 & 0 \\ 0 & -1 \end{bmatrix},$$

and so the matrix of T is the product

$$\begin{bmatrix} 1 & 0 \\ 0 & -1 \end{bmatrix}\begin{bmatrix} \cos(\theta) & -\sin(\theta) \\ \sin(\theta) & \cos(\theta) \end{bmatrix} = \begin{bmatrix} \cos(\theta) & -\sin(\theta) \\ -\sin(\theta) & -\cos(\theta) \end{bmatrix}.$$ ▲

QA #11: (a) 2×3 (b) undefined (c) undefined (d) 3×2 (e) 2×2 (f) 3×3

Quick Exercise #12. Let $S : \mathbb{R}^2 \to \mathbb{R}^2$ be the map which first reflects in the x-axis and then rotates counterclockwise by θ radians. Find the matrix of S. Under what circumstances is it the same as the matrix of T above?

The arithmetic of multiplication and addition of matrices works just as one would expect; in particular, there are the following associative and distributive laws.

Theorem 2.10 *Let* $\mathrm{A}, \mathrm{A}_1, \mathrm{A}_2 \in M_{m,n}(\mathbb{F})$, $\mathrm{B}, \mathrm{B}_1, \mathrm{B}_2 \in M_{n,p}(\mathbb{F})$, *and* $\mathrm{C} \in M_{p,q}(\mathbb{F})$. *Then*

1. $\mathrm{A}(\mathrm{BC}) = (\mathrm{AB})\mathrm{C}$,
2. $\mathrm{A}(\mathrm{B}_1 + \mathrm{B}_2) = \mathrm{AB}_1 + \mathrm{AB}_2$,
3. $(\mathrm{A}_1 + \mathrm{A}_2)\mathrm{B} = \mathrm{A}_1\mathrm{B} + \mathrm{A}_2\mathrm{B}$.

Proof For the first part, the corresponding fact for linear maps is a result of the definition of composition; see Lemma A.1 in the appendix. The fact that we defined matrix multiplication to correspond to composition of linear maps thus makes the first part automatic.

For parts 2 and 3, the corresponding facts for linear maps were proved in Theorem 2.6. By Theorem 2.9, the sum of two matrices is the matrix of the sum of the corresponding linear maps, and so parts 2 and 3 follow immediately. ▲

Other Ways of Looking at Matrix Multiplication

We saw in Section 2.1 that it's often useful to view matrix–vector multiplication in terms of the columns of the matrix rather than in terms of individual entries; this is also true in the setting of matrix–matrix multiplication, where the corresponding viewpoint is the following.

Lemma 2.11 *Let* $\mathrm{A} \in \mathrm{M}_{m,n}(\mathbb{F})$ *and let* $\mathrm{B} = \begin{bmatrix} | & & | \\ \mathbf{b}_1 & \cdots & \mathbf{b}_p \\ | & & | \end{bmatrix}$ *be an* $n \times p$ *matrix with columns* $\mathbf{b}_1, \ldots, \mathbf{b}_p \in \mathbb{F}^n$. *Then*

$$\mathrm{AB} = \begin{bmatrix} | & & | \\ \mathrm{A}\mathbf{b}_1 & \cdots & \mathrm{A}\mathbf{b}_p \\ | & & | \end{bmatrix}. \tag{2.12}$$

QA #12: $\begin{bmatrix} \cos(\theta) & -\sin(\theta) \\ \sin(\theta) & \cos(\theta) \end{bmatrix} \begin{bmatrix} 1 & 0 \\ 0 & -1 \end{bmatrix} = \begin{bmatrix} \cos(\theta) & \sin(\theta) \\ \sin(\theta) & -\cos(\theta) \end{bmatrix}$. This is only the same for $\theta = 0, \pi$.

Proof We could prove this from equation (2.11), but it's actually easier to work directly from our definition of the matrix product. By definition, for each j,

$$(\mathrm{AB})\mathbf{e}_j = \mathrm{A}(\mathrm{B}\mathbf{e}_j). \tag{2.13}$$

Remembering that for any matrix C, $\mathrm{C}\mathbf{e}_j$ is the jth column of C, equation (2.13) says exactly that the jth column of AB is $\mathrm{A}\mathbf{b}_j$. ▲

Lemma 2.11 can be summarized as:

> The jth column of AB is A times the jth column of B.

An immediate consequence of this observation is that if A is an $m \times n$ matrix and $\mathbf{v} \in \mathbb{F}^n = \mathrm{M}_{n,1}(\mathbb{F})$, then regardless of whether we think of $\mathbf{v}$ as an n-dimensional vector or as an $n \times 1$ matrix, $\mathrm{A}\mathbf{v}$ is the same thing.

Another important consequence is the following.

Corollary 2.12 *If the matrix product* AB *is defined, then every column of* AB *is a linear combination of the columns of* A.

Proof The jth column of AB is the matrix–vector product $\mathrm{A}\mathbf{b}_j$. By (2.3), if $\mathbf{a}_k$ are the columns of A, then

$$\mathrm{A}\mathbf{b}_j = \sum_k [\mathbf{b}_j]_k \mathbf{a}_k,$$

so $\mathrm{A}\mathbf{b}_j$ is a linear combination of the $\mathbf{a}_k$. ▲

We can similarly describe AB in terms of B and the rows of A.

Definition A row vector is a matrix $\mathbf{v} \in \mathrm{M}_{1,m}(\mathbb{F})$.

By the definition of matrix multiplication, if $\mathbf{v} = \begin{bmatrix} v_1 & \cdots & v_m \end{bmatrix} \in \mathrm{M}_{1,m}(\mathbb{F})$ and $\mathrm{B} \in \mathrm{M}_{m,n}(\mathbb{F})$, then the product $\mathbf{v}\mathrm{B}$ is defined and is a row vector in $\mathrm{M}_{1,n}(\mathbb{F})$, whose jth (really $(1,j)$) entry is

$$\sum_{k=1}^{m} v_k b_{kj}.$$

Notice that this is very similar (but not identical!) to the formula (2.2) for the entries of $\mathrm{A}\mathbf{x}$ when $\mathbf{x}$ is a column vector. We could just as well have chosen to use row vectors instead of column vectors as our standard representation of n-tuples in $\mathbb{F}^n$; it's purely a matter of historical convention that we mainly use column vectors.

Lemma 2.13 *Let* $A = \begin{bmatrix} -a_1- \\ \vdots \\ -a_m- \end{bmatrix}$ *be an* $m \times n$ *matrix with rows* $a_1, \ldots, a_m \in M_{1,n}(\mathbb{F})$ *and* $B \in M_{n,p}(\mathbb{F})$. *Then*

$$AB = \begin{bmatrix} -a_1B- \\ \vdots \\ -a_mB- \end{bmatrix}.$$

Proof See Exercise 2.3.14. ▲

That is:

The ith row of AB is the ith row of A, times B.

Quick Exercise #13. Prove that if AB is defined, then every row of AB is a linear combination of the rows of B.

There is one last viewpoint on matrix multiplication, which essentially combines the viewpoints of Lemmas 2.11 and 2.13. To do this, first note that if $v \in M_{1,n}(\mathbb{F})$ is a row vector and $w \in \mathbb{F}^n = M_{n,1}(\mathbb{F})$ is a column vector with the same number of entries, then vw is the scalar*

$$vw = \sum_{k=1}^{n} v_k w_k.$$

Using this observation, we can rewrite equation (2.11) as in the following lemma.

Lemma 2.14 *Let* $A = \begin{bmatrix} -a_1- \\ \vdots \\ -a_m- \end{bmatrix} \in M_{m,n}(\mathbb{F})$ *and* $B = \begin{bmatrix} | & & | \\ b_1 & \cdots & b_p \\ | & & | \end{bmatrix} \in M_{n,p}(\mathbb{F})$.

Then

$$AB = \begin{bmatrix} a_1b_1 & \cdots & a_1b_p \\ \vdots & \ddots & \vdots \\ a_mb_1 & \cdots & a_mb_p \end{bmatrix}.$$

*Technically, vw is the 1×1 matrix $\left[\sum_{k=1}^{n} v_k w_k\right]$, but there's generally no harm in treating a 1×1 matrix as the same thing as its single entry.

QA #13: Modify the proof of Corollary 2.12.

That is:

> The (i, j) entry of $\mathbf{AB}$ is the ith row of $\mathbf{A}$ times the jth column of $\mathbf{B}$.

Quick Exercise #14. Compute

$$\begin{bmatrix} 0 & -1 \\ 2 & 1 \end{bmatrix} \begin{bmatrix} 1 & -2 & 0 \\ 0 & 1 & 3 \end{bmatrix}$$

using any perspective on matrix multiplication you like (or better yet, use all of them as a way to check yourself).

The Transpose

The operation that switches the roles of rows and columns is called **transposition** and is defined as follows.*

Definition Given an $m \times n$ matrix $\mathrm{A} = \left[a_{ij}\right]_{\substack{1 \le i \le m \\ 1 \le j \le n}}$ over $\mathbb{F}$, we define the **transpose** A^{T} of A to be the $n \times m$ matrix whose (i, j) entry is

$$[\mathrm{A}^{\mathrm{T}}]_{ij} := a_{ji}.$$

In particular, the transpose of a row vector is a column vector and the transpose of a column vector is a row vector.

It is useful to visualize the definition as follows:

$$\mathrm{A} = \begin{bmatrix} \text{—}\mathbf{a}_1\text{—} \\ \vdots \\ \text{—}\mathbf{a}_m\text{—} \end{bmatrix} \implies \mathrm{A}^{\mathrm{T}} = \begin{bmatrix} | & & | \\ \mathbf{a}_1^{\mathrm{T}} & \cdots & \mathbf{a}_m^{\mathrm{T}} \\ | & & | \end{bmatrix}.$$

The following lemma is immediate from the definition.

*The transpose of A is also sometimes denoted A′, or by many obvious variations on our notation: A^T, A^{t}, A^t, $\mathrm{A}^{\top}$, etc. You may also have seen A^* or $\mathrm{A}^{\dagger}$ described as a notation for the transpose, but we use those for something different.

QA #14: $\begin{bmatrix} 0 & -1 & -3 \\ 2 & -3 & 3 \end{bmatrix}$.

Lemma 2.15 *Let* $A \in M_{m,n}(\mathbb{F})$. *Then* $(A^T)^T = A$.

The following proposition shows how transposition and matrix multiplication interact.

Proposition 2.16 *Let* $A \in M_{m,n}(\mathbb{F})$ *and* $B \in M_{n,p}(\mathbb{F})$. *Then*

$$(AB)^T = B^T A^T.$$

Proof The (i,j) entry of $(AB)^T$ is

$$[AB]_{ji} = \sum_{k=1}^{n} a_{jk} b_{ki},$$

and the (i,j) entry of $B^T A^T$ is

$$\left[B^T A^T\right]_{ij} = \sum_{k=1}^{n} \left[B^T\right]_{ik} \left[A^T\right]_{kj} = \sum_{k=1}^{n} b_{ki} a_{jk}.$$ ▲

The way the order changes in Proposition 2.16 is typical in situations in which there is a product which is not commutative; notice that $A^T \in M_{n,m}(\mathbb{F})$ and $B^T \in M_{p,n}(\mathbb{F})$, so the change in order is necessary for the multiplication even to make sense in general.

In the case that $\mathbb{F} = \mathbb{C}$, it is often more natural to consider the **conjugate transpose** A^*:

Definition The **conjugate transpose** of a matrix $B \in M_{m,n}(\mathbb{C})$ is the matrix $B^* \in M_{n,m}(\mathbb{C})$ such that $[B^*]_{jk} = \overline{b_{kj}}$. That is, it is the matrix whose entries are the complex conjugates of the entries of B^T (if the entries of B are real, then $B^* = B^T$).

By Proposition 2.16 and the fact that $\overline{zw} = (\overline{z})(\overline{w})$, it follows that

$$(AB)^* = B^* A^*.$$

Matrix Inverses

Definition Let $A \in M_n(\mathbb{F})$. The matrix $B \in M_n(\mathbb{F})$ is an **inverse (matrix)** of A if

$$AB = BA = I_n.$$

If a matrix A has an inverse, we say that A is **invertible** or **nonsingular.** If A is not invertible, then we say that A is **singular.**

Note that if you happen to know an inverse matrix B of A, then you can solve $\mathrm{Ax} = \mathrm{b}$ by matrix multiplication:

$$\mathrm{x} = \mathrm{BAx} = \mathrm{Bb}.$$

Lemma 2.17 *If a matrix* $\mathrm{A} \in \mathrm{M}_n(\mathbb{F})$ *has an inverse, then its inverse is unique and we denote it* A^{-1}.

Proof Suppose that B and C are both inverses to A. Then

$$\mathrm{B} = \mathrm{B(AC)} = \mathrm{(BA)C} = \mathrm{C}.$$

▲

Quick Exercise #15. Suppose that $a, b, c, d \in \mathbb{F}$ and that $ad \neq bc$. Verify that $\begin{bmatrix} a & b \\ c & d \end{bmatrix}$ is invertible, and that

$$\begin{bmatrix} a & b \\ c & d \end{bmatrix}^{-1} = \frac{1}{ad - bc} \begin{bmatrix} d & -b \\ -c & a \end{bmatrix}. \qquad (2.14)$$

Lemma 2.18 *Suppose that* $\mathrm{A}, \mathrm{B} \in \mathrm{M}_n(\mathbb{F})$ *are invertible. Then* AB *is invertible and*

$$(\mathrm{AB})^{-1} = \mathrm{B}^{-1}\mathrm{A}^{-1}.$$

Proof If $\mathrm{A}, \mathrm{B} \in \mathrm{M}_n(\mathbb{F})$ are invertible, then

$$\left(\mathrm{B}^{-1}\mathrm{A}^{-1}\right)\mathrm{AB} = \mathrm{B}^{-1}\left(\mathrm{A}^{-1}\mathrm{A}\right)\mathrm{B} = \mathrm{B}^{-1}\mathrm{B} = \mathrm{I}_n,$$

and similarly,

$$\mathrm{AB}\left(\mathrm{B}^{-1}\mathrm{A}^{-1}\right) = \mathrm{A}\left(\mathrm{BB}^{-1}\right)\mathrm{A}^{-1} = \mathrm{AA}^{-1} = \mathrm{I}_n.$$

▲

Note the appearance of the Rat Poison Principle: we have shown that $(\mathrm{AB})^{-1} = \mathrm{B}^{-1}\mathrm{A}^{-1}$ by showing that $\mathrm{B}^{-1}\mathrm{A}^{-1}$ does the job that defines $(\mathrm{AB})^{-1}$.

QA #15: Remember the Rat Poison Principle! We haven't yet discussed how to *find* the inverse of a matrix, but since you've been given a putative inverse, you can just multiply it by the original matrix.

Lemma 2.19 *Let* $A \in M_n(\mathbb{F})$ *be invertible. Then* A^{-1} *is also invertible and*

$$(A^{-1})^{-1} = A.$$

Proof See Exercise 2.3.15. ▲

Example Consider the $n \times n$ matrices

$$A = \begin{bmatrix} 1 & 1 & 1 & \cdots & \cdots & 1 \\ 0 & 1 & 1 & \cdots & \cdots & 1 \\ 0 & 0 & 1 & \ddots & & 1 \\ \vdots & & \ddots & \ddots & \ddots & \vdots \\ 0 & & & \ddots & 1 & 1 \\ 0 & \cdots & \cdots & \cdots & 0 & 1 \end{bmatrix} \text{ and } B = \begin{bmatrix} 1 & -1 & 0 & \cdots & \cdots & 0 \\ 0 & 1 & -1 & 0 & & \vdots \\ 0 & 0 & 1 & -1 & \ddots & \vdots \\ \vdots & & \ddots & \ddots & \ddots & 0 \\ 0 & & & \ddots & 1 & -1 \\ 0 & \cdots & \cdots & \cdots & 0 & 1 \end{bmatrix}. \tag{2.15}$$

Then we can check that $AB = BA = I_n$. This shows that A and B are both invertible, and that $A^{-1} = B$ and $B^{-1} = A$. ▲

We have only defined invertibility for square matrices. This is not an accident; it is impossible for a non-square matrix to be "invertible." To see this, let $A \in M_{m,n}(\mathbb{F})$ be **left-invertible**, meaning that there is an $n \times m$ matrix B such that $BA = I_n$. Consider the homogeneous $m \times n$ linear system

$$Ax = 0 \tag{2.16}$$

over $\mathbb{F}$. The system (2.16) is consistent (why?), and we can multiply both sides of (2.16) on the left by B to get

$$x = BAx = B0 = 0.$$

That is, $x = 0$ is the *unique* solution to the system $Ax = 0$. By Corollary 1.4 (and its analog for systems over an arbitrary field, whose proof is identical), this can only happen if $m \geq n$.

Similarly, if A is **right-invertible**, then $m \leq n$ (see Exercise 2.3.11). Thus, an $m \times n$ matrix can only hope to be reasonably called "invertible" if $m = n$.

As expected, invertibility of matrices and linear operators is the same thing:

Theorem 2.20 *Let* $A \in M_n(\mathbb{F})$ *and let* $T : \mathbb{F}^n \to \mathbb{F}^n$ *be the operator defined by multiplication by* A. *Then* A *is invertible if and only if* T *is an invertible linear map; if* A *and* T *are invertible, then the matrix of* T^{-1} *is* A^{-1}.

Proof Suppose that T is invertible with inverse S. Then $T \circ S = S \circ T = I$. Let B be the matrix of S. Then by definition of matrix multiplication, $\mathrm{AB} = \mathrm{BA} = \mathrm{I}_n$, so A is invertible and $\mathrm{B} = \mathrm{A}^{-1}$ is the matrix of T^{-1}.

Conversely, suppose that A is invertible, and let S be the operator defined by multiplication by A^{-1}. Then the matrices of $S \circ T$ and $T \circ S$ are both I_n; that is, $S = T^{-1}$. ▲

KEY IDEAS

- The product of matrices is defined so that $\mathrm{A}(\mathrm{B}\mathbf{v}) = \mathrm{AB}\mathbf{v}$ for every $\mathbf{v}$.
- There are various formulas for AB (they are all collected in the Perspectives at the end of this chapter).
- Taking the transpose of a matrix swaps its rows and columns.
- A matrix A is invertible if it has an inverse matrix A^{-1} so that

$$\mathrm{AA}^{-1} = \mathrm{A}^{-1}\mathrm{A} = \mathrm{I}_n.$$

- When matrix inverses exist, they are unique.
- Only square matrices have inverses.

EXERCISES

2.3.1 Compute each of the following matrix products, or state that it is undefined.

(a) $\begin{bmatrix} 1 & 2 \\ 3 & 4 \end{bmatrix}\begin{bmatrix} 4 & 3 \\ 2 & 1 \end{bmatrix}$

(b) $\begin{bmatrix} 3 & 1-i & 0 \\ 2 & 4 & -5i \end{bmatrix}\begin{bmatrix} -1+2i & 0 \\ 0 & 2 \\ 3 & 1 \end{bmatrix}$

(c) $\begin{bmatrix} 1 & -5 & 2 \\ 7 & 0 & -3 \end{bmatrix}\begin{bmatrix} 0 & 2 \\ -3 & 6 \end{bmatrix}$

(d) $\begin{bmatrix} -2 & 3 & 1 \end{bmatrix}\begin{bmatrix} 5 \\ -2 \\ 4 \end{bmatrix}$

(e) $\begin{bmatrix} -3 & 1 \\ -1 & 2 \end{bmatrix}\begin{bmatrix} 0 & 4 \\ -2 & 1 \end{bmatrix}\begin{bmatrix} 5 & -1 \\ 0 & -2 \\ 1 & 3 \end{bmatrix}$

2.3.2 Compute each of the following matrix products, or state that it is undefined.

(a) $\begin{bmatrix} -1+2i & 0 \\ 0 & 2 \\ 3 & 1 \end{bmatrix}\begin{bmatrix} 3 & 1-i & 0 \\ 2 & 4 & -5i \end{bmatrix}$

(b) $\begin{bmatrix} 0 & 2 \\ -3 & 6 \end{bmatrix}\begin{bmatrix} 1 & -5 & 2 \\ 7 & 0 & -3 \end{bmatrix}$

(c) $\begin{bmatrix} -2 \\ 3 \\ 1 \end{bmatrix}\begin{bmatrix} 5 \\ -2 \\ 4 \end{bmatrix}$

(d) $\begin{bmatrix} -2 \\ 3 \\ 1 \end{bmatrix}\begin{bmatrix} 5 & -2 & 4 \end{bmatrix}$

(e) $\begin{bmatrix} -3 & 1 \\ -1 & 2 \end{bmatrix}\begin{bmatrix} 0 & 4 \\ -2 & 1 \end{bmatrix}\begin{bmatrix} 5 & 0 & 1 \\ -1 & -2 & 3 \end{bmatrix}$

2.3.3 Write out carefully how to compute

$$\begin{bmatrix} 0 & -1 \\ 2 & 1 \end{bmatrix} \begin{bmatrix} 1 & -2 & 0 \\ 0 & 1 & 3 \end{bmatrix}$$

three ways, using:

(a) Lemma 2.11,

(b) Lemma 2.13,

(c) Lemma 2.14.

2.3.4 Let $T : \mathbb{R}^2 \to \mathbb{R}^2$ be the map defined by first rotating counterclockwise by θ and then reflecting across the line $y = x$. Find the matrix of T.

2.3.5 Find the matrix of the map $T : \mathbb{R}^2 \to \mathbb{R}^2$ which first reflects across the y-axis, then stretches by a factor of 2 in the y-direction, then rotates counterclockwise by $\frac{\pi}{4}$.

2.3.6 Use the multiplication rule in (2.11) to prove that for $\mathrm{A} \in \mathrm{M}_{m,n}(\mathbb{F})$,

$$\mathrm{AZ}_{n\times p} = \mathrm{Z}_{m\times p}, \qquad \mathrm{Z}_{p\times m}\mathrm{A} = \mathrm{Z}_{p\times n},$$

and

$$\mathrm{AI}_n = \mathrm{A}, \qquad \mathrm{I}_m\mathrm{A} = \mathrm{A}.$$

2.3.7 Suppose that $\mathrm{A} = \mathbf{diag}(a_1, \ldots, a_n)$ and $\mathrm{B} = \mathbf{diag}(b_1, \ldots, b_n)$. What is AB?

2.3.8 The **norm** of a vector $\mathrm{v} \in \mathbb{R}^n$ is defined (in analogy with $\mathbb{R}^2$ and $\mathbb{R}^3$) as $\|\mathrm{v}\| = \sqrt{v_1^2 + \cdots + v_n^2}$. Show that if $\mathrm{v} \in \mathbb{R}^n$, then $\|\mathrm{v}\|^2 = \mathrm{v}^{\mathrm{T}}\mathrm{v}$.

2.3.9 When is a 1×1 matrix invertible? What is its inverse?

2.3.10 Suppose that $\mathrm{A} \in \mathrm{M}_n(\mathbb{F})$ is invertible. Show that A^{T} is also invertible, and that $\left(\mathrm{A}^{\mathrm{T}}\right)^{-1} = \left(\mathrm{A}^{-1}\right)^{\mathrm{T}}$.

2.3.11 Suppose that $\mathrm{A} \in \mathrm{M}_{m,n}(\mathbb{F})$ is **right-invertible**, meaning that there is a $\mathrm{B} \in \mathrm{M}_{n,m}(\mathbb{F})$ such that $\mathrm{AB} = \mathrm{I}_m$. Show that $m \le n$.
Hint: Show that given any $\mathbf{b} \in \mathbb{F}^m$, the $m \times n$ linear system

$$\mathrm{Ax} = \mathrm{b}$$

is consistent, and use Theorem 1.2.

2.3.12 An $n \times n$ matrix A is called **upper triangular** if $a_{ij} = 0$ whenever $i > j$, and is called **lower triangular** if $a_{ij} = 0$ whenever $i < j$.

(a) Suppose that $\mathrm{A}, \mathrm{B} \in \mathrm{M}_n(\mathbb{F})$ are both upper triangular. Prove that AB is also upper triangular. What are the diagonal entries of AB?

(b) Do the same for lower triangular matrices.

Warning: Don't be tricked by this into thinking that $\mathrm{AB} = \mathrm{BA}$ for triangular matrices!

2.3.13 Use equation (2.11) to give another proof of Theorem 2.10.

2.3.14 Prove Lemma 2.13.

2.3.15 Prove Lemma 2.19.

2.4 Row Operations and the LU Decomposition

Row Operations and Matrix Multiplication

The row operations **R1**, **R2**, and **R3** we've used for solving linear systems can be interpreted in terms of matrix multiplication, using the following special set of matrices.

Definition Given $1 \le i,j \le m$ and $c \in \mathbb{F}$, define the following matrices in $\mathrm{M}_m(\mathbb{F})$.

- For $i \neq j$,

$$\mathrm{P}_{c,i,j} := \begin{bmatrix} 1 & 0 & & 0 \\ 0 & \ddots & c & \\ & & \ddots & 0 \\ 0 & & 0 & 1 \end{bmatrix}.$$

  The (i,j) entry is c, all diagonal entries are 1, and all off-diagonal entries other than the (i,j) entry are 0.

- For $c \neq 0$,

$$\mathrm{Q}_{c,i} := \begin{bmatrix} 1 & 0 & & & 0 \\ 0 & \ddots & & & \\ & & c & & \\ & & & \ddots & 0 \\ 0 & & & 0 & 1 \end{bmatrix}.$$

  The (i,i) entry is c, all other diagonal entries are 1, and all the off-diagonal entries are 0.

- For $i \neq j$,

$$\mathrm{R}_{i,j} := \begin{bmatrix} 1 & & & & & \\ & \ddots & & & & \\ & & 0 & & 1 & \\ & & & \ddots & & \\ & & 1 & & 0 & \\ & & & & & \ddots & \\ & & & & & & 1 \end{bmatrix}.$$

The (i,j) and (j,i) entries are both 1, all the other off-diagonal entries are 0, the i,i and j,j entries are both 0, and all the other diagonal entries are 1.

These are called **elementary matrices.**

Observe that the elementary matrices are exactly what you get if you perform a single row operation on the identity matrix I_m:

- $P_{c,i,j}$ is the result of adding c times the jth row of I_m to the ith row.
- $Q_{c,i}$ is the result of multiplying the ith row of I_m by c.
- $R_{i,j}$ is the result of switching the ith and jth rows of I_m.

Theorem 2.21 *Given $1 \leq i,j \leq m$ with $i \neq j$ and $c \in \mathbb{F}$, let $P_{c,i,j}$, $Q_{c,i}$, and $R_{i,j}$ be the $m \times m$ elementary matrices defined above. Then for any $A \in M_{m,n}(\mathbb{F})$:*

- *$P_{c,i,j}A$ is the result of adding c times the jth row of A to the ith row.*
- *$Q_{c,i}A$ is the result of multiplying the ith row of A by c.*
- *$R_{i,j}A$ is the result of switching the ith and jth rows of A.*

Proof We will prove the first statement only; the other two are left as Exercises 2.4.21 and 2.4.22.

Let $P_{c,i,j} = \left[p_{ij}\right]_{1 \leq i,j \leq m}$ and $A = \left[a_{ij}\right]_{\substack{1 \leq i \leq m \\ 1 \leq j \leq n}}$. Then the (k,ℓ) entry of $P_{c,i,j}A$ is

$$\left[P_{c,i,j}A\right]_{k\ell} = \sum_{r=1}^{m} p_{kr}a_{r\ell}. \tag{2.17}$$

Since

$$p_{kr} = \begin{cases} c & \text{if } k = i, r = j, \\ 1 & \text{if } k = r, \\ 0 & \text{otherwise,} \end{cases}$$

equation (2.17) becomes

$$\left[P_{c,i,j}A\right]_{k\ell} = \begin{cases} a_{k\ell} & \text{if } k \neq i, \\ a_{i\ell} + ca_{j\ell} & \text{if } k = i, \end{cases}$$

which says exactly that $P_{c,i,j}A$ is the result of adding c times the jth row of A to the ith row. ▲

The point of Theorem 2.21 is that we can think of performing row operations as simply multiplying on the left by a series of elementary matrices; having encoded the algorithm of row operations so neatly as a simple algebraic operation will turn out to be quite useful.

As a first step, we state the following theorem, which is just a restatement of Theorem 1.6 of Section 1.4, with the elementary matrices $E_1, \ldots, E_k$ corresponding, in order, to the row operations used to put A into RREF.

Theorem 2.22 *Given any* $A \in M_{m,n}(\mathbb{F})$, *there is a matrix* $B \in M_{m,n}(\mathbb{F})$ *in RREF and elementary matrices* $E_1, \ldots, E_k \in M_m(\mathbb{F})$ *such that*

$$B = E_k \cdots E_1 A.$$

Quick Exercise #16. The matrix $A = \begin{bmatrix} 1 & 3 \\ -1 & -1 \end{bmatrix}$ can be put into RREF via Gaussian elimination as follows:

$$\begin{bmatrix} 1 & 3 \\ -1 & -1 \end{bmatrix} \to \begin{bmatrix} 1 & 3 \\ 0 & 2 \end{bmatrix} \to \begin{bmatrix} 1 & 3 \\ 0 & 1 \end{bmatrix} \to \begin{bmatrix} 1 & 0 \\ 0 & 1 \end{bmatrix}.$$

Write the corresponding factorization of I_2 in terms of A and elementary matrices.

Theorem 2.23 *Given* $1 \le i, j \le m$ *with* $i \ne j$ *and* $c \in \mathbb{F}$ *with* $c \ne 0$, *let* $P_{c,i,j}$, $Q_{c,i}$, *and* $R_{i,j}$ *be the* $m \times m$ *elementary matrices defined above. All of the elementary matrices are invertible, and moreover:*

- $P_{c,i,j}^{-1} = P_{-c,i,j}$,
- $Q_{c,i}^{-1} = Q_{c^{-1},i}$,
- $R_{i,j}^{-1} = R_{i,j}$.

Proof We could check these statements using formula (2.11) for matrix multiplication, but it's simpler to use the fact that we know which row operations the elementary matrices correspond to.

Firstly, given $c \in \mathbb{F}$, the matrix $P_{-c,i,j}P_{c,i,j}$ corresponds to adding c times the jth row to the ith row, and then adding $-c$ times the jth row to the ith row: that is, adding 0 to the ith row.

Similarly, $Q_{c^{-1},i}Q_{c,i}$ corresponds to multiplication of the ith row by c, then by c^{-1}, so $Q_{c^{-1},i}Q_{c,i} = Q_{c,i}Q_{c^{-1},i} = I_n$.

Finally, $R_{i,j}$ switches the ith and jth rows and another application of $R_{i,j}$ switches them back. ▲

Because the elementary matrices are invertible, we can rearrange the conclusion of Theorem 2.22 as a factorization of the matrix A as follows.

QA #16: $P_{-3,1,2}Q_{1/2,2}P_{1,2,1}A$.

Theorem 2.24 *Given any* $A \in M_{m,n}(\mathbb{F})$*, there is a matrix* $B \in M_{m,n}(\mathbb{F})$ *in RREF and elementary matrices* $E_1, \ldots, E_k \in M_m(\mathbb{F})$ *such that*

$$A = E_1 \cdots E_k B.$$

Proof By Theorem 2.22, there are elementary matrices $F_1, \ldots, F_k$ such that

$$B = F_k \cdots F_1 A.$$

Since each F_i is invertible, it follows by repeated applications of Lemma 2.18 that $F_k \cdots F_1$ is invertible and

$$A = (F_k \cdots F_1)^{-1} B = F_1^{-1} \cdots F_k^{-1} B.$$

By Theorem 2.23, each $E_i := F_i^{-1}$ is an elementary matrix. ▲

Quick Exercise #17. Write $A = \begin{bmatrix} 1 & 3 \\ -1 & -1 \end{bmatrix}$ as a product of elementary matrices and a matrix in RREF.
Hint: In Quick Exercise #16, you found that $I_2 = P_{-3,1,2} Q_{1/2,2} P_{1,2,1} A$.

Inverting Matrices via Row Operations

Our observations about the elementary matrices allow us to give an algorithm to find the inverse of an invertible matrix via row operations, as follows.

Algorithm 2.25 To determine whether a square matrix A is invertible and, if so, find A^{-1}:

- Form the augmented matrix $[A \mid I_n]$ and put it into RREF.
- If the RREF has the form $[I_n \mid B]$, then A is invertible and $B = A^{-1}$.
- If the matrix in the left half of the RREF is not I_n, then A is singular.

Proof If the RREF of $[A \mid I_n]$ has the form $[I_n \mid B]$, then in particular the RREF of A is I_n. By Theorem 2.22, this means that there are elementary matrices $E_1, \ldots, E_k$ such that

$$I_n = E_k \cdots E_1 A.$$

Since elementary matrices are invertible, this implies that

$$A = (E_k \cdots E_1)^{-1},$$

QA #17: $A = P_{-1,2,1} Q_{2,2} P_{3,1,2} I_2$.

and so by Lemma 2.19, A is invertible and

$$A^{-1} = \left[(E_k \cdots E_1)^{-1}\right]^{-1} = E_k \cdots E_1.$$

Thus

$$E_k \cdots E_1 [A \mid I_n] = A^{-1}[A \mid I_n] = [I_n \mid A^{-1}];$$

that is, the RREF of $[A \mid I_n]$ is $[I_n \mid A^{-1}]$.

It remains to show that if the RREF of A is not I_n, then A is singular. We do this by proving the contrapositive: suppose that A is invertible. Then the linear system

$$Ax = b$$

has a unique solution for any choice of **b**. By Theorem 1.3 (or rather its analog over general fields, whose proof is identical), this means that the RREF of A has a pivot in each column. Since A is a square matrix, this means that the RREF of A is I_n. ▲

Tucked into the algorithm is the following result, which is important enough to be stated separately.

Corollary 2.26 *Let* $A \in M_n(\mathbb{F})$. *Then* A *is invertible if and only if the RREF of* A *is* I_n.

This observation is useful in proving the following.

Proposition 2.27 *Let* $A \in M_n(\mathbb{F})$. *If* $BA = I_n$ *or* $AB = I_n$, *then* $B = A^{-1}$.

Proof Suppose that $BA = I_n$. Then the argument in the second half of the proof of Algorithm 2.25 still applies: $Ax = b$ has unique solutions, so the RREF of A is I_n. But then by Corollary 2.26, A has an inverse, and so

$$B = BAA^{-1} = I_n A^{-1} = A^{-1}.$$

If $AB = I_n$, then we have just shown that $A = B^{-1}$. Inverting both sides gives that

$$A^{-1} = (B^{-1})^{-1} = B.$$ ▲

The point is that if B is only known to be a one-sided inverse of A, it must actually be a two-sided inverse. This is not automatic in settings where multiplication is not commutative.

Quick Exercise #18. Let $A = \begin{bmatrix} 1 & 1 \\ 2 & -1 \\ 1 & 1 \end{bmatrix}$ and $B = \frac{1}{6}\begin{bmatrix} 1 & 2 & 1 \\ 2 & -2 & 2 \end{bmatrix}$. Show that $BA = I_2$, but that A is singular. Why does this not contradict Proposition 2.27?

The LU Decomposition

Theorem 2.24 gives a factorization of a matrix as a product of elementary matrices times a matrix in RREF. In practice, a slightly different version of Gaussian elimination as a matrix factorization is normally used; this is the so-called **LU decomposition.** Here, L stands for lower triangular and U stands for upper triangular, defined as follows.

Definition A matrix $A \in M_{m,n}(\mathbb{F})$ is **lower triangular** if $a_{ij} = 0$ whenever $i < j$. A matrix $A \in M_{m,n}(\mathbb{F})$ is **upper triangular** if $a_{ij} = 0$ whenever $i > j$.

Notice that we do not require upper and lower triangular matrices to be square.

Now, suppose for the moment that when we perform the Gaussian elimination algorithm (as described in the proof of Theorem 1.1) on an $m \times n$ matrix A, we never use row operation R3. If we do not insist on normalizing all pivot entries to be 1, we can transform A into an $m \times n$ upper triangular matrix U using only operation R1, and only ever adding a multiple of row j to row i for $j < i$. If we encode this process as multiplication by $m \times m$ elementary matrices, we have

$$P_{c_k,i_k,j_k} \cdots P_{c_1,i_1,j_1} A = U.$$

Multiplying through by the inverses to the $P_{c,i,j}$ in sequence, we have

$$A = P^{-1}_{c_1,i_1,j_1} \cdots P^{-1}_{c_k,i_k,j_k} U = P_{-c_1,i_1,j_1} \cdots P_{-c_k,i_k,j_k} U.$$

Because the matrices $P_{c,i,j}$ above always have $i > j$, it follows that the $m \times m$ matrix

$$L := P_{-c_1,i_1,j_1} \cdots P_{-c_k,i_k,j_k}$$

is lower triangular, since it is the product of lower triangular matrices (see Exercise 2.3.12). Moreover, L turns out to be very easy to compute. When we perform the row operations of Gaussian elimination, we use matrices of the form $P_{c,i,1}$ first, then matrices of the form $P_{c,i,2}$, and so on. That is, $j_1 \le j_2 \le \cdots \le j_k$. Now, we can think of L as a sequence of row operations performed on I_m. Observe that for each $\ell < m$, we have $j_\ell \le j_m < i_m$, and so $j_\ell \ne i_m$. This means that when we come to multiply by $P_{-c_\ell,i_\ell,j_\ell}$, the j_ℓth row of our matrix is still untouched: it has a 1 on the diagonal and zeroes everywhere else. Moreover, since we started from I_m and

QA #18: A is not square.

have not yet multiplied by a matrix of the form $\mathbf{P}_{c,i_\ell,j_\ell}$, the (i_ℓ,j_ℓ) entry is still zero. This means that when we do multiply by $\mathbf{P}_{-c_\ell,i_\ell,j_\ell}$, the only change is to put a $-c_\ell$ in the (i_ℓ,j_ℓ) entry.

All together then, we have the following algorithm.

Algorithm 2.28 To find the LU decomposition of A:

- Try to reduce A to an upper triangular matrix using only row operation R1, and only ever adding a multiple of row j to row i if $j < i$. If this succeeds, call the result U. If it fails, stop; the algorithm fails to produce an LU decomposition for A.
- If the previous step succeeded, define L to be the lower triangular matrix with 1s on the diagonal and $-c$ in the (i,j) entry if, during the elimination process, c times row j was added to row i.
- $\mathbf{A} = \mathbf{LU}$.

Example Let $\mathbf{A} = \begin{bmatrix} 1 & 2 & 3 \\ 4 & 5 & 6 \\ 7 & 8 & 9 \end{bmatrix}$. Performing Gaussian elimination using only **R1** yields

$$\begin{bmatrix} 1 & 2 & 3 \\ 4 & 5 & 6 \\ 7 & 8 & 9 \end{bmatrix} \rightsquigarrow \begin{bmatrix} 1 & 2 & 3 \\ 0 & -3 & -6 \\ 0 & -6 & -12 \end{bmatrix} \rightsquigarrow \begin{bmatrix} 1 & 2 & 3 \\ 0 & -3 & -6 \\ 0 & 0 & 0 \end{bmatrix}.$$

Quick Exercise #19. Show that the Gaussian elimination above leads to

$$\mathbf{L} = \begin{bmatrix} 1 & 0 & 0 \\ 4 & 1 & 0 \\ 7 & 2 & 1 \end{bmatrix}.$$

We obtain

$$\mathbf{A} = \mathbf{LU} = \begin{bmatrix} 1 & 0 & 0 \\ 4 & 1 & 0 \\ 7 & 2 & 1 \end{bmatrix} \begin{bmatrix} 1 & 2 & 3 \\ 0 & -3 & -6 \\ 0 & 0 & 0 \end{bmatrix}.$$

▲

An important numerical use for the LU decomposition is that it makes it possible to solve a linear system $\mathbf{Ax} = \mathbf{b}$ by solving two triangular systems via substitution; this method is both fast and not as susceptible to round-off errors as other methods.

QA #19: We added −4 times row 1 to row 2, −7 times row 1 to row 3, and −2 times (the new) row 2 to row 3.

Example To solve

$$\begin{bmatrix} 1 & 2 & 3 \\ 4 & 5 & 6 \\ 7 & 8 & 9 \end{bmatrix} \mathbf{x} = \begin{bmatrix} 0 \\ 3 \\ 6 \end{bmatrix},$$

we use the LU decomposition found in the previous example to write the system as

$$\begin{bmatrix} 1 & 0 & 0 \\ 4 & 1 & 0 \\ 7 & 2 & 1 \end{bmatrix} \begin{bmatrix} 1 & 2 & 3 \\ 0 & -3 & -6 \\ 0 & 0 & 0 \end{bmatrix} \mathbf{x} = \mathbf{LUx} = \begin{bmatrix} 0 \\ 3 \\ 6 \end{bmatrix}.$$

First, let $\mathbf{y} = \mathbf{Ux}$ and solve

$$\begin{bmatrix} 1 & 0 & 0 \\ 4 & 1 & 0 \\ 7 & 2 & 1 \end{bmatrix} \begin{bmatrix} y_1 \\ y_2 \\ y_3 \end{bmatrix} = \mathbf{Ly} = \begin{bmatrix} 0 \\ 3 \\ 6 \end{bmatrix}$$

by substitution: $y_1 = 0$, $y_2 = 3$, and $y_3 = 0$. The system

$$\mathbf{y} = \begin{bmatrix} 0 \\ 3 \\ 0 \end{bmatrix} = \mathbf{Ux} = \begin{bmatrix} 1 & 2 & 3 \\ 0 & -3 & -6 \\ 0 & 0 & 0 \end{bmatrix} \begin{bmatrix} x_1 \\ x_2 \\ x_3 \end{bmatrix}$$

has x_3 as a free variable; solving for x_1 and x_2 in terms of x_3 yields

$$x_2 = -1 - 2x_3$$

and

$$x_1 = -2x_2 - 3x_3 = 2 + x_3,$$

so the set of all solutions to the system is

$$\left\{ \begin{bmatrix} 2 \\ -1 \\ 0 \end{bmatrix} + t \begin{bmatrix} 1 \\ -2 \\ 1 \end{bmatrix} \middle| \, t \in \mathbb{R} \right\}.$$

▲

If A cannot be put into upper triangular form with operation R1 only, then Algorithm 2.28 will not produce an LU decomposition for A. Such a matrix might still have an LU decomposition, although it typically doesn't. The issue is that the need for row swaps in order to arrive at an upper triangular matrix prevents what would play the role of L from being upper triangular. However, it is always possible to multiply A by a matrix P (a so-called permutation matrix) so that PA does have an LU decomposition.

Theorem 2.29 *For every* $A \in M_{m,n}(\mathbb{F})$, *there is a matrix* P *which is a product of elementary matrices of type* $R_{i,j}$, *an upper triangular matrix* U, *and a lower triangular matrix* L *such that*

$$PA = LU.$$

This is referred to as an **LUP decomposition** of A, or LU decomposition with **pivoting.***

Note that multiplication on the left by a sequence of matrices of type $R_{i,j}$ simply rearranges the rows of A; the theorem says that it is always possible to rearrange the rows of A so that the result has an LU decomposition.

Proof From the proof of Theorem 1.6, it is clear that every matrix over a field $\mathbb{F}$ can be put into upper triangular form using row operations R1 and R3 (where, again, upper triangular means only that $a_{ij} = 0$ if $i < j$ and does not imply that the matrix is square). There are thus elementary matrices $E_1, \ldots, E_k$ of types $P_{c,i,j}$ and $R_{i,j}$ such that

$$E_k \cdots E_1 A = U,$$

where U is upper triangular. It is a general fact (see Exercise 2.4.15) that

$$R_{i,j}P_{c,k,\ell} = \begin{cases} P_{c,k,\ell}R_{i,j} & \text{if } \{i,j\} \cap \{k,\ell\} = \emptyset, \\ P_{c,j,\ell}R_{j,k} & \text{if } i = k, j \neq \ell, \\ P_{c,k,j}R_{\ell,j} & \text{if } i = \ell, j \neq k, \\ P_{c,\ell,k}R_{k,\ell} & \text{if } i = k, j = \ell. \end{cases}$$

We can thus move every matrix of type R_{ij} among the E_i to the right until it is past all matrices of type $P_{c,i,j}$, changing indices on the way as needed. At that point, we simply multiply both sides by the inverses of the $P_{c,i,j}$ as before. ▲

KEY IDEAS

- The row operations can be encoded as multiplication by the elementary matrices.
- You can find the inverse of a matrix (or find that it isn't invertible) using row operations: row reduce $[A \mid I]$; if you get $[I \mid B]$, then $B = A^{-1}$. If not, A is not invertible.
- If A can be put in upper triangular form using only row operation R1, then you can factor A as $A = LU$, with U the upper triangular matrix resulting from the

*Yes, the letters in LUP appear to be in the wrong order, but that's what people call it.

applications of **R1** and L being lower triangular with 1s on the diagonal and below-diagonal entries determined by the row operations used to get U.

- Writing A = LU lets you solve $A\mathbf{x} = \mathbf{b}$ by solving two triangular systems: $L\mathbf{y} = \mathbf{b}$, then $U\mathbf{x} = \mathbf{y}$.

EXERCISES

2.4.1 Write each of the following matrices as a product of elementary matrices and a matrix in RREF.

(a) $\begin{bmatrix} 2 & -1 \\ 0 & 3 \end{bmatrix}$ (b) $\begin{bmatrix} 1 & 2 \\ 3 & 4 \end{bmatrix}$ (c) $\begin{bmatrix} 1 & -2 & -1 \\ -2 & 4 & 2 \end{bmatrix}$ (d) $\begin{bmatrix} 1 & -2 & 0 & 2 \\ 2 & -4 & 1 & 3 \\ -1 & 2 & 1 & -3 \end{bmatrix}$

2.4.2 Write each of the following matrices as a product of elementary matrices and a matrix in RREF.

(a) $\begin{bmatrix} 1 & 2 \\ 0 & 3 \end{bmatrix}$ (b) $\begin{bmatrix} 1 & 1 \\ 2 & 2 \end{bmatrix}$ (c) $\begin{bmatrix} 2 & 0 & 1 \\ 1 & -1 & 3 \end{bmatrix}$ (d) $\begin{bmatrix} 2 & 1 & 3 & 0 \\ 1 & 1 & 1 & -1 \\ 1 & -1 & 3 & 1 \end{bmatrix}$

2.4.3 Determine whether each of the following matrices is singular or invertible. For the invertible matrices, find the inverse.

(a) $\begin{bmatrix} 1 & 2 \\ 3 & 4 \end{bmatrix}$ (b) $\begin{bmatrix} i & -3 \\ 2+i & 0 \end{bmatrix}$ (c) $\begin{bmatrix} 1 & 1 & 0 \\ 1 & 0 & 1 \\ 0 & 1 & 1 \end{bmatrix}$, over $\mathbb{R}$

(d) $\begin{bmatrix} 1 & 1 & 0 \\ 1 & 0 & 1 \\ 0 & 1 & 1 \end{bmatrix}$, over $\mathbb{F}_2$ (e) $\begin{bmatrix} 2 & -1 & 4 \\ -3 & -1 & -1 \\ 1 & -2 & 5 \end{bmatrix}$ (f) $\begin{bmatrix} 1 & 1 & 0 & -1 \\ 2 & 0 & -2 & -1 \\ -1 & 0 & 2 & 1 \\ 1 & -1 & -1 & 0 \end{bmatrix}$

2.4.4 Determine whether each of the following matrices is singular or invertible. For the invertible matrices, find the inverse.

(a) $\begin{bmatrix} 1 & -2 \\ -2 & 4 \end{bmatrix}$ (b) $\begin{bmatrix} 1+2i & 5 \\ i & 2+i \end{bmatrix}$ (c) $\begin{bmatrix} 1 & 1 & 1 \\ 1 & 2 & 2 \\ 1 & 2 & 3 \end{bmatrix}$ (d) $\begin{bmatrix} 1 & 2 & 3 \\ 4 & 5 & 6 \\ 7 & 8 & 9 \end{bmatrix}$

(e) $\begin{bmatrix} 1 & 0 & -1 \\ 2 & 1 & 1 \\ 1 & -2 & -3 \end{bmatrix}$ (f) $\begin{bmatrix} 1 & 0 & 2 & -1 \\ 0 & -1 & -2 & 1 \\ 2 & 0 & 1 & 0 \\ 0 & 2 & -3 & 2 \end{bmatrix}$

2.4.5 Compute $\begin{bmatrix} 1 & 0 & 3 \\ 0 & 1 & 1 \\ -1 & 0 & 2 \end{bmatrix}^{-1}$, and use your answer to solve each of the following linear systems.

(a) $\begin{aligned} x + 3z &= 1 \\ y + z &= 2 \\ -x + 2z &= 3 \end{aligned}$ (b) $\begin{aligned} x + 3z &= -7 \\ y + z &= 3 \\ -x + 2z &= \frac{1}{2} \end{aligned}$ (c) $\begin{aligned} x + 3z &= \sqrt{2} \\ y + z &= \pi \\ -x + 2z &= 83 \end{aligned}$

(d) $\begin{aligned} x + 3z &= a \\ y + z &= b \\ -x + 2z &= c \end{aligned}$

2.4.6 Compute $\begin{bmatrix} 1 & 0 & 1 \\ 3 & 2 & 0 \\ 0 & 1 & -2 \end{bmatrix}^{-1}$, and use your answer to solve each of the following linear systems.

(a) $\begin{aligned} x + z &= 1 \\ 3x + 2y &= 2 \\ y - 2z &= 3 \end{aligned}$ (b) $\begin{aligned} x + z &= 0 \\ 3x + 2y &= -\frac{2}{3} \\ y - 2z &= \frac{5}{2} \end{aligned}$

(c) $\begin{aligned} x + z &= -2 \\ 3x + 2y &= 1 \\ y - 2z &= -1 \end{aligned}$ (d) $\begin{aligned} x + z &= 1 \\ 3x + 2y &= 0 \\ y - 2z &= 0 \end{aligned}$

2.4.7 Find the LU factorization of each of the following matrices.

(a) $\begin{bmatrix} 2 & 3 \\ -2 & 1 \end{bmatrix}$ (b) $\begin{bmatrix} 2 & -1 \\ 4 & -3 \end{bmatrix}$ (c) $\begin{bmatrix} 2 & 0 & -1 \\ -2 & -1 & 3 \\ 0 & 2 & -3 \end{bmatrix}$ (d) $\begin{bmatrix} -1 & 1 & 2 \\ -1 & 3 & 1 \end{bmatrix}$

2.4.8 Find the LU factorization of each of the following matrices.

(a) $\begin{bmatrix} -1 & 3 \\ 2 & -4 \end{bmatrix}$ (b) $\begin{bmatrix} 1 & -1 & 2 \\ 3 & -1 & 5 \\ -1 & 3 & -4 \end{bmatrix}$ (c) $\begin{bmatrix} 2 & 1 & -3 \\ 2 & -1 & 0 \\ -4 & -2 & 3 \end{bmatrix}$

(d) $\begin{bmatrix} -1 & 1 \\ -1 & -1 \\ 1 & -3 \end{bmatrix}$

2.4.9 Solve each of the following linear systems using LU decompositions.

(a) $2x+3y=1$, $-2x+y=2$ (b) $2x-y=0$, $4x-3y=-2$

(c) $2x-z=-2$, $-2x-y+3z=3$, $2y-3z=0$ (d) $-x+y+2z=5$, $-x+3y+z=-2$

2.4.10 Solve each of the following linear systems using LU decompositions.

(a) $-x+3y=-5$, $2x-4y=6$ (b) $x-y+2z=3$, $3x-y+5z=10$, $-x+3y-4z=-3$

(c) $2x+y-3z=1$, $2x-y=-5$, $-4x-2y+3z=-10$ (d) $-x+y=1$, $-x-y=3$, $x-3y=1$

2.4.11 Find an LUP decomposition of each of the following matrices.

(a) $\begin{bmatrix} 0 & 1 \\ 2 & 3 \end{bmatrix}$ (b) $\begin{bmatrix} 2 & -1 & 2 \\ -2 & 1 & -1 \\ 0 & 2 & -1 \end{bmatrix}$

2.4.12 Find an LUP decomposition of each of the following matrices.

(a) $\begin{bmatrix} 0 & 2 \\ -1 & 4 \end{bmatrix}$ (b) $\begin{bmatrix} 1 & 2 & -2 \\ 3 & 6 & -4 \\ 1 & 5 & -5 \end{bmatrix}$

2.4.13 Use an LUP decomposition to solve the linear system

$$\begin{aligned} 2x-y+2z&=5 \\ -2x+y-z&=-2 \\ 2y-z&=2. \end{aligned}$$

2.4.14 Use an LUP decomposition to solve the linear system

$$\begin{aligned} x+2y-2z&=2 \\ 3x+6y-4z&=4 \\ x+5y-5z&=2. \end{aligned}$$

2.4.15 Show that

$$\mathbf{R}_{i,j}\mathbf{P}_{c,k,\ell} = \begin{cases} \mathbf{P}_{c,k,\ell}\mathbf{R}_{i,j} & \text{if } \{i,j\}\cap\{k,\ell\}=\emptyset, \\ \mathbf{P}_{c,j,\ell}\mathbf{R}_{j,k} & \text{if } i=k, j\neq\ell, \\ \mathbf{P}_{c,k,j}\mathbf{R}_{\ell,j} & \text{if } i=\ell, j\neq k, \\ \mathbf{P}_{c,\ell,k}\mathbf{R}_{k,\ell} & \text{if } i=k, j=\ell. \end{cases}$$

2.4.16 Suppose that $A \in M_n(\mathbb{F})$ and $a_{11} = 0$.

(a) Show that if A has the LU decomposition $A = LU$, then either $\ell_{11} = 0$ or $u_{11} = 0$.

(b) Show that if $u_{11} = 0$ then U must be singular, and that if $\ell_{11} = 0$ then L must be singular.
Hint: For the second part, Exercise 2.3.10 may be helpful.

(c) Show that in this case, $A = LU$ must be singular.

(d) Give an example of an invertible matrix $A \in M_n(\mathbb{F})$ with $a_{11} = 0$.

This shows that for certain matrices, pivoting is necessary in order to find an LU decomposition.

2.4.17 Suppose that $A \in M_{m,n}(\mathbb{F})$ has an LU decomposition. Prove that we can write $A = LDU$, where:

- $L \in M_m(\mathbb{F})$ is lower triangular with $\ell_{i,i} = 1$ for each i,
- $D \in M_m(\mathbb{F})$ is diagonal,
- $U \in M_{m,n}(\mathbb{F})$ is upper triangular with $u_{i,i} = 1$ for each i. (This U is not necessarily the same as in the LU decomposition of A.)

This is called the **LDU decomposition** of A.

2.4.18 Find an LDU factorization of each of the following matrices (see Exercise 2.4.17).

(a) $\begin{bmatrix} -1 & 3 \\ 2 & 4 \end{bmatrix}$ (b) $\begin{bmatrix} 1 & -1 & 2 \\ 3 & -1 & 5 \\ -1 & 3 & -4 \end{bmatrix}$ (c) $\begin{bmatrix} 2 & 1 & -3 \\ 2 & -1 & 0 \\ -4 & -2 & 3 \end{bmatrix}$

(d) $\begin{bmatrix} -1 & 1 \\ -1 & -1 \\ 1 & -3 \end{bmatrix}$

2.4.19 Suppose that $A \in M_n(\mathbb{F})$ is upper triangular and invertible. Prove that A^{-1} is also upper triangular.
Hint: Think about the row operations used in computing A^{-1} via Gaussian elimination.

2.4.20 Let $A, B \in M_n(\mathbb{F})$ be the matrices given in equation (2.15). Use Algorithm 2.25 to show that:

(a) B is the inverse of A,

(b) A is the inverse of B.

2.4.21 Prove the second statement of Theorem 2.21.

2.4.22 Prove the third statement of Theorem 2.21.

2.5 Range, Kernel, and Eigenspaces

In this section we introduce some important subspaces associated with linear maps.

Range

Recall that if $T \in \mathcal{L}(V, W)$ and $A \subseteq V$, then

$$T(A) = \{T(v) \mid v \in A\}$$

is called the **image** of A under T.

Definition Let $T \in \mathcal{L}(V, W)$.

The vector space W is called the **codomain** of T.

The **range** of T is the image of V under T and is denoted range T:

$$\operatorname{range} T := T(V) = \{T(v) \mid v \in V\} \subseteq W.$$

Quick Exercise #20. Let $T : \mathbb{R}^2 \to \mathbb{R}^2$ be the projection onto the line $y = x$. What is range T?

Note that the range and the codomain of T are the same if and only if T is surjective. However, if the range is smaller than the codomain, it is not just an arbitrary subset.

Theorem 2.30 *If $T \in \mathcal{L}(V, W)$, then* range T *is a subspace of W.*

Proof By Theorem 2.3, $T(0) = 0$, so $0 \in \operatorname{range} T$.

Let $w_1, w_2 \in \operatorname{range} T$. Then by definition of range T, there are $v_1, v_2 \in V$ so that

$$T(v_1) = w_1 \qquad \text{and} \qquad T(v_2) = w_2.$$

Then by linearity,

$$T(v_1 + v_2) = T(v_1) + T(v_2) = w_1 + w_2,$$

so $w_1 + w_2 \in \operatorname{range} T$.

Similarly, if $w \in \operatorname{range} T$, then there is $v \in V$ such that $T(v) = w$. Thus if $c \in \mathbb{F}$, then

$$cw = cT(v) = T(cv) \in \operatorname{range} T. \qquad \blacktriangle$$

Theorem 2.31 *Suppose that $T \in \mathcal{L}(\mathbb{F}^n, \mathbb{F}^m)$ is represented by $\mathrm{A} \in \mathrm{M}_{m,n}(\mathbb{F})$. Then the range of T is exactly the span of the columns of* A.

QA #20: See Exercise 2.1.2 for the formal definition of orthogonal projection. The range is the line $y = x$.

Proof Recall that if $\mathbf{a}_j$ denotes the jth column of A, then

$$\mathbf{a}_j = T(\mathbf{e}_j).$$

It follows that each of the columns of A is in the range of T, and since the range of T is a vector space, it also includes the span of the $\mathbf{a}_j$.

Conversely, if $\mathbf{w} \in T(\mathbb{F}^n)$, then there is a $\mathbf{v} = \begin{bmatrix} v_1 \\ \vdots \\ v_n \end{bmatrix} \in \mathbb{F}^n$ such that $T(\mathbf{v}) = \mathbf{w}$.

Then by linearity

$$\mathbf{w} = T(v_1\mathbf{e}_1 + \cdots + v_n\mathbf{e}_n) = v_1 T(\mathbf{e}_1) + \cdots + v_n T(\mathbf{e}_1) = v_1\mathbf{a}_1 + \cdots + v_n\mathbf{a}_n,$$

so $\mathbf{w} \in \langle \mathbf{a}_1, \ldots, \mathbf{a}_n \rangle$.

We have shown that range $T \supseteq \langle \mathbf{a}_1, \ldots, \mathbf{a}_n \rangle$ and range $T \subseteq \langle \mathbf{a}_1, \ldots, \mathbf{a}_n \rangle$, so in fact range $T = \langle \mathbf{a}_1, \ldots, \mathbf{a}_n \rangle$. ▲

Definition The **column space** $C(\mathrm{A})$ of a matrix A is the span of the columns of A. That is, if $\mathrm{A} = \begin{bmatrix} | & & | \\ \mathbf{a}_1 & \cdots & \mathbf{a}_n \\ | & & | \end{bmatrix}$, then $C(\mathrm{A}) := \langle \mathbf{a}_1, \ldots, \mathbf{a}_n \rangle$.

In this language, Theorem 2.31 says that the range of $T \in \mathcal{L}(\mathbb{F}^n, \mathbb{F}^m)$ is the same as the column space of the matrix of T.

Corollary 2.32 *If the matrix product* AB *is defined, then* $C(\mathrm{AB}) \subseteq C(\mathrm{A})$.

Proof Observe that if $T \in \mathcal{L}(U, V)$ and $S \in \mathcal{L}(V, W)$, then range $ST \subseteq$ range S: if $w \in$ range ST, then there is a $u \in U$ so that

$$w = ST(u) = S\big(T(u)\big) \in \text{range}\, S.$$

If T is the operator on $\mathbb{F}^m$ given by multiplication by B and S is the operator on $\mathbb{F}^n$ given by multiplication by A, then the range of S is $C(\mathrm{A})$ and the range of ST is $C(\mathrm{AB})$, so by the observation above, $C(\mathrm{AB}) \subseteq C(\mathrm{A})$. ▲

The column space of the coefficient matrix plays an important role in solving linear systems:

Corollary 2.33 *Let* $\mathrm{A} \in \mathrm{M}_{m,n}(\mathbb{F})$. *The linear system* $\mathrm{A}\mathbf{x} = \mathbf{b}$ *is consistent if and only if* $\mathbf{b}$ *is in the column space of* A.

Quick Exercise #21. Prove Corollary 2.33.

This new perspective on the consistency of linear systems also provides us with an algorithm for determining when a list of vectors in $\mathbb{F}^m$ spans the whole space.

Corollary 2.34 *A list of vectors* $(\mathbf{v}_1, \ldots, \mathbf{v}_n)$ *in* $\mathbb{F}^m$ *spans* $\mathbb{F}^m$ *if and only if the RREF of the matrix* $\mathrm{A} := \begin{bmatrix} | & & | \\ \mathbf{v}_1 & \cdots & \mathbf{v}_n \\ | & & | \end{bmatrix}$ *has a pivot in every row.*

Proof The span of the vectors $(\mathbf{v}_1, \ldots, \mathbf{v}_n)$ in $\mathbb{F}^m$ is all of $\mathbb{F}^m$ if and only if each $\mathbf{b} \in \mathbb{F}^m$ is in the column space of A. By Corollary 2.33, this is true if and only if the linear system with augmented matrix

$$\left[\begin{array}{ccc|c} | & & | & b_1 \\ \mathbf{v}_1 & \cdots & \mathbf{v}_n & \vdots \\ | & & | & b_m \end{array}\right] \tag{2.18}$$

is consistent for each choice of $\mathbf{b} \in \mathbb{F}^m$.

Now, by Theorem 1.2 (and its analog for systems over arbitrary fields), a linear system is consistent if and only if there is no pivot in the last column of the augmented matrix. If the RREF of A has a pivot in every row, then for any choice of $\mathbf{b}$, there is no pivot in the final column of the RREF of (2.18) because each row already has a pivot coming from the RREF A.

Conversely, if the RREF of A does not have a pivot in every row, it has some rows of all zeroes, and then there are choices of $\mathbf{b}$ that put a pivot in the final column of (2.18). Indeed, imagine the row-reduced form of (2.18); in the first row without a pivot, there are all zeroes to the left of the bar. Put a 1 to the right of the bar (and whatever you like in the rest of the column). Now reverse the row-reducing to get back to A to the left of the bar; what is left is the augmented matrix of a system $\mathrm{A}\mathbf{x} = \mathbf{b}$ which is inconsistent. ▲

Quick Exercise #22. Is every vector in $\mathbb{R}^3$ in the span of the list

$$\left(\begin{bmatrix} 1 \\ 0 \\ 1 \end{bmatrix}, \begin{bmatrix} 0 \\ 2 \\ 2 \end{bmatrix}, \begin{bmatrix} 2 \\ -1 \\ 1 \end{bmatrix} \right)?$$

QA #21: The system is consistent if and only if there is an $\mathbf{x} \in \mathbb{F}^n$ such that $\mathrm{A}\mathbf{x} = \mathbf{b}$: that is, if and only if $\mathbf{b}$ is in the range of the operator T represented by A. By Theorem 2.31, the range of T is the column space of A.

QA #22: The RREF of $\begin{bmatrix} 1 & 0 & 2 \\ 0 & 2 & -1 \\ 1 & 2 & 1 \end{bmatrix}$ is $\begin{bmatrix} 1 & 0 & 2 \\ 0 & 1 & -\frac{1}{2} \\ 0 & 0 & 0 \end{bmatrix}$, so no.

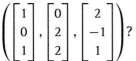

As we've observed before, if a matrix is just put into REF via row operations, then the pivots are in the same positions as if it is put all the way into RREF. Therefore in using Corollary 2.34 it's enough to put A into REF.

Examples

1. Consider the list of vectors $(\mathbf{v}_1, \ldots, \mathbf{v}_n)$ in $\mathbb{F}^n$ given by

$$\mathbf{v}_j = \mathbf{e}_1 + \cdots + \mathbf{e}_j$$

for $1 \le j \le n$. The matrix whose jth column is $\mathbf{v}_j$ is the matrix A in equation (2.15). Since A is already in REF, we can see that each row contains a pivot, so by Corollary 2.34, the list $(\mathbf{v}_1, \ldots, \mathbf{v}_n)$ spans $\mathbb{F}^n$.

2. Now consider the list of vectors $(\mathbf{w}_1, \ldots, \mathbf{w}_n)$ in $\mathbb{F}^n$ given by $\mathbf{w}_1 = \mathbf{e}_1$ and

$$\mathbf{w}_j = \mathbf{e}_j - \mathbf{e}_{j-1}$$

for $2 \le j \le n$. The matrix whose jth column is $\mathbf{w}_j$ is the matrix B in equation (2.15). Again, since B is already in REF, we can see that each row contains a pivot, so by Corollary 2.34, the list $(\mathbf{w}_1, \ldots, \mathbf{w}_n)$ spans $\mathbb{F}^n$. ▲

An important general consequence of Corollary 2.34 is the following result.

Corollary 2.35 *If $n < m$, then no list $(\mathbf{v}_1, \ldots, \mathbf{v}_n)$ of n vectors in $\mathbb{F}^m$ spans $\mathbb{F}^m$.*

Proof See Exercise 2.5.20. ▲

Kernel

Definition Let $T \in \mathcal{L}(V, W)$. The **kernel*** or **null space** of T is the set

$$\ker T := \{v \in V \mid T(v) = 0\}.$$

The **kernel** or **null space** of a matrix $\mathrm{A} \in \mathrm{M}_{m,n}(\mathbb{F})$ is the kernel of the linear map from $\mathbb{F}^n$ to $\mathbb{F}^m$ represented by A.

That is, the null space of a matrix A is exactly the set of solutions to the homogeneous linear system

$$\mathrm{A}\mathbf{x} = \mathbf{0}.$$

*This is completely different from the the kernel of an integral operator, as discussed in the example on page 87; it's just an unfortunate coincidence of terminology.

Example (Finding the kernel of a matrix) Let A be the matrix

$$A = \begin{bmatrix} 1 & 2 & 1 & -1 \\ 2 & 4 & 1 & -1 \\ 1 & 2 & 0 & 0 \end{bmatrix}.$$

The RREF of A is

$$\begin{bmatrix} 1 & 2 & 0 & 0 \\ 0 & 0 & 1 & -1 \\ 0 & 0 & 0 & 0 \end{bmatrix},$$

and so $\mathbf{x}$ is a solution to the linear system $\mathbf{A}\mathbf{x} = \mathbf{0}$ if and only if $x = -2y$ and $z = w$. That is,

$$\ker \mathbf{A} = \left\{ \begin{bmatrix} x \\ y \\ z \\ w \end{bmatrix} \middle| \, x = -2y, z = w \right\} = \left\{ \begin{bmatrix} -2y \\ y \\ w \\ w \end{bmatrix} \middle| \, y, w \in \mathbb{R} \right\}$$

$$= \left\{ y \begin{bmatrix} -2 \\ 1 \\ 0 \\ 0 \end{bmatrix} + w \begin{bmatrix} 0 \\ 0 \\ 1 \\ 1 \end{bmatrix} \middle| \, y, w \in \mathbb{R} \right\} = \left\langle \begin{bmatrix} -2 \\ 1 \\ 0 \\ 0 \end{bmatrix}, \begin{bmatrix} 0 \\ 0 \\ 1 \\ 1 \end{bmatrix} \right\rangle. \quad \blacktriangle$$

Quick Exercise #23. Suppose that $\mathbf{D} = \mathbf{diag}(d_1, \ldots, d_n)$. Show that $\ker \mathbf{D}$ is the span of all the $\mathbf{e}_i$ such that $d_i = 0$.

In both the example above and Quick Exercise #23, we found that the kernel of a matrix could be expressed as the span of some vectors. Since we know that the span of any list of vectors in V is a subspace of V, the following theorem should come as no surprise.

Theorem 2.36 *Let $T \in \mathcal{L}(V, W)$. Then $\ker T$ is a subspace of V.*

Proof Since $T(0) = 0$, we know $0 \in \ker T$.

If $v_1, v_2 \in \ker T$, then $T(v_1 + v_2) = T(v_1) + T(v_2) = 0$, so $v_1 + v_2 \in \ker T$.

Similarly, $v \in \ker T$ and $c \in \mathbb{F}$, then $T(cv) = cT(v) = 0$. ▲

We proved the matrix analog of Theorem 2.36, that the set of solutions to $\mathbf{A}\mathbf{x} = \mathbf{0}$ is a subspace of $\mathbb{F}^m$, in the example on page 53. The proofs are the same, although this one takes up rather less space.

QA #23: $\mathbf{D}\mathbf{x} = \begin{bmatrix} d_1 x_1 \\ \vdots \\ d_n x_n \end{bmatrix}$, which is 0 iff $d_i x_i = 0$ for each i; that is, iff $x_i = 0$ whenever $d_i \neq 0$.

One of the most important facts about the kernel of a linear map is its relationship to injectivity.

Theorem 2.37 *A linear map $T : V \to W$ is injective iff** $\ker T = \{0\}$.

Proof Suppose that $T \in \mathcal{L}(V, W)$ is injective. By linearity, $T(0) = 0$. By injectivity, if $T(v) = 0$ for some $v \in V$, then in fact $v = 0$. Therefore $\ker T = \{0\}$.

Conversely, suppose that $T \in \mathcal{L}(V, W)$ and $\ker T = \{0\}$. Let $v_1, v_2 \in V$ be such that $T(v_1) = T(v_2)$. Then

$$T(v_1 - v_2) = T(v_1) - T(v_2) = 0,$$

so $v_1 - v_2 \in \ker T = \{0\}$; that is, $v_1 = v_2$. ▲

The kernel of the coefficient matrix also plays an important role in solving linear systems:

Corollary 2.38 *Suppose that the linear system* $\mathrm{A}\mathbf{x} = \mathbf{b}$ *is consistent. The system has a unique solution if and only if* $\ker \mathrm{A} = \{\mathbf{0}\}$.

Proof If $\ker \mathrm{A} = \{0\}$, then the linear map given by A is injective by Theorem 2.37, which implies there is only one $\mathbf{x}$ with $\mathrm{A}\mathbf{x} = \mathbf{b}$.

Conversely, if $\mathrm{A}\mathbf{x} = \mathbf{b}$ and $\mathbf{0} \neq \mathbf{y} \in \ker \mathrm{A}$, then $\mathrm{A}(\mathbf{x} + \mathbf{y}) = \mathbf{b}$, and so the system has at least two distinct solutions. ▲

Eigenspaces

The kernel of a linear map $T \in \mathcal{L}(V)$ is a special case of the following more general kind of subspace associated to T.

Definition Let $T \in \mathcal{L}(V)$ and suppose that $\lambda \in \mathbb{F}$ is an eigenvalue of T. The **λ-eigenspace** of T is the subspace $\mathrm{Eig}_\lambda(T) \subseteq V$ defined by

$$\mathrm{Eig}_\lambda(T) := \{v \in V \mid T(v) = \lambda v\}.$$

If $\lambda \in \mathbb{F}$ is an eigenvalue of $\mathrm{A} \in \mathrm{M}_n(\mathbb{F})$, the **$\lambda$-eigenspace** of A, denoted $\mathrm{Eig}_\lambda(\mathrm{A})$ is the λ-eigenspace of the linear map given by A.

*"iff" is short for "if and only if."

That is, the λ-eigenspace of T consists of all the eigenvectors of T corresponding to the eigenvalue λ, together with the zero vector. (Remember that we don't count 0 itself as an eigenvector.)

This definition is an example of a generally frowned upon but frequently used practice, namely including in a definition a claim that requires verification; we have claimed that the set $\mathrm{Eig}_\lambda(T)$ is in fact a subspace of V. The quickest way to confirm this fact is via the following theorem.

Theorem 2.39 *Let* $T \in \mathcal{L}(V)$ *and suppose that* $\lambda \in \mathbb{F}$ *is an eigenvalue of* T*. Then* $\mathrm{Eig}_\lambda(T) = \ker(T - \lambda I)$.

Proof For $v \in V$,

$$\begin{aligned} v \in \mathrm{Eig}_\lambda(T) \quad &\Longleftrightarrow \quad Tv = \lambda v \\ &\Longleftrightarrow \quad Tv = \lambda I v \\ &\Longleftrightarrow \quad Tv - \lambda I v = 0 \\ &\Longleftrightarrow \quad (T - \lambda I)(v) = 0 \\ &\Longleftrightarrow \quad v \in \ker(T - \lambda I). \end{aligned}$$ ▲

Since we've already proved that the kernel of a linear map is a vector space, we immediately get that eigenspaces are indeed vector spaces:

Corollary 2.40 *Let* $T \in \mathcal{L}(V)$ *and suppose that* $\lambda \in \mathbb{F}$ *is an eigenvalue of* T*. Then* $\mathrm{Eig}_\lambda(T)$ *is a subspace of* V.

Example (Identifying an eigenspace) Suppose you are told that -2 is an eigenvalue of

$$\mathrm{B} = \begin{bmatrix} 0 & -1 & 1 \\ 1 & 0 & 3 \\ -1 & 1 & -2 \end{bmatrix}$$

and you wish to find the eigenspace $\mathrm{Eig}_{-2}(\mathrm{B})$. By Theorem 2.39,

$$\mathrm{Eig}_{-2}(\mathrm{B}) = \ker(\mathrm{B} + 2\mathrm{I}_3) = \ker \begin{bmatrix} 2 & -1 & 1 \\ 1 & 2 & 3 \\ -1 & 1 & 0 \end{bmatrix}.$$

To find the kernel of

$$\mathrm{A} = \begin{bmatrix} 2 & -1 & 1 \\ 1 & 2 & 3 \\ -1 & 1 & 0 \end{bmatrix},$$

we need to solve the homogeneous linear system $\mathbf{A}\mathbf{x} = \mathbf{0}$. By Gaussian elimination, the RREF of the augmented matrix $[\mathbf{A}|\mathbf{0}]$ is

$$\left[\begin{array}{ccc|c} 1 & 0 & 1 & 0 \\ 0 & 1 & 1 & 0 \\ 0 & 0 & 0 & 0 \end{array}\right],$$

which has solutions $\begin{bmatrix} x \\ y \\ z \end{bmatrix}$ with $x = y = -z$, where z is a free variable. Therefore

$$\operatorname{Eig}_{-2}(\mathbf{B}) = \ker \mathbf{A} = \left\{ \begin{bmatrix} -z \\ -z \\ z \end{bmatrix} \middle| \, z \in \mathbb{F} \right\} = \left\langle \begin{bmatrix} -1 \\ -1 \\ 1 \end{bmatrix} \right\rangle.$$ ▲

The following restatement of the definition of eigenvalue is often useful.

Theorem 2.41 *Let $T \in \mathcal{L}(V)$ and $\lambda \in \mathbb{F}$. Then λ is an eigenvalue of T iff* $\ker(T - \lambda I) \neq \{0\}$.

Proof By definition, λ is an eigenvalue of T if and only if there is an eigenvector of T with eigenvalue λ: that is, if and only if there is a vector $v \neq 0$ with $T(v) = \lambda v$. This is exactly the statement that $\ker(T - \lambda I) \neq \{0\}$. ▲

Quick Exercise #24. Use Theorem 2.41 to show that 2 is an eigenvalue of $\mathbf{A} = \begin{bmatrix} 3 & 2 \\ -1 & 0 \end{bmatrix}$.

Example We've already seen (see Example 4 on page 71) that if $\mathbf{D} = \operatorname{diag}(d_1, \ldots, d_n)$, then each of the d_j is an eigenvalue of $\mathbf{D}$.

Now, by Theorem 2.41, λ is an eigenvalue of $\mathbf{D}$ if and only if

$$\ker(\mathbf{D} - \lambda \mathbf{I}_n) = \ker\left(\operatorname{diag}(d_1 - \lambda, \ldots, d_n - \lambda)\right)$$

is nonzero. By Quick Exercise #23, $\ker\left(\operatorname{diag}(d_1 - \lambda, \ldots, d_n - \lambda)\right)$ is the span of all the $\mathbf{e}_i$ such that $d_i - \lambda = 0$: that is, the span of all the $\mathbf{e}_i$ for which $d_i = \lambda$. So $\ker(\mathbf{D} - \lambda \mathbf{I}_n)$ is nonzero if and only if there is at least one i for which $\lambda = d_i$. That is, $d_1, \ldots, d_n$ are the *only* eigenvalues of $\mathbf{D}$. ▲

QA #24: $\ker(\mathbf{A} - 2\mathbf{I}_2) = \ker \begin{bmatrix} 1 & 2 \\ -1 & -2 \end{bmatrix} = \left\langle \begin{bmatrix} 2 \\ -1 \end{bmatrix} \right\rangle$.

Solution Spaces

Theorem 2.36 shows that the set of solutions of an $m \times n$ homogeneous linear system

$$\mathrm{A}\mathbf{x} = \mathbf{0} \tag{2.19}$$

over $\mathbb{F}$ is a subspace of $\mathbb{F}^n$. The set of solutions to a non-homogeneous system

$$\mathrm{A}\mathbf{x} = \mathbf{b} \tag{2.20}$$

with $\mathbf{b} \neq \mathbf{0}$ is not a vector space; see Exercise 1.5.2. However, it does have a related structure.

Suppose that the system (2.20) is consistent, and let $\mathbf{x}_0$ be a solution. Suppose that $\mathbf{v} \in \ker \mathrm{A}$. Then

$$\mathrm{A}(\mathbf{x}_0 + \mathbf{v}) = \mathrm{A}\mathbf{x}_0 + \mathrm{A}\mathbf{v} = \mathbf{b} + \mathbf{0} = \mathbf{b},$$

so $\mathbf{x}_0 + \mathbf{v}$ is also a solution of (2.20).

On the other hand, if $\mathbf{x}$ is any solution of (2.20), then

$$\mathrm{A}(\mathbf{x} - \mathbf{x}_0) = \mathrm{A}\mathbf{x} - \mathrm{A}\mathbf{x}_0 = \mathbf{b} - \mathbf{b} = \mathbf{0}.$$

We can thus write $\mathbf{x} = \mathbf{x}_0 + \mathbf{y}$ with $\mathbf{y} = \mathbf{x} - \mathbf{x}_0 \in \ker \mathrm{A}$.

Together then, we have that the set of solutions of (2.20) is

$$\{\mathbf{x}_0 + \mathbf{y} \mid \mathbf{y} \in \ker A\}.$$

While this set isn't a subspace of $\mathbb{F}^n$, it is a translation of the subspace $\ker \mathrm{A}$ by the vector $\mathbf{x}_0$. A set of this form is called an **affine subspace.***

Example In Section 1.3, we saw that the solutions of the system

$$\begin{bmatrix} 1 & \frac{1}{4} & 0 & \frac{1}{2} \\ 2 & \frac{1}{4} & 0 & 0 \\ 1 & 0 & 8 & 1 \end{bmatrix} \begin{bmatrix} x \\ y \\ z \\ w \end{bmatrix} = \begin{bmatrix} 20 \\ 36 \\ 176 \end{bmatrix}$$

could be written in the form

$$\begin{bmatrix} 16 \\ 16 \\ 20 \\ 0 \end{bmatrix} + w \begin{bmatrix} \frac{1}{2} \\ -4 \\ -\frac{3}{16} \\ 1 \end{bmatrix}.$$

*It's slightly unfortunate that an "affine subspace" is not actually a type of subspace.

In this situation, $\mathbf{x}_0 = \begin{bmatrix} 16 \\ 16 \\ 20 \\ 0 \end{bmatrix}$ is one particular solution, and

$$\ker \begin{bmatrix} 1 & \frac{1}{4} & 0 & \frac{1}{2} \\ 2 & \frac{1}{4} & 0 & 0 \\ 1 & 0 & 8 & 1 \end{bmatrix} = \left\langle \begin{bmatrix} \frac{1}{2} \\ -4 \\ -\frac{3}{16} \\ 1 \end{bmatrix} \right\rangle .$$ ▲

The argument above works for general linear maps between vector spaces to produce the following.

Proposition 2.42 *Let $T \in \mathcal{L}(V, W)$, and suppose that $T(v_0) = w$. Then*

$$\{v \in V \mid T(v) = w\} = \{v_0 + y \mid y \in \ker T\} .$$

Proof See Exercise 2.5.22. ▲

You may have come across this kind of structure in another context if you have taken a course in ordinary differential equations, or in calculus-based classical mechanics. A **second-order linear differential equation** is an equation of the form

$$a(t)\frac{d^2f}{dt^2} + b(t)\frac{df}{dt} + c(t)f(t) = g(t), \tag{2.21}$$

where $a(t)$, $b(t)$, $c(t)$, and $g(t)$ are all known functions of a variable $t \in \mathbb{R}$, and $f(t)$ is an unknown function of t. If V and W are appropriately chosen spaces of functions, then we can define a map $T : V \to W$ by

$$(T(f))(t) = a(t)\frac{d^2f}{dt^2} + b(t)\frac{df}{dt} + c(t)f(t).$$

(Since $T(f) \in W$ is itself a function of t, to say what $T(f)$ is, we need to say what it does to t.) The map T is linear; it is built up out of typical examples of linear maps on function spaces of the kinds we saw in Section 2.2.

The differential equation (2.21) can now be more simply written as

$$T(f) = g.$$

Proposition 2.42 now tells us that if $f_p \in V$ is any particular solution of the differential equation (2.21), then *every* solution $f \in V$ can be written in the form

$$f(t) = f_p(t) + f_h(t)$$

for some $f_h \in \ker T$. That is, f_h satisfies the homogeneous differential equation

$$a(t)\frac{d^2 f_h}{dt^2} + b(t)\frac{df_h}{dt} + c(t)f_h(t) = 0.$$

Differential equations of the form in equation (2.21) in which $a(t)$, $b(t)$, and $c(t)$ are all positive constants are used in physics to model **forced damped harmonic oscillators.** The function $f(t)$ is the displacement of a block on the end of a spring from its resting position; the constant $a(t) = a$ is the mass of the block, $b(t) = b$ has to do with the friction of the surface on which the block is sitting, and $c(t) = c$ describes the strength of the spring. The function $g(t)$ represents a time-dependent external force being exerted on the block.

KEY IDEAS

- The range of T is the subspace $\{Tv \mid v \in V\} \subseteq W$.
- If A is the matrix of T, then the range of T is the column space of A.
- The kernel of T is the subspace $\{v \in V \mid Tv = 0\} \subseteq V$.
- A linear map T is injective if and only if $\ker T = \{0\}$.
- The λ-eigenspace of T is the collection of all eigenvectors of T with eigenvalue λ, together with the zero vector.

EXERCISES

2.5.1 Express the kernel of each of the following matrices as the span of some list of vectors.

(a) $\begin{bmatrix} 1 & 2 & 3 \\ 4 & 5 & 6 \end{bmatrix}$ (b) $\begin{bmatrix} 2 & -1 & 3 \\ 1 & 2 & 0 \\ 1 & -3 & 3 \end{bmatrix}$ (c) $\begin{bmatrix} -1 & 0 & 2 & 1 \\ 2 & 1 & -1 & 0 \\ 1 & 1 & 1 & 1 \\ 0 & 1 & 3 & 2 \end{bmatrix}$

(d) $\begin{bmatrix} 1 & 1 & 0 \\ 1 & 0 & 1 \\ 0 & 1 & 1 \end{bmatrix}$, over $\mathbb{R}$ (e) $\begin{bmatrix} 1 & 1 & 0 \\ 1 & 0 & 1 \\ 0 & 1 & 1 \end{bmatrix}$, over $\mathbb{F}_2$

2.5.2 Express the kernel of each of the following matrices as the span of some list of vectors.

(a) $\begin{bmatrix} 1 & -2 \\ 4 & -8 \end{bmatrix}$ (b) $\begin{bmatrix} -2 & 0 & 1 \\ 3 & 2 & -4 \end{bmatrix}$ (c) $\begin{bmatrix} 3 & 1 & 4 \\ 1 & 5 & 9 \\ -2 & 4 & 5 \end{bmatrix}$

(d) $\begin{bmatrix} 2 & 0 & -1 & 3 \\ -1 & 2 & 3 & -2 \\ 0 & 4 & 5 & -1 \end{bmatrix}$ (e) $\begin{bmatrix} 2 & -1 & 1 & -3 \\ -3 & 1 & -2 & 2 \\ 0 & 2 & 2 & 2 \\ -1 & 0 & -1 & 1 \end{bmatrix}$

2.5.3 Determine whether each of the following lists of vectors spans $\mathbb{F}^n$.

(a) $\left(\begin{bmatrix}1\\2\end{bmatrix},\begin{bmatrix}-3\\4\end{bmatrix}\right)$ (b) $\left(\begin{bmatrix}-2\\1\end{bmatrix},\begin{bmatrix}1\\1\end{bmatrix},\begin{bmatrix}3\\-1\end{bmatrix}\right)$ (c) $\left(\begin{bmatrix}1\\0\\2\end{bmatrix},\begin{bmatrix}3\\-1\\1\end{bmatrix}\right)$

(d) $\left(\begin{bmatrix}1\\0\\2\end{bmatrix},\begin{bmatrix}3\\-1\\1\end{bmatrix},\begin{bmatrix}2\\-1\\-1\end{bmatrix}\right)$ (e) $\left(\begin{bmatrix}1\\0\\2\end{bmatrix},\begin{bmatrix}3\\-1\\1\end{bmatrix},\begin{bmatrix}0\\1\\-1\end{bmatrix}\right)$

2.5.4 Determine whether each list of vectors spans $\mathbb{F}^n$.

(a) $\left(\begin{bmatrix}1\\-1\end{bmatrix},\begin{bmatrix}-1\\1\end{bmatrix}\right)$ (b) $\left(\begin{bmatrix}1\\-1\end{bmatrix},\begin{bmatrix}-1\\1\end{bmatrix},\begin{bmatrix}2\\-1\end{bmatrix}\right)$

(c) $\left(\begin{bmatrix}1\\0\\2\end{bmatrix},\begin{bmatrix}3\\-1\\1\end{bmatrix},\begin{bmatrix}2\\-1\\-1\end{bmatrix},\begin{bmatrix}0\\1\\-1\end{bmatrix}\right)$ (d) $\left(\begin{bmatrix}1\\1\\0\end{bmatrix},\begin{bmatrix}1\\0\\1\end{bmatrix},\begin{bmatrix}0\\1\\1\end{bmatrix}\right)$, $\mathbb{F}=\mathbb{R}$

(e) $\left(\begin{bmatrix}1\\1\\0\end{bmatrix},\begin{bmatrix}1\\0\\1\end{bmatrix},\begin{bmatrix}0\\1\\1\end{bmatrix}\right)$, $\mathbb{F}=\mathbb{F}_2$

2.5.5 For each matrix A, determine whether λ is an eigenvalue, and if so, express the eigenspace $\operatorname{Eig}_\lambda(A)$ as the span of some list of vectors.

(a) $A=\begin{bmatrix}2&2\\2&-1\end{bmatrix}$, $\lambda=3$ (b) $A=\begin{bmatrix}2&2\\2&-1\end{bmatrix}$, $\lambda=-1$

(c) $A=\begin{bmatrix}-1&1&-1\\1&0&-2\\0&1&-3\end{bmatrix}$, $\lambda=-2$ (d) $A=\begin{bmatrix}7&-2&4\\0&1&0\\-6&2&-3\end{bmatrix}$, $\lambda=1$

(e) $A=\begin{bmatrix}-2&0&1&0\\0&-2&0&1\\1&0&-2&0\\0&1&0&-2\end{bmatrix}$, $\lambda=-3$

2.5.6 For each matrix A, determine whether λ is an eigenvalue, and if so, express the eigenspace $\operatorname{Eig}_\lambda(A)$ as the span of some list of vectors.

(a) $A=\begin{bmatrix}1&2\\3&4\end{bmatrix}$, $\lambda=1$ (b) $A=\begin{bmatrix}-1&1&-1\\1&0&-2\\0&1&-3\end{bmatrix}$, $\lambda=0$

(c) $A=\begin{bmatrix}-1&1&-1\\1&0&-2\\0&1&-3\end{bmatrix}$, $\lambda=1$ (d) $A=\begin{bmatrix}7&-2&4\\0&1&0\\-6&2&-3\end{bmatrix}$, $\lambda=3$

(e) $A=\begin{bmatrix}1&4&4&0\\1&0&2&1\\-2&-3&-5&-1\\1&1&2&0\end{bmatrix}$, $\lambda=-1$

2.5.7 Find all solutions for each of the following linear systems.

(a) $$\begin{aligned} x + 3y - z &= 2 \\ 2x + 2y + z &= -3 \end{aligned}$$

(b) $$\begin{aligned} 2x - 3y &= -1 \\ -x + 2y + z &= 1 \\ y + 2z &= 1 \end{aligned}$$

(c) $$\begin{aligned} x - 4y - 2z - w &= 0 \\ -x + 2y - w &= 2 \\ y + z + w &= 1 \end{aligned}$$

(d) $$\begin{aligned} x + y - z &= 2 \\ y + z &= 1 \\ x - 2z &= 1 \\ x + 3y + z &= 4 \end{aligned}$$

(e) $$\begin{bmatrix} 1 & 0 & 0 & -1 \\ 0 & 2 & 1 & 0 \\ 2 & 1 & 0 & -1 \\ 1 & 1 & 1 & -2 \end{bmatrix} \begin{bmatrix} x \\ y \\ z \\ w \end{bmatrix} = \begin{bmatrix} 1 \\ 1 \\ 3 \\ 1 \end{bmatrix}$$

2.5.8 Find all solutions for each of the following linear systems.

(a) $$\begin{aligned} x - 2y &= -3 \\ -2x + 4y &= 6 \end{aligned}$$

(b) $$\begin{aligned} 2x - y + z &= -1 \\ x + z &= 0 \\ 3x + y + 4z &= 1 \end{aligned}$$

(c) $$\begin{aligned} x + 2y - 3z &= 0 \\ 2x + y - z + w &= 4 \\ -x + y - 2z - w &= -3 \\ 3y - 5z - 2w &= -5 \end{aligned}$$

(d) $$\begin{bmatrix} 2 & 0 & 1 & 0 \\ 1 & -1 & 0 & 3 \\ 0 & 2 & 0 & -1 \end{bmatrix} \begin{bmatrix} x \\ y \\ z \\ w \end{bmatrix} = \begin{bmatrix} -2 \\ 0 \\ 3 \end{bmatrix}$$

(e) $$\begin{aligned} 2x + 2y + z &= 5 \\ x - y &= -3 \\ 4y + z &= 11 \\ 3x + y + z &= 2 \end{aligned}$$

2.5.9 It is a fact (which you don't have to prove) that every solution of the homogeneous differential equation

$$\frac{d^2 f}{dt^2} + f(t) = 0$$

is a linear combination of $f_1(t) = \sin(t)$ and $f_2(t) = \cos(t)$.

(a) Show that $f_p(t) = 2e^{-t}$ is a solution of the differential equation

$$\frac{d^2 f}{dt^2} + f(t) = 4e^{-t}. \tag{2.22}$$

(b) Use Proposition 2.42 to show that every solution of the differential equation (2.22) is of the form

$$f(t) = 2e^{-t} + k_1 \sin(t) + k_2 \cos(t)$$

for constants $k_1, k_2 \in \mathbb{R}$.

(c) Determine all solutions $f(t)$ of the differential equation (2.22) which satisfy $f(0) = 1$.

(d) Show that for any $a, b \in \mathbb{R}$ there is a unique solution of the differential equation (2.22) which satisfies $f(0) = a$ and $f(\pi/2) = b$.

2.5.10 Let $T : V \to W$ be a linear map. Prove that if U is a subspace of V, then $T(U)$ is a subspace of W.

2.5.11 (a) Show that if $S, T \in \mathcal{L}(V, W)$, then

$$\operatorname{range}(S + T) \subseteq \operatorname{range} S + \operatorname{range} T,$$

where the sum of two subspaces is as defined in Exercise 1.5.11.

(b) Show that if $\mathrm{A}, \mathrm{B} \in \mathrm{M}_{m,n}(\mathbb{F})$, then

$$C(\mathrm{A} + \mathrm{B}) \subseteq C(\mathrm{A}) + C(\mathrm{B}).$$

2.5.12 Let D be the differentiation operator on $C^\infty(\mathbb{R})$. What is the kernel of D?

2.5.13 Let $\mathrm{A} \in \mathrm{M}_{m,n}(\mathbb{F})$ and $\mathrm{B} \in \mathrm{M}_{n,p}(\mathbb{F})$. Show that $\mathrm{AB} = 0$ if and only if $C(\mathrm{B}) \subseteq \ker \mathrm{A}$.

2.5.14 Show that $T \in \mathcal{L}(V)$ is injective if and only if 0 is not an eigenvalue of T.

2.5.15 Suppose that $\lambda \in \mathbb{F}$ is an eigenvalue of $T \in \mathcal{L}(V)$, and $k \in \mathbb{N}$. Show that λ^k is an eigenvalue of T^k, and $\mathrm{Eig}_\lambda(T) \subseteq \mathrm{Eig}_{\lambda^k}(T^k)$.

2.5.16 Give an example of a linear map $T : V \to V$ with an eigenvalue λ, such that $\mathrm{Eig}_\lambda(T) \neq \mathrm{Eig}_{\lambda^2}(T^2)$.

2.5.17 Is it true that the set of all eigenvectors of a linear map $T \in \mathcal{L}(V)$, together with the zero vector, must be a subspace of V? Give a proof or a counterexample.

2.5.18 Suppose that $T : V \to W$ is linear, and that U is a subspace of W. Let

$$X = \{v \in V \mid Tv \in U\}.$$

Show that X is a subspace of V.

2.5.19 Use Corollary 2.12 to give a different proof of Corollary 2.32.

2.5.20 Prove Corollary 2.35.

2.5.21 Prove Corollary 2.40 directly from the definition of eigenspace.

2.5.22 Prove Proposition 2.42.

2.6 Error-correcting Linear Codes

Linear Codes

In this section we'll see an application of some of the machinery we've built up – subspaces, matrices, column spaces, and kernels – to a situation which doesn't initially appear to have anything to do either with solving linear systems or geometry. This application is also an important example of how it can be useful to work with this machinery over the field $\mathbb{F}_2$.

As mentioned before, the field $\mathbb{F}_2$ which only has two elements, 0 and 1, is a natural mathematical tool in computer science, since computers internally represent everything as strings of 0s and 1s. Each 0 or 1 is referred to as a **bit**; if we group the data into chunks of k bits, then each chunk of data is the same as a vector $\mathbf{x} \in \mathbb{F}_2^k$.

We will be interested in ways of turning our chunks of data into different chunks of data, in general with a different number of bits. To discuss how we do this, we need the following definition.

Definition A **binary code** is a subset $\mathcal{C} \subseteq \mathbb{F}_2^n$ for some n. An **encoding function** is a bijective function

$$f : \mathbb{F}_2^k \to \mathcal{C}.$$

A binary code $\mathcal{C}$ should be thought of as a set of "code words," which get used in place of the original chunks of data from $\mathbb{F}_2^k$. The encoding function f tells us which code word represents each original chunk of data. One reason we might want to do this is suggested by the "code" terminology, namely encryption. Instead of storing your data

$$\mathbf{x}_1, \ldots, \mathbf{x}_m \in \mathbb{F}_2^k,$$

you apply this function f and store

$$f(\mathbf{x}_1), \ldots, f(\mathbf{x}_m) \in \mathcal{C}.$$

If someone else comes along who wants to read your data but doesn't know the function f you used, they'll have a much harder time.

The requirement that the encoding function f is surjective means that every code word is actually used, and is included only for convenience. If f were not surjective, we could just replace $\mathcal{C}$ with the smaller set $f(\mathbb{F}_2^k)$. Injectivity, on the other hand, is more important.

Quick Exercise #25. Why is it important for the encoding function f to be injective?

In this section, we'll be interested in encoding data not for encryption, but for error detection and error correction. If you're working with data in the real world (as opposed to in a math textbook), then mistakes can happen when data is transmitted or stored. Richard Hamming, who invented the Hamming code that we'll discuss below, was concerned with errors that resulted from the fact that 1950s-era computers read data, not always reliably, from a pattern of holes in cardboard punch cards. Modern RAM chips are so small that data can be corrupted by cosmic rays. In every era, mistakes are made, and so it would be nice to be able to detect them. Error-detecting codes give a way to tell that a mistake has occurred; error-correcting codes go one step further, by identifying what the mistake was, so it can be fixed automatically.

Before getting into those aspects, let's first see how to get linear algebra involved. We'll start in the most obvious way: stick linearity requirements into the definition above.

Definition A **binary linear code** is a *subspace* $\mathcal{C} \subseteq \mathbb{F}_2^n$, and a **linear encoding** is a bijective linear map $T : \mathbb{F}_2^k \to \mathcal{C}$.

Since $\mathcal{C}$ is a subspace of $\mathbb{F}_2^n$, another way to say this is that T is an injective linear map $\mathbb{F}_2^k \to \mathbb{F}_2^n$ such that range $T = \mathcal{C}$.

We know that T is represented by a matrix $\mathrm{A} \in \mathrm{M}_{n,k}(\mathbb{F}_2)$, which we call the **encoding matrix**. We then have that $\ker \mathrm{A} = \{0\}$ since T is injective, and

$$C(\mathrm{A}) = \operatorname{range} T = \mathcal{C}.$$

So if your original chunk of data is $\mathbf{x} \in \mathbb{F}_2^k$, your encoded data is $\mathbf{y} = \mathrm{A}\mathbf{x} \in \mathbb{F}_2^n$.

Now suppose that Alice has some data $\mathbf{x} \in \mathbb{F}_2^k$ which she encodes as $\mathbf{y} = \mathrm{A}\mathbf{x}$, and then she transmits the encoded data $\mathbf{y}$ to her friend Bob. If everything works properly, then Bob will receive $\mathbf{y}$ itself, and, if he knows the encoding matrix A that Alice used, he can recover the original data $\mathbf{x}$. But if something goes wrong, Bob will get some other vector $\mathbf{z} \in \mathbb{F}_2^n$. It would be nice if Bob had some way of telling, just by looking at the received vector $\mathbf{z}$, that he'd received the wrong vector, so that he and Alice could sort out the situation instead of decoding the wrong thing.

Error-detecting Codes

One way that Bob could tell there must be a problem is if $\mathbf{z} \notin C(\mathrm{A})$; since $\mathbf{y} = \mathrm{A}\mathbf{x} \in C(\mathrm{A})$, that would certainly imply that $\mathbf{z} \neq \mathbf{y}$.

QA #25: If $f(\mathbf{x}) = f(\mathbf{y})$, then you yourself won't be able to tell whether your original chunk of data was $\mathbf{x}$ or $\mathbf{y}$.

Definition A linear encoding with encoding matrix $A \in M_{n,k}(\mathbb{F}_2)$ is **error-detecting** if, whenever $y \in C(A)$ and $z \in \mathbb{F}_2^n$ differ in exactly one entry, $z \notin C(A)$.

Suppose that Alice and Bob are using an error-detecting code. That means that, if Alice transmits y to Bob but something changes one of the bits, so that Bob receives a vector z which differs from y in exactly one entry, then Bob will definitely be able to tell that something went wrong by checking whether $z \in C(A)$. If $z \in C(A)$, then Bob can tell that either $y = z$ and nothing went wrong, or that multiple bits were changed in transmission. (We assume in practice that multiple errors are rare enough to ignore.) If, on the other hand, $z \notin C(A)$, then something definitely went wrong. Bob could ask Alice to send that signal again.

So now Bob needs a way to check whether the vector z he receives is in $C(A)$.

Definition A **parity-check matrix** for a linear encoding with encoding matrix $A \in M_{n,k}(\mathbb{F}_2)$ is a matrix $B \in M_{m,n}(\mathbb{F}_2)$ such that $\ker B = C(A)$.

So if Bob knows a parity-check matrix for Alice's encoding, when he receives $z \in \mathbb{F}_2^n$, he should compute Bz. If $Bz \neq 0$, then $z \notin \ker B = C(A)$, and so some error definitely occurred. The following proposition shows that, if B doesn't have any zero columns, this process will always let Bob know that an error occurred, as long as there is at most one.

Proposition 2.43 *If a linear encoding has a parity-check matrix whose columns are all nonzero, then the encoding is error-detecting.*

Proof Let A be the encoding matrix and let B be a parity-check matrix with nonzero columns. If y and z differ only in the ith entry, then $z_i = y_i + 1$ (since we're working over $\mathbb{F}_2$) and $z_j = y_j$ for $j \neq i$. That is, $z = y + e_i$. So if $y \in C(A) = \ker B$, then

$$Bz = B(y + e_i) = By + Be_i = b_i.$$

Since $b_i \neq 0$, this means that $z \notin \ker B = C(A)$. ▲

The simplest example of an error-detecting code is simply called a **parity bit code** and is the source of the name of the parity-check matrix. We'll describe it first without matrices, then see how it fits into the framework we've been discussing.

For the parity-check, the encoded vector y consists of just the original data vector x, plus one more bit which tells whether x contains an even or an odd number of 1s. (The *parity* of an integer is whether it is even or odd.) That is, $n = k + 1$,

$$\begin{aligned} y_i &= x_i \text{ for each } i = 1, \ldots, k, \text{ and} \\ y_{k+1} &= x_1 + \cdots + x_k, \end{aligned} \tag{2.23}$$

using $\mathbb{F}_2$ arithmetic. (In $\mathbb{F}_2$, the sum of an even number of 1s is 0, and the sum of an odd number of 1s is 1.) Now if Bob receives $\mathbf{z} \in \mathbb{F}_2^{k+1}$, he can check whether z_{k+1} properly reflects the number of 1s in $z_1, \ldots, z_k$: that is, whether

$$z_{k+1} = z_1 + \cdots + z_k \tag{2.24}$$

in $\mathbb{F}_2$. If not, then something went wrong.

There's one important point we've ignored so far:

Quick Exercise #26. What if two errors occur in transmission? What if there are more?

Now let's see how to express what's going on here in terms of matrices. First of all, we can tell what the encoding matrix A is by finding its columns $\mathbf{A}\mathbf{e}_i$. By (2.23),

$$\mathbf{A}\mathbf{e}_i = \mathbf{e}_i + \mathbf{e}_{k+1}.$$

(Watch out that $\mathbf{e}_i \in \mathbb{F}_2^k$ on the left-hand side and $\mathbf{e}_i \in \mathbb{F}_2^{k+1}$ on the right-hand side.) That is,

$$\mathbf{A} = \begin{bmatrix} 1 & 0 & \cdots & 0 \\ 0 & 1 & & \vdots \\ \vdots & & \ddots & \vdots \\ 0 & \cdots & \cdots & 1 \\ 1 & 1 & \cdots & 1 \end{bmatrix}.$$

As for the parity-check matrix, we use one of our favorite tricks and rewrite equation (2.24) with everything on one side:

$$z_1 + \cdots + z_k + z_{k+1} = 0$$

(remember that in $\mathbb{F}_2$ addition and subtraction are the same thing). So a parity-check matrix for the parity bit code has as its kernel all those $\mathbf{z} \in \mathbb{F}_2^{k+1}$ whose coordinates add up to 0. The simplest way to achieve that is just to use

$$\mathbf{B} = \begin{bmatrix} 1 & \cdots & 1 \end{bmatrix} \in \mathrm{M}_{1,k+1}(\mathbb{F}_2),$$

since then

$$\mathbf{B}\mathbf{z} = [z_1 + \cdots + z_k + z_{k+1}].$$

QA #26: If there were exactly two errors, Bob won't be able to tell. This is true regardless of whether both errors occur in the "data bits," that is, the first k bits, or if one takes place in a data bit and one in the "parity bit" at the end. More generally, Bob will know if an odd number of errors occur, but not if an even number occur.

Error-correcting Codes

The ability to catch errors automatically, without having any idea what's actually going on in the transmission process, is pretty useful. But by being a bit cleverer, we can do an even better trick.

Definition An error-detecting linear encoding with encoding matrix $A \in M_{n,k}(\mathbb{F}_2)$ is **error-correcting** if, whenever $z \in \mathbb{F}_2^n$ differs in at most one entry from some $y \in C(A)$, there is a *unique* such $y \in C(A)$.

Quick Exercise #27. Show that the parity bit code is *not* error-correcting.
Hint: It's enough to check this for $k = 2$.

As this Quick Exercise shows, with the parity bit code, Bob can tell that an error occurred in transmission, so he knows that the z he received is not the y that Alice sent, but that's all he knows. On the other hand, if they were using an error-correcting code, then for a given z there would only be one possible y. That way Bob wouldn't have to ask Alice to go to the trouble of sending y again – he could figure out what it was on his own. (To make things simpler, we will assume from now on that at most one error can occur.)

The simplest error-correcting code is the **triple repetition code**. This code works bit-by-bit (so $k = 1$), and simply repeats each bit three times. If Bob receives a $z \in \mathbb{F}_2^3$ whose three entries are not all the same, then he knows an error has occurred. Even better, if all three entries are not the same, then two of the entries of z must be the same as each other, so Alice's transmitted vector y must have had all three entries equal to those two, and that tells Bob what the original bit $x \in \mathbb{F}_2$ was.

In terms of matrices, the encoding matrix for the triple repetition code is

$$A = \begin{bmatrix} 1 \\ 1 \\ 1 \end{bmatrix},$$

which has column space

$$C(A) = \left\{ y \in \mathbb{F}_2^3 \mid y_1 = y_2 = y_3 \right\}.$$

QA #27: The vector $z = \begin{bmatrix} 1 \\ 1 \\ 1 \end{bmatrix}$ differs in one entry from all three of the valid code words $\begin{bmatrix} 1 \\ 0 \\ 1 \end{bmatrix}$, $\begin{bmatrix} 0 \\ 1 \\ 1 \end{bmatrix}$, and $\begin{bmatrix} 1 \\ 1 \\ 0 \end{bmatrix}$.

A parity-check matrix B must therefore have

$$\begin{aligned}\ker B &= \{y \in \mathbb{F}_2^3 \mid y_1 = y_2 = y_3\} \\ &= \{y \in \mathbb{F}_2^3 \mid y_1 = y_3 \text{ and } y_2 = y_3\} \\ &= \{y \in \mathbb{F}_2^3 \mid y_1 + y_3 = 0 \text{ and } y_2 + y_3 = 0\},\end{aligned}$$

remembering again that addition and subtraction are the same in $\mathbb{F}_2$. Now notice that the linear system in this last description is in RREF, so we can tell immediately that

$$B = \begin{bmatrix} 1 & 0 & 1 \\ 0 & 1 & 1 \end{bmatrix} \tag{2.25}$$

is such a matrix.

The matrix formulation of error correction works as follows. Suppose that when Alice transmits $y \in \mathbb{F}_2^n$, an error occurs in the ith bit. That means that Bob receives the vector $z = y + e_i$, and when he applies the parity-check matrix $B \in M_{m,n}(\mathbb{F}_2)$, he finds

$$Bz = B(y + e_i) = By + Be_i = b_i,$$

since $y = Ax \in C(A) \subseteq \ker B$. The code is error-correcting if one can tell from the result of this computation exactly where the error occurred, and thereby recover y. In other words, the code is error-correcting if the columns $b_1, \ldots, b_n$ of B are distinct and nonzero.

This argument proves the following proposition.

Proposition 2.44 *If a linear encoding has a parity-check matrix whose columns are all distinct and nonzero, then the encoding is error-correcting.*

In particular, we can see that the triple repetition code is error-correcting because the columns of the parity-check matrix in (2.25) are distinct.

The Hamming Code

The triple repetition code is a nice proof of concept, but it has the major drawback that Alice needs to transmit three times as many bits as in her original data. For Richard Hamming, who transmitted data by physically feeding pieces of cardboard into a computer in 1950, this was not a great trade-off in terms of how he spent his time. Even in more modern systems, the more bits you store or transmit, the more likely it is that more than one error will occur (remember that the codes we're discussing can only correct one error). Hamming wanted to find a more efficient error-correcting code; the one he found is now known as the **Hamming code**.

The starting point is to notice that the columns of the parity-check matrix $B \in M_{2,3}(\mathbb{F}_2)$ for the triple repetition code in (2.25) are not simply distinct; they actually are all of the nonzero vectors in $\mathbb{F}_2^2$. To build the Hamming code, we'll start not with the encoding matrix but with a parity-check matrix

$$B = \begin{bmatrix} 1 & 1 & 1 & 0 & 1 & 0 & 0 \\ 1 & 1 & 0 & 1 & 0 & 1 & 0 \\ 1 & 0 & 1 & 1 & 0 & 0 & 1 \end{bmatrix} \in M_{3,7}(\mathbb{F}_2) \tag{2.26}$$

whose columns are all seven of the distinct nonzero vectors in $\mathbb{F}_2^3$.

Now we need an encoding matrix $A \in M_{k,7}(\mathbb{F}_2)$ for some k such that $\ker B = C(A)$. Notice that B can be written as

$$B = \begin{bmatrix} M & I_3 \end{bmatrix}, \qquad \text{where} \qquad M = \begin{bmatrix} 1 & 1 & 1 & 0 \\ 1 & 1 & 0 & 1 \\ 1 & 0 & 1 & 1 \end{bmatrix} \in M_{3,4}(\mathbb{F}_2).$$

Matrix multiplication of matrices written in blocks like this is a simple extension of the usual multiplication: if we define

$$A = \begin{bmatrix} I_4 \\ M \end{bmatrix} = \begin{bmatrix} 1 & 0 & 0 & 0 \\ 0 & 1 & 0 & 0 \\ 0 & 0 & 1 & 0 \\ 0 & 0 & 0 & 1 \\ 1 & 1 & 1 & 0 \\ 1 & 1 & 0 & 1 \\ 1 & 0 & 1 & 1 \end{bmatrix} \in M_{7,4}(\mathbb{F}_2), \tag{2.27}$$

then BA is defined and can be computed by

$$BA = \begin{bmatrix} M & I_3 \end{bmatrix} \begin{bmatrix} I_4 \\ M \end{bmatrix} = MI_4 + I_3M = M + M = \mathbf{0},$$

since we're working over $\mathbb{F}_2$. This implies (see Exercise 2.5.13) that $C(A) \subseteq \ker B$. We also need $\ker A = \{\mathbf{0}\}$; this is easy to check since the first four entries of Ax consist of the entries of x itself.

It's actually true that for these matrices $C(A) = \ker B$ (see Exercise 2.6.6), but what we've just seen is already enough to make everything work: If Alice has a chunk of data $x \in \mathbb{F}_2^4$, then she can transmit $y = Ax \in \mathbb{F}_2^7$. If an error occurs in the ith bit then Bob can compute

$$Bz = B(y + e_i) = By + Be_i = b_i,$$

and he'll be able to tell from this where the error occurred and thus what vector y Alice transmitted. He can then recover the original vector x by solving the linear system $Ax = y$. As noted above, this is trivial since the first four entries of Ax consist of x itself; that is, x is given by the first four entries of y.

Furthermore, with this code Alice only needs to transmit $7/4$ as many bits as in her original data, making the Hamming code much more efficient than the triple repetition code.

Example Suppose Alice wants to send Bob the vector $\mathbf{x} = \begin{bmatrix} 0 & 1 & 1 & 0 \end{bmatrix}^{\mathrm{T}}$. She encodes it using the Hamming code and transmits

$$\mathbf{y} = \mathbf{A}\mathbf{x} = \begin{bmatrix} 0 & 1 & 1 & 0 & 0 & 1 & 1 \end{bmatrix}^{\mathrm{T}}$$

to Bob. Because stuff happens, Bob actually receives the vector

$$\mathbf{z} = \begin{bmatrix} 0 & 1 & 0 & 0 & 0 & 1 & 1 \end{bmatrix}^{\mathrm{T}}.$$

Bob computes

$$\mathbf{B}\mathbf{z} = \begin{bmatrix} 1 \\ 0 \\ 1 \end{bmatrix} = \mathbf{b}_3,$$

and concludes that an error occurred in the transmission of the third bit (always assuming that at most one error occurred). He therefore recovers $\mathbf{y} = \mathbf{z} + \mathbf{e}_3$, and taking just the first four entries of $\mathbf{y}$, recovers $\mathbf{x} = \begin{bmatrix} 0 & 1 & 1 & 0 \end{bmatrix}^{\mathrm{T}}$. ▲

KEY IDEAS

- A binary linear code encodes vectors of bits (0s and 1s) by multiplication by a matrix over $\mathbb{F}_2$.
- A parity-check matrix lets you check whether a message may have been corrupted, because its kernel is the same as the image of the encoding matrix. It can miss errors, but doesn't give false positives.
- Error-detecting codes have parity-check matrices that always detect one-bit errors.
- Error-correcting codes not only detect one-bit errors, but can tell you which bit is wrong.

EXERCISES

2.6.1 How would the vector $\begin{bmatrix} 1 \\ 0 \\ 1 \\ 1 \end{bmatrix} \in \mathbb{F}_2^4$ be encoded by each of the following linear encodings?

(a) Parity bit.

(b) Triple repetition.

(c) Hamming code.

2.6.2 Alice encodes several vectors $\mathbf{x}$ using a parity bit encoding, and after transmission Bob receives the following vectors $\mathbf{z}$. Assuming that at most one error has occurred for each vector, determine whether or not an error has occurred, and what all the possible original x vectors are.

(a) $\begin{bmatrix} 0 \\ 1 \\ 0 \end{bmatrix}$ (b) $\begin{bmatrix} 1 \\ 1 \\ 0 \end{bmatrix}$ (c) $\begin{bmatrix} 1 \\ 1 \\ 1 \\ 1 \end{bmatrix}$ (d) $\begin{bmatrix} 1 \\ 0 \\ 1 \\ 1 \end{bmatrix}$

2.6.3 Alice encodes several vectors **x** using a parity bit encoding, and after transmission Bob receives the following vectors **z**. Assuming that at most one error has occurred for each vector, determine whether or not an error has occurred, and what all the possible original **x** vectors are.

(a) $\begin{bmatrix} 1 \\ 0 \\ 1 \end{bmatrix}$ (b) $\begin{bmatrix} 1 \\ 1 \\ 1 \end{bmatrix}$ (c) $\begin{bmatrix} 1 \\ 0 \\ 0 \\ 1 \end{bmatrix}$ (d) $\begin{bmatrix} 1 \\ 1 \\ 1 \\ 0 \end{bmatrix}$

2.6.4 Alice encodes several vectors $\mathbf{x} \in \mathbb{F}_2^4$ using the Hamming code, and after transmission Bob receives the following vectors **z**. Assuming that at most one error has occurred for each vector, determine what the original **x** vectors are.

(a) $\begin{bmatrix} 0 \\ 1 \\ 1 \\ 0 \\ 0 \\ 0 \\ 1 \end{bmatrix}$ (b) $\begin{bmatrix} 0 \\ 1 \\ 0 \\ 0 \\ 0 \\ 0 \\ 1 \end{bmatrix}$ (c) $\begin{bmatrix} 1 \\ 0 \\ 0 \\ 1 \\ 1 \\ 0 \\ 0 \end{bmatrix}$ (d) $\begin{bmatrix} 1 \\ 1 \\ 1 \\ 1 \\ 1 \\ 1 \\ 1 \end{bmatrix}$ (e) $\begin{bmatrix} 1 \\ 0 \\ 1 \\ 0 \\ 1 \\ 0 \\ 1 \end{bmatrix}$

2.6.5 Alice encodes several vectors $\mathbf{x} \in \mathbb{F}_2^4$ using the Hamming code, and after transmission Bob receives the following vectors **z**. Assuming that at most one error has occurred for each vector, determine what the original **x** vectors are.

(a) $\begin{bmatrix} 1 \\ 0 \\ 1 \\ 1 \\ 0 \\ 1 \\ 1 \end{bmatrix}$ (b) $\begin{bmatrix} 1 \\ 0 \\ 1 \\ 1 \\ 1 \\ 1 \\ 0 \end{bmatrix}$ (c) $\begin{bmatrix} 1 \\ 1 \\ 0 \\ 1 \\ 0 \\ 0 \\ 1 \end{bmatrix}$ (d) $\begin{bmatrix} 0 \\ 1 \\ 0 \\ 1 \\ 1 \\ 1 \\ 1 \end{bmatrix}$ (e) $\begin{bmatrix} 1 \\ 0 \\ 0 \\ 0 \\ 0 \\ 1 \\ 0 \end{bmatrix}$

2.6.6 Show that $C(\mathbf{A}) = \ker \mathbf{B}$ for the Hamming code (where $\mathbf{A}$ and $\mathbf{B}$ are given in equations (2.27) and (2.26)).

2.6.7 Suppose that when using the Hamming code, two or more errors occur in a single 7-bit block. What will $\mathbf{Bz}$ look like in that case? What does this imply about the usefulness of the Hamming code for a very noisy channel?

2.6.8 It would be nice if we could do the error checking and decoding for the Hamming code linearly in one step. That is, we'd like to have a matrix $\mathbf{C} \in \mathrm{M}_{4,7}(\mathbb{F}_2)$ such that

$$\mathbf{Cz} = \mathbf{x},$$

both when $\mathbf{z} = \mathbf{y} = \mathbf{Ax}$, and whenever $\mathbf{z} = \mathbf{y} + \mathbf{e}_i$ for $i = 1, \ldots, 7$. Prove that there is no such matrix $\mathbf{C}$.

Hint: Think first about the case $\mathbf{x} = \mathbf{0}$. What can you conclude about the columns $\mathbf{Ce}_i$? Then think about when $\mathbf{x} \neq \mathbf{0}$.

2.6.9 Show that every vector $\mathbf{z} \in \mathbb{F}_2^7$ can occur by encoding a vector $\mathbf{x} \in \mathbb{F}_2^4$ as $\mathbf{y} = \mathbf{Ax}$ with the Hamming code, then changing at most one bit.

Hint: This is most easily done by simply *counting*: show that the number of distinct vectors you can get is the same as the total number of vectors in $\mathbb{F}_2^7$.

2.6.10 Extend the idea of the Hamming code to construct an error-correcting code with an encoding matrix $\mathbf{A} \in \mathrm{M}_{15,11}(\mathbb{F}_2)$.

Hint: Start with a parity-check matrix $\mathbf{B} \in \mathrm{M}_{4,15}(\mathbb{F}_2)$.

2.6.11 The Hamming code can be modified by adding a parity bit: $\mathbf{x} \in \mathbb{F}_2^4$ gets encoded by the vector $\begin{bmatrix} \mathbf{Ax} \\ x_1 + x_2 + x_3 + x_4 \end{bmatrix} \in \mathbb{F}_2^8$, where $\mathbf{A}$ is the encoding matrix for the Hamming code, given in equation (2.27).

(a) Find an encoding matrix $\tilde{\mathbf{A}} \in \mathrm{M}_{4,8}(\mathbb{F}_2)$ for the Hamming-plus-parity code.

(b) Find a parity-check matrix $\tilde{\mathbf{B}} \in \mathrm{M}_{8,4}(\mathbb{F}_2)$ for the Hamming-plus-parity code.

PERSPECTIVES: Matrix multiplication

Let $\mathrm{A} = \begin{bmatrix} -\mathbf{a}_1- \\ \vdots \\ -\mathbf{a}_m- \end{bmatrix} \in \mathrm{M}_{m,n}(\mathbb{F})$ and $\mathrm{B} = \begin{bmatrix} | & & | \\ \mathbf{b}_1 & \cdots & \mathbf{b}_p \\ | & & | \end{bmatrix} \in \mathrm{M}_{n,p}(\mathbb{F})$.

- $[\mathrm{AB}]_{ij} = \sum_{k=1}^{n} a_{ik} b_{kj} = \begin{bmatrix} -\mathbf{a}_i- \end{bmatrix} \begin{bmatrix} | \\ \mathbf{b}_j \\ | \end{bmatrix}$.
- $\mathrm{AB} = \begin{bmatrix} | & & | \\ \mathrm{A}\mathbf{b}_1 & \cdots & \mathrm{A}\mathbf{b}_p \\ | & & | \end{bmatrix}$.
- $\mathrm{AB} = \begin{bmatrix} -\mathbf{a}_1\mathrm{B}- \\ \vdots \\ -\mathbf{a}_m\mathrm{B}- \end{bmatrix}$.
- If $T_{\mathrm{A}} \in \mathcal{L}(\mathbb{F}^n, \mathbb{F}^m)$ has matrix A and $T_{\mathrm{B}} \in \mathcal{L}(\mathbb{F}^p, \mathbb{F}^n)$ has matrix B, then AB is the matrix of $T_{\mathrm{A}} \circ T_{\mathrm{B}}$.

3

Linear Independence, Bases, and Coordinates

3.1 Linear (In)dependence

Redundancy

We've come across the idea of redundancy in a couple of ways now. For example, recall the system of equations (1.3):

$$\begin{aligned} x + \frac{1}{4}y &= 20 \\ 2x + \frac{1}{4}y &= 36 \\ x + 8z &= 176 \\ 2x + \frac{1}{4}y + z &= 56. \end{aligned} \tag{3.1}$$

We observed that the system was redundant, in that we could uniquely solve the first three equations, and then the fourth was automatically satisfied. In fact, we explicitly demonstrated this redundancy by observing that if the first three equations are satisfied, then

$$\begin{aligned} 2x + \frac{1}{4}y + z &= \frac{1}{8}\left(x + \frac{1}{4}y\right) + \frac{7}{8}\left(2x + \frac{1}{4}y\right) + \frac{1}{8}(x + 8z) \\ &= \frac{1}{8}(20) + \frac{7}{8}(36) + \frac{1}{8}(176) \\ &= 56. \end{aligned} \tag{3.2}$$

Another type of redundancy we've seen is geometric: in Section 1.3, we found that solving a linear system was the same thing as finding a way of writing one vector as a linear combination of two others. We saw that to write $\begin{bmatrix} 20 \\ 36 \end{bmatrix}$ as a linear combination of $\begin{bmatrix} 1 \\ 2 \end{bmatrix}$ and $\begin{bmatrix} \frac{1}{4} \\ \frac{1}{4} \end{bmatrix}$, we really need both vectors, but if we want

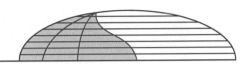

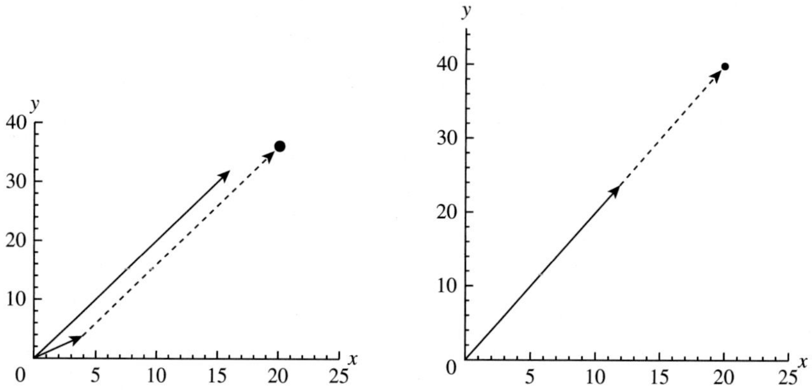

Figure 3.1 Forming $\begin{bmatrix} 20 \\ 40 \end{bmatrix}$ as a linear combination of $\begin{bmatrix} 1 \\ 2 \end{bmatrix}$ and $\begin{bmatrix} \frac{1}{4} \\ \frac{1}{4} \end{bmatrix}$ (left) and as a linear combination of $\begin{bmatrix} 1 \\ 2 \end{bmatrix}$ and $\begin{bmatrix} \frac{1}{4} \\ \frac{1}{2} \end{bmatrix}$ (right). In the combination on the left, both vectors are needed, but on the right, the two pieces are in the same direction.

to write $\begin{bmatrix} 20 \\ 40 \end{bmatrix}$ as a linear combination of $\begin{bmatrix} 1 \\ 2 \end{bmatrix}$ and $\begin{bmatrix} \frac{1}{4} \\ \frac{1}{2} \end{bmatrix}$ we don't need both; either of the vectors $\begin{bmatrix} 1 \\ 2 \end{bmatrix}$ and $\begin{bmatrix} \frac{1}{4} \\ \frac{1}{2} \end{bmatrix}$ would do on its own.

In both of these redundant situations, something (an equation, a vector) is unnecessary because we can get it as a linear combination of other things we already have.

Definition A list of vectors $(v_1, \ldots, v_n)$ in V (with $n \geq 2$) is **linearly dependent** if for some i, v_i is a linear combination of $\{v_j \mid j \neq i\}$.

A single vector v in V is **linearly dependent** if and only if $v = 0$.

Quick Exercise #1. Show that $\left(\begin{bmatrix} 1 \\ 0 \end{bmatrix}, \begin{bmatrix} 0 \\ 1 \end{bmatrix}, \begin{bmatrix} 1 \\ 1 \end{bmatrix}\right)$ is linearly dependent.

Notice that by the definition of a linear combination, for $n \geq 2$, $(v_1 \ldots, v_n)$ is linearly dependent if, for some i, there are scalars $a_j \in \mathbb{F}$ for all $j \neq i$ such that

$$v_i = a_1 v_1 + \cdots + a_{i-1} v_{i-1} + a_{i+1} v_{i+1} + \cdots + a_n v_n.$$

Subtracting v_i from both sides, this is the same as

$$0 = a_1 v_1 + \cdots + a_{i-1} v_{i-1} + (-1) v_i + a_{i+1} v_{i+1} + \cdots + a_n v_n.$$

QA #1: $\begin{bmatrix} 1 \\ 0 \end{bmatrix} + \begin{bmatrix} 0 \\ 1 \end{bmatrix} = \begin{bmatrix} 1 \\ 1 \end{bmatrix}$

So we've written $0 \in V$ as a linear combination of the vectors $v_1, \dots, v_n$ in a *non-trivial* way; that is, the coefficients are not all zero.

On the other hand, suppose we are given vectors $v_1, \dots, v_n$ in V $(n \geq 2)$ and $b_1, \dots, b_n \in \mathbb{F}$ such that

$$\sum_{j=1}^{n} b_j v_j = 0.$$

If, say, $b_i \neq 0$, then we can solve for v_i:

$$v_i = \left(-\frac{b_1}{b_i}\right) v_1 + \cdots + \left(-\frac{b_{i-1}}{b_i}\right) v_{i-1} + \left(-\frac{b_{i+1}}{b_i}\right) v_{i+1} + \cdots + \left(-\frac{b_n}{b_i}\right) v_n,$$

and so v_i is a linear combination of the vectors $v_1, \dots, v_{i-1}, v_{i+1}, \dots, v_n$.

This discussion shows that we could just as well have taken the following as our definition of linear dependence.

Definition A list of vectors $(v_1, \dots, v_n)$ in V is **linearly dependent** if there exist scalars $a_1, \dots, a_n \in \mathbb{F}$ which are not all 0 such that

$$\sum_{j=1}^{n} a_j v_j = 0.$$

The two definitions of linear dependence we've just seen are *equivalent*. That is, a list of vectors is linearly dependent according to the first definition if and only if it is linearly dependent according to the second definition. (This is exactly what we proved in the discussion between the two definitions.) The first definition is what our examples led us to, but the second one is frequently easier to work with.

Quick Exercise #2. Verify that the two definitions of linear dependence are equivalent for $n = 1$. (We omitted that case in the discussion above.)

When we want to emphasize the base field, we sometimes say "linearly dependent over $\mathbb{F}$" rather than just "linearly dependent." This becomes important in some cases, because we may have multiple possible base fields, and whether a list of vectors is linearly dependent may depend on which field we are working over. (See Exercise 3.1.8 for an example of this phenomenon.)

Linear Independence

Definition A list of vectors $(v_1, \dots, v_n)$ in V is **linearly independent** if it is not linearly dependent.

QA #2: If v is linearly dependent then $0 = v = 1v$. If $av = 0$ and $a \neq 0$, then $v = a^{-1}0 = 0$.

Thus, $(v_1, \ldots, v_n)$ is linearly independent iff the *only* choice of scalars $a_1, \ldots, a_n \in \mathbb{F}$ for which

$$\sum_{j=1}^{n} a_j v_j = 0 \tag{3.3}$$

is $a_1 = \cdots = a_n = 0$.

Quick Exercise #3. Show that the list $(\mathbf{e}_1, \ldots, \mathbf{e}_m)$ in $\mathbb{F}^m$ is linearly independent. (Recall that $\mathbf{e}_i$ has ith entry equal to 1 and all other entries 0.)

The next result is a first indication of the central role that linear independence plays in the study of linear systems.

Proposition 3.1 *The columns of a matrix* $\mathrm{A} \in \mathrm{M}_{m,n}(\mathbb{F})$ *are linearly independent iff* $\mathbf{x} = \mathbf{0}$ *is the only solution of the* $m \times n$ *linear system* $\mathrm{A}\mathbf{x} = \mathbf{0}$. *That is, the columns of* A *are linearly independent iff* $\ker \mathrm{A} = \{\mathbf{0}\}$.

Proof Let $\mathbf{v}_1, \ldots, \mathbf{v}_n \in \mathbb{F}^m$ be the columns of A. Then the linear system $\mathrm{A}\mathbf{x} = \mathbf{0}$ has the vector form

$$\sum_{j=1}^{n} x_j \mathbf{v}_j = \mathbf{0}.$$

Thus $\mathbf{v}_1, \ldots, \mathbf{v}_n$ are linearly independent if and only if the only solution to the system is $x_1 = \cdots = x_n = 0$. ▲

Corollary 3.2 *A linear map* $T : \mathbb{F}^n \to \mathbb{F}^m$ *is injective iff the columns of its matrix are linearly independent.*

Proof See Exercise 3.1.19. ▲

Corollary 3.3 *Suppose that* $(\mathbf{v}_1, \ldots, \mathbf{v}_n)$ *is a list of vectors in* $\mathbb{F}^m$, *with* $n > m$. *Then* $(\mathbf{v}_1, \ldots, \mathbf{v}_n)$ *is linearly dependent.*

Proof By Theorem 1.4, the $m \times n$ linear system $\begin{bmatrix} \mathbf{v}_1 & \cdots & \mathbf{v}_n \end{bmatrix} \mathbf{x} = \mathbf{0}$ cannot have a unique solution because $m < n$. By Proposition 3.1, this means that $(\mathbf{v}_1, \ldots, \mathbf{v}_n)$ is linearly dependent. ▲

QA #3: If $a_1\mathbf{e}_1 + \cdots + a_m\mathbf{e}_m = \begin{bmatrix} a_1 \\ \vdots \\ a_m \end{bmatrix} = \mathbf{0}$, then $a_1 = \cdots = a_m = 0$, so $(\mathbf{e}_1, \ldots, \mathbf{e}_m)$ is linearly independent.

Proposition 3.1 provides us with an algorithm for checking linear (in)dependence of vectors in $\mathbb{F}^m$.

Algorithm 3.4 To check whether a list $(\mathbf{v}_1, \ldots, \mathbf{v}_n)$ in $\mathbb{F}^m$ is linearly independent:

- Put the matrix $\mathbf{A} = \begin{bmatrix} \mathbf{v}_1 & \cdots & \mathbf{v}_n \end{bmatrix}$ in REF.
- $(\mathbf{v}_1, \ldots, \mathbf{v}_n)$ in $\mathbb{F}^m$ is linearly independent if and only if there is a pivot in every column of the REF.

Proof By Proposition 3.1, $(\mathbf{v}_1, \ldots, \mathbf{v}_n)$ is linearly independent iff the linear system $\mathbf{A}\mathbf{x} = \mathbf{0}$ has the unique solution $\mathbf{x} = \mathbf{0}$. The system is consistent since $\mathbf{x} = \mathbf{0}$ is a solution automatically, so by Theorem 1.3, $\mathbf{x} = \mathbf{0}$ is the unique solution iff the REF of A has a pivot in each column. ▲

Quick Exercise #4. Use Algorithm 3.4 to show (again) that $\left(\begin{bmatrix} 1 \\ 0 \end{bmatrix}, \begin{bmatrix} 0 \\ 1 \end{bmatrix}, \begin{bmatrix} 1 \\ 1 \end{bmatrix} \right)$ is linearly dependent.

Examples

1. Consider the list of vectors $(\mathbf{v}_1, \ldots, \mathbf{v}_n)$ in $\mathbb{F}^n$ given by

$$\mathbf{v}_j = \mathbf{e}_1 + \cdots + \mathbf{e}_j$$

for $1 \leq j \leq n$. Since the matrix $\begin{bmatrix} \mathbf{v}_1 & \cdots & \mathbf{v}_n \end{bmatrix}$ is already in REF, we can see that each column contains a pivot, so by Algorithm 3.4, the list $(\mathbf{v}_1, \ldots, \mathbf{v}_n)$ is linearly independent.

2. Now consider the list of vectors $(\mathbf{w}_1, \ldots, \mathbf{w}_n)$ in $\mathbb{F}^n$ given by $\mathbf{w}_1 = \mathbf{e}_1$ and

$$\mathbf{w}_j = \mathbf{e}_j - \mathbf{e}_{j-1}$$

for $2 \leq j \leq n$. Again, since $\begin{bmatrix} \mathbf{w}_1 & \cdots & \mathbf{w}_n \end{bmatrix}$ is already in REF, we can see that each column contains a pivot, so by Algorithm 3.4, the list $(\mathbf{w}_1, \ldots, \mathbf{w}_n)$ is linearly independent. ▲

The following consequence shows a very special property of lists of exactly n vectors in $\mathbb{F}^n$.

Corollary 3.5 *Suppose that $(\mathbf{v}_1, \ldots, \mathbf{v}_n)$ is a list of n vectors in $\mathbb{F}^n$. Then $(\mathbf{v}_1, \ldots, \mathbf{v}_n)$ is linearly independent if and only if $(\mathbf{v}_1, \ldots, \mathbf{v}_n)$ spans $\mathbb{F}^n$.*

QA #4: $\begin{bmatrix} 1 & 0 & 1 \\ 0 & 1 & 1 \end{bmatrix}$ is in REF, and has no pivot in the third column.

Proof Let $\mathrm{A} = \begin{bmatrix} \mathbf{v}_1 & \cdots & \mathbf{v}_n \end{bmatrix} \in \mathrm{M}_n(\mathbb{F})$. By Corollary 2.34, $(\mathbf{v}_1, \ldots, \mathbf{v}_n)$ spans $\mathbb{F}^n$ iff the RREF of A has a pivot in each row. By Algorithm 3.4, $(\mathbf{v}_1, \ldots, \mathbf{v}_n)$ is linearly independent iff the RREF of A has a pivot in each column. Since A is square, each of these happens precisely when the RREF is I_n. ▲

You may have noticed that the examples above (showing that two lists of n vectors in $\mathbb{F}^n$ are linearly independent) were very similar to the examples on page 118 (showing that the same two lists each span $\mathbb{F}^n$). Corollary 3.5 shows that this is not a coincidence. In fact, if we'd done just one set of the examples, the others would follow immediately from the corollary.

The Linear Dependence Lemma

The following lemma is hardly more than a restatement of the definition of linear dependence, but turns out to be quite a useful one.

Theorem 3.6 (Linear Dependence Lemma) *Suppose that $(v_1, \ldots, v_n)$ is linearly dependent and that $v_1 \neq 0$. Then there is a $k \in \{2, \ldots, n\}$ such that $v_k \in \langle v_1, \ldots, v_{k-1} \rangle$.*

Proof If $(v_1, \ldots, v_n)$ is linearly dependent, then there are scalars $b_1, \ldots, b_n \in \mathbb{F}$ which are not all 0 such that

$$\sum_{j=1}^{n} b_j v_j = 0.$$

Let k be the largest index such that $b_k \neq 0$. If $k = 1$, then we have $b_1 v_1 = 0$, which is impossible, because $b_1 \neq 0$ and $v_1 \neq 0$. So $k \geq 2$, and then

$$b_k v_k = -b_1 v_1 - \cdots - b_{k-1} v_{k-1},$$

so

$$v_k = \sum_{j=1}^{k-1} \left(-\frac{b_j}{b_k} \right) v_j \in \langle v_1, \ldots, v_{k-1} \rangle .$$ ▲

Corollary 3.7 *Given a list $(v_1, \ldots, v_n)$ in V with $v_1 \neq 0$, suppose that for each $k \in \{2, \ldots, n\}$, $v_k \notin \langle v_1, \ldots, v_{k-1} \rangle$. Then $(v_1, \ldots, v_n)$ is linearly independent.*

The corollary is simply a restatement of Theorem 3.6 in the logically equivalent contrapositive form.

Quick Exercise #5. Show that if $(v_1, \ldots, v_n)$ is linearly independent, and $v_{n+1} \notin \langle v_1, \ldots, v_n \rangle$, then $(v_1, \ldots, v_n, v_{n+1})$ is also linearly independent.

Example Let $n \geq 1$ be an integer, and consider the functions $1, x, x^2, \ldots, x^n$ in the function space $C(\mathbb{R})$. For $1 \leq k \leq n$, every function $f \in \langle 1, x, \ldots, x^{k-1} \rangle$ is a polynomial of degree at most $k - 1$, so the kth derivative of f is 0. Since the kth derivative of x^k is not 0 (it is the constant function $k!$), this implies that $x^k \notin \langle 1, x, \ldots, x^{k-1} \rangle$. By Corollary 3.7, the list of functions $(1, x, \ldots, x^n)$ is linearly independent. ▲

Linear Independence of Eigenvectors

The following theorem is quite important in its own right; its proof is also a nice illustration of how to show that vectors in an abstract vector space are linearly independent.

Theorem 3.8 *Suppose that $v_1, \ldots, v_k$ are eigenvectors of $T \in \mathcal{L}(V)$ with distinct eigenvalues $\lambda_1, \ldots, \lambda_k$, respectively. Then $(v_1, \ldots, v_k)$ is linearly independent.*

Proof Suppose that $(v_1, \ldots, v_k)$ is linearly dependent. Since $v_1 \neq 0$ (eigenvectors are never 0), by the Linear Dependence Lemma (Theorem 3.6), there is some $j \geq 2$ such that $v_j \in \langle v_1, \ldots, v_{j-1} \rangle$. Pick the smallest such j. Then $(v_1, \ldots, v_{j-1})$ is linearly independent, and we can write

$$v_j = \sum_{i=1}^{j-1} a_i v_i \tag{3.4}$$

for $a_1, \ldots, a_{j-1} \in \mathbb{F}$.

Apply T and use linearity and the fact that for each i, v_i is an eigenvector of T with eigenvalue λ_i, to get

$$\lambda_j v_j = T(v_j) = \sum_{i=1}^{j-1} a_i T(v_i) = \sum_{i=1}^{j-1} a_i \lambda_i v_i. \tag{3.5}$$

Multiplying both sides of equation (3.4) by λ_j and subtracting from equation (3.5) gives

$$0 = \sum_{i=1}^{j-1} a_i (\lambda_i - \lambda_j) v_i.$$

Since $(v_1, \ldots, v_{j-1})$ is linearly independent, this means that $a_i(\lambda_i - \lambda_j) = 0$ for each $i = 1, \ldots, j-1$. Since $\lambda_1, \ldots, \lambda_k$ are distinct, $\lambda_i - \lambda_j \neq 0$ for each i, so $a_i = 0$ for

QA #5: By linear independence, $v_1 \neq 0$ and $v_k \notin \langle v_1, \ldots, v_{k-1} \rangle$ for $2 \leq k \leq n$. By assumption, this is also true for $k = n + 1$, so the claim follows from Corollary 3.7.

$i = 1, \dots, j-1$. But by equation (3.4), this means that $v_j = 0$, which is impossible since v_j is an eigenvector. Therefore the assumption that $(v_1, \dots, v_k)$ is linearly dependent must be false; i.e., $(v_1, \dots, v_k)$ is linearly independent. ▲

Theorem 3.8 is false without the assumption that the eigenvalues corresponding to the eigenvectors are distinct; see Exercise 3.1.17.

KEY IDEAS

- The vectors $v_1, \dots, v_n$ are linearly independent if the only way to have $c_1v_1 + \cdots + c_nv_n = 0$ is to take all the $c_i = 0$.
- The vectors $v_1, \dots, v_n$ are linearly dependent if one of the v_i is in the span of the others (or if $n = 1$ and $v_1 = 0$).
- The vectors $\mathbf{v}_1, \dots, \mathbf{v}_n \in \mathbb{F}^m$ are linearly independent if and only if the RREF of the matrix with those columns has a pivot in every column.
- Eigenvectors corresponding to distinct eigenvalues are linearly independent.

EXERCISES

3.1.1 Determine whether each of the following lists of vectors is linearly independent.

(a) $\left(\begin{bmatrix} 1 \\ 2 \\ 3 \end{bmatrix}, \begin{bmatrix} 2 \\ 3 \\ 1 \end{bmatrix}, \begin{bmatrix} 1 \\ 1 \\ 2 \end{bmatrix} \right)$

(b) $\left(\begin{bmatrix} 1 \\ 2 \\ 3 \end{bmatrix}, \begin{bmatrix} 2 \\ 3 \\ 1 \end{bmatrix}, \begin{bmatrix} 1 \\ 1 \\ -2 \end{bmatrix} \right)$

(c) $\left(\begin{bmatrix} 1 \\ 1 \\ 0 \end{bmatrix}, \begin{bmatrix} 1 \\ 0 \\ 1 \end{bmatrix}, \begin{bmatrix} 0 \\ 1 \\ 1 \end{bmatrix} \right)$, in $\mathbb{R}^3$

(d) $\left(\begin{bmatrix} 1 \\ 1 \\ 0 \end{bmatrix}, \begin{bmatrix} 1 \\ 0 \\ 1 \end{bmatrix}, \begin{bmatrix} 0 \\ 1 \\ 1 \end{bmatrix} \right)$, in $\mathbb{F}_2^3$

(e) $\left(\begin{bmatrix} 1 \\ -2 \\ -1 \\ 0 \end{bmatrix}, \begin{bmatrix} 2 \\ 1 \\ 3 \\ -2 \end{bmatrix}, \begin{bmatrix} 0 \\ 5 \\ 5 \\ -2 \end{bmatrix} \right)$

3.1.2 Determine whether each of the following lists of vectors is linearly independent.

(a) $\left(\begin{bmatrix} 1 \\ 1+i \end{bmatrix}, \begin{bmatrix} 1-i \\ 2 \end{bmatrix}\right)$, in $\mathbb{C}^2$

(b) $\left(\begin{bmatrix} -1 \\ 2 \\ 0 \end{bmatrix}, \begin{bmatrix} 2 \\ -3 \\ 1 \end{bmatrix}, \begin{bmatrix} 0 \\ 4 \\ -5 \end{bmatrix}, \begin{bmatrix} 1 \\ -2 \\ -1 \end{bmatrix}\right)$

(c) $\left(\begin{bmatrix} 1 \\ 1 \\ 2 \end{bmatrix}, \begin{bmatrix} 2 \\ 1 \\ 3 \end{bmatrix}, \begin{bmatrix} 1 \\ 0 \\ 1 \end{bmatrix}\right)$

(d) $\left(\begin{bmatrix} 2 \\ -1 \\ 1 \\ 1 \end{bmatrix}, \begin{bmatrix} 1 \\ 2 \\ -1 \\ 3 \end{bmatrix}, \begin{bmatrix} 1 \\ -8 \\ 5 \\ -7 \end{bmatrix}, \begin{bmatrix} 6 \\ 0 \\ -1 \\ 2 \end{bmatrix}\right)$

(e) $\left(\begin{bmatrix} 1 \\ 0 \\ 1 \\ 0 \end{bmatrix}, \begin{bmatrix} 0 \\ 1 \\ 0 \\ 1 \end{bmatrix}, \begin{bmatrix} 1 \\ 0 \\ 0 \\ 1 \end{bmatrix}, \begin{bmatrix} 1 \\ 1 \\ 0 \\ 1 \end{bmatrix}\right)$

3.1.3 Solve a linear system to find where the coefficients $\frac{1}{8}$, $\frac{7}{8}$, and $\frac{1}{8}$ came from in equation (3.2).

3.1.4 Show that whether a list of vectors is linearly independent does not depend on the order in which the vectors are given.

3.1.5 Show that any list of vectors in V which contains 0 is linearly dependent.

3.1.6 Show that any list of vectors which contains a linearly dependent list is linearly dependent.

3.1.7 Two vectors $v, w \in V$ are called **collinear** if there is a scalar $c \in \mathbb{F}$ such that either $v = cw$ or $w = cv$.

(a) Given two vectors $v, w \in V$, prove that (v, w) is linearly dependent if and only if v and w are collinear.

(b) Give an example of a list of three linearly dependent vectors, no two of which are collinear.

3.1.8 Show that the list of vectors $\left(\begin{bmatrix} 1 \\ i \end{bmatrix}, \begin{bmatrix} i \\ -1 \end{bmatrix}\right)$ in $\mathbb{C}^2$ is linearly independent over $\mathbb{R}$, but is linearly dependent over $\mathbb{C}$.

3.1.9 Suppose that $T \in \mathcal{L}(V, W)$ is injective and that $(v_1, \ldots, v_n)$ is linearly independent in V. Prove that $(Tv_1, \ldots, Tv_n)$ is linearly independent in W.

3.1.10 (a) Let $A \in M_n(\mathbb{F})$. Show that $\ker A = \{0\}$ if and only if $C(A) = \mathbb{F}^n$.

(b) Let $T \in \mathcal{L}(\mathbb{F}^n)$. Show that T is injective if and only if it is surjective.

3.1.11 Suppose that $A \in M_n(\mathbb{F})$ is upper triangular (see Exercise 2.3.12) and that all the diagonal entries are nonzero (i.e., $a_{ii} \neq 0$ for each $i = 1, \ldots, n$). Use Algorithm 3.4 to prove that the columns of A are linearly independent.

3.1.12 Suppose that $A \in M_n(\mathbb{F})$ is upper triangular (see Exercise 2.3.12) and that all the diagonal entries are nonzero (i.e., $a_{ii} \neq 0$ for each $i = 1, \ldots, n$). Use Corollary 3.7 to prove that the columns of A are linearly independent.

3.1.13 Suppose that $\mathbb{F}$ is a subfield of $\mathbb{K}$ (see Exercise 1.5.7), so that every vector in the vector space $\mathbb{F}^m$ over $\mathbb{F}$ is also a vector in the vector space $\mathbb{K}^m$ over $\mathbb{K}$. Show that a list of vectors $(\mathrm{v}_1, \ldots, \mathrm{v}_n)$ in $\mathbb{F}^n$ is linearly independent when thought of as vectors in $\mathbb{F}^n$ if and only if it is linearly independent when thought of as vectors in $\mathbb{K}^n$.

3.1.14 Let $n \geq 1$ be an integer, and suppose that there are constants $a_1, \ldots, a_n \in \mathbb{R}$ such that

$$\sum_{k=1}^{n} a_k \sin(kx) = 0$$

for every $x \in \mathbb{R}$. Prove that $a_1 = \cdots = a_n = 0$.
Hint: Consider the linear map $D^2 : C^\infty(\mathbb{R}) \to C^\infty(\mathbb{R})$ given by $D^2 f = f''$, where $C^\infty(\mathbb{R})$ is the space of infinitely differentiable functions on $\mathbb{R}$, and use Theorem 3.8.

3.1.15 Let $n \geq 1$ be an integer, and suppose that there are constants $a_1, \ldots, a_n \in \mathbb{R}$ and distinct nonzero constants $c_1, \ldots, c_n \in \mathbb{R}$ such that

$$\sum_{k=1}^{n} a_k e^{c_k x} = 0$$

for every $x \in \mathbb{R}$. Prove that $a_1 = \cdots = a_n = 0$.
Hint: Use Theorem 3.8.

3.1.16 Let V be a vector space over the field $\mathbb{Q}$ of rational numbers. Show that a list $(v_1, \ldots, v_n)$ in V is linearly dependent if and only if there are *integers* $a_1, \ldots, a_n \in \mathbb{Z}$, not all equal to 0, such that

$$a_1 v_1 + \cdots + a_n v_n = 0.$$

3.1.17 Give an example of a linear map $T \in \mathcal{L}(V)$ with two linearly *dependent* eigenvectors v and w.
Hint: Don't overthink this.

3.1.18 Consider the set of real numbers $\mathbb{R}$ as a vector space over $\mathbb{Q}$ (see Exercise 1.5.8). Show that if $p_1, \ldots, p_n$ are distinct prime numbers, then the list

$$(\log p_1, \ldots, \log p_n)$$

in $\mathbb{R}$ is linearly independent over $\mathbb{Q}$.

3.1.19 Prove Corollary 3.2.

3.1.20 Use Algorithm 3.4 to give another proof of Corollary 3.3.

3.2 Bases

Bases of Vector Spaces

Recall that the **span** of $v_1, v_2, \ldots, v_k \in V$ is the set $\langle v_1, v_2, \ldots, v_k\rangle$ of all linear combinations of $v_1, v_2, \ldots, v_k$. We also say that the list of vectors $(v_1, v_2, \ldots, v_k)$ **spans** a subspace $W \subseteq V$ if $\langle v_1, v_2, \ldots, v_k\rangle = W$.

Definition A vector space V is **finite-dimensional** if there is a finite list of vectors $(v_1, \ldots, v_n)$ in V such that $V \subseteq \langle v_1, \ldots, v_n\rangle$. A vector space V is **infinite-dimensional** if it is not finite-dimensional.

Quick Exercise #6. Show that $\mathbb{F}^n$ is finite-dimensional.

Of course if $v_1, \ldots, v_n \in V$, then $\langle v_1, \ldots, v_n\rangle \subseteq V$ automatically, so it would be equivalent to require $V = \langle v_1, \ldots, v_n\rangle$ in the definition above. We wrote the definition the way we did because it's formally simpler.

Remember that if $(v_1, \ldots, v_n)$ spans V, then we can write any vector in V as a linear combination of $(v_1, \ldots, v_n)$; i.e., $(v_1, \ldots, v_n)$ suffices to describe all of V. On the other hand, we can think of a linearly independent list as a list with no redundancy. We would like to combine the virtues of each. That is, we would like to describe the whole space with no redundancy.

Definition Let V be a finite-dimensional vector space. A list $(v_1, \ldots, v_n)$ in V is a **basis** for V if $(v_1, \ldots, v_n)$ is linearly independent and $V \subseteq \langle v_1, \ldots, v_n\rangle$.

Examples

1. Recall that $\mathbf{e}_i \in \mathbb{F}^m$ denotes the vector with 1 in the ith position and zeroes elsewhere. The list $(\mathbf{e}_1, \ldots, \mathbf{e}_m)$ in $\mathbb{F}^m$ is called the **standard basis** of $\mathbb{F}^m$. It is indeed a basis: We saw in Quick Exercise #3 of Section 3.1 that it is linearly independent, and in Quick Exercise #6 above that it spans $\mathbb{F}^m$.

QA #6: Given $\mathbf{v} = \begin{bmatrix} v_1 \\ \vdots \\ v_n \end{bmatrix} \in \mathbb{F}^n$, $\mathbf{v} = v_1\mathbf{e}_1 + \cdots + v_n\mathbf{e}_n$. Thus $\mathbb{F}^n = \langle \mathbf{e}_1, \ldots, \mathbf{e}_n\rangle$.

2. Let $\mathcal{P}_n(\mathbb{R}) := \{a_0 + a_1x + \cdots + a_nx^n \mid a_0, a_1, \ldots, a_n \in \mathbb{R}\}$ be the vector space over $\mathbb{R}$ of polynomials in x of degree n or less. Then it is similarly easy to see, with the help of the example on page 146, that $(1, x, \ldots, x^n)$ forms a basis of $\mathcal{P}_n(\mathbb{R})$.
3. Let $\mathbf{v}_1 = \begin{bmatrix} 1 \\ 0 \end{bmatrix}$ and $\mathbf{v}_2 = \begin{bmatrix} 1 \\ 2 \end{bmatrix}$ in $\mathbb{R}^2$. The two vectors are not collinear, and so they're linearly independent (see Exercise 3.1.7). We've already seen that $\langle \mathbf{v}_1, \mathbf{v}_2 \rangle = \mathbb{R}^2$ (see pp. 26 and 30).
4. Consider the two lists of vectors in $\mathbb{F}^n$, $(\mathbf{v}_1, \ldots, \mathbf{v}_n)$ given by

$$\mathbf{v}_j = \mathbf{e}_1 + \cdots + \mathbf{e}_j$$

for $1 \leq j \leq n$, and $(\mathbf{w}_1, \ldots, \mathbf{w}_n)$ given by $\mathbf{w}_1 = \mathbf{e}_1$ and

$$\mathbf{w}_j = \mathbf{e}_j - \mathbf{e}_{j-1}$$

for $2 \leq j \leq n$. We saw in the examples on page 118 that both these lists span $\mathbb{F}^n$, and in the examples on page 144 that they are linearly independent. Therefore each of these lists is a basis of $\mathbb{F}^n$.
5. **(Finding the basis of the kernel)** Consider the matrix

$$\mathbf{A} = \begin{bmatrix} 2 & 1 & -1 & 3 \\ 1 & 0 & -1 & 2 \\ 1 & 1 & 0 & 1 \end{bmatrix} \in \mathrm{M}_{3,4}(\mathbb{R}).$$

The RREF of $[\mathbf{A} \mid \mathbf{0}]$ is

$$\left[\begin{array}{cccc|c} 1 & 0 & -1 & 2 & 0 \\ 0 & 1 & 1 & -1 & 0 \\ 0 & 0 & 0 & 0 & 0 \end{array}\right],$$

and so every vector $\mathbf{x} \in \ker \mathbf{A}$ can be written in the form

$$\begin{bmatrix} x \\ y \\ z \\ w \end{bmatrix} = \begin{bmatrix} z - 2w \\ -z + w \\ z \\ w \end{bmatrix} = z \begin{bmatrix} 1 \\ -1 \\ 1 \\ 0 \end{bmatrix} + w \begin{bmatrix} -2 \\ 1 \\ 0 \\ 1 \end{bmatrix},$$

where z and w are free variables. It follows that

$$\ker \mathbf{A} = \left\langle \begin{bmatrix} 1 \\ -1 \\ 1 \\ 0 \end{bmatrix}, \begin{bmatrix} -2 \\ 1 \\ 0 \\ 1 \end{bmatrix} \right\rangle.$$

Furthermore, these two vectors are linearly independent: the only way for a linear combination to have 0s in the last two entries is for both coefficients to be 0. Therefore $\left(\begin{bmatrix}1\\-1\\1\\0\end{bmatrix}, \begin{bmatrix}-2\\1\\0\\1\end{bmatrix}\right)$ is a basis of $\ker \mathrm{A}$. ▲

The following proposition gives a simple condition on the RREF of a matrix that determines whether or not a list of vectors in $\mathbb{F}^n$ is a basis of $\mathbb{F}^n$.

Proposition 3.9 *The list* $(\mathbf{v}_1, \ldots, \mathbf{v}_n)$ *of vectors in* $\mathbb{F}^m$ *forms a basis for* $\mathbb{F}^m$ *if and only if the RREF of the matrix* $\mathrm{A} := \begin{bmatrix} | & & | \\ \mathbf{v}_1 & \cdots & \mathbf{v}_n \\ | & & | \end{bmatrix}$ *is* I_m.

Proof We saw in Corollary 2.34 that $(\mathbf{v}_1, \ldots, \mathbf{v}_n)$ spans $\mathbb{F}^m$ if and only if the RREF of A has a pivot in every row, and we saw in Algorithm 3.4 that $(\mathbf{v}_1, \ldots, \mathbf{v}_n)$ is linearly independent if and only if the RREF of A has a pivot in every column. A matrix in RREF has a pivot in every row and every column if and only if it is the identity matrix. ▲

This condition gives an easy alternative proof that $\left(\begin{bmatrix}1\\0\end{bmatrix}, \begin{bmatrix}1\\2\end{bmatrix}\right)$ spans $\mathbb{R}^2$: the RREF of $\mathrm{A} = \begin{bmatrix}1 & 1\\0 & 2\end{bmatrix}$ is clearly I_2.

Quick Exercise #7. Show that every basis of $\mathbb{F}^m$ contains precisely m vectors.

Properties of Bases

The main reason that bases* are nice to work with is the following.

Theorem 3.10 *A list of vectors* $\mathcal{B} = (v_1, \ldots, v_n)$ *is a basis of* V *iff each element of* V *can be uniquely represented as a linear combination of the vectors in* $\mathcal{B}$.

*That's the plural of basis – it's pronounced *bay-seez*.

QA #7: Proposition 3.9 implies that if $(\mathbf{v}_1, \ldots, \mathbf{v}_n)$ is a basis of $\mathbb{F}^m$, then the matrix with columns $\mathbf{v}_j$ is $m \times m$, and thus in fact $n = m$.

Proof Suppose first that $\mathcal{B} = (v_1, \ldots, v_n)$ is a basis of V. Since $\langle v_1, \ldots, v_n \rangle = V$, each vector $v \in V$ can be represented as a linear combination of $(v_1, \ldots, v_n)$. We need to show that if

$$v = \sum_{i=1}^{n} a_i v_i \qquad \text{and} \qquad v = \sum_{i=1}^{n} b_i v_i,$$

then in fact $a_i = b_i$ for each i. Subtracting one representation from the other,

$$0 = v - v = \sum_{i=1}^{n} (a_i - b_i) v_i.$$

Since $(v_1, \ldots, v_n)$ is linearly independent, this implies that $a_i = b_i$ for each i.

Now suppose that each element of V can be uniquely represented as a linear combination of the vectors in $\mathcal{B}$. Then in particular, each element of V *can* be represented as a linear combination of vectors in $\mathcal{B}$, so $V \subseteq \langle v_1, \ldots, v_n \rangle$. On the other hand, the fact that $0 \in V$ can be uniquely represented as a linear combination of $(v_1, \ldots, v_n)$ means that if

$$\sum_{i=1}^{n} a_i v_i = 0,$$

then $a_i = 0$ for each i. In other words, $\mathcal{B}$ is linearly independent. ▲

Theorem 3.11 *Suppose that V is a nonzero vector space and $V = \langle \mathcal{B} \rangle$. Then there is a sublist of $\mathcal{B}$ which is a basis of V.*

Proof Let $(v_1, \ldots, v_n)$ be the list of all nonzero vectors in $\mathcal{B}$, and define $\mathcal{B}'$ to be the list containing, in order, each v_j such that

$$v_j \notin \langle v_1, \ldots, v_{j-1} \rangle.$$

Then $\mathcal{B}'$ is linearly independent by Corollary 3.7.

Now, since $\mathcal{B}$ spans V, for $v \in V$ given, we can write

$$v = \sum_{i=1}^{n} a_i v_i.$$

Pick the largest k such that v_k is not in $\mathcal{B}'$. By construction of $\mathcal{B}'$, this means that

$$v_k = \sum_{i=1}^{k-1} b_i v_i$$

for some $b_1, \ldots, b_{k-1} \in \mathbb{F}$. Therefore

$$v = \sum_{i=1}^{k-1} a_i v_i + a_k v_k + \sum_{i=k+1}^{n} a_i v_i = \sum_{i=1}^{k-1} (a_i + a_k b_i) v_i + \sum_{i=k+1}^{n} a_i v_i,$$

so $v \in \langle v_1, \ldots, v_{k-1}, v_{k+1}, \ldots, v_n \rangle$. Now iterate this construction: pick the largest $\ell < k$ such that v_ℓ is not in $\mathcal{B}'$, and show in the same way that

$$v \in \langle v_1, \ldots, v_{\ell-1}, v_{\ell+1}, \ldots, v_{k-1}, v_{k+1}, \ldots, v_n \rangle .$$

Continue in this way until v has been written as a linear combination of the vectors in $\mathcal{B}'$. Since v was arbitrary, we have shown that $\mathcal{B}'$ spans V and is thus a basis. ▲

Corollary 3.12 *If V is a nonzero finite-dimensional vector space, then there is a basis for V.*

Proof Since V is finite-dimensional, $V = \langle \mathcal{B} \rangle$ for some finite list $\mathcal{B}$ of vectors in V. By Theorem 3.11, there is then a sublist of $\mathcal{B}$ which is a basis for V. ▲

The proof of Theorem 3.11 implicitly contains an algorithm for finding a sublist of a spanning list which forms a basis, but it's not terribly practical. For each vector v_j in the list, you have to determine whether $v_j \in \langle v_1, \ldots, v_{j-1} \rangle$; depending on the space V, this may or may not be feasible. In $\mathbb{F}^m$, though, we can use row operations to give a practical way to find a linearly independent spanning sublist of a list of vectors.

Algorithm 3.13 Let $(\mathbf{v}_1, \ldots, \mathbf{v}_n)$ be a list of vectors in $\mathbb{F}^m$. To find a basis $\mathcal{B}$ for $\langle \mathbf{v}_1, \ldots, \mathbf{v}_n \rangle$:

- Put $\mathbf{A} = [\mathbf{v}_1 \cdots \mathbf{v}_n]$ in RREF.
- If the ith column contains a pivot, include $\mathbf{v}_i$ in $\mathcal{B}$.

Proof We saw in the proof of Theorem 3.11 that we can find a basis of $\langle \mathbf{v}_1, \ldots, \mathbf{v}_n \rangle$ by first removing any vectors which are $\mathbf{0}$ and then removing any vectors $\mathbf{v}_j$ such that $\mathbf{v}_j \in \langle \mathbf{v}_1, \ldots, \mathbf{v}_{j-1} \rangle$. Observe that if $[\mathbf{v}_1 \cdots \mathbf{v}_n]$ is already in RREF, then the pivot columns are exactly the ones that get put in the basis. We can thus complete the proof by showing that whether or not a column of a matrix is a linear combination of preceding columns is unchanged by the row operations **R1–R3**. But in fact, we already know this: $\mathbf{v}_j = \sum_{k=1}^{j-1} c_k \mathbf{v}_k$ if and only if the vector $\begin{bmatrix} c_1 \\ \vdots \\ c_{j-1} \end{bmatrix}$ is a solution to the linear system with augmented matrix

$$\left[\begin{array}{ccc|c} | & & | & | \\ \mathbf{v}_1 & \cdots & \mathbf{v}_{j-1} & \mathbf{v}_j \\ | & & | & | \end{array}\right].$$

Since the set of solutions to a linear system are not changed by the row operations, we are done. ▲

Quick Exercise #8. Find a basis for the subspace

$$\left\langle \begin{bmatrix} 1 \\ 0 \\ 1 \\ -1 \end{bmatrix}, \begin{bmatrix} -2 \\ 0 \\ -2 \\ 2 \end{bmatrix}, \begin{bmatrix} 0 \\ 1 \\ 0 \\ 1 \end{bmatrix}, \begin{bmatrix} 0 \\ 0 \\ 1 \\ 0 \end{bmatrix}, \begin{bmatrix} 1 \\ 1 \\ 3 \\ 0 \end{bmatrix} \right\rangle \subseteq \mathbb{R}^4.$$

Bases and Linear Maps

An important fact about linear maps is that they are completely determined by what they do to the elements of a basis.

Theorem 3.14 *Suppose that $(v_1, \ldots, v_n)$ is a basis for V, and let $w_1, \ldots, w_n \in W$ be any vectors. Then there is a unique linear map $T : V \to W$ such that $T(v_i) = w_i$ for each i.*

Proof We first prove the existence of such a map by explicitly constructing one. Given $v \in V$, by Theorem 3.10, there is a unique representation

$$v = \sum_{i=1}^{n} c_i v_i$$

with $c_1, \ldots, c_n \in \mathbb{F}$. We define

$$T(v) := \sum_{i=1}^{n} c_i w_i.$$

Then clearly $T(v_i) = w_i$ for each i (write it out carefully for yourself), so we just have to show that T is in fact a linear map. If $c \in \mathbb{F}$, then

$$T(cv) = T\left(\sum_{i=1}^{n} (cc_i)v_i\right) = \sum_{i=1}^{n} (cc_i)w_i = c\sum_{i=1}^{n} c_i w_i = cT(v),$$

QA #8: The RREF of the matrix with those columns is $\begin{bmatrix} 1 & -2 & 0 & 0 & 1 \\ 0 & 0 & 1 & 0 & -1 \\ 0 & 0 & 0 & 1 & 2 \\ 0 & 0 & 0 & 0 & 0 \end{bmatrix}$, with pivots in the first, third, and fourth columns. Thus $\left(\begin{bmatrix} 1 \\ 0 \\ 1 \\ -1 \end{bmatrix}, \begin{bmatrix} 0 \\ 1 \\ 0 \\ 1 \end{bmatrix}, \begin{bmatrix} 0 \\ 0 \\ 1 \\ 0 \end{bmatrix}\right)$ is a basis for the subspace.

so T is homogeneous. If

$$u = \sum_{i=1}^{n} d_i v_i,$$

then

$$T(u+v) = T\left(\sum_{i=1}^{n}(c_i + d_i)v_i\right) = \sum_{i=1}^{n}(c_i + d_i)w_i = \sum_{i=1}^{n} c_i w_i + \sum_{i=1}^{n} d_i w_i = T(v) + T(u).$$

To show uniqueness, observe that if $S \in \mathcal{L}(V, W)$ and $S(v_i) = w_i$ for each i, then by linearity,

$$S\left(\sum_{i=1}^{n} c_i v_i\right) = \sum_{i=1}^{n} c_i S(v_i) = \sum_{i=1}^{n} c_i w_i,$$

so S is the same as the map T defined above. ▲

The construction of T in the proof above is very common, and is referred to as "extending by linearity."

Quick Exercise #9. Let $\mathbf{w}_1, \ldots, \mathbf{w}_n \in \mathbb{F}^m$ be given. Since $(\mathbf{e}_1, \ldots, \mathbf{e}_n)$ is a basis of $\mathbb{F}^n$, Theorem 3.14 says there is a unique linear map $T : \mathbb{F}^n \to \mathbb{F}^m$ such that $T\mathbf{e}_j = \mathbf{w}_j$ for each j. What is the matrix of T?

From a slightly different perspective, Theorem 3.14 points to a potential use of the Rat Poison Principle: since there is one and only one linear map which sends a basis to a specified list of vectors, to show that two maps are the same it suffices to check that they do the same thing to a basis.

Theorem 3.15 *Let $T \in \mathcal{L}(V, W)$ and let $(v_1, \ldots, v_n)$ be a basis of V. Then T is an isomorphism iff $(T(v_1), \ldots, T(v_n))$ is a basis of W.*

Proof Suppose first that T is an isomorphism. We need to show that $(T(v_1), \ldots, T(v_n))$ is linearly independent and spans W.

Suppose that

$$\sum_{i=1}^{n} c_i T(v_i) = 0.$$

By linearity, this means that

$$T\left(\sum_{i=1}^{n} c_i v_i\right) = 0.$$

QA #9: $\begin{bmatrix} | & & | \\ \mathbf{w}_1 & \cdots & \mathbf{w}_n \\ | & & | \end{bmatrix}$. This is almost a trick question – we've already known this since Section 2.2!

Since T is injective, by Theorem 2.37,

$$\sum_{i=1}^{n} c_i v_i = 0,$$

and since $(v_1, \dots, v_n)$ is linearly independent, this means that $c_1 = \cdots = c_n = 0$. Thus $(T(v_1), \dots, T(v_n))$ is linearly independent.

Now suppose that $w \in W$. Since T is surjective, there is a $v \in V$ such that $T(v) = w$. Since $V = \langle v_1, \dots, v_n \rangle$, we can write

$$v = \sum_{i=1}^{n} d_i v_i$$

for some $d_1, \dots, d_n \in \mathbb{F}$. By linearity,

$$w = T(v) = \sum_{i=1}^{n} d_i T(v_i).$$

Therefore $W \subseteq \langle T(v_1), \dots, T(v_n) \rangle$, and so $(T(v_1), \dots, T(v_n))$ is a basis for W.

Now suppose that $(T(v_1), \dots, T(v_n))$ is a basis for W. We need to show that T is injective and surjective.

Suppose that $T(v) = 0$. Since $V = \langle v_1, \dots, v_n \rangle$, we can write

$$v = \sum_{i=1}^{n} d_i v_i$$

for some $d_1, \dots, d_n \in \mathbb{F}$. By linearity,

$$0 = T(v) = \sum_{i=1}^{n} d_i T(v_i).$$

Since $(T(v_1), \dots, T(v_n))$ is linearly independent, this implies that $d_1 = \cdots = d_n = 0$, and so $v = 0$. By Theorem 2.37, this means that T is injective.

Now suppose that $w \in W$. Since $W = \langle T(v_1), \dots, T(v_n) \rangle$, we can write

$$w = \sum_{i=1}^{n} c_i T(v_i)$$

for some $c_1, \dots, c_n \in \mathbb{F}$. By linearity,

$$w = T\left(\sum_{i=1}^{n} c_i v_i\right) \in \operatorname{range} T,$$

and thus T is surjective. ▲

Corollary 3.16 *Let* $A \in M_{m,n}(\mathbb{F})$. *The following statements are equivalent.*

- A *is invertible.*
- *The columns of* A *are a basis of* $\mathbb{F}^m$.
- *The RREF of* A *is* I_m.

Proof Let $T \in \mathcal{L}(\mathbb{F}^n, \mathbb{F}^m)$ be the linear map represented by A; then A is invertible if and only if T is an isomorphism. By Theorem 3.15, T is an isomorphism if and only if $(Te_1, \ldots, Te_n)$ are a basis of $\mathbb{F}^m$. But $(Te_1, \ldots, Te_n)$ are exactly the columns of A.

By Proposition 3.9, the columns of A are a basis of $\mathbb{F}^m$ if and only if the RREF of A is I_m. ▲

Quick Exercise #10. Use Corollary 3.16 to show that a matrix $A \in M_{m,n}(\mathbb{F})$ can only be invertible if $m = n$.

Example The example on page 99 showed that the two matrices

$$A = \begin{bmatrix} 1 & 1 & 1 & \cdots & \cdots & 1 \\ 0 & 1 & 1 & \cdots & \cdots & 1 \\ 0 & 0 & 1 & \ddots & & 1 \\ \vdots & & \ddots & \ddots & \ddots & \vdots \\ 0 & & & \ddots & 1 & 1 \\ 0 & \cdots & \cdots & \cdots & 0 & 1 \end{bmatrix} \quad \text{and } B = \begin{bmatrix} 1 & -1 & 0 & \cdots & \cdots & 0 \\ 0 & 1 & -1 & 0 & & \vdots \\ 0 & 0 & 1 & -1 & \ddots & \vdots \\ \vdots & & \ddots & \ddots & \ddots & 0 \\ 0 & & & \ddots & 1 & -1 \\ 0 & \cdots & \cdots & \cdots & 0 & 1 \end{bmatrix}.$$

are invertible. Corollary 3.16 thus gives another way to see that the columns of A and the columns of B are two different bases of $\mathbb{F}^n$ (this was shown first in Example 4 on page 151). ▲

KEY IDEAS

- A basis of a finite-dimensional vector space is a linearly independent list of vectors which spans the space.
- Finite-dimensional vector spaces always have bases: you can start with any spanning list, and eliminate unnecessary elements to get a basis.
- A list $(v_1, \ldots, v_m)$ is a basis of $\mathbb{F}^n$ if and only if the RREF of the matrix with those columns is I_n (so you need $m = n$).

QA #10: If the RREF of A is I_m, then A must be square.

- You can define a linear map by defining it on a basis and extending by linearity.
- If two linear maps do the same thing to a basis, they are the same.
- A map is invertible if and only if it takes a basis to a basis.
- A matrix is invertible if and only if its columns are a basis.

EXERCISES

3.2.1 Determine whether each of the following lists of vectors is a basis for $\mathbb{R}^n$.

(a) $\left(\begin{bmatrix}1\\-3\end{bmatrix},\begin{bmatrix}2\\6\end{bmatrix}\right)$

(b) $\left(\begin{bmatrix}1\\-3\end{bmatrix},\begin{bmatrix}2\\6\end{bmatrix},\begin{bmatrix}4\\-1\end{bmatrix}\right)$

(c) $\left(\begin{bmatrix}1\\0\\2\end{bmatrix},\begin{bmatrix}1\\2\\3\end{bmatrix},\begin{bmatrix}2\\1\\-1\end{bmatrix}\right)$

(d) $\left(\begin{bmatrix}1\\1\\1\\1\end{bmatrix},\begin{bmatrix}1\\2\\2\\2\end{bmatrix},\begin{bmatrix}1\\2\\3\\3\end{bmatrix},\begin{bmatrix}1\\2\\3\\4\end{bmatrix}\right)$

3.2.2 Determine whether each of the following lists of vectors is a basis for $\mathbb{R}^n$.

(a) $\left(\begin{bmatrix}1\\-3\end{bmatrix},\begin{bmatrix}-2\\6\end{bmatrix}\right)$

(b) $\left(\begin{bmatrix}1\\2\\3\end{bmatrix},\begin{bmatrix}2\\3\\1\end{bmatrix}\right)$

(c) $\left(\begin{bmatrix}1\\2\\3\end{bmatrix},\begin{bmatrix}2\\3\\1\end{bmatrix},\begin{bmatrix}3\\1\\2\end{bmatrix}\right)$

(d) $\left(\begin{bmatrix}1\\2\\3\end{bmatrix},\begin{bmatrix}2\\3\\1\end{bmatrix},\begin{bmatrix}1\\1\\-2\end{bmatrix}\right)$

3.2.3 Find a basis for each of the following spaces.

(a) $\left\langle \begin{bmatrix} 1 \\ 0 \\ 2 \end{bmatrix}, \begin{bmatrix} 0 \\ -3 \\ 1 \end{bmatrix}, \begin{bmatrix} 2 \\ 3 \\ 3 \end{bmatrix}, \begin{bmatrix} 1 \\ -3 \\ 3 \end{bmatrix} \right\rangle$

(b) $\left\langle \begin{bmatrix} 4 \\ -2 \\ 2 \end{bmatrix}, \begin{bmatrix} -2 \\ 1 \\ -1 \end{bmatrix}, \begin{bmatrix} 0 \\ 3 \\ -2 \end{bmatrix}, \begin{bmatrix} 4 \\ 1 \\ 0 \end{bmatrix} \right\rangle$

(c) $\left\langle \begin{bmatrix} 2 \\ 0 \\ -1 \\ 3 \end{bmatrix}, \begin{bmatrix} 1 \\ 1 \\ 0 \\ 1 \end{bmatrix}, \begin{bmatrix} 0 \\ -2 \\ -1 \\ 1 \end{bmatrix}, \begin{bmatrix} 2 \\ 1 \\ -2 \\ 0 \end{bmatrix}, \begin{bmatrix} -1 \\ 0 \\ 2 \\ 1 \end{bmatrix} \right\rangle$

(d) $\left\langle \begin{bmatrix} 1 \\ -2 \\ 3 \\ -4 \end{bmatrix}, \begin{bmatrix} 4 \\ -3 \\ 2 \\ -1 \end{bmatrix}, \begin{bmatrix} 1 \\ 0 \\ 1 \\ 0 \end{bmatrix}, \begin{bmatrix} 0 \\ 1 \\ 0 \\ 1 \end{bmatrix}, \begin{bmatrix} 1 \\ 1 \\ 1 \\ 1 \end{bmatrix} \right\rangle$

3.2.4 Find a basis for each of the following spaces.

(a) $\left\langle \begin{bmatrix} 1 \\ 2 \\ 3 \end{bmatrix}, \begin{bmatrix} 2 \\ 3 \\ 1 \end{bmatrix}, \begin{bmatrix} 1 \\ 1 \\ -2 \end{bmatrix} \right\rangle$

(b) $\left\langle \begin{bmatrix} 1 \\ -2 \\ 1 \\ 0 \end{bmatrix}, \begin{bmatrix} 2 \\ 0 \\ 3 \\ -1 \end{bmatrix}, \begin{bmatrix} 1 \\ 2 \\ 2 \\ -1 \end{bmatrix}, \begin{bmatrix} 0 \\ 3 \\ -1 \\ 2 \end{bmatrix}, \begin{bmatrix} 1 \\ 1 \\ -3 \\ 2 \end{bmatrix} \right\rangle$

(c) $\left\langle \begin{bmatrix} 2 \\ -1 \\ 0 \\ -3 \end{bmatrix}, \begin{bmatrix} 1 \\ 0 \\ 2 \\ -1 \end{bmatrix}, \begin{bmatrix} 1 \\ -1 \\ -2 \\ -2 \end{bmatrix} \right\rangle$

(d) $\left\langle \begin{bmatrix} 1 \\ 1 \\ 0 \\ 0 \end{bmatrix}, \begin{bmatrix} 1 \\ 0 \\ 1 \\ 0 \end{bmatrix}, \begin{bmatrix} 0 \\ 1 \\ 0 \\ 1 \end{bmatrix}, \begin{bmatrix} 0 \\ 0 \\ 1 \\ 1 \end{bmatrix} \right\rangle$

3.2.5 Find a basis for the kernel of each of the following matrices.

(a) $\begin{bmatrix} 1 & 2 & 3 & 4 \\ 5 & 6 & 7 & 8 \end{bmatrix}$

(b) $\begin{bmatrix} 1 & 0 & 2 & -3 \\ 2 & 1 & -1 & -3 \\ 1 & 1 & -3 & 0 \end{bmatrix}$

(c) $\begin{bmatrix} 1 & -2 & 0 & 0 & 1 \\ 0 & 0 & 1 & 0 & 1 \\ 1 & -2 & 0 & 1 & 3 \\ -1 & 2 & 1 & 0 & 0 \end{bmatrix}$

(d) $\begin{bmatrix} 2 & 0 & 1 & -1 & 0 & 3 \\ 1 & 1 & -2 & 0 & 3 & 0 \\ -1 & 1 & 0 & 0 & 2 & -1 \end{bmatrix}$

3.2.6 Find a basis for the kernel of each of the following matrices.

(a) $\begin{bmatrix} 3 & -1 & 4 & -1 \\ -5 & 9 & -2 & 6 \end{bmatrix}$ (b) $\begin{bmatrix} 2 & 0 & 1 & -1 \\ 1 & -3 & 0 & 4 \\ 3 & -3 & 1 & 3 \end{bmatrix}$

(c) $\begin{bmatrix} 2 & 1 & 0 & -4 & 1 \\ 0 & 3 & 1 & 0 & -2 \\ 1 & -2 & 1 & 2 & 0 \end{bmatrix}$ (d) $\begin{bmatrix} 1 & 0 & 2 & 4 & -1 & 3 \\ 0 & 1 & 4 & 2 & 0 & -2 \\ -1 & 2 & 0 & -2 & 0 & 1 \end{bmatrix}$

3.2.7 Write $\begin{bmatrix} 1 \\ 2 \end{bmatrix}$ as a linear combination of each of the following lists of vectors in $\mathbb{R}^2$, or else show that it is not possible to do so.

(a) $\left(\begin{bmatrix} 1 \\ 1 \end{bmatrix}, \begin{bmatrix} 1 \\ -1 \end{bmatrix}\right)$ (b) $\left(\begin{bmatrix} 2 \\ 3 \end{bmatrix}, \begin{bmatrix} 4 \\ 5 \end{bmatrix}\right)$ (c) $\left(\begin{bmatrix} -1 \\ 2 \end{bmatrix}, \begin{bmatrix} 2 \\ 1 \end{bmatrix}\right)$

(d) $\left(\begin{bmatrix} 2 \\ -1 \end{bmatrix}, \begin{bmatrix} 4 \\ 2 \end{bmatrix}\right)$

3.2.8 Write $\begin{bmatrix} 1 \\ 2 \\ 3 \end{bmatrix}$ as a linear combination of each of the following lists of vectors in $\mathbb{R}^3$, or else show that it is not possible to do so.

(a) $\left(\begin{bmatrix} 1 \\ 1 \\ 0 \end{bmatrix}, \begin{bmatrix} 1 \\ 0 \\ 1 \end{bmatrix}, \begin{bmatrix} 0 \\ 1 \\ 1 \end{bmatrix}\right)$ (b) $\left(\begin{bmatrix} 1 \\ 0 \\ 0 \end{bmatrix}, \begin{bmatrix} 2 \\ -1 \\ 0 \end{bmatrix}, \begin{bmatrix} 1 \\ -2 \\ 1 \end{bmatrix}\right)$

(c) $\left(\begin{bmatrix} 3 \\ -2 \\ 1 \end{bmatrix}, \begin{bmatrix} 1 \\ 0 \\ -1 \end{bmatrix}, \begin{bmatrix} 0 \\ -2 \\ 4 \end{bmatrix}\right)$

3.2.9 Show that $(1, i)$ is a basis of $\mathbb{C}$, thought of as a vector space over $\mathbb{R}$.

3.2.10 Find the unique matrix $\mathrm{A} \in \mathrm{M}_2(\mathbb{R})$ such that $\mathrm{A}\begin{bmatrix} 1 \\ 2 \end{bmatrix} = \begin{bmatrix} 0 \\ -1 \end{bmatrix}$ and $\mathrm{A}\begin{bmatrix} 2 \\ 1 \end{bmatrix} = \begin{bmatrix} -3 \\ 2 \end{bmatrix}$.

3.2.11 Suppose that $T \in \mathcal{L}(\mathbb{R}^3, \mathbb{R}^2)$ satisfies

$$T\begin{bmatrix} 1 \\ 1 \\ 0 \end{bmatrix} = \begin{bmatrix} -1 \\ 2 \end{bmatrix}, \quad T\begin{bmatrix} 1 \\ 0 \\ 1 \end{bmatrix} = \begin{bmatrix} 2 \\ 3 \end{bmatrix}, \quad \text{and} \quad T\begin{bmatrix} 0 \\ 1 \\ 1 \end{bmatrix} = \begin{bmatrix} -3 \\ 1 \end{bmatrix}.$$

Find $T\begin{bmatrix} 1 \\ 2 \\ 3 \end{bmatrix}$.

3.2.12 Show that if A is any matrix, then ker A is finite-dimensional.

3.2.13 Fix $k \in \mathbb{N}$, and let V be the subspace of $\mathbb{F}^\infty$ consisting of all sequences $(a_i)_{i \geq 1}$ such that $a_{i+k} = a_i$ for every i. Prove that V is finite-dimensional.

3.2.14 Let D be the derivative operator on $C^\infty(\mathbb{R})$. Show that the eigenspaces of D are all finite-dimensional.

3.2.15 Suppose that $(v_1, \ldots, v_n)$ is a basis of V. Prove that

$$(v_1 + v_2, v_2 + v_3, \ldots, v_{n-1} + v_n, v_n)$$

is also a basis of V.

3.2.16 Find a sublist of $(x + 1, x^2 - 1, x^2 + 2x + 1, x^2 - x)$ which is a basis for $\mathcal{P}_2(\mathbb{R})$.

3.2.17 Show that if $\mathcal{B}$ is a linearly independent list of vectors in V, then $\mathcal{B}$ is a basis for the subspace $\langle \mathcal{B} \rangle$ spanned by $\mathcal{B}$.

3.2.18 Suppose that $\mathrm{A} \in \mathrm{M}_n(\mathbb{F})$ is upper triangular (see Exercise 2.3.12) and that all the diagonal entries are nonzero (i.e., $a_{ii} \neq 0$ for each $i = 1, \ldots, n$). Show that the columns of A form a basis of $\mathbb{F}^n$.

3.2.19 Let $(\mathbf{v}_1, \ldots, \mathbf{v}_n)$ be a basis of $\mathbb{F}^n$, and let $\mathbf{b} \in \mathbb{F}^n$. Show that

$$\mathbf{b} = x_1\mathbf{v}_1 + \cdots + x_n\mathbf{v}_n,$$

where $\mathbf{x} = \mathrm{V}^{-1}\mathbf{b}$, and $\mathrm{V} \in \mathrm{M}_n(\mathbb{F})$ has columns $\mathbf{v}_1, \ldots, \mathbf{v}_n$.

3.2.20 Suppose that $(v_1, \ldots, v_n)$ is a basis of V, and let $w_i = \sum_{j=1}^n a_{ij}v_j$ for $i = 1, \ldots, n$ and scalars $a_{ij} \in \mathbb{F}$. Show that if the matrix $\mathrm{A} \in \mathrm{M}_n(\mathbb{F})$ with entries a_{ij} is invertible, then $(w_1, \ldots, w_n)$ is a basis of V.

3.2.21 Let $(v_1, \ldots, v_n)$ be a basis of V, and let $a_1, \ldots, a_n \in \mathbb{F}$. Prove that there is a unique linear map $T : V \to \mathbb{F}$ such that $Tv_j = a_j$ for each j.

3.2.22 Suppose that $(v_1, \ldots, v_n)$ and $(w_1, \ldots, w_n)$ are both bases of V. Show that there is an isomorphism $T \in \mathcal{L}(V)$ such that $Tv_i = w_i$ for each $i = 1, \ldots, n$.

3.2.23 Show that Theorem 3.14 fails if the vectors $(v_1, \ldots, v_n)$ are not linearly independent. That is, give an example of a list $(v_1, \ldots, v_n)$ which spans V and a list $(w_1, \ldots, w_n)$ in W, such that there is *no* linear map $T \in \mathcal{L}(V, W)$ with $Tv_j = w_j$ for each j.

3.2.24 Suppose that $(v_1, \ldots, v_n)$ is a basis for V, $T : V \to W$ is linear, and $(Tv_1, \ldots, Tv_n)$ is a basis for W. Prove that if $(u_1, \ldots, u_n)$ is another basis for V, then $(Tu_1, \ldots, Tu_n)$ is another basis for W.

3.2.25 Use Proposition 3.9 to give an alternative proof of Theorem 3.10 when $V = \mathbb{F}^n$.

3.3 Dimension

In this section we will begin making finer distinctions among finite-dimensional spaces.

The Dimension of a Vector Space

Lemma 3.17 *Suppose that $V \subseteq \langle v_1, \ldots, v_m \rangle$, and that $(w_1, \ldots, w_n)$ is a linearly independent list in V. Then $m \geq n$.*

Proof For each $j = 1, \ldots, n$, $w_j \in \langle v_1, \ldots, v_m \rangle$, and so there are scalars $a_{1j}, \ldots, a_{mj} \in \mathbb{F}$ such that

$$w_j = \sum_{i=1}^{m} a_{ij} v_i.$$

Now consider the homogeneous $m \times n$ linear system $\mathrm{A}\mathbf{x} = \mathbf{0}$, where $\mathrm{A} = [a_{ij}] \in \mathrm{M}_{m,n}(\mathbb{F})$. If $\mathbf{x}$ is a solution of this system, then

$$\sum_{j=1}^{n} a_{ij} x_j = 0$$

for each $i = 1, \ldots, m$, and so

$$0 = \sum_{i=1}^{m} \left(\sum_{j=1}^{n} a_{ij} x_j \right) v_i = \sum_{j=1}^{n} x_j \left(\sum_{i=1}^{m} a_{ij} v_i \right) = \sum_{j=1}^{n} x_j w_j.$$

Since $(w_1, \ldots, w_n)$ is linearly independent, this implies that $x_j = 0$ for each j. Therefore, $\mathbf{x} = \mathbf{0}$ is the only solution of the $m \times n$ system $\mathrm{A}\mathbf{x} = \mathbf{0}$, which by Corollary 1.4 means that $m \geq n$. ▲

Quick Exercise #11. Show that any linearly independent list of polynomials in $\mathcal{P}_n(\mathbb{R})$ consists of at most $n + 1$ polynomials.

Intuitively, Lemma 3.17 says that a list of vectors in V with no redundancy is necessarily shorter than (or the same length as) any list that describes all of V. One easy consequence is the following way of recognizing infinite-dimensional spaces.

Theorem 3.18 *Suppose that, for every $n \geq 1$, V contains a linearly independent list of n vectors. Then V is infinite-dimensional.*

Proof We will prove the contrapositive: if V is finite-dimensional, then there is some n such that V does *not* contain any linearly independent list of n vectors. Indeed, suppose that $V \subseteq \langle v_1, \ldots, v_m \rangle$. By Lemma 3.17, for any $n > m$, V cannot contain a linearly independent list of n vectors. ▲

QA #11: $\mathcal{P}_n(\mathbb{R})$ is spanned by the $n + 1$ polynomials $(1, x, \ldots, x^n)$.

In fact, the converse of Theorem 3.18 also holds (see Exercise 3.3.21).

Example For each n, the space $\mathcal{P}(\mathbb{R})$ of *all* polynomials over $\mathbb{R}$ contains the linearly independent list $(1, x, \ldots, x^n)$. Therefore $\mathcal{P}(\mathbb{R})$ is infinite-dimensional. ▲

The most important consequence of Lemma 3.17 is the following fact.

Theorem 3.19 *Suppose that $(v_1, \ldots, v_m)$ and $(w_1, \ldots, w_n)$ are both bases of V. Then $m = n$.*

Proof Since $V \subseteq \langle v_1, \ldots, v_m \rangle$ and $(w_1, \ldots, w_n)$ is a linearly independent list in V, Lemma 3.17 implies that $m \geq n$. On the other hand, since $V \subseteq \langle w_1, \ldots, w_n \rangle$ and $(v_1, \ldots, v_n)$ is a linearly independent list in V, Lemma 3.17 implies that $n \geq m$. ▲

Theorem 3.19 tells us that, even though a given (finite-dimensional) vector space has many different bases, they all have something important in common – their length. So given a finite-dimensional vector space V, "the length of a basis of V" is a single number, which only depends on V itself, and not on which basis you consider. This means that the following definition makes sense.

Definition Let V be a finite-dimensional vector space.

- If $V \neq \{0\}$, then the **dimension** of V, written $\dim V$, is the length of any basis of V. If $\dim V = n$, we say that V is **n-dimensional**.
- If $V = \{0\}$, then we define the dimension of V to be 0.

If it were possible for V to have two (or more) bases with different lengths, then our definition of dimension would have serious problems: what we thought $\dim V$ was would depend on which basis we were using. We would say in that case that the dimension of V was not **well-defined.** In general when making mathematical definitions, one has to be careful about well-definedness whenever the definition appears to depend on a choice. Here, Theorem 3.19 tells us that even though the definition of dimension *appears* to depend on a choice of basis, it really doesn't – we'd get the same number by choosing a different basis.

Examples

1. We have seen that $(\mathbf{e}_1, \ldots, \mathbf{e}_n)$ forms a basis for $\mathbb{F}^n$, so $\dim \mathbb{F}^n = n$.
2. We observed in the example on page 146 that $(1, x, \ldots, x^n)$ is a basis of $\mathcal{P}_n(\mathbb{R})$ the space of polynomials over $\mathbb{R}$ with degree n or less, so $\dim \mathcal{P}_n(\mathbb{R}) = n + 1$. ▲

We next record some simple consequences of Lemma 3.17 for the dimension of a vector space.

Proposition 3.20 *If* $V = \langle v_1, \ldots, v_n \rangle$, *then* $\dim V \leq n$.

Quick Exercise #12. Prove Proposition 3.20.

Proposition 3.21 *If* V *is finite-dimensional and* $(v_1, \ldots, v_n)$ *is a linearly independent list in* V, *then* $\dim V \geq n$.

Proof By Corollary 3.12, V has a basis $\mathcal{B}$, and by Lemma 3.17, $\mathcal{B}$ is at least as long as the linearly independent list $(v_1, \ldots, v_n)$. ▲

The following lemma gives a good example of the way we can sometimes learn something non-trivial by counting dimensions of spaces.

Lemma 3.22 *Suppose that* U_1 *and* U_2 *are subspaces of a finite-dimensional vector space* V, *and that* $\dim U_1 + \dim U_2 > \dim V$. *Then* $U_1 \cap U_2 \neq \{0\}$.

Proof We will prove the contrapositive: if $U_1 \cap U_2 = \{0\}$, then $\dim U_1 + \dim U_2 \leq \dim V$. Let $(u_1, \ldots, u_p)$ be a basis of U_1 and $(v_1, \ldots, v_q)$ be a basis of U_2. (Theorem 3.29 below will show that U_1 and U_2 are finite-dimensional and therefore have bases.) Suppose that

$$\sum_{i=1}^{p} a_i u_i + \sum_{j=1}^{q} b_j v_j = 0$$

for scalars $a_1, \ldots, a_p, b_1, \ldots, b_q$. Then

$$\sum_{i=1}^{p} a_i u_i = -\sum_{j=1}^{q} b_j v_j,$$

and this vector must be in $U_1 \cap U_2$, so by assumption it is 0. Since $(u_1, \ldots, u_p)$ is linearly independent, $a_i = 0$ for each i, and since $(v_1, \ldots, v_q)$ is linearly independent, $b_j = 0$ for each j. Therefore $(u_1, \ldots, u_p, v_1, \ldots, v_q)$ is linearly independent, which implies that $\dim V \geq p + q$. ▲

The following result shows that, up to isomorphism, there is only one n-dimensional vector space over $\mathbb{F}$.

Theorem 3.23 *Let* V *and* W *be finite-dimensional vector spaces. Then* $\dim V = \dim W$ *if and only if* V *is isomorphic to* W.

QA #12: $(v_1, \ldots, v_n)$ has a sublist which is a basis of V.

Proof Suppose first that $\dim V = \dim W = n$. Pick bases $(v_1, \ldots, v_n)$ of V and $(w_1, \ldots, w_n)$ of W. By Theorem 3.14, there is a linear map $T \in \mathcal{L}(V, W)$ such that $T(v_i) = w_i$ for each i. So by Theorem 3.15, T is an isomorphism.

Now suppose that V and W are isomorphic. Then there is an isomorphism $T \in \mathcal{L}(V, W)$. If $(v_1, \ldots, v_n)$ is any basis of V, then by Theorem 3.15, $(T(v_1), \ldots, T(v_n))$ is a basis of W, and so $\dim W = n = \dim V$. ▲

Theorem 3.23 gives us the abstract statement that every n-dimensional vector space over $\mathbb{F}$ is isomorphic to every other one, but it is helpful to have a concrete example of such a space.

Corollary 3.24 *If* $\dim V = n$*, then* V *is isomorphic to* $\mathbb{F}^n$.

Quick Exercise #13. Suppose that $\mathbf{v}_1, \mathbf{v}_2 \in \mathbb{R}^3$ are noncollinear. Show that the plane $U = \langle \mathbf{v}_1, \mathbf{v}_2 \rangle$ is isomorphic to $\mathbb{R}^2$.

Algorithm 3.25 Let $(\mathbf{v}_1, \ldots, \mathbf{v}_n)$ be a list of vectors in $\mathbb{F}^m$. To find the dimension of $\langle \mathbf{v}_1, \ldots, \mathbf{v}_n \rangle$:

- Put $\mathrm{A} = \begin{bmatrix} | & & | \\ \mathbf{v}_1 & \cdots & \mathbf{v}_n \\ | & & | \end{bmatrix}$ in RREF.
- $\dim \langle \mathbf{v}_1, \ldots, \mathbf{v}_n \rangle$ is the number of pivots.

Proof Algorithm 3.13 shows the pivot columns of the RREF of A form a basis for $\langle \mathbf{v}_1, \ldots, \mathbf{v}_n \rangle$, so the dimension is the number of pivot columns. ▲

Quick Exercise #14. Find the dimension of the subspace

$$\left\langle \begin{bmatrix} 1 \\ 0 \\ 1 \\ -1 \end{bmatrix}, \begin{bmatrix} -2 \\ 0 \\ -2 \\ 2 \end{bmatrix}, \begin{bmatrix} 0 \\ 1 \\ 0 \\ 1 \end{bmatrix}, \begin{bmatrix} 0 \\ 0 \\ 1 \\ 0 \end{bmatrix}, \begin{bmatrix} 1 \\ 1 \\ 3 \\ 0 \end{bmatrix} \right\rangle \subseteq \mathbb{R}^4$$

(from Quick Exercise #8 on page 155).

QA #13: Since $\mathbf{v}_1, \mathbf{v}_2$ are noncollinear, $(\mathbf{v}_1, \mathbf{v}_2)$ is linearly independent and hence a basis of U. Therefore $\dim U = 2$, so U is isomorphic to $\mathbb{R}^2$.

QA #14: There are three pivot columns of the matrix with those columns (see page 155), so the subspace is three-dimensional.

Dimension, Bases, and Subspaces

The next result gives one way that understanding dimension can save you some work. It says that to check whether a list in V is a basis, as long as it has the right length, you only need to check that the list spans V, and you get linear independence for free.

Theorem 3.26 *Suppose that* $\dim V = n \neq 0$ *and that* $V = \langle \mathcal{B} \rangle$, *where* $\mathcal{B} = (v_1, \ldots, v_n)$. *Then* $\mathcal{B}$ *is a basis for* V.

Proof By Theorem 3.11, some sublist $\mathcal{B}'$ of $\mathcal{B}$ is a basis for V. Since $\dim V = n$, $\mathcal{B}'$ must have length n, so in fact $\mathcal{B}' = \mathcal{B}$. Thus $\mathcal{B}$ is a basis for V. ▲

Example Consider the two lists of vectors in $\mathbb{F}^n$, $(\mathbf{v}_1, \ldots, \mathbf{v}_n)$ in $\mathbb{F}^n$ given by

$$\mathbf{v}_j = \mathbf{e}_1 + \cdots + \mathbf{e}_j$$

for $1 \leq j \leq n$, and $(\mathbf{w}_1, \ldots, \mathbf{w}_n)$ given by $\mathbf{w}_1 = \mathbf{e}_1$ and

$$\mathbf{w}_j = \mathbf{e}_j - \mathbf{e}_{j-1}$$

for $2 \leq j \leq n$. We saw in the examples on page 118 that both these lists span $\mathbb{F}^n$. Since each of these lists consists of n vectors and $\dim \mathbb{F}^n = n$, this is enough to show that each of them is a basis of $\mathbb{F}^n$, without further consideration of linear independence. ▲

Theorem 3.26 says that if a list is the right length to be a basis, you only need to check that it spans the space. It is also enough to only check linear independence; to prove this, we first need the following counterpart to Theorem 3.11.

Theorem 3.27 *Suppose that* V *is a finite-dimensional vector space and* $\mathcal{B}$ *is a linearly independent list of vectors in* V. *Then* $\mathcal{B}$ *can be extended to a basis of* V. *That is, there is a basis for* V *which contains* $\mathcal{B}$.

Proof Suppose that $\mathcal{B} = (v_1, \ldots, v_m)$. If $V \subseteq \langle v_1, \ldots, v_m \rangle$, then $\mathcal{B}$ is a basis and we are done. Otherwise, there is some vector $v_{m+1} \in V \setminus \langle v_1, \ldots, v_m \rangle$. By the Linear Dependence Lemma (Theorem 3.6), the list $(v_1, \ldots, v_{m+1})$ is linearly independent.

Now if $V \subseteq \langle v_1, \ldots, v_{m+1} \rangle$, then we are done. Otherwise, there is some vector $v_{m+2} \in V \setminus \langle v_1, \ldots, v_{m+1} \rangle$. By the Linear Dependence Lemma, the list $(v_1, \ldots, v_{m+2})$ is linearly independent.

We continue in this way, adding vectors to the list one at a time. At some step, it must be the case that $V \subseteq \langle v_1, \ldots, v_{m+k} \rangle$, since otherwise Theorem 3.18 would imply that V is infinite-dimensional. ▲

Theorem 3.28 *Suppose that* $\dim V = n$ *and* $\mathcal{B} = (v_1, \ldots, v_n)$ *is a linearly independent list in* V. *Then* $\mathcal{B}$ *is a basis for* V.

Proof See Exercise 3.3.23. ▲

Example Consider (again) the two lists of vectors $(\mathbf{v}_1, \ldots, \mathbf{v}_n)$ and $(\mathbf{w}_1, \ldots, \mathbf{w}_n)$ from the last example. We saw in Example 2 on page 144 that both these lists are linearly independent. Since each of these lists consists of n vectors and $\dim \mathbb{F}^n = n$, this is enough to show that each of them is a basis of $\mathbb{F}^n$, without considering whether they span $\mathbb{F}^n$. ▲

Quick Exercise #15. Suppose that $(\mathbf{v}_1, \mathbf{v}_2)$ are noncollinear vectors in $\mathbb{R}^2$ (see Exercise 3.1.7). Prove that $(\mathbf{v}_1, \mathbf{v}_2)$ spans $\mathbb{R}^2$.

The following facts about finite-dimensional vector spaces are fundamental and will be used constantly throughout the rest of this book, often without comment.

Theorem 3.29 *Suppose that* V *is a finite-dimensional vector space and* U *is a subspace of* V. *Then*

1. U *is finite-dimensional,*
2. $\dim U \leq \dim V$,
3. *if* $\dim U = \dim V$, *then* $U = V$.

Proof 1. The proof is essentially the same as the proof of Theorem 3.27.

If $U = \{0\}$, then there is nothing to prove. Otherwise, there is some vector $u_1 \in U \setminus \{0\}$.

If $U = \langle u_1 \rangle$, then we are done. Otherwise, there is some vector $u_2 \in U \setminus \langle u_1 \rangle$. By the Linear Dependence Lemma, the list (u_1, u_2) is linearly independent.

We continue in this way, adding vectors in U to the list one at a time. Since each successive list $(u_1, \ldots, u_k)$ is linearly independent in the finite-dimensional vector space V, we cannot add more vectors indefinitely. If we are forced to stop after adding u_k, it must be that $U \subseteq \langle u_1, \ldots, u_k \rangle$.

QA #15: Since $(\mathbf{v}_1, \mathbf{v}_2)$ are two noncollinear vectors, they are linearly independent by Exercise 3.1.7, and since $\dim \mathbb{R}^2 = 2$, Theorem 3.28 implies that $(\mathbf{v}_1, \mathbf{v}_2)$ is a basis of $\mathbb{R}^2$.

2. The space U is finite-dimensional, so it has a basis $\mathcal{B}$. Since $\mathcal{B}$ is a linearly independent list in V, Theorem 3.27 implies that it extends to a basis $\mathcal{B}'$ of V. There are at least as many vectors in $\mathcal{B}'$ as in $\mathcal{B}$, so $\dim V \geq \dim U$.
3. Let $(u_1, \ldots, u_k)$ be a basis for U. If $U \neq V$, then there is a vector $v \in V$ such that $v \notin U = \langle u_1, \ldots, u_k \rangle$. Then $(u_1, \ldots, u_k, v)$ is linearly independent by the Linear Dependence Lemma, and so $\dim V \geq k + 1 > \dim U$. ▲

Example Within $\mathbb{R}^3$, we are familiar with subspaces in the form of lines through the origin, planes containing the origin, the zero subspace, and all of $\mathbb{R}^3$. Using Theorem 3.29, we can tell that we haven't missed anything more exotic: any subspace of $\mathbb{R}^3$ must have dimension 0, 1, 2, or 3. If it has dimension 1, it is spanned by one nonzero vector; i.e., it is a line through the origin. If it has dimension 2, it is spanned by two linearly independent vectors; i.e., it is a plane (containing the origin). The only zero-dimensional subspace is $\{0\}$, and the only three-dimensional subspace is all of $\mathbb{R}^3$. ▲

Remark Thanks to Corollary 3.12 every finite-dimensional vector space has a dimension (and its dimension is indeed finite). For convenience, we sometimes write $\dim V = \infty$ when V is infinite-dimensional and $\dim V < \infty$ when V is finite-dimensional. It's important to note that we have not shown that an infinite-dimensional vector space possesses a basis with infinite length. Such things come up in more advanced courses, and depend on subtle issues of set theory or analysis. ▲

KEY IDEAS

- All bases of a finite-dimensional vector space have the same length; we call this the dimension of the space.
- Two finite-dimensional vector spaces are isomorphic if and only if they have the same dimension.
- Linearly independent lists can be extended to bases.

EXERCISES

3.3.1 Determine the dimension of each of the following vector spaces.

(a) $\left\langle \begin{bmatrix} 2 \\ -1 \\ 3 \end{bmatrix}, \begin{bmatrix} 1 \\ 0 \\ 1 \end{bmatrix}, \begin{bmatrix} 0 \\ 1 \\ -1 \end{bmatrix} \right\rangle$

(b) $\left\langle \begin{bmatrix} 1 \\ 2 \\ 3 \\ 4 \end{bmatrix}, \begin{bmatrix} 1 \\ 2 \\ 3 \\ 3 \end{bmatrix}, \begin{bmatrix} 1 \\ 2 \\ 2 \\ 2 \end{bmatrix} \right\rangle$

(c) $\left\langle \begin{bmatrix} 1 \\ 2 \end{bmatrix}, \begin{bmatrix} 3 \\ 4 \end{bmatrix}, \begin{bmatrix} 5 \\ 6 \end{bmatrix} \right\rangle$

(d) $C\left(\begin{bmatrix} 1 & 0 & 2 & -1 \\ 2 & 1 & 0 & 1 \\ 1 & 1 & -2 & 2 \end{bmatrix}\right)$

(e) $C\left(\begin{bmatrix} 1 & 1 & 0 \\ 1 & 0 & 1 \\ 0 & 1 & 1 \end{bmatrix}\right)$, over $\mathbb{R}$

(f) $C\left(\begin{bmatrix} 1 & 1 & 0 \\ 1 & 0 & 1 \\ 0 & 1 & 1 \end{bmatrix}\right)$, over $\mathbb{F}_2$

(g) $\ker \begin{bmatrix} 1 & 0 & 2 & -1 \\ 2 & 1 & 0 & 1 \\ 1 & 1 & -2 & 2 \end{bmatrix}$

3.3.2 Determine the dimension of each of the following vector spaces.

(a) $\left\langle \begin{bmatrix} 1 \\ -2 \end{bmatrix}, \begin{bmatrix} -2 \\ 4 \end{bmatrix} \right\rangle$

(b) $\left\langle \begin{bmatrix} 1 \\ 2 \\ 3 \end{bmatrix}, \begin{bmatrix} 3 \\ 2 \\ 1 \end{bmatrix}, \begin{bmatrix} 1 \\ 1 \\ 1 \end{bmatrix}, \begin{bmatrix} 1 \\ 0 \\ 1 \end{bmatrix} \right\rangle$

(c) $\left\langle \begin{bmatrix} 1 \\ 0 \\ -2 \\ 1 \end{bmatrix}, \begin{bmatrix} -2 \\ 5 \\ 0 \\ -1 \end{bmatrix}, \begin{bmatrix} 0 \\ 1 \\ 1 \\ 2 \end{bmatrix}, \begin{bmatrix} -1 \\ 4 \\ -3 \\ -2 \end{bmatrix}, \begin{bmatrix} -2 \\ 3 \\ -2 \\ -5 \end{bmatrix} \right\rangle$

(d) $C\left(\begin{bmatrix} 3 & 2 & 1 & 0 \\ 2 & 1 & 0 & 3 \\ 1 & 0 & 3 & 2 \end{bmatrix}\right)$

(e) $\ker \begin{bmatrix} 3 & 2 & 1 & 0 \\ 2 & 1 & 0 & 3 \\ 1 & 0 & 3 & 2 \end{bmatrix}$

(f) $\left\langle \begin{bmatrix} 1 \\ 1+i \\ i \end{bmatrix}, \begin{bmatrix} i \\ -1+i \\ -1 \end{bmatrix} \right\rangle$, over $\mathbb{C}$

(g) $\left\langle \begin{bmatrix} 1 \\ 1+i \\ i \end{bmatrix}, \begin{bmatrix} i \\ -1+i \\ -1 \end{bmatrix} \right\rangle$, over $\mathbb{R}$

3.3.3 Use Exercise 3.1.14 to prove that $C(\mathbb{R})$ is infinite-dimensional.

3.3.4 Prove that the space $\mathbb{F}^\infty$ of infinite sequences with entries in $\mathbb{F}$ is infinite-dimensional.

3.3.5 Suppose that $v_1 \neq 0$, and that for each $i = 2, \ldots, n$, $v_i \notin \langle v_1, \ldots, v_{i-1} \rangle$. Show that $\dim \langle v_1, \ldots, v_n \rangle = n$.

3.3.6 Suppose that $\lambda_1, \ldots, \lambda_n \in \mathbb{R}$ are distinct numbers, and let $f_i(x) = e^{\lambda_i x}$. Prove that

$$\dim\langle f_1, \ldots, f_n \rangle = n.$$

Hint: Think about what the derivative operator $D : C^\infty(\mathbb{R}) \to C^\infty(\mathbb{R})$ does to the f_i.

3.3.7 Show that if U_1 and U_2 are subspaces of a vector space V, then $\dim(U_1 + U_2) \leq \dim U_1 + \dim U_2$, where $U_1 + U_2$ is as defined in Exercise 1.5.11.

3.3.8 Show that if U_1 and U_2 are subspaces of a vector space V, then

$$\dim(U_1 + U_2) = \dim U_1 + \dim U_2 - \dim(U_1 \cap U_2),$$

where $U_1 + U_2$ is as defined in Exercise 1.5.11.
Hint: Start with a basis of $U_1 \cap U_2$. Extend it to a basis of U_1 and a basis of U_2, and show that both these bases together form a basis of $U_1 + U_2$.

3.3.9 Suppose that $T \in \mathcal{L}(V, W)$, and that V is finite-dimensional. Prove that $\dim \operatorname{range} T \leq \dim V$.

3.3.10 Show that if $(v_1, \ldots, v_k)$ is a linearly independent list in V and $w_1, \ldots, w_k \in W$ then there exists a linear map $T \in \mathcal{L}(V, W)$ such that $T(v_i) = w_i$ for each i.

3.3.11 Show that $\mathbb{C}^2$ is two-dimensional as a vector space over $\mathbb{C}$, but four-dimensional as a vector space over $\mathbb{R}$.

3.3.12 Suppose that $\dim V = n$ and that $T \in \mathcal{L}(V)$. Prove that T has at most n distinct eigenvalues.

3.3.13 Suppose that $\mathrm{A} \in \mathrm{M}_n(\mathbb{F})$ has n distinct eigenvalues. Show that there is a basis of $\mathbb{F}^n$ consisting of eigenvectors of A.

3.3.14 Let D be the derivative operator on $C^\infty(\mathbb{R})$. Show that the eigenspaces of D are all one-dimensional.

3.3.15 Recall Exercise 1.5.14, in which it was shown that the simplex

$$V = \left\{ (p_1, \ldots, p_n) \,\middle|\, p_i > 0,\ i = 1, \ldots, n;\ \sum_{i=1}^{n} p_i = 1 \right\}$$

is a real vector space, with non-standard versions of addition and scalar multiplication. Describe the one-dimensional subspaces of V.

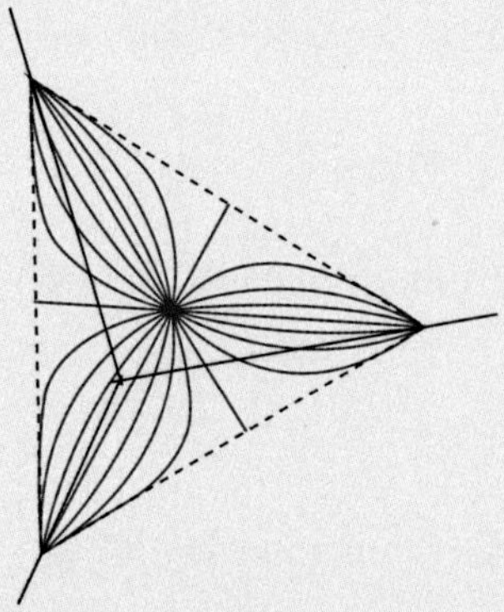

Some one-dimensional subspaces of the vector space V from Exercise 3.3.15.

3.3.16 Prove that $\mathbb{R}$, considered as a vector space over $\mathbb{Q}$, is infinite-dimensional.
Hint: Use Exercise 3.1.18.

3.3.17 Show that the intersection of any two planes containing the origin in $\mathbb{R}^3$ must contain a line through the origin.

3.3.18 Suppose that $\mathbb{F}$ is a finite field with q elements. Show that if V is a finite-dimensional vector space over $\mathbb{F}$, then V has exactly q^n elements for some n.

3.3.19 Suppose that $\mathbb{F}$ is a finite field with q elements and that $1 \leq k \leq n$.

(a) Show that there are exactly $(q^n - 1)(q^n - q) \cdots (q^n - q^{k-1})$ distinct linearly independent lists of length k of vectors in $\mathbb{F}^n$. In particular, there are $(q^n - 1)(q^n - q) \cdots (q^n - q^{n-1})$ distinct bases of $\mathbb{F}^n$.
Hint: Do this by induction on k. For the basis step, how many linearly independent lists of length 1 are there in $\mathbb{F}^n$? For the inductive step, every linearly independent list of length k can be formed by taking a linearly independent list of length $k - 1$ and adding a vector not in its span (why?).

(b) Show that there are q^{n^2} matrices in $\mathrm{M}_n(\mathbb{F})$, and that exactly $(q^n - 1)(q^n - q) \cdots (q^n - q^{n-1})$ of them are invertible.

3.3.20 Use Theorem 3.27 to give another proof of Corollary 3.12.

3.3.21 Prove the converse of Theorem 3.18. That is, show that if V is infinite-dimensional, then for every $n \geq 1$, V contains a linearly independent list of n vectors.
Hint: Mine the proof of part 1 of Theorem 3.29.

3.3.22 Use Exercise 3.3.8 to give a new proof of Lemma 3.22.

3.3.23 Use Theorem 3.27 to prove Theorem 3.28.

3.4 Rank and Nullity

The Rank and Nullity of Maps and Matrices

In Section 2.5 we introduced some important subspaces associated with linear operators (or matrices): the range (or column space) and the kernel (or null space).

It turns out that many conclusions can be drawn just by knowing the dimensions of these subspaces. We begin as before with the range.

Definition The **rank** of a linear map T is the dimension of its range, i.e.,

$$\operatorname{rank} T := \dim \operatorname{range} T;$$

the **rank** of a matrix A is the dimension of its column space, i.e.,

$$\operatorname{rank} \mathrm{A} := \dim C(\mathrm{A}).$$

Recall that, by Theorem 2.31, range T is exactly the column space of the matrix of T, so these two uses of the term "rank" are consistent.

Quick Exercise #16. Suppose that $\mathrm{A} \in \mathrm{M}_{m,n}(\mathbb{F})$ has rank 1. Show that $\mathrm{A} = \mathrm{v}\mathrm{w}^{\mathrm{T}}$ (equivalently, $a_{ij} = v_i w_j$) for some $\mathrm{v} \in \mathbb{F}^m$ and $\mathrm{w} \in \mathbb{F}^n$.
Hint: If $\operatorname{rank} \mathrm{A} = 1$, then $C(\mathrm{A}) = \langle \mathrm{v} \rangle$ for some $\mathrm{v} \in \mathbb{F}^m$.

We start with a few easy observations.

Proposition 3.30 *If* $T \in \mathcal{L}(V, W)$, *then* $\operatorname{rank} T \le \min\{\dim V, \dim W\}$.

Proof Since range $T \subseteq W$ by definition, Theorem 3.29 implies that $\operatorname{rank} T \le \dim W$. On the other hand, Exercise 3.3.9 shows that $\operatorname{rank} T \le \dim V$. ▲

Proposition 3.31 *If* $T \in \mathcal{L}(V, W)$ *and* $\dim W < \infty$, *then* T *is surjective iff* $\operatorname{rank} T = \dim W$.

Proof See Exercise 3.4.6. ▲

Theorem 3.32 *If* A *is an* $m \times n$ *matrix, then* $\operatorname{rank} \mathrm{A} \le \min\{m, n\}$.

Proof Since $C(\mathrm{A}) \subseteq \mathbb{F}^m$ by definition, $\operatorname{rank} \mathrm{A} \le \dim \mathbb{F}^m = m$. On the other hand, $C(\mathrm{A})$ is spanned by the n columns of A, $\operatorname{rank} \mathrm{A} \le n$. ▲

Quick Exercise #17. In the special case that $V = \mathbb{F}^n$ and $W = \mathbb{F}^m$, use Theorem 3.32 to give another proof of Theorem 3.30.

QA #17: If $T : \mathbb{F}^n \to \mathbb{F}^m$, then rank T is the same as the rank of its $(m \times n)$ matrix.

QA #16: The jth column of A is $w_j \mathrm{v}$ for some scalar w_j.

We can use row operations to find the rank of a matrix, as follows.

Algorithm 3.33 To find the rank of a matrix A:

- put A into RREF,
- rank A is the number of pivots.

Proof Recall from Algorithm 3.25 that the pivot columns in the RREF of A can be taken as a basis for $C(\mathrm{A})$. The result is therefore immediate from the definition of rank A. ▲

The connection between the column space of a matrix and the range of the corresponding operator makes the rank a natural thing to look at. But when working with matrices, we often think in terms of the rows rather than the columns, and we can perfectly well define *row space* and *row rank*: given a matrix

$$\mathrm{A} = \begin{bmatrix} -\mathbf{a}_1- \\ \vdots \\ -\mathbf{a}_m- \end{bmatrix} \in \mathrm{M}_{m,n}(\mathbb{F}),$$

the **row space** of A is

$$R(\mathrm{A}) := \langle \mathbf{a}_1, \ldots, \mathbf{a}_m \rangle \subseteq \mathrm{M}_{1,n}(\mathbb{F}),$$

and the **row rank** of A is $\dim R(\mathrm{A})$. Notice that this is exactly the same thing as $\operatorname{rank} \mathrm{A}^{\mathrm{T}}$.

The reason that this terminology is less standard is that there is in fact no need to separately refer to the row rank of a matrix, thanks to the following theorem.

Theorem 3.34 *Let* A *be a matrix. Then the row rank of* A *is the same as the rank of* A. *In other words,* $\operatorname{rank} \mathrm{A}^{\mathrm{T}} = \operatorname{rank} \mathrm{A}$.

Proof If the matrix A is in RREF, it is clear that the pivot rows form a basis of the row space: non-pivot rows are all zero, and working up from the bottom, each pivot row has a 1 where previous rows had zeroes, so the pivot rows are linearly independent by the Linear Dependence Lemma. That is, the row rank of A is equal to the number of pivots. But by Algorithm 3.33, the (column) rank of A is the number of pivots in the RREF of A.

It is thus enough to see that the row operations **R1**–**R3** do not change the row space of a matrix. Clearly, multiplying a row by a nonzero constant or swapping two rows does not change the row space. The span of the rows after operation **R1** is a subspace of the span of the original rows; since we can also get back to the original matrix via row operation **R1**, the two row spaces are equal. ▲

Quick Exercise #18. Find the rank of $A = \begin{bmatrix} 1 & 0 & 0 \\ 0 & 1 & 0 \\ -2 & 3 & 0 \\ 0 & 0 & 1 \end{bmatrix}$ using Theorem 3.34.

We next move on to the kernel.

Definition The **nullity** of a linear map T is

$$\operatorname{null} T := \dim \ker T,$$

and the **nullity** of a matrix A is

$$\operatorname{null} A := \dim \ker A.$$

Again, the nullity of $T \in \mathcal{L}(\mathbb{F}^n, \mathbb{F}^m)$ is the same as the nullity of its matrix.

Quick Exercise #19. Let $P \in \mathcal{L}(\mathbb{R}^3)$ be the projection onto the x–y plane. What is null P?

The Rank–Nullity Theorem

The following simple statement has far-reaching consequences in linear algebra and its applications.

Theorem 3.35 (The Rank–Nullity Theorem) *If* $A \in M_{m,n}(\mathbb{F})$, *then*

$$\operatorname{rank} A + \operatorname{null} A = n.$$

If $T \in \mathcal{L}(V, W)$, *then*

$$\operatorname{rank} T + \operatorname{null} T = \dim V.$$

The statement for linear maps implies the statement for matrices, simply by specializing to the case $V = \mathbb{F}^n$ and $W = \mathbb{F}^m$. Nevertheless, we will give separate proofs of the two statements in order to illustrate different ideas, one for matrices based on Gaussian elimination, and a more abstract one for maps.

QA #18: $A^T = \begin{bmatrix} 1 & 0 & -2 & 0 \\ 0 & 1 & 3 & 0 \\ 0 & 0 & 0 & 1 \end{bmatrix}$, which is in RREF and has three pivots, so rank A = 3.

QA #19: ker P consists of the z-axis in $\mathbb{R}^3$, which is one-dimensional, so null $P = 1$.

Proof of Theorem 3.35, matrix version By Algorithm 3.33, rank A is the number of pivots in the RREF of A, i.e., it is the number of pivot variables in the system $\mathrm{Ax} = \mathbf{0}$. If we prove that null A is the number of free variables, then we will be done since the total number of variables is n.

Denote the free variables of the system $\mathrm{Ax} = \mathbf{0}$ by $x_{i_1}, \ldots, x_{i_k}$; our goal is to show that $\operatorname{null} \mathrm{A} = k$. Recall our discussion at the end of Section 1.3 of the geometry of solutions: if we find the RREF of the augmented matrix $\left[\mathrm{A} \mid \mathbf{0}\right]$, we can use each row that contains a pivot to solve for the corresponding pivot variable in terms of the free variables. Then by including the trivial equations $x_{i_j} = x_{i_j}$ for each of the free variables, we can write the solutions of the system as linear combinations of a set of fixed vectors, where the coefficients are exactly the free variables:

$$\begin{bmatrix} x_1 \\ \vdots \\ x_n \end{bmatrix} = x_{i_1} \begin{bmatrix} c_{11} \\ \vdots \\ c_{1n} \end{bmatrix} + \cdots + x_{i_k} \begin{bmatrix} c_{k1} \\ \vdots \\ c_{kn} \end{bmatrix}. \tag{3.6}$$

Since the null space of A is exactly the set of solutions of $\mathrm{Ax} = \mathbf{0}$, this formula makes it clear that

$$\ker \mathrm{A} \subseteq \left\langle \begin{bmatrix} c_{11} \\ \vdots \\ c_{1n} \end{bmatrix}, \ldots, \begin{bmatrix} c_{k1} \\ \vdots \\ c_{kn} \end{bmatrix} \right\rangle; \tag{3.7}$$

that is, $\operatorname{null} \mathrm{A} \leq k$. But recall that the equations for the free variables in (3.6) are just $x_{i_j} = x_{i_j}$; i.e., the vector of coefficients for x_{i_j} above has a 1 in the i_jth position, and zeroes in the positions corresponding to all the other free variables. From this it is easy to see that the vectors in (3.7) are linearly independent, and so in fact $\operatorname{null} \mathrm{A} = k$ and we are done. ▲

Proof of Theorem 3.35, linear map version Since $\ker T \subseteq V$, $\ker T$ is finite-dimensional by Theorem 3.29. Suppose that $\ker T \neq \{0\}$, so that it has a basis $(u_1, \ldots, u_k)$. (For the case when $\ker T = \{0\}$, see Exercise 3.4.20.) By Theorem 3.27, this list can be extended to a basis

$$\mathcal{B} = (u_1, \ldots, u_k, v_1, \ldots, v_\ell)$$

of V. We claim that

$$\mathcal{B}' = (T(v_1), \ldots, T(v_\ell))$$

is a basis of range T. Assuming that this claim is true, we then have

$$\operatorname{rank} T + \operatorname{null} T = \ell + k = \dim V.$$

We therefore need only to show that $\mathcal{B}'$ is linearly independent and spans range T. Suppose then that

$$\sum_{i=1}^{\ell} a_i T(v_i) = 0$$

for $a_1, \ldots, a_\ell \in \mathbb{F}$. By linearity, we have

$$T\left(\sum_{i=1}^{\ell} a_i v_i\right) = 0,$$

so

$$\sum_{i=1}^{\ell} a_i v_i \in \ker T.$$

Since $(u_1, \ldots, u_k)$ is a basis for $\ker T$, we must have

$$\sum_{i=1}^{\ell} a_i v_i = \sum_{j=1}^{k} b_j u_j$$

for some $b_1, \ldots, b_k \in \mathbb{F}$; i.e.,

$$\sum_{j=1}^{k} (-b_j) u_j + \sum_{i=1}^{\ell} a_i v_i = 0.$$

Since the whole list $\mathcal{B}$ is linearly independent, we must have $a_1 = \cdots = a_\ell = b_1 = \cdots = b_k = 0$; in particular, since it must be that $a_1 = \cdots = a_\ell = 0$, $\mathcal{B}'$ is linearly independent.

Now suppose that $w \in \operatorname{range} T$. By definition, this means that $w = T(v)$ for some $v \in V$. Since $\mathcal{B}$ spans V,

$$v = \sum_{i=1}^{k} c_i u_i + \sum_{j=1}^{\ell} d_j v_j$$

for some $c_1, \ldots, c_k, d_1, \ldots, d_\ell \in \mathbb{F}$. By linearity,

$$w = T(v) = \sum_{i=1}^{k} c_i T(u_i) + \sum_{j=1}^{\ell} d_j T(v_j) = \sum_{j=1}^{\ell} d_j T(v_j)$$

since $u_i \in \ker T$ for each i. So $w \in \langle \mathcal{B}' \rangle$. ▲

Quick Exercise #20. Prove that for any square matrix A, $\operatorname{null} A = \operatorname{null} A^T$.

QA #20: If A is $n \times n$, then $\operatorname{null} A = n - \operatorname{rank} A = n - \operatorname{rank} A^T = \operatorname{null} A^T$ by the Rank–Nullity Theorem and Theorem 3.34.

Example In Section 2.6 on the Hamming code, we considered the encoding matrix $A \in M_{7,4}(\mathbb{F}_2)$ given in equation (2.27) and the parity check matrix $B \in M_{3,7}(\mathbb{F}_2)$ given in equation (2.26). It was easy to check that $C(A) \subseteq \ker B$ and $\ker A = \{0\}$; it was left as an exercise to show that actually $C(A) = \ker B$. In fact, this follows with almost no additional work using the Rank–Nullity Theorem.

Since $\operatorname{null} A = 0$ and A is 7×4, the Rank–Nullity Theorem implies that $\operatorname{rank} A = 4$. Since the last three columns of B are the three standard basis vectors, $\operatorname{rank} B = 3$, and the Rank–Nullity Theorem implies that $\operatorname{null} B = 7 - 3 = 4$. So $C(A)$ is a four-dimensional subspace of the four-dimensional space $\ker B$, which means that $C(A) = \ker B$. ▲

Consequences of the Rank–Nullity Theorem

The Rank–Nullity Theorem has some important consequences for linear maps, starting with the following.

Corollary 3.36 *Suppose that $T \in \mathcal{L}(V, W)$. If $\dim V = \dim W$, then the following are equivalent:*

1. *T is injective.*
2. *T is surjective.*
3. *T is an isomorphism.*

Proof Recall (see Theorem 2.37) that T is injective if and only if $\ker T = \{0\}$, i.e., if and only if $\operatorname{null} T = 0$.

We thus have the following equivalences:

$$\begin{aligned} T \text{ is injective} &\iff \operatorname{null} T = 0 \\ &\iff \operatorname{rank} T = \dim V \quad \text{(by the Rank–Nullity Theorem)} \\ &\iff \operatorname{rank} T = \dim W \quad (\text{since } \dim V = \dim W) \\ &\iff T \text{ is surjective} \quad \text{(by Proposition 3.31).} \end{aligned}$$

It follows that if T is either injective or surjective, then in fact T is both injective and surjective, and thus an isomorphism. Conversely, we already know that if T is an isomorphism then it is both injective and surjective. ▲

This last corollary is an example of what's known as a "dimension-counting" argument. We showed that injectivity and surjectivity of T are equivalent, not directly, but by using the Rank–Nullity Theorem to relate the dimensions of the image and kernel of T.

In addition to possibly being injective but not surjective or vice versa, in general it is possible for linear maps to have only a left-inverse or a right-inverse, i.e., to have $S \in \mathcal{L}(V, W)$ and $T \in \mathcal{L}(W, V)$ so that, for example, $ST = I$ but $TS \neq I$.

However, it follows from the previous result that this cannot happen in the setting of linear maps if $\dim V = \dim W$. Similarly, we have already seen that while some matrices have only one-sided inverses, this is not possible for square matrices; the result above gives a quick proof of this fact.

Corollary 3.37

1. *Suppose that* $\dim V = \dim W$ *and that* $S \in \mathcal{L}(V, W)$ *and* $T \in \mathcal{L}(W, V)$. *If* $ST = I$, *then* $TS = I$ *as well. In particular,* S *and* T *are invertible, and* $T^{-1} = S$.
2. *Suppose that* $A, B \in M_n(\mathbb{F})$. *If* $AB = I_n$, *then* $BA = I_n$ *as well. In particular,* A *and* B *are invertible, and* $A^{-1} = B$.

Proof

1. If $ST = I$, then for every $w \in W$, $w = Iw = S(Tw)$, which implies that S is surjective. By Corollary 3.36, S is invertible, and so
$$T = S^{-1}ST = S^{-1}.$$
2. By Corollary 2.32,
$$\mathbb{F}^n = C(I_n) = C(AB) \subseteq C(A),$$
so $\operatorname{rank} A \geq n$. Since $A \in M_n(\mathbb{F})$, this means that $\operatorname{rank} A = n$; i.e., the RREF of A contains n pivots. Since A is an $n \times n$ square matrix, this means that the RREF of A is I_n, which by Algorithm 2.25 means that A is invertible. Then, exactly as above,
$$B = A^{-1}AB = A^{-1}.$$
▲

Quick Exercise #21. Let $A = \begin{bmatrix} 1 & 0 & 0 \\ 0 & 1 & 0 \end{bmatrix}$ and $B = A^T = \begin{bmatrix} 1 & 0 \\ 0 & 1 \\ 0 & 0 \end{bmatrix}$. Compute AB and BA. Why does this not contradict Corollary 3.37?

Dimension counting lets us make the following interesting observation about the eigenvalues of square matrices.

Corollary 3.38 *Let* $A \in M_n(\mathbb{F})$. *Then* $\lambda \in \mathbb{F}$ *is an eigenvalue of* A *if and only if* λ *is an eigenvalue of* A^T.

QA #21: $AB = I_2$ and $BA = \begin{bmatrix} 1 & 0 & 0 \\ 0 & 1 & 0 \\ 0 & 0 & 0 \end{bmatrix}$. Corollary 3.37 doesn't apply because A and B aren't square matrices.

Proof Recall that $\lambda \in \mathbb{F}$ is an eigenvalue of A if and only if $\ker(\mathrm{A} - \lambda \mathrm{I}_n) \neq \{0\}$. By the Rank–Nullity Theorem, this is equivalent to $\operatorname{rank}(\mathrm{A} - \lambda \mathrm{I}_n) < n$. But by Theorem 3.34,

$$\operatorname{rank}(\mathrm{A} - \lambda \mathrm{I}_n) = \operatorname{rank}(\mathrm{A} - \lambda \mathrm{I}_n)^{\mathrm{T}} = \operatorname{rank}(\mathrm{A}^{\mathrm{T}} - \lambda \mathrm{I}_n),$$

so $\operatorname{rank}(\mathrm{A} - \lambda \mathrm{I}_n) < n$ if and only if $\operatorname{rank}(\mathrm{A}^{\mathrm{T}} - \lambda \mathrm{I}_n) < n$, i.e., if and only if λ is an eigenvalue of A^{T}. ▲

An important thing to notice here is that to say that λ is an eigenvalue of A means that there is a nonzero vector $\mathrm{v} \in \mathbb{F}^n$ such that $\mathrm{Av} = \lambda \mathrm{v}$. The result above tells us that such a vector v exists if and only if there is a vector $\mathrm{w} \in \mathbb{F}^n$ such that $\mathrm{A}^{\mathrm{T}}\mathrm{w} = \lambda \mathrm{w}$, but it tells us nothing at all about any relationship between v and w. In particular, it does not tell us how to find eigenvectors for A^{T} if we know eigenvectors of A.

Proposition 3.39 *Let $T \in \mathcal{L}(V, W)$.*

- *If* $\dim V > \dim W$, *then T cannot be injective.*
- *If* $\dim W > \dim V$, *then T cannot be surjective.*

Proof of the first half of Proposition 3.39 Suppose that $\dim V > \dim W$. By Theorem 3.29, $\operatorname{rank} T \leq \dim W$, and so the Rank–Nullity Theorem implies

$$\operatorname{null} T = \dim V - \operatorname{rank} T \geq \dim V - \dim W \geq 1.$$

Therefore, $\ker T \neq \{0\}$, and so T is not injective. ▲

Quick Exercise #22. Complete the proof of Proposition 3.39. That is, prove that if $T \in \mathcal{L}(V, W)$ and $\dim W > \dim V$, then T cannot be surjective.

This result relates directly to observations we've already made about linear systems: let A be an $m \times n$ matrix over $\mathbb{F}$, and consider the linear system

$$\mathrm{Ax} = 0. \tag{3.8}$$

Then the operator $T_A : \mathbb{F}^n \to \mathbb{F}^m$ defined by multiplication by A is injective if and only if (3.8) has 0 as its unique solution. In fact, we already observed (see Corollary 1.4) that if $m < n$, then it is impossible for (3.8) to have a unique solution.

To relate the second part to linear systems, note that $T_A : \mathbb{F}^n \to \mathbb{F}^m$ is surjective if and only if the $m \times n$ linear system

$$\mathrm{Ax} = \mathrm{b} \tag{3.9}$$

QA #22: $\operatorname{rank} T = \dim V - \operatorname{null} T \leq \dim V \leq \dim W - 1$.

is consistent for every $\mathbf{b} \in \mathbb{F}^m$. Recall (Theorem 1.2) that (3.9) is consistent if and only if there is no pivot in the last column of its RREF. But if $m > n$ (i.e., there are more rows than columns), then the RREF of A cannot have a pivot in every row, since each column can contain at most one pivot. Therefore, there are some rows of all zeroes. This means that for some choices of $\mathbf{b}$, there will be a pivot in the final column of the RREF of (3.9), making the system inconsistent.

Proposition 3.39 also has important geometric consequences for linear maps between finite-dimensional vector spaces. Firstly, it reproves the fact that two vector spaces of different dimensions cannot be isomorphic, but it refines that observation. On the one hand, a linear map cannot stretch out a vector space to manufacture more dimensions: when mapping V into W, a linear map cannot make the dimension go up, only down. On the other hand, linear maps cannot squeeze vector spaces down into spaces of smaller dimension without collapsing some vectors to zero. One can put these ideas together to justify visualizing a linear map as collapsing part of a space to zero and putting the rest down in the codomain pretty much as it was. This kind of restricted behavior is a very special property of linear maps; non-linear maps between vector spaces can do all kinds of strange things.*

Linear Constraints

In many applications of linear algebra, one encounters sets which are subsets of an ambient vector space, subject to certain linear constraints; i.e., elements are required to satisfy some linear equations. For example, one might encounter a set like

$$S := \left\{ \begin{bmatrix} x_1 \\ x_2 \\ x_3 \\ x_4 \\ x_5 \end{bmatrix} \middle| \begin{array}{c} 3x_1 + 2x_4 + x_5 = 10 \\ x_2 - x_3 - 5x_5 = 7 \end{array} \right\}.$$

The set S is not a subspace of $\mathbb{R}^5$, and so we have not defined the notion of dimension for S, but it is not hard to do so: recall from Section 2.5 that S is a so-called "affine subspace" of $\mathbb{R}^5$, meaning that it is a subspace which has been shifted so that it no longer contains the origin. It is thus reasonable to declare the dimension of S to be the dimension of the subspace $S - \mathbf{v}_0$ of $\mathbb{R}^5$, where $\mathbf{v}_0 \in \mathbb{R}^5$ is a vector such that $S - \mathbf{v}_0$ contains $\mathbf{0}$.

It is easy to read off from S which subspace S is a translation of, and how to translate it back to containing the origin. Indeed, if $\mathbf{v}_0$ is any element of S (assuming that S is nonempty), then trivially $S - \mathbf{v}_0$ contains $\mathbf{0}$. Moreover,

*For example, you might enjoy googling "space-filling curve."

$$S - \mathbf{v}_0 = \left\{ \begin{bmatrix} x_1 \\ x_2 \\ x_3 \\ x_4 \\ x_5 \end{bmatrix} \middle| \begin{array}{c} 3x_1 + 2x_4 + x_5 = 0 \\ x_2 - x_3 - 5x_5 = 0 \end{array} \right\}.$$

(See Exercise 3.4.7.)

So if we define $T : \mathbb{R}^5 \to \mathbb{R}^2$ by

$$T \begin{bmatrix} x_1 \\ x_2 \\ x_3 \\ x_4 \\ x_5 \end{bmatrix} = \begin{bmatrix} 3x_1 + 2x_4 + x_5 \\ x_2 - x_3 - 5x_5 \end{bmatrix} = \begin{bmatrix} 3 & 0 & 0 & 2 & 1 \\ 0 & 1 & -1 & 0 & -5 \end{bmatrix} \begin{bmatrix} x_1 \\ x_2 \\ x_3 \\ x_4 \\ x_5 \end{bmatrix},$$

then $S - \mathbf{v}_0$ is exactly the kernel of T. Since the rank of T is obviously 2, this means that the nullity of T, i.e., the dimension of S, is $5 - 2 = 3$.

The same argument works in general: if S is the subset of an n-dimensional vector space defined by k (*linearly independent* and *consistent*) constraints, then the dimension of S is $n - k$.

KEY IDEAS

- The rank of a map is the dimension of its image; the rank of a matrix is the dimension of its column space (which is the number of pivots in its RREF).
- The nullity of a map or matrix is the dimension of its kernel.
- Row rank = column rank.
- The Rank–Nullity Theorem: If $T \in \mathcal{L}(V, W)$, then $\operatorname{rank} T + \operatorname{null} T = \dim(V)$. If $\mathrm{A} \in \mathrm{M}_{m,n}(\mathbb{F})$, then $\operatorname{rank} \mathrm{A} + \operatorname{null} \mathrm{A} = n$.

EXERCISES

3.4.1 Find the rank and nullity of each of the following matrices.

(a) $\begin{bmatrix} 1 & 2 & 3 \\ 4 & 5 & 6 \end{bmatrix}$ (b) $\begin{bmatrix} -2 & 1 & 3 & -1 \\ 1 & 0 & 2 & 1 \\ 3 & -1 & 2 & 2 \end{bmatrix}$

(c) $\begin{bmatrix} 2 & 0 & 1 & -3 & 4 \\ -1 & 2 & 4 & 0 & 3 \\ 3 & -2 & -3 & -3 & 1 \end{bmatrix}$ (d) $\begin{bmatrix} 1 & 1 & 0 \\ 1 & 0 & 1 \\ 0 & 1 & 1 \end{bmatrix}$, over $\mathbb{R}$

(e) $\begin{bmatrix} 1 & 1 & 0 \\ 1 & 0 & 1 \\ 0 & 1 & 1 \end{bmatrix}$, over $\mathbb{F}_2$

3.4.2 Find the rank and nullity of each of the following matrices.

(a) $\begin{bmatrix} 3 & -1 & 4 & -1 \\ -9 & 2 & -6 & 5 \end{bmatrix}$ (b) $\begin{bmatrix} 2 & -3 & 1 \\ 4 & -6 & 2 \\ -2 & 3 & -1 \end{bmatrix}$ (c) $\begin{bmatrix} 2 & -1 & 0 \\ 1 & 3 & -1 \\ 3 & 2 & -1 \end{bmatrix}$

(d) $\begin{bmatrix} 5 & -3 & 2 \\ 1 & -2 & 1 \\ 2 & 3 & -1 \\ 4 & -1 & 1 \end{bmatrix}$ (e) $\begin{bmatrix} 3 & 1 & 0 & -2 \\ -2 & 0 & 4 & -3 \\ 1 & 1 & 4 & -5 \end{bmatrix}$

3.4.3 Let D be the differentiation operator on the space $\mathcal{P}_n(\mathbb{R})$ of polynomials over $\mathbb{R}$ of degree at most n. Find the rank and nullity of D.

3.4.4 Let $\mathrm{A} \in \mathrm{M}_{m,n}(\mathbb{R})$ be given by $a_{ij} = ij$ for $1 \le i \le m$ and $1 \le j \le n$. Find the rank and nullity of A.

3.4.5 Prove that $\operatorname{rank} \mathrm{AB} \le \min\{\operatorname{rank} \mathrm{A}, \operatorname{rank} \mathrm{B}\}$.

3.4.6 Prove that if $T \in \mathcal{L}(V, W)$ and $\dim W$ is finite-dimensional, then T is surjective if and only if $\operatorname{rank} T = \dim W$.

3.4.7 Show that if

$$S := \left\{ \begin{bmatrix} x_1 \\ x_2 \\ x_3 \\ x_4 \\ x_5 \end{bmatrix} \middle| \begin{array}{c} 3x_1 + 2x_4 + x_5 = 10 \\ x_2 - x_3 - 5x_5 = 7 \end{array} \right\}$$

and $\mathbf{v}_0 \in S$, then

$$S - \mathbf{v}_0 = \left\{ \begin{bmatrix} x_1 \\ x_2 \\ x_3 \\ x_4 \\ x_5 \end{bmatrix} \middle| \begin{array}{c} 3x_1 + 2x_4 + x_5 = 0 \\ x_2 - x_3 - 5x_5 = 0 \end{array} \right\}.$$

3.4.8 Prove that if $\mathrm{A} \in \mathrm{M}_{m,n}(\mathbb{F})$ has rank r, then there exist $\mathbf{v}_1, \ldots, \mathbf{v}_r \in \mathbb{F}^m$ and $\mathbf{w}_1, \ldots, \mathbf{w}_r \in \mathbb{F}^n$ such that $\mathrm{A} = \sum_{i=1}^r \mathbf{v}_i \mathbf{w}_i^{\mathrm{T}}$.
Hint: Write the columns of A as linear combinations of a basis $(\mathbf{v}_1, \ldots, \mathbf{v}_r)$ of $C(\mathrm{A})$.

3.4.9 (a) Prove that if $S, T \in \mathcal{L}(V, W)$, then $\operatorname{rank}(S + T) \le \operatorname{rank} S + \operatorname{rank} T$.
(b) Prove that if $\mathrm{A}, \mathrm{B} \in \mathrm{M}_{m,n}(\mathbb{F})$, then $\operatorname{rank}(\mathrm{A} + \mathrm{B}) \le \operatorname{rank} \mathrm{A} + \operatorname{rank} \mathrm{B}$.
Hint: Use Exercises 2.5.11 and 3.3.7.

3.4.10 Prove that if $T \in \mathcal{L}(V, W)$ and V is finite-dimensional, then T is injective iff $\operatorname{rank} T = \dim V$.

3.4.11 Suppose that $\mathrm{A}\mathbf{x} = \mathbf{b}$ is a 5×5 linear system which is consistent but does *not* have a unique solution. Prove that there must be a $\mathbf{c} \in \mathbb{F}^5$ such that the system $\mathrm{A}\mathbf{x} = \mathbf{c}$ is inconsistent.

3.4.12 Suppose that, for a given $A \in M_3(\mathbb{R})$, there is a plane P through the origin in $\mathbb{R}^3$ such that the linear system $Ax = b$ is consistent if and only if $b \in P$. Prove that the set of solutions of the homogeneous system $Ax = 0$ is a line through the origin in $\mathbb{R}^3$.

3.4.13 Show that if $T \in \mathcal{L}(V, W)$ and $\operatorname{rank} T = \min\{\dim V, \dim W\}$, then T is either injective or surjective.

3.4.14 Suppose that $\dim V = n$ and $S, T \in \mathcal{L}(V)$.

(a) Show that if $\operatorname{rank} ST < n$, then $\operatorname{rank} TS < n$.

Hint: Prove the contrapositive.

(b) Show that if 0 is an eigenvalue of ST, then 0 is an eigenvalue of TS.

Remark: Together with Exercise 2.1.16, this shows that ST and TS have all the same eigenvalues. Notice, though, that this doesn't say anything about their eigen*vectors*, which might be completely different!

3.4.15 Prove that the rank of $T \in \mathcal{L}(V)$ is greater than or equal to the number of distinct nonzero eigenvalues of T.

3.4.16 Show that the set of matrices

$$\left\{ A \in M_n(\mathbb{R}) \;\middle|\; \sum_{i=1}^{n} a_{ij} = 0 \text{ for every } j \text{ and } \sum_{j=1}^{n} a_{ij} = 0 \text{ for every } i \right\}$$

has dimension $n^2 - 2n + 1$.

3.4.17 Suppose that $\mathbb{F}$ is a subfield of $\mathbb{K}$ (see Exercise 1.5.7) and $A \in M_{m,n}(\mathbb{F})$. Show that the rank and nullity of A are the same whether A is thought of as a matrix over $\mathbb{F}$ or as a matrix over $\mathbb{K}$.

3.4.18 (a) Use Exercise 3.4.9 to show that if

$$A = \sum_{i=1}^{r} v_i w_i^T$$

for some $v_1, \ldots, v_r \in \mathbb{F}^m$ and $w_1, \ldots, w_r \in \mathbb{F}^n$, then $\operatorname{rank} A \leq r$.

(b) Give another proof of Theorem 3.34 using Exercise 3.4.9 and the above fact.

3.4.19 (a) Suppose that $A \in M_{m,n}(\mathbb{F})$ has rank r. Let $(c_1, \ldots, c_r)$ be a basis of $C(A)$, and let $C \in M_{m,r}(\mathbb{F})$ be the matrix with those columns. Show that there is a matrix $B \in M_{r,n}(\mathbb{F})$ such that $A = CB$.

(b) Use this fact, Corollary 2.32, and Theorem 3.32 to show that $\operatorname{rank} A^T \leq r = \operatorname{rank} A$.

(c) Use the above fact to give another proof of Theorem 3.34.

3.4.20 Prove the Rank–Nullity Theorem (Theorem 3.35) in the case that $\ker T = \{0\}$.

3.4.21 State and prove the matrix analog of Proposition 3.39.

3.5 Coordinates

Coordinate Representations of Vectors

We have sometimes given proofs specific to $\mathbb{F}^n$, then showed by a different argument that the same result holds in a general vector space. We have seen, though, that all n-dimensional vector spaces are isomorphic to $\mathbb{F}^n$, so there must be a sense in which anything that we can do in $\mathbb{F}^n$ works the same way in an arbitrary n-dimensional vector space. The definitions and results of this section allow us to see exactly how that works.*

Let V be a finite-dimensional vector space. Then we know that V has a basis, say $\mathcal{B} = (v_1, \ldots, v_n)$. Because $\mathcal{B}$ is a basis, each vector $v \in V$ can be uniquely represented as a linear combination of the elements of $\mathcal{B}$ (see Theorem 3.10): there is a unique list of scalars $a_1, \ldots, a_n \in \mathbb{F}$ such that

$$v = a_1 v_1 + \cdots + a_n v_n.$$

There's a lot of unnecessary notation in the expression above. If we've already established that we're thinking in terms of the basis $\mathcal{B}$, then the whole content of the expression for v above is the (ordered) list of coefficients a_i. We therefore sometimes associate the vector v with its list of coefficients and write it "in coordinates" as follows.

Definition The **coordinates** of $v \in V$ with respect to the basis $\mathcal{B}$ are the (unique ordered list of) scalars $a_1, \ldots, a_n$ such that

$$v = a_1 v_1 + \cdots + a_n v_n.$$

The **coordinate representation** $[v]_{\mathcal{B}} \in \mathbb{F}^n$ of v with respect to $\mathcal{B}$ is

$$[v]_{\mathcal{B}} := \begin{bmatrix} a_1 \\ \vdots \\ a_n \end{bmatrix}.$$

If V is an n-dimensional vector space over $\mathbb{F}$ with basis $\mathcal{B}$, representing vectors in coordinates gives an explicit isomorphism between V and $\mathbb{F}^n$:

---

*There has historically been a certain class of mathematician very resistant to the idea of making every space into $\mathbb{F}^n$ and every map into a matrix, sometimes going through preposterous contortions in order to do everything "intrinsically." The UNIX fortune cookie program sometimes returns a quote from such a mathematician: "A gentleman never takes bases unless he really has to."

Proposition 3.40 *Let V be an n-dimensional vector space over $\mathbb{F}$. Let $\mathcal{B}$ be a basis for V. Then the map $C_{\mathcal{B}} : V \to \mathbb{F}^n$ defined by $C_{\mathcal{B}}(v) = [v]_{\mathcal{B}}$ is an isomorphism.*

Proof The map $C_{\mathcal{B}}$ is clearly linear, and maps the basis $\mathcal{B}$ to the standard basis of $\mathbb{F}^n$, and so by Theorem 3.15, $C_{\mathcal{B}}$ is invertible. ▲

Examples

1. If $V = \mathbb{F}^n$, the elements of V come to us as linear combinations of the standard basis vectors $\mathbf{e}_1, \ldots, \mathbf{e}_n$:

$$\begin{bmatrix} a_1 \\ \vdots \\ a_n \end{bmatrix} = a_1\mathbf{e}_1 + \cdots + a_n\mathbf{e}_n,$$

and so vectors in $\mathbb{F}^n$ are already written as their coordinate representations in this basis.

2. Consider the subspace $V \subseteq \mathbb{R}^3$ defined by

$$V := \left\{ \begin{bmatrix} x \\ y \\ z \end{bmatrix} \middle| \, x + 2y - z = 0 \right\}.$$

The set $\mathcal{B} = \left(\begin{bmatrix} 0 \\ 1 \\ 2 \end{bmatrix}, \begin{bmatrix} 2 \\ 1 \\ 4 \end{bmatrix} \right)$ is a basis of V.

Quick Exercise #23. Verify that $\mathcal{B}$ is a basis of V.

The vector $\mathbf{v} = \begin{bmatrix} 1 \\ 0 \\ 1 \end{bmatrix} \in V$. To find its coordinates, we solve the linear system

$$a_1 \begin{bmatrix} 0 \\ 1 \\ 2 \end{bmatrix} + a_2 \begin{bmatrix} 2 \\ 1 \\ 4 \end{bmatrix} = \begin{bmatrix} 1 \\ 0 \\ 1 \end{bmatrix}$$

to find that $a_1 = -\frac{1}{2}$ and $a_2 = \frac{1}{2}$, so

$$[\mathbf{v}]_{\mathcal{B}} = \frac{1}{2}\begin{bmatrix} -1 \\ 1 \end{bmatrix}.$$

QA #23: The two elements of $\mathcal{B}$ are linearly independent. $V \subsetneq \mathbb{R}^3$ since $\mathbf{e}_1 \notin V$, so $\dim(V) \leq 2$.

Note that even though $\mathbf{v}$ came to us in (some) coordinates, its coordinate representation with respect to the basis $\mathcal{B}$ looks completely different from its representation with respect to the standard basis of $\mathbb{R}^3$ – it is even a different length. ▲

Matrix Representations of Linear Maps

Recall Theorem 2.8: if $T : \mathbb{F}^n \to \mathbb{F}^m$ is a linear map, then there is a unique $m \times n$ matrix A over $\mathbb{F}$ so that for any $\mathbf{v} \in \mathbb{F}^n$, $T(\mathbf{v}) = \mathrm{A}\mathbf{v}$; i.e., the map T is just multiplication by the matrix A. Moreover, we had a simple expression for A: the ith column of A is $T(\mathbf{e}_i)$, where $(\mathbf{e}_1, \ldots, \mathbf{e}_n)$ is the standard basis of $\mathbb{F}^n$.

Working with coordinate representations of vectors in arbitrary finite-dimensional vector spaces lets us represent all linear maps this way. Specifically, we have the following.

Definition Let V and W be n- and m-dimensional vector spaces over $\mathbb{F}$, respectively, and let $T : V \to W$ be a linear map. Suppose that V has basis $\mathcal{B}_V = (v_1, \ldots, v_n)$ and W has basis $\mathcal{B}_W = (w_1, \ldots, w_m)$. For each $1 \leq j \leq n$, define the scalars a_{ij} for $1 \leq i \leq m$ to be the coordinates of Tv_j in the basis $\mathcal{B}_W$; i.e.,

$$Tv_j = a_{1j}w_1 + \cdots + a_{mj}w_m.$$

The **matrix of T with respect to $\mathcal{B}_V$ and $\mathcal{B}_W$** is the matrix $\mathrm{A} = \left[a_{ij}\right]_{\substack{1 \leq i \leq m \\ 1 \leq j \leq n}}$; we denote this matrix as $[T]_{\mathcal{B}_V,\mathcal{B}_W}$.

When $V = W$ and $\mathcal{B}_V = \mathcal{B}_W = \mathcal{B}$, we simply write $[T]_{\mathcal{B}}$ and refer to it as the **matrix of T with respect to $\mathcal{B}$.**

In particular, the definition means that

the jth column of $[T]_{\mathcal{B}_V,\mathcal{B}_W}$ is the coordinate representation $\left[Tv_j\right]_{\mathcal{B}_W}$.

The whole point, of course, is to represent T by multiplication by a matrix, as follows.

Lemma 3.41 *Let $T : V \to W$ be a linear map. Suppose that V has basis $\mathcal{B}_V$ and W has basis $\mathcal{B}_W$, and let $\mathrm{A} = [T]_{\mathcal{B}_V,\mathcal{B}_W}$. Then for any $v \in V$,*

$$[Tv]_{\mathcal{B}_W} = \mathrm{A}\,[v]_{\mathcal{B}_V}.$$

That is, when vectors in V and W are expressed in coordinates with respect to the bases $\mathcal{B}_V$ and $\mathcal{B}_W$, the map T is given by multiplication by the matrix $[T]_{\mathcal{B}_V,\mathcal{B}_W}$.

Proof Write $\mathcal{B}_V = (v_1, \ldots, v_n)$ and $\mathcal{B}_W = (w_1, \ldots, w_m)$. Let $[v]_{\mathcal{B}_V} = \begin{bmatrix} c_1 \\ \vdots \\ c_n \end{bmatrix}$; in other words,

$$v = c_1 v_1 + \cdots + c_n v_n.$$

Then by linearity,

$$Tv = \sum_{j=1}^{n} c_j T v_j = \sum_{j=1}^{n} c_j \left(\sum_{i=1}^{m} a_{ij} w_i \right) = \sum_{i=1}^{m} \left(\sum_{j=1}^{n} a_{ij} c_j \right) w_i.$$

That is, the ith coordinate of Tv with respect to $\mathcal{B}_W$ is $\sum_{j=1}^{n} a_{ij} c_j$, which is the same as the ith entry of $A\,[v]_{\mathcal{B}_V}$. ▲

Of course, whenever we have an $m \times n$ matrix A, it defines a linear map in $\mathcal{L}(\mathbb{F}^n, \mathbb{F}^m)$ via matrix–vector multiplication. In general, if V and W are n- and m-dimensional vector spaces over $\mathbb{F}$, with bases $\mathcal{B}_V$ and $\mathcal{B}_W$ respectively, and $A \in M_{m,n}(\mathbb{F})$, then we can define a map $T_{A,\mathcal{B}_V,\mathcal{B}_W} : V \to W$ by

$$\left[T_{A,\mathcal{B}_V,\mathcal{B}_W} v\right]_{\mathcal{B}_W} := A\,[v]_{\mathcal{B}_V}. \tag{3.10}$$

The rather unwieldy notation $T_{A,\mathcal{B}_V,\mathcal{B}_W}$ is essential: which linear map a given matrix defines depends completely on the bases we're working in. For example, consider the 2×2 matrix $\begin{bmatrix} 1 & 0 \\ 0 & -1 \end{bmatrix}$. If we let the domain and codomain be $\mathbb{R}^2$ with the standard basis, the map defined by multiplication by this matrix gives the left-most picture in Figure 3.2; if we simply write the standard basis in the non-standard order (e_2, e_1) in both the domain and the codomain, we get the picture in the middle. If we use the standard basis (in the standard order) in the domain, but we use the non-standard order (e_2, e_1) in the codomain, we get the right-most picture.

Examples

1. Let I be the identity operator on a vector space V. If we use the same basis $\mathcal{B}$ for V in both the domain and codomain of I, then the matrix of I is the (appropriately sized) identity matrix. (Check carefully that this is really true!) However, if $\mathcal{B}_1$ and $\mathcal{B}_2$ are two different bases, then $[I]_{\mathcal{B}_1,\mathcal{B}_2}$ is not the identity matrix; all we can say about it is that it is invertible, but aside from that the entries could be anything.
2. Let $\mathcal{P}_n(\mathbb{R})$ denote the polynomials in x over $\mathbb{R}$ of degree n or less, and let $D : \mathcal{P}_n(\mathbb{R}) \to \mathcal{P}_{n-1}(\mathbb{R})$ be the derivative operator:

$$Df(x) := f'(x).$$

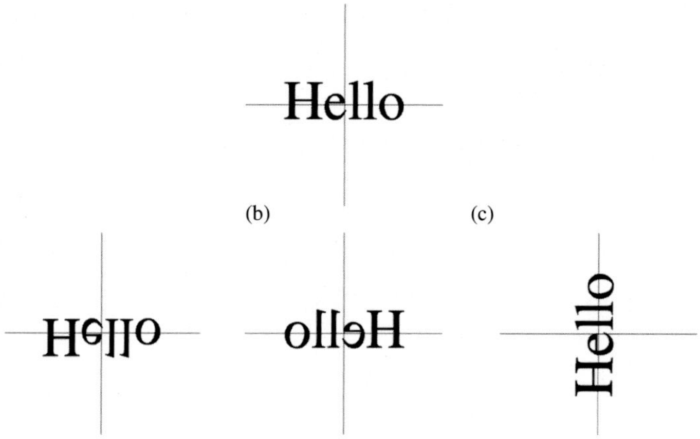

Figure 3.2 Multiplication by the matrix $\begin{bmatrix} 1 & 0 \\ 0 & -1 \end{bmatrix}$ when the bases in the domain and codomain are: (a) both $(\mathbf{e}_1, \mathbf{e}_2)$; (b) both $(\mathbf{e}_2, \mathbf{e}_1)$; (c) $(\mathbf{e}_1, \mathbf{e}_2)$ in the domain and $(\mathbf{e}_2, \mathbf{e}_1)$ in the codomain.

We saw in Section 2.2 that D is linear. The matrix of D with respect to the bases $\{1, x, x^2, \ldots, x^n\}$ and $\{1, x, x^2, \ldots, x^{n-1}\}$ is the $n \times (n+1)$ matrix

$$\begin{bmatrix} 0 & 1 & 0 & & 0 \\ 0 & 0 & 2 & & 0 \\ & & & \ddots & \\ 0 & 0 & 0 & & n \end{bmatrix}.$$

3. Let $T : \mathbb{R}^2 \to \mathbb{R}^2$ be given by reflection across the line $y = 2x$.

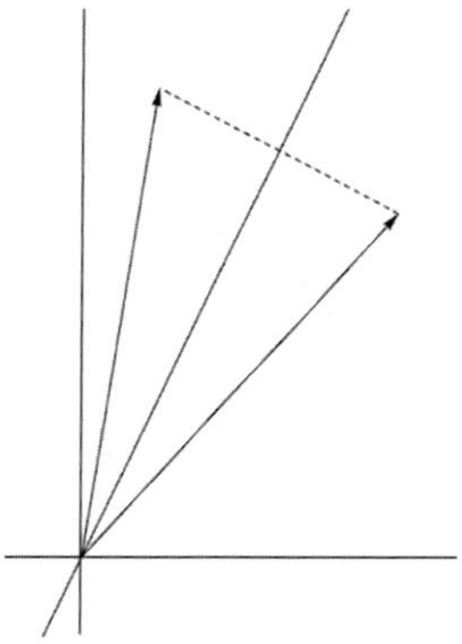

Figure 3.3 Reflection across $y = 2x$.

The matrix of T with respect to the standard basis is $\begin{bmatrix} -\frac{3}{5} & \frac{4}{5} \\ \frac{4}{5} & \frac{3}{5} \end{bmatrix}$.

Quick Exercise #24. Convince yourself that this is a non-trivial claim.

However, things look much simpler if we consider the basis $\mathcal{B} := \left(\begin{bmatrix} 1 \\ 2 \end{bmatrix}, \begin{bmatrix} -2 \\ 1 \end{bmatrix} \right)$. We have that $T\left(\begin{bmatrix} 1 \\ 2 \end{bmatrix} \right) = \begin{bmatrix} 1 \\ 2 \end{bmatrix}$ and $T\left(\begin{bmatrix} -2 \\ 1 \end{bmatrix} \right) = \begin{bmatrix} 2 \\ -1 \end{bmatrix}$, so the matrix of T with respect to $\mathcal{B}$ is

$$\begin{bmatrix} 1 & 0 \\ 0 & -1 \end{bmatrix}.$$

The matrix in this basis makes it easy to see that applying T twice gives the identity map, which is geometrically obvious but rather obscured by the form of the matrix in the standard basis. ▲

Our representation of linear maps as matrices with respect to arbitrary bases has the same nice properties that our representation with respect to the standard basis had. In particular, we have the following generalization of Theorem 2.9.

Theorem 3.42 *Suppose that* $\dim V = n$, $\dim W = m$, *and* $\mathcal{B}_V$ *and* $\mathcal{B}_W$ *are bases of* V *and* W, *respectively. Then the map* $C_{\mathcal{B}_V,\mathcal{B}_W} : \mathcal{L}(V, W) \to \mathrm{M}_{\dim W, \dim V}(\mathbb{F})$ *defined by*

$$C_{\mathcal{B}_V,\mathcal{B}_W}(T) = [T]_{\mathcal{B}_V,\mathcal{B}_W}$$

is an isomorphism.

Proof The proof is essentially identical to that of Theorem 2.9. Let $\mathcal{B}_V = (v_1, \ldots, v_n)$ and $\mathcal{B}_W = (w_1, \ldots, w_m)$. Let $T_1, T_2 \in \mathcal{L}(V, W)$. Then the jth column of the matrix $C_{\mathcal{B}_V,\mathcal{B}_W}(T_1 + T_2)$ is given by

$$\left[(T_1 + T_2)(v_j)\right]_{\mathcal{B}_W} = \left[T_1 v_j\right]_{\mathcal{B}_W} + \left[T_2 v_j\right]_{\mathcal{B}_W},$$

which is the sum of the jth columns of $C_{\mathcal{B}_V,\mathcal{B}_W}(T_1)$ and $C_{\mathcal{B}_V,\mathcal{B}_W}(T_2)$, so

$$C_{\mathcal{B}_V,\mathcal{B}_W}(T_1 + T_2) = C_{\mathcal{B}_V,\mathcal{B}_W}(T_1) + C_{\mathcal{B}_V,\mathcal{B}_W}(T_2).$$

Similarly, if $a \in \mathbb{F}$ and $T \in \mathcal{L}(V, W)$, then the jth column of $C_{\mathcal{B}_V,\mathcal{B}_W}(aT)$ is given by

$$\left[(aT)(v_j)\right]_{\mathcal{B}_W} = a\left[T v_j\right]_{\mathcal{B}_W},$$

so that

$$C_{\mathcal{B}_V,\mathcal{B}_W}(aT) = aC_{\mathcal{B}_V,\mathcal{B}_W}(T).$$

The map $C_{\mathcal{B}_V,\mathcal{B}_W}$ is surjective, since given a matrix $\mathrm{A} \in \mathrm{M}_{\dim V, \dim W}(\mathbb{F})$, one can define a linear map $T_{\mathrm{A},\mathcal{B}_V,\mathcal{B}_W} \in \mathcal{L}(V, W)$ as in (3.10), and then $C_{\mathcal{B}_V,\mathcal{B}_W}(T_{\mathrm{A},\mathcal{B}_V,\mathcal{B}_W}) = \mathrm{A}$.

If $C_{\mathcal{B}_V,\mathcal{B}_W}(T) = 0$, then for every $v_j \in \mathcal{B}_V$, $T(v_j) = 0$, and so T is the zero map. The map $C_{\mathcal{B}_V,\mathcal{B}_W}$ is therefore injective and we are done. ▲

Theorem 3.42 in particular immediately tells us the dimension of the vector space $\mathcal{L}(V, W)$:

Corollary 3.43 *Let V and W be finite-dimensional vector spaces. Then*

$$\dim \mathcal{L}(V, W) = (\dim V)(\dim W).$$

We have already observed that the matrix of a given linear map looks completely different depending on which bases are used. In particular, it is always possible to choose bases so that the matrix of an operator is extremely simple, as in the following result.

Theorem 3.44 *Let V and W be finite-dimensional vector spaces and $T \in \mathcal{L}(V, W)$. Then there are bases $\mathcal{B}_V$ of V and $\mathcal{B}_W$ of W such that if $\mathrm{A} = [T]_{\mathcal{B}_V,\mathcal{B}_W}$, then $a_{ii} = 1$ for $1 \le i \le \operatorname{rank} T$ and all other entries of A are 0.*

Proof Recall that in the linear map version of the Rank–Nullity Theorem, we showed that if $(u_1, \ldots, u_k)$ is a basis of $\ker(T)$ and $(u_1, \ldots, u_k, v_1, \ldots, v_\ell)$ is a basis of all of V, then $(T(v_1), \ldots, T(v_\ell))$ is a basis for $\operatorname{range}(T)$; in particular, $\ell = \operatorname{rank}(T)$. Take $\mathcal{B}_V := (v_1, \ldots, v_\ell, u_1, \ldots, u_k)$ and extend $(T(v_1), \ldots, T(v_\ell))$ to a basis

$$\mathcal{B}_W := (T(v_1), \ldots, T(v_\ell), w_1, \ldots, w_m)$$

of all of W. With respect to these bases, $[T]_{\mathcal{B}_V,\mathcal{B}_W}$ has the desired form. ▲

This result may look useful, but in most respects it really isn't – it essentially just shifts all the complication from the map to the bases. In particular, there is no clear connection between the bases used in the domain and in the codomain, other than to say they are related through the map T.

Eigenvectors and Diagonalizability

Recall the example of reflection in the line $y = 2x$; in the basis $\mathcal{B} = \left(\begin{bmatrix} 1 \\ 2 \end{bmatrix}, \begin{bmatrix} -2 \\ 1 \end{bmatrix}\right)$, the matrix of T is

$$\begin{bmatrix} 1 & 0 \\ 0 & -1 \end{bmatrix}.$$

This is an example of a **diagonal** matrix; i.e., one whose only nonzero entries are those in the (i, i) entries.

Diagonal matrices are nice to work with from a computational point of view, as demonstrated for example by Quick Exercise #3 in Section 2.1. There is also the following important connection to eigenvalues and eigenvectors.

Proposition 3.45 *Let $\mathcal{B} = (v_1, \dots, v_n)$ be a basis for V and let $T \in \mathcal{L}(V)$. The matrix $[T]_{\mathcal{B}}$ is diagonal if and only if for each $i = 1, \dots, n$, v_i is an eigenvector of T. Moreover, in this case the ith diagonal entry of $[T]_{\mathcal{B}}$ is the eigenvalue of T corresponding to v_i.*

Proof Suppose first that

$$[T]_{\mathcal{B}} = \mathrm{A} = \begin{bmatrix} \lambda_1 & & 0 \\ & \ddots & \\ 0 & & \lambda_n \end{bmatrix}.$$

By definition, this means that, for each j,

$$Tv_j = a_{1j}v_1 + \cdots + a_{nj}v_n = a_{jj}v_j = \lambda_j v_j,$$

and thus v_j is an eigenvector of T with eigenvalue λ_j.

Suppose now that, for each j, v_j is an eigenvector of T, say with eigenvalue $\lambda_j \in \mathbb{F}$. By definition, this means that $Tv_j = \lambda_j v_j$. Thus the jth column of $[T]_{\mathcal{B}}$ has jth entry equal to λ_j, and all other entries 0. ▲

Quick Exercise #25. Suppose that $T \in \mathcal{L}(V)$, and that $\mathcal{B} = (v_1, \dots, v_n)$ is a basis of V consisting of eigenvectors of T with eigenvalues $\lambda_1, \dots, \lambda_n$ respectively. Show that if $[v]_{\mathcal{B}} = \begin{bmatrix} c_1 \\ \vdots \\ c_n \end{bmatrix}$, then $[Tv]_{\mathcal{B}} = \begin{bmatrix} \lambda_1 c_1 \\ \vdots \\ \lambda_n c_n \end{bmatrix}$.

Definition A linear map $T \in \mathcal{L}(V)$ is **diagonalizable** if there exists a basis $\mathcal{B}$ of V such that $[T]_{\mathcal{B}}$ is diagonal. We refer to finding such a basis as **diagonalizing** the map.

It's important to note the distinction between the definition above and Theorem 3.44. We observed that that theorem was not especially useful because it shifted the complication from the map to the bases. The key difference here is that diagonalizable maps have diagonal matrices with *the same basis* being used in the domain and the codomain.

Corollary 3.46 *A linear map $T \in \mathcal{L}(V)$ is* **diagonalizable** *if and only if there exists a basis $\mathcal{B}$ of V consisting entirely of eigenvectors of T.*

QA #25: Solution 1: $v = \sum_{i=1}^{n} c_i v_i$, so $Tv = \sum_{i=1}^{n} c_i T v_i = \sum_{i=1}^{n} c_i \lambda_i v_i$.
Solution 2: Combine Proposition 3.45 with Lemma 3.41 and Quick Exercise #3 from Section 2.1.

Examples

1. We saw above that reflection in the line $y = 2x$ is a diagonalizable operator; $\mathcal{B} = \left(\begin{bmatrix} 1 \\ 2 \end{bmatrix}, \begin{bmatrix} -2 \\ 1 \end{bmatrix} \right)$ is a basis of eigenvectors (with corresponding eigenvalues 1 and -1).
2. Consider the map $T : \mathbb{R}^2 \to \mathbb{R}^2$ defined by $T\begin{bmatrix} x \\ y \end{bmatrix} := \begin{bmatrix} 0 \\ x \end{bmatrix}$. If $\begin{bmatrix} x \\ y \end{bmatrix}$ is an eigenvector of T with eigenvalue λ, then

$$\begin{bmatrix} 0 \\ x \end{bmatrix} = \begin{bmatrix} \lambda x \\ \lambda y \end{bmatrix},$$

so that either $x = 0$ or $\lambda = 0$. If $\lambda \neq 0$, then it must be that $x = 0$, from which the equality of the second entries gives that $y = 0$. Since the zero vector is never an eigenvector, this means that the only possible eigenvalue of T is $\lambda = 0$.

For $\begin{bmatrix} x \\ y \end{bmatrix}$ to be an eigenvector with eigenvalue 0, we need

$$\begin{bmatrix} 0 \\ x \end{bmatrix} = \begin{bmatrix} 0 \\ 0 \end{bmatrix};$$

i.e., $x = 0$. In particular, the zero eigenspace is spanned by $\begin{bmatrix} 0 \\ 1 \end{bmatrix}$, and so we cannot have a basis of eigenvectors; T is not diagonalizable. ▲

Matrix Multiplication and Coordinates

We introduced matrix multiplication in Section 2.3 as the operation of matrices that corresponded to composition of the associated linear maps; the following result shows that, even in arbitrary coordinates, matrix multiplication corresponds to composition.

Theorem 3.47 *Let U, V, and W be finite-dimensional vector spaces with bases $\mathcal{B}_U$, $\mathcal{B}_V$, and $\mathcal{B}_W$ respectively, and suppose that $T \in \mathcal{L}(U, V)$ and $S \in \mathcal{L}(V, W)$. Then*

$$[ST]_{\mathcal{B}_U,\mathcal{B}_W} = [S]_{\mathcal{B}_V,\mathcal{B}_W}[T]_{\mathcal{B}_U,\mathcal{B}_V}.$$

Proof Write $\mathcal{B}_U = (u_1, \ldots, u_p)$, $\mathcal{B}_V = (v_1, \ldots, v_n)$, and $\mathcal{B}_W = (w_1, \ldots, w_m)$, and let $\mathrm{A} = [S]_{\mathcal{B}_V,\mathcal{B}_W}$, $\mathrm{B} = [T]_{\mathcal{B}_U,\mathcal{B}_V}$, and $\mathrm{C} = [ST]_{\mathcal{B}_U,\mathcal{B}_W}$. We need to show that $\mathrm{C} = \mathrm{AB}$.

By the definition of A, for each $k = 1, \ldots, n$, the kth column of A is the coordinate representation of Sv_k with respect to $\mathcal{B}_W$. That is,

$$Sv_k = \sum_{i=1}^{m} a_{ik} w_i. \tag{3.11}$$

Similarly, for each $j = 1, \ldots, p$,

$$Tu_j = \sum_{k=1}^{n} b_{kj} v_k. \tag{3.12}$$

By linearity, (3.12) and (3.11) imply that

$$STu_j = \sum_{k=1}^{n} b_{kj} Sv_k = \sum_{k=1}^{n} b_{kj} \left(\sum_{i=1}^{m} a_{ik} w_i \right) = \sum_{i=1}^{m} \left(\sum_{k=1}^{n} a_{ik} b_{kj} \right) w_i,$$

which means that

$$c_{ij} = \sum_{k=1}^{n} a_{ik} b_{kj}$$

for each i and j, which is the same as $\mathbf{C} = \mathbf{AB}$. ▲

Example Consider the map $T : \mathbb{R}^2 \to \mathbb{R}^2$ given by first reflecting across the line $y = 2x$, then rotating counterclockwise by an angle of 2θ, where θ is the angle the line $y = 2x$ makes with the y-axis. We have seen (see the example on page 189) that in the basis $\mathcal{B} = \left(\begin{bmatrix} 1 \\ 2 \end{bmatrix}, \begin{bmatrix} -2 \\ 1 \end{bmatrix} \right)$, the matrix of the reflection is $\begin{bmatrix} 1 & 0 \\ 0 & -1 \end{bmatrix}$.

Quick Exercise #26. Show that if we use the basis $\mathcal{B}$ in the domain and the standard basis $\mathcal{E} = (\mathbf{e}_1, \mathbf{e}_2)$ in the codomain, the matrix of the rotation by 2θ is

$$\begin{bmatrix} -1 & -2 \\ 2 & -1 \end{bmatrix}.$$

It follows that

$$[T]_{\mathcal{B},\mathcal{E}} = \begin{bmatrix} -1 & -2 \\ 2 & -1 \end{bmatrix} \begin{bmatrix} 1 & 0 \\ 0 & -1 \end{bmatrix} = \begin{bmatrix} -1 & 2 \\ 2 & 1 \end{bmatrix}.$$ ▲

Corollary 3.48 *Let V and W be finite-dimensional vector spaces with bases $\mathcal{B}_V$ and $\mathcal{B}_W$, respectively. Suppose that $T \in \mathcal{L}(V, W)$ and let $\mathrm{A} = [T]_{\mathcal{B}_V,\mathcal{B}_W}$. Then A is invertible if and only if T is invertible, and in that case $\left[T^{-1}\right]_{\mathcal{B}_W,\mathcal{B}_V} = \mathrm{A}^{-1}$.*

QA #26: Since θ is exactly the angle the line $y = 2x$ makes with the y-axis, if R denotes the rotation, then $R\begin{bmatrix} 1 \\ 2 \end{bmatrix} = \begin{bmatrix} -1 \\ 2 \end{bmatrix}$ and $R\begin{bmatrix} -2 \\ 1 \end{bmatrix} = \begin{bmatrix} -2 \\ -1 \end{bmatrix}$.

Proof Suppose first that T is invertible, and let $\mathrm{B} = \left[T^{-1}\right]_{\mathcal{B}_W,\mathcal{B}_V}$. Since T is invertible, V and W have the same dimension, say n. By Theorem 3.47,

$$\mathrm{AB} = \left[TT^{-1}\right]_{\mathcal{B}_W,\mathcal{B}_W} = [I]_{\mathcal{B}_W,\mathcal{B}_W} = \mathrm{I}_n,$$

and so by Corollary 3.37, A is invertible and $\mathrm{B} = \mathrm{A}^{-1}$.

Now suppose that A is invertible, which implies that A is square and so $\dim V = \dim W$. Let $S \in \mathcal{L}(W, V)$ be the linear map with $[S]_{\mathcal{B}_W,\mathcal{B}_V} = \mathrm{A}^{-1}$. Then by Theorem 3.47,

$$[ST]_{\mathcal{B}_V,\mathcal{B}_V} = \mathrm{A}^{-1}\mathrm{A} = \mathrm{I}_n,$$

and so $ST = I$. Thus by Corollary 3.37, T is invertible and $S = T^{-1}$. ▲

KEY IDEAS

- Saying that the vector v has coordinates $[v]_{\mathcal{B}} = \begin{bmatrix} c_1 \\ \vdots \\ c_n \end{bmatrix}$ means exactly that $v = c_1 v_1 + \cdots + c_n v_n$, where $\mathcal{B} = (v_1, \ldots, v_n)$.
- The matrix of T with respect to bases $\mathcal{B}_V, \mathcal{B}_W$ is denoted $[T]_{\mathcal{B}_V,\mathcal{B}_W}$ and defined so that $[T]_{\mathcal{B}_V,\mathcal{B}_W} [v]_{\mathcal{B}_V} = [Tv]_{\mathcal{B}_W}$.
- The matrix $[T]_{\mathcal{B}_1,\mathcal{B}_2}$ has jth column $\left[Tv_j\right]_{\mathcal{B}_2}$.
- A linear map $T \in \mathcal{L}(V)$ is diagonalizable if there is a basis of V in which the matrix of T is diagonal. In this case, all of the basis vectors are eigenvectors of T.
- $[S]_{\mathcal{B}_V,\mathcal{B}_W} [T]_{\mathcal{B}_U,\mathcal{B}_V} = [ST]_{\mathcal{B}_U,\mathcal{B}_W}$.
- $\left[T^{-1}\right]_{\mathcal{B}_W,\mathcal{B}_V} = [T]^{-1}_{\mathcal{B}_V,\mathcal{B}_W}$.

EXERCISES

3.5.1 Find the coordinate representations of each of the following vectors with respect to the basis $\mathcal{B} = \left(\begin{bmatrix} 2 \\ -3 \end{bmatrix}, \begin{bmatrix} -1 \\ 2 \end{bmatrix} \right)$ of $\mathbb{R}^2$.

(a) $\begin{bmatrix} 1 \\ 0 \end{bmatrix}$ (b) $\begin{bmatrix} 0 \\ 1 \end{bmatrix}$ (c) $\begin{bmatrix} 2 \\ 3 \end{bmatrix}$ (d) $\begin{bmatrix} 4 \\ -5 \end{bmatrix}$ (e) $\begin{bmatrix} -6 \\ 1 \end{bmatrix}$

3.5.2 Find the coordinate representations of each of the following vectors with respect to the basis $\mathcal{B} = \left(\begin{bmatrix} 1 \\ 2 \end{bmatrix}, \begin{bmatrix} 3 \\ 4 \end{bmatrix} \right)$ of $\mathbb{R}^2$.

(a) $\begin{bmatrix} 1 \\ 0 \end{bmatrix}$ (b) $\begin{bmatrix} 0 \\ 1 \end{bmatrix}$ (c) $\begin{bmatrix} 2 \\ 3 \end{bmatrix}$ (d) $\begin{bmatrix} 4 \\ -5 \end{bmatrix}$ (e) $\begin{bmatrix} -6 \\ 1 \end{bmatrix}$

3.5.3 Find the coordinate representations of each of the following vectors with respect to the basis $\mathcal{B} = \left(\begin{bmatrix} 1 \\ -1 \\ 0 \end{bmatrix}, \begin{bmatrix} 1 \\ 0 \\ 1 \end{bmatrix}, \begin{bmatrix} 0 \\ 1 \\ -1 \end{bmatrix} \right)$ of $\mathbb{R}^3$.

(a) $\begin{bmatrix} 1 \\ 0 \\ 0 \end{bmatrix}$ (b) $\begin{bmatrix} 0 \\ 0 \\ 1 \end{bmatrix}$ (c) $\begin{bmatrix} -4 \\ 0 \\ 2 \end{bmatrix}$ (d) $\begin{bmatrix} 1 \\ 2 \\ -3 \end{bmatrix}$ (e) $\begin{bmatrix} 6 \\ -4 \\ 1 \end{bmatrix}$

3.5.4 Find the coordinate representations of each of the following vectors with respect to the basis $\mathcal{B} = \left(\begin{bmatrix} 1 \\ -1 \\ 1 \end{bmatrix}, \begin{bmatrix} 1 \\ 2 \\ -1 \end{bmatrix}, \begin{bmatrix} 0 \\ 2 \\ -1 \end{bmatrix} \right)$ of $\mathbb{R}^3$.

(a) $\begin{bmatrix} 1 \\ 0 \\ 0 \end{bmatrix}$ (b) $\begin{bmatrix} 0 \\ 0 \\ 1 \end{bmatrix}$ (c) $\begin{bmatrix} -4 \\ 0 \\ 2 \end{bmatrix}$ (d) $\begin{bmatrix} 1 \\ 2 \\ -3 \end{bmatrix}$ (e) $\begin{bmatrix} 6 \\ -4 \\ 1 \end{bmatrix}$

3.5.5 Consider the bases $\mathcal{B} = \left(\begin{bmatrix} 2 \\ 1 \end{bmatrix}, \begin{bmatrix} 1 \\ 1 \end{bmatrix} \right)$ of $\mathbb{R}^2$ and

$\mathcal{C} = \left(\begin{bmatrix} 1 \\ 3 \\ 0 \end{bmatrix}, \begin{bmatrix} 0 \\ 2 \\ 3 \end{bmatrix}, \begin{bmatrix} 1 \\ 0 \\ -4 \end{bmatrix} \right)$ of $\mathbb{R}^3$, and the linear maps $S \in \mathcal{L}(\mathbb{R}^2, \mathbb{R}^3)$ and $T \in \mathcal{L}(\mathbb{R}^3, \mathbb{R}^2)$ given (with respect to the standard bases) by

$$[S]_{\mathcal{E},\mathcal{E}} = \begin{bmatrix} 0 & 1 \\ 3 & -3 \\ 4 & -8 \end{bmatrix} \quad \text{and} \quad [T]_{\mathcal{E},\mathcal{E}} = \begin{bmatrix} 4 & -1 & 1 \\ 3 & -1 & 1 \end{bmatrix}.$$

Find each of the following coordinate representations.
(a) $[S]_{\mathcal{B},\mathcal{E}}$ (b) $[S]_{\mathcal{E},\mathcal{C}}$ (c) $[S]_{\mathcal{B},\mathcal{C}}$ (d) $[T]_{\mathcal{C},\mathcal{E}}$ (e) $[T]_{\mathcal{E},\mathcal{B}}$ (f) $[T]_{\mathcal{C},\mathcal{B}}$

3.5.6 Consider the bases $\mathcal{B} = \left(\begin{bmatrix} 2 \\ 3 \end{bmatrix}, \begin{bmatrix} 3 \\ 5 \end{bmatrix} \right)$ of $\mathbb{R}^2$ and

$\mathcal{C} = \left(\begin{bmatrix} 1 \\ 1 \\ 0 \end{bmatrix}, \begin{bmatrix} 1 \\ 0 \\ 1 \end{bmatrix}, \begin{bmatrix} 0 \\ 1 \\ 1 \end{bmatrix} \right)$ of $\mathbb{R}^3$, and the linear maps $S \in \mathcal{L}(\mathbb{R}^2, \mathbb{R}^3)$ and $T \in \mathcal{L}(\mathbb{R}^3, \mathbb{R}^2)$ given (with respect to the standard bases) by

$$[S]_{\mathcal{E},\mathcal{E}} = \begin{bmatrix} 2 & -1 \\ 5 & -3 \\ -3 & 2 \end{bmatrix} \quad \text{and} \quad [T]_{\mathcal{E},\mathcal{E}} = \begin{bmatrix} 1 & -1 & 1 \\ 1 & 1 & -1 \end{bmatrix}.$$

Find each of the following coordinate representations.
(a) $[S]_{\mathcal{B},\mathcal{E}}$ (b) $[S]_{\mathcal{E},\mathcal{C}}$ (c) $[S]_{\mathcal{B},\mathcal{C}}$ (d) $[T]_{\mathcal{C},\mathcal{E}}$ (e) $[T]_{\mathcal{E},\mathcal{B}}$ (f) $[T]_{\mathcal{C},\mathcal{B}}$

3.5.7 Consider the bases $\mathcal{B} = \left(\begin{bmatrix} 1 \\ 1 \end{bmatrix}, \begin{bmatrix} 1 \\ -1 \end{bmatrix} \right)$ of $\mathbb{R}^2$ and $\mathcal{C} = \left(\begin{bmatrix} 1 \\ 1 \\ 0 \end{bmatrix}, \begin{bmatrix} 1 \\ 0 \\ 1 \end{bmatrix}, \begin{bmatrix} 0 \\ 1 \\ 1 \end{bmatrix} \right)$ of $\mathbb{R}^3$, and let $T \in \mathcal{L}(\mathbb{R}^3, \mathbb{R}^2)$ be given by

$$T \begin{bmatrix} x \\ y \\ z \end{bmatrix} = \begin{bmatrix} z \\ y \end{bmatrix}.$$

(a) Find each of the following coordinate representations.

(i) $\left[\begin{bmatrix} 1 \\ 2 \end{bmatrix} \right]_{\mathcal{B}}$ (ii) $\left[\begin{bmatrix} -2 \\ 3 \end{bmatrix} \right]_{\mathcal{B}}$ (iii) $\left[\begin{bmatrix} 1 \\ 2 \\ 3 \end{bmatrix} \right]_{\mathcal{C}}$ (iv) $\left[\begin{bmatrix} 3 \\ -2 \\ 1 \end{bmatrix} \right]_{\mathcal{C}}$

(v) $[T]_{\mathcal{C},\mathcal{E}}$ (vi) $[T]_{\mathcal{E},\mathcal{B}}$ (vii) $[T]_{\mathcal{C},\mathcal{B}}$

(b) Compute each of the following in two ways: by finding $T\mathbf{x}$ directly and finding the coordinate representation, and by using your answers to the previous parts and Lemma 3.41.

(i) $\left[T \begin{bmatrix} 1 \\ 2 \\ 3 \end{bmatrix} \right]_{\mathcal{E}}$ (ii) $\left[T \begin{bmatrix} 3 \\ -2 \\ 1 \end{bmatrix} \right]_{\mathcal{B}}$

3.5.8 Consider the bases $\mathcal{B} = \left(\begin{bmatrix} 1 \\ 1 \end{bmatrix}, \begin{bmatrix} 1 \\ -1 \end{bmatrix} \right)$ of $\mathbb{R}^2$ and $\mathcal{C} = \left(\begin{bmatrix} 1 \\ 1 \\ 0 \end{bmatrix}, \begin{bmatrix} 1 \\ 0 \\ 1 \end{bmatrix}, \begin{bmatrix} 0 \\ 1 \\ 1 \end{bmatrix} \right)$ of $\mathbb{R}^3$, and let $T \in \mathcal{L}(\mathbb{R}^3, \mathbb{R}^2)$ be given by

$$T \begin{bmatrix} x \\ y \\ z \end{bmatrix} = \begin{bmatrix} y \\ x \end{bmatrix}.$$

(a) Find each of the following coordinate representations.

(i) $\left[\begin{bmatrix} -2 \\ 1 \end{bmatrix} \right]_{\mathcal{B}}$ (ii) $\left[\begin{bmatrix} 2 \\ 3 \end{bmatrix} \right]_{\mathcal{B}}$ (iii) $\left[\begin{bmatrix} 1 \\ -2 \\ 3 \end{bmatrix} \right]_{\mathcal{C}}$ (iv) $\left[\begin{bmatrix} 3 \\ 2 \\ 1 \end{bmatrix} \right]_{\mathcal{C}}$

(v) $[T]_{\mathcal{C},\mathcal{E}}$ (vi) $[T]_{\mathcal{E},\mathcal{B}}$ (vii) $[T]_{\mathcal{C},\mathcal{B}}$

(b) Compute each of the following in two ways: by finding $T\mathbf{x}$ directly and finding the coordinate representation, and by using your answers to the previous parts and Lemma 3.41.

(i) $\left[T \begin{bmatrix} 1 \\ -2 \\ 3 \end{bmatrix} \right]_{\mathcal{E}}$ (ii) $\left[T \begin{bmatrix} 3 \\ 2 \\ 1 \end{bmatrix} \right]_{\mathcal{B}}$

3.5.9 Let P be the plane

$$\left\{ \begin{bmatrix} x \\ y \\ z \end{bmatrix} \in \mathbb{R}^3 \,\middle|\, x - 2y + 3z = 0 \right\}.$$

(a) Find a basis for P.

(b) Determine whether each of the following vectors is in P, and for each one i.e., give its coordinate representation in terms of your basis.

(i) $\begin{bmatrix} 1 \\ -1 \\ -1 \end{bmatrix}$ (ii) $\begin{bmatrix} 2 \\ 3 \\ 1 \end{bmatrix}$ (iii) $\begin{bmatrix} 5 \\ -2 \\ -3 \end{bmatrix}$

3.5.10 Let P be the plane

$$\left\{ \begin{bmatrix} x \\ y \\ z \end{bmatrix} \in \mathbb{R}^3 \,\middle|\, 4x + y - 2z = 0 \right\}.$$

(a) Find a basis for P.

(b) Determine whether each of the following vectors is in P, and for each one i.e., give its coordinate representation in terms of your basis.

(i) $\begin{bmatrix} 1 \\ 1 \\ 1 \end{bmatrix}$ (ii) $\begin{bmatrix} 0 \\ 2 \\ 1 \end{bmatrix}$ (iii) $\begin{bmatrix} 1 \\ -2 \\ 1 \end{bmatrix}$

3.5.11 Let $D : \mathcal{P}_n(\mathbb{R}) \to \mathcal{P}_{n-1}(\mathbb{R})$ be the derivative operator, and let $T : \mathcal{P}_{n-1}(\mathbb{R}) \to \mathcal{P}_n(\mathbb{R})$ be the linear map given by multiplication by x:

$$(Tp)(x) = xp(x).$$

Find the matrices of the maps T, DT, and TD with respect to the bases $(1, x, \ldots, x^n)$ and $(1, x, \ldots, x^{n-1})$.

3.5.12 (a) Show that $\mathcal{B} = \left(1, x, \frac{3}{2}x^2 - \frac{1}{2}\right)$ is a basis of $\mathcal{P}_2(\mathbb{R})$.

(b) Find the coordinate representation of x^2 with respect to $\mathcal{B}$.

(c) Let $D : \mathcal{P}_2(\mathbb{R}) \to \mathcal{P}_2(\mathbb{R})$ be the derivative operator. Find the coordinate representation of D with respect to $\mathcal{B}$ (i.e., with the same basis $\mathcal{B}$ on both the domain and the codomain).

(d) Use your answers to the last two parts to calculate $\frac{d}{dx}(x^2)$.

3.5.13 Suppose that $T \in \mathcal{L}(V)$ is invertible, and that for some basis $\mathcal{B}$ of V, $[T]_{\mathcal{B}} = \operatorname{diag}(\lambda_1, \ldots, \lambda_n)$. What is $\left[T^{-1}\right]_{\mathcal{B}}$?

3.5.14 Let $R \in \mathcal{L}\left(\mathbb{R}^2\right)$ be the counterclockwise rotation of the plane by $\pi/2$ radians.

(a) Show that R is not diagonalizable.

(b) Show that R^2 is diagonalizable.

3.5.15 Show that the projection $P \in \mathcal{L}(\mathbb{R}^3)$ onto the x–y plane is diagonalizable.

3.5.16 Let L be a line through the origin in $\mathbb{R}^2$, and let $P \in \mathcal{L}(\mathbb{R}^2)$ be the orthogonal projection onto L. (See Exercise 2.1.2.) Show that P is diagonalizable.

3.5.17 Suppose that $\mathcal{B} = (\mathbf{v}_1, \ldots, \mathbf{v}_n)$ is a linearly independent list in $\mathbb{F}^m$, so that it is a basis for $U = \langle \mathcal{B} \rangle$. Let $\mathbf{A} = \begin{bmatrix} \mathbf{v}_1 & \cdots & \mathbf{v}_n \end{bmatrix} \in \mathrm{M}_{m,n}(\mathbb{F})$ be the matrix whose jth column is $\mathbf{v}_j$. Show that, for any $\mathbf{u} \in U$, $\mathbf{u} = \mathbf{A}\,[\mathbf{u}]_{\mathcal{B}}$.

3.5.18 Suppose that $\mathcal{B} = (\mathbf{v}_1, \ldots, \mathbf{v}_n)$ is a basis for $\mathbb{F}^n$, and let $\mathbf{A} = \begin{bmatrix} \mathbf{v}_1 & \cdots & \mathbf{v}_n \end{bmatrix} \in \mathrm{M}_n(\mathbb{F})$ be the matrix whose jth column is $\mathbf{v}_j$. Show that, for any $\mathbf{x} \in \mathbb{F}^n$, $[\mathbf{x}]_{\mathcal{B}} = \mathbf{A}^{-1}\mathbf{x}$.

3.5.19 Show that if $S, T \in \mathcal{L}(V)$ are both diagonalized by a basis $\mathcal{B}$, then ST is also diagonalized by $\mathcal{B}$.

3.5.20 Show that if $\mathcal{B}_1$ and $\mathcal{B}_2$ are bases of V, then $[I]_{\mathcal{B}_1,\mathcal{B}_2}$ is an invertible matrix, and find its inverse.

3.6 Change of Basis

Change of Basis Matrices

Say $\mathcal{B} = (v_1, \ldots, v_n)$ and $\mathcal{B}' = (v'_1, \ldots, v'_n)$ are two different bases of V. In an abstract sense, at least, knowing the coordinate representation $[v]_{\mathcal{B}}$ of a vector $v \in V$ is equivalent to knowing the vector v itself, and so it determines the coordinate representation $[v]_{\mathcal{B}'}$. How can we actually express $[v]_{\mathcal{B}'}$ in terms of $[v]_{\mathcal{B}}$?

Theorem 3.49 *Let $\mathcal{B}$ and $\mathcal{B}'$ be two different bases of V. Then for any $v \in V$,*

$$[v]_{\mathcal{B}'} = [I]_{\mathcal{B},\mathcal{B}'}\,[v]_{\mathcal{B}}\,.$$

Proof Applying Lemma 3.41 to the identity map $I : V \to V$,

$$[v]_{\mathcal{B}'} = [Iv]_{\mathcal{B}'} = [I]_{\mathcal{B},\mathcal{B}'}\,[v]_{\mathcal{B}}\,.$$

▲

Definition Let $\mathcal{B}$ and $\mathcal{B}'$ be two different bases of V. The matrix $\mathbf{S} = [I]_{\mathcal{B},\mathcal{B}'}$ is called the **change of basis matrix** from $\mathcal{B}$ to $\mathcal{B}'$.

Example Let $\mathcal{B} = \left(\begin{bmatrix} 1 \\ 2 \end{bmatrix}, \begin{bmatrix} 2 \\ -1 \end{bmatrix} \right)$ and $\mathcal{B}' = \left(\begin{bmatrix} 1 \\ 1 \end{bmatrix}, \begin{bmatrix} 1 \\ -1 \end{bmatrix} \right)$. Then to find $[I]_{\mathcal{B},\mathcal{B}'}$, we need to find $[\mathbf{v}]_{\mathcal{B}'}$ for each of the elements $\mathbf{v} \in \mathcal{B}$; i.e., we need to identify the coefficients needed to express $\mathbf{v}$ as a linear combination of the elements of $\mathcal{B}'$. This is equivalent to solving the linear systems with augmented matrices

$$\left[\begin{array}{cc|c} 1 & 1 & 1 \\ 1 & -1 & 2 \end{array}\right] \quad \text{and} \quad \left[\begin{array}{cc|c} 1 & 1 & 2 \\ 1 & -1 & -1 \end{array}\right].$$

The solution of the first system is $\frac{1}{2}\begin{bmatrix} 3 \\ -1 \end{bmatrix}$; that is,

$$\begin{bmatrix} 1 \\ 2 \end{bmatrix} = \frac{3}{2}\begin{bmatrix} 1 \\ 1 \end{bmatrix} - \frac{1}{2}\begin{bmatrix} 1 \\ -1 \end{bmatrix} \iff \left[\begin{bmatrix} 1 \\ 2 \end{bmatrix}\right]_{\mathcal{B}'} = \frac{1}{2}\begin{bmatrix} 3 \\ -1 \end{bmatrix}.$$

Solving the second system gives that the change of basis matrix $[I]_{\mathcal{B},\mathcal{B}'}$ is

$$[I]_{\mathcal{B},\mathcal{B}'} = \frac{1}{2}\begin{bmatrix} 3 & 1 \\ -1 & 3 \end{bmatrix}. \tag{3.13}$$

▲

It turns out that there is a more systematic way to find change of basis matrices when $V = \mathbb{F}^n$, using the following two lemmas.

Lemma 3.50 *Let $\mathcal{E}$ be the standard basis of $\mathbb{F}^n$, and let $\mathcal{B}$ be any other basis of $\mathbb{F}^n$. Then the columns of the change of basis matrix $\mathrm{S} = [I]_{\mathcal{B},\mathcal{E}}$ are the column vectors in $\mathcal{B}$, in order.*

Proof Let $\mathcal{B} = (\mathbf{v}_1, \ldots, \mathbf{v}_n)$. The jth column of S is the coordinate representation of $I\mathbf{v}_j = \mathbf{v}_j$ with respect to $\mathcal{E}$. Since $\mathcal{E}$ is the standard basis of $\mathbb{F}^n$, the jth column of S is $\mathbf{v}_j$ itself. ▲

Lemma 3.51 *Let $\mathcal{B}$ and $\mathcal{B}'$ be two bases of V, and let S be the change of basis matrix from $\mathcal{B}$ to $\mathcal{B}'$. Then the change of basis matrix from $\mathcal{B}'$ to $\mathcal{B}$ is S^{-1}.*

Proof By definition, $\mathrm{S} = [I]_{\mathcal{B},\mathcal{B}'}$, and so by Theorem 3.47,

$$\mathrm{S}\,[I]_{\mathcal{B}',\mathcal{B}} = [I]_{\mathcal{B},\mathcal{B}'}\,[I]_{\mathcal{B}',\mathcal{B}} = [I]_{\mathcal{B}',\mathcal{B}'} = \mathrm{I}_n.$$

Since S is a square matrix, this proves that $[I]_{\mathcal{B}',\mathcal{B}} = \mathrm{S}^{-1}$ by Corollary 3.37. ▲

Quick Exercise #27. Let $\mathcal{B} = \left(\begin{bmatrix} 1 \\ 2 \end{bmatrix}, \begin{bmatrix} 2 \\ -1 \end{bmatrix}\right)$ and $\mathcal{B}' = \left(\begin{bmatrix} 1 \\ 1 \end{bmatrix}, \begin{bmatrix} 1 \\ -1 \end{bmatrix}\right)$. Find the change of basis matrix from $\mathcal{B}'$ to $\mathcal{B}$.

QA #27: Invert the change of basis matrix from (3.13) to get $\frac{1}{5}\begin{bmatrix} 3 & -1 \\ 1 & 3 \end{bmatrix}$.

Theorem 3.47 and Lemma 3.51 now show that if $\mathcal{B}$ and $\mathcal{C}$ are two bases of $\mathbb{F}^n$, then the change of basis matrix $[I]_{\mathcal{B},\mathcal{C}}$ is

$$[I]_{\mathcal{B},\mathcal{C}} = [I]_{\mathcal{E},\mathcal{C}} [I]_{\mathcal{B},\mathcal{E}} = \left([I]_{\mathcal{C},\mathcal{E}}\right)^{-1} [I]_{\mathcal{B},\mathcal{E}}, \tag{3.14}$$

and Lemma 3.50 tells what both the matrices in this last expression are.

Example We saw in Section 3.2 that the list of vectors $\mathcal{B} = (\mathbf{v}_1, \dots, \mathbf{v}_n)$ in $\mathbb{F}^n$ given by

$$\mathbf{v}_j = \mathbf{e}_1 + \cdots + \mathbf{e}_j$$

for $1 \le j \le n$ is a basis of $\mathbb{F}^n$. By Lemma 3.50, the change of basis matrix $[I]_{\mathcal{B},\mathcal{E}}$ is just $\mathbf{A} = \begin{bmatrix} \mathbf{v}_1 & \cdots & \mathbf{v}_n \end{bmatrix}$, and the change of basis matrix $[I]_{\mathcal{E},\mathcal{B}}$ is $\mathbf{A}^{-1}$, which we saw in the example on page 99 is

$$\mathbf{B} = \begin{bmatrix} 1 & -1 & 0 & \cdots & \cdots & 0 \\ 0 & 1 & -1 & 0 & & \vdots \\ 0 & 0 & 1 & -1 & \ddots & \vdots \\ \vdots & & \ddots & \ddots & \ddots & 0 \\ 0 & & & \ddots & 1 & -1 \\ 0 & \cdots & \cdots & \cdots & 0 & 1 \end{bmatrix}.$$

Therefore, for example, the coordinate representation of $\begin{bmatrix} 1 \\ 2 \\ \vdots \\ n \end{bmatrix}$ with respect to $\mathcal{B}$ is

$$\mathbf{B} \begin{bmatrix} 1 \\ 2 \\ \vdots \\ n \end{bmatrix} = \begin{bmatrix} 1-2 \\ 2-3 \\ \vdots \\ (n-1)-n \\ n \end{bmatrix} = \begin{bmatrix} -1 \\ -1 \\ \vdots \\ -1 \\ n \end{bmatrix}.$$

Similarly, if $\mathcal{C}$ is the basis $(\mathbf{w}_1, \dots, \mathbf{w}_n)$ of $\mathbb{F}^n$ given by $\mathbf{w}_1 = \mathbf{e}_1$ and

$$\mathbf{w}_j = \mathbf{e}_j - \mathbf{e}_{j-1}$$

for $2 \le j \le n$, then $[I]_{\mathcal{C},\mathcal{E}} = \mathbf{B}$ and $[I]_{\mathcal{E},\mathcal{C}} = \mathbf{A}$, where $\mathbf{A}$ and $\mathbf{B}$ are the same matrices as above. By formula (3.14),

$$[I]_{\mathcal{B},\mathcal{C}} = \left([I]_{\mathcal{E},\mathcal{C}}\right)^{-1} [I]_{\mathcal{B},\mathcal{E}} = \mathrm{A}^2 = \begin{bmatrix} 1 & 2 & 3 & \cdots & \cdots & n \\ 0 & 1 & 2 & \cdots & \cdots & n-1 \\ 0 & 0 & 1 & \ddots & & n-2 \\ \vdots & & \ddots & \ddots & \ddots & \vdots \\ 0 & & & \ddots & 1 & 2 \\ 0 & \cdots & \cdots & \cdots & 0 & 1 \end{bmatrix}.$$

▲

Theorem 3.52 *Let $\mathcal{B}_V, \mathcal{B}'_V$ be two bases of V, let $\mathcal{B}_W, \mathcal{B}'_W$ be two bases of W, and let $T \in \mathcal{L}(V, W)$. Then*

$$[T]_{\mathcal{B}'_V,\mathcal{B}'_W} = [I]_{\mathcal{B}_W,\mathcal{B}'_W} [T]_{\mathcal{B}_V,\mathcal{B}_W} [I]_{\mathcal{B}'_V,\mathcal{B}_V}.$$

Proof Using Theorem 3.47 twice,

$$[T]_{\mathcal{B}'_V,\mathcal{B}'_W} = [ITI]_{\mathcal{B}'_V,\mathcal{B}'_W} = [I]_{\mathcal{B}_W,\mathcal{B}'_W} [T]_{\mathcal{B}_V,\mathcal{B}_W} [I]_{\mathcal{B}'_V,\mathcal{B}_V}.$$ ▲

The most important special case of Theorem 3.52 is when $V = W$, $\mathcal{B}_V = \mathcal{B}_W$, and $\mathcal{B}'_V = \mathcal{B}'_W$.

Corollary 3.53 *Let $\mathcal{B}$ and $\mathcal{B}'$ be two bases of V, let S be the change of basis matrix from $\mathcal{B}$ to $\mathcal{B}'$. Given $T \in \mathcal{L}(V)$, let $\mathrm{A} = [T]_{\mathcal{B}}$. Then*

$$[T]_{\mathcal{B}'} = \mathrm{SAS}^{-1}.$$

Proof By Theorem 3.52 and Lemma 3.51,

$$[T]_{\mathcal{B}'} = [I]_{\mathcal{B},\mathcal{B}'} [T]_{\mathcal{B}} [I]_{\mathcal{B}',\mathcal{B}} = \mathrm{SAS}^{-1}.$$ ▲

In summary, we have the following:

The Change of Basis Matrix

If V has bases $\mathcal{B}$ and $\mathcal{B}'$, the change of basis matrix is $\mathrm{S} := [I]_{\mathcal{B},\mathcal{B}'}$; i.e.,

$$[v]_{\mathcal{B}'} = \mathrm{S}\,[v]_{\mathcal{B}}, \qquad [T]_{\mathcal{B}'} = \mathrm{S}\,[T]_{\mathcal{B}}\,\mathrm{S}^{-1}.$$

Example Recall the example on page 189: let $T : \mathbb{R}^2 \to \mathbb{R}^2$ be reflection across the line $y = 2x$. We saw that if $\mathcal{B} = \left(\begin{bmatrix} 1 \\ 2 \end{bmatrix}, \begin{bmatrix} -2 \\ 1 \end{bmatrix} \right)$, then

$$[T]_{\mathcal{B}} = \begin{bmatrix} 1 & 0 \\ 0 & -1 \end{bmatrix}.$$

To recover the matrix of T with respect to the standard basis $\mathcal{E}$, we first compute the change of basis matrix

$$\mathrm{S} = [I]_{\mathcal{B},\mathcal{E}} = \begin{bmatrix} 1 & -2 \\ 2 & 1 \end{bmatrix},$$

and its inverse (by Gaussian elimination):

$$\left[\begin{array}{cc|cc} 1 & -2 & 1 & 0 \\ 2 & 1 & 0 & 1 \end{array}\right] \rightsquigarrow \left[\begin{array}{cc|cc} 1 & 0 & \frac{1}{5} & \frac{2}{5} \\ 0 & 1 & -\frac{2}{5} & \frac{1}{5} \end{array}\right],$$

so

$$\mathrm{S}^{-1} = \frac{1}{5} \begin{bmatrix} 1 & 2 \\ -2 & 1 \end{bmatrix}.$$

It then follows that

$$[T]_{\mathcal{E}} = \begin{bmatrix} 1 & -2 \\ 2 & 1 \end{bmatrix} \begin{bmatrix} 1 & 0 \\ 0 & -1 \end{bmatrix} \left(\frac{1}{5} \begin{bmatrix} 1 & 2 \\ -2 & 1 \end{bmatrix} \right) = \frac{1}{5} \begin{bmatrix} -3 & 4 \\ 4 & 3 \end{bmatrix}.$$

▲

The example gives one illustration of why changing bases is desirable: it is very easy to write the matrix of T above in the non-standard basis $\mathcal{B}$, and it is moreover very easy to manipulate that matrix. However, if we wanted to actually evaluate $T(v)$ in specific cases, then in order to make use of $[T]_{\mathcal{B}}$, we would need to express all vectors in the basis $\mathcal{B}$. It is much more efficient to take our easily obtained matrix with respect to $\mathcal{B}$ and use the change of basis formula to get a matrix with respect to the standard basis.

Similarity and Diagonalizability

Definition A matrix $\mathrm{A} \in \mathrm{M}_n(\mathbb{F})$ is **similar** to $\mathrm{B} \in \mathrm{M}_n(\mathbb{F})$ if there is an invertible $\mathrm{S} \in \mathrm{M}_n(\mathbb{F})$ such that $\mathrm{B} = \mathrm{SAS}^{-1}$.

Quick Exercise #28. Show that if A is similar to B, then B is similar to A. Thus we can simply say that A and B are similar.

QA #28: If $\mathrm{B} = \mathrm{SAS}^{-1}$, then $\mathrm{A} = \mathrm{S}^{-1}\mathrm{BS} = \mathrm{S}^{-1}\mathrm{B}(\mathrm{S}^{-1})^{-1}$.

Theorem 3.54 *If $\mathcal{B}$ and $\mathcal{B}'$ are both bases of V, $T \in \mathcal{L}(V)$, $A = [T]_{\mathcal{B}}$, and $B = [T]_{\mathcal{B}'}$, then A and B are similar.*

Conversely, suppose that $A, B \in M_n(\mathbb{F})$ are similar, $\mathcal{B}$ is a basis of V, and $A = [T]_{\mathcal{B}}$. Then there is a basis $\mathcal{B}'$ of V such that $B = [T]_{\mathcal{B}'}$.

The point is that two matrices A and B are similar exactly when they represent the same map, with respect to different bases.

Proof The first statement follows immediately from Corollary 3.53 and the definition of similarity.

Now suppose that $A, B \in M_n(\mathbb{F})$ are similar, $\mathcal{B} = (v_1, \ldots, v_n)$ is a basis of V, and $A = [T]_{\mathcal{B}}$. Let $B = SAS^{-1}$, where S is invertible, and define $R = S^{-1}$. For each j, set

$$w_j = \sum_{i=1}^{n} r_{ij} v_i,$$

so that $\mathbf{r}_j = \left[w_j\right]_{\mathcal{B}}$. Since R is invertible, $(\mathbf{r}_1, \ldots, \mathbf{r}_n)$ is a basis of $\mathbb{F}^n$, and therefore $\mathcal{B}' = (w_1, \ldots, w_n)$ is a basis of V. Furthermore,

$$[I]_{\mathcal{B}',\mathcal{B}}\, \mathbf{e}_j = [I]_{\mathcal{B}',\mathcal{B}} \left[w_j\right]_{\mathcal{B}'} = \left[w_j\right]_{\mathcal{B}} = \mathbf{r}_j$$

for each j, so $[I]_{\mathcal{B}',\mathcal{B}} = R$. Therefore $S = [I]_{\mathcal{B},\mathcal{B}'}$ by Lemma 3.51, and so by Corollary 3.53, $[T]_{\mathcal{B}'} = SAS^{-1} = B$. ▲

Definition A matrix $A \in M_n(\mathbb{F})$ is **diagonalizable** if A is similar to a diagonal matrix.

The following result is immediate from Corollary 3.46 and Theorem 3.54, and justifies our *a priori* different definitions of diagonalizability for maps and for matrices.

Proposition 3.55 *Let $T \in \mathcal{L}(V)$, let $\mathcal{B}$ be a basis for V, and let $A = [T]_{\mathcal{B}}$. Then A is a diagonalizable matrix if and only if T is a diagonalizable linear map.*

Examples

1. We have seen (see the example on page 203) that if $T : \mathbb{R}^2 \to \mathbb{R}^2$ is the operator given by reflection across the line $y = 2x$, then

$$[T]_{\mathcal{E}} = \begin{bmatrix} -\frac{3}{5} & \frac{4}{5} \\ \frac{4}{5} & \frac{3}{5} \end{bmatrix}, \qquad [T]_{\mathcal{B}} = \begin{bmatrix} 1 & 0 \\ 0 & -1 \end{bmatrix},$$

where $\mathcal{E}$ is the standard basis and $\mathcal{B} = \left(\begin{bmatrix} 1 \\ 2 \end{bmatrix}, \begin{bmatrix} -2 \\ 1 \end{bmatrix} \right)$. This means that the matrix $\begin{bmatrix} -\frac{3}{5} & \frac{4}{5} \\ \frac{4}{5} & \frac{3}{5} \end{bmatrix}$ is diagonalizable, and that

$$\begin{bmatrix} -\frac{3}{5} & \frac{4}{5} \\ \frac{4}{5} & \frac{3}{5} \end{bmatrix} = S \begin{bmatrix} 1 & 0 \\ 0 & -1 \end{bmatrix} S^{-1},$$

where $S = \begin{bmatrix} 1 & -2 \\ 2 & 1 \end{bmatrix}$ is the change of basis matrix from $\mathcal{B}$ to $\mathcal{E}$.

2. Consider the matrix $A = \begin{bmatrix} 1 & -2 \\ 1 & -1 \end{bmatrix}$. If we view A as an element of $M_2(\mathbb{C})$, then A is diagonalizable:

$$A = S \begin{bmatrix} i & 0 \\ 0 & -i \end{bmatrix} S^{-1}, \qquad S = \begin{bmatrix} 1+i & 1-i \\ 1 & 1 \end{bmatrix}.$$

(Check for yourself!)

On the other hand, if we view A as an element of $M_2(\mathbb{R})$, then A is not diagonalizable. It's not *a priori* obvious that just because our diagonalization over $\mathbb{C}$ uses complex numbers, there isn't a diagonalization over $\mathbb{R}$; the easiest way to prove that diagonalizing A over $\mathbb{R}$ is impossible uses Theorem 3.56 below. ▲

We could prove the next result by combining Proposition 3.45 with Theorem 3.54, but it's actually easier to just work directly from definitions.

Theorem 3.56 *A matrix* $A \in M_n(\mathbb{F})$ *is diagonalizable if and only if there is a basis* $(v_1, \ldots, v_n)$ *of* $\mathbb{F}^n$ *such that for each* $i = 1, \ldots, n$, v_i *is an eigenvector of* A. *In that case,*

$$A = S \begin{bmatrix} \lambda_1 & & 0 \\ & \ddots & \\ 0 & & \lambda_n \end{bmatrix} S^{-1},$$

where $Av_i = \lambda_i v_i$ *and*

$$S = \begin{bmatrix} | & & | \\ v_1 & \cdots & v_n \\ | & & | \end{bmatrix}.$$

Proof Suppose first that A is diagonalizable. Then there is an invertible $S \in M_n(\mathbb{F})$ and a diagonal matrix

$$B = \begin{bmatrix} \lambda_1 & & 0 \\ & \ddots & \\ 0 & & \lambda_n \end{bmatrix}$$

such that $B = S^{-1}AS$. Since S is invertible, its columns $(v_1, \ldots, v_n)$ form a basis of $\mathbb{F}^n$. Furthermore,

$$\begin{bmatrix} | & & | \\ Av_1 & \cdots & Av_n \\ | & & | \end{bmatrix} = AS = SB = \begin{bmatrix} | & & | \\ \lambda_1 v_1 & \cdots & \lambda_n v_n \\ | & & | \end{bmatrix},$$

and so $Av_i = \lambda_i v_i$ for each i.

Suppose now that $(v_1, \ldots, v_n)$ is a basis of $\mathbb{F}^n$ and that $Av_i = \lambda_i v_i$ for each i. Define

$$S = \begin{bmatrix} | & & | \\ v_1 & \cdots & v_n \\ | & & | \end{bmatrix} \quad \text{and} \quad B = \begin{bmatrix} \lambda_1 & & 0 \\ & \ddots & \\ 0 & & \lambda_n \end{bmatrix}.$$

Since the columns of S form a basis of $\mathbb{F}^n$, S is invertible, and

$$AS = \begin{bmatrix} | & & | \\ Av_1 & \cdots & Av_n \\ | & & | \end{bmatrix} = \begin{bmatrix} | & & | \\ \lambda_1 v_1 & \cdots & \lambda_n v_n \\ | & & | \end{bmatrix} = SB,$$

and thus $A = SBS^{-1}$. ▲

Quick Exercise #29. (a) Show that if an $n \times n$ matrix A has n distinct eigenvalues, then A is diagonalizable.
(b) Give an example of an $n \times n$ diagonalizable matrix with fewer than n distinct eigenvalues.

Invariants

Consider a matrix $A \in M_n(\mathbb{F})$, and suppose that it has eigenvalue $\lambda \in \mathbb{F}$; i.e., there is a vector $v \in \mathbb{F}^n$ with $v \neq \mathbf{0}$, such that

$$Av = \lambda v.$$

Suppose now that $B \in M_n(\mathbb{F})$ is similar to A, i.e., there is some $S \in M_n(\mathbb{F})$ with $B = SAS^{-1}$. If $w = Sv$, then $w \neq \mathbf{0}$ (since S is invertible), and

$$Bw = SAS^{-1}(Sv) = SAv = S(\lambda v) = \lambda Sv = \lambda w,$$

so λ is an eigenvalue of B as well. This proves the following lemma.

QA #29: (a) A list of n eigenvectors corresponding to the n distinct eigenvalues is linearly independent, hence a basis of $\mathbb{F}^n$. (b) I_n is diagonal, hence clearly diagonalizable, but has only the single eigenvalue 1.

Lemma 3.57 *Similar matrices have the same eigenvalues.*

Recall that two matrices are similar if and only if they are matrix representations of the same map, with respect to different bases. When a property of a matrix is preserved by similarity, as the set of eigenvalues is, we call it an **invariant** of the matrix.

Further invariants of a matrix are its rank and nullity:

Theorem 3.58 *Suppose V and W are finite-dimensional vector spaces and $T \in \mathcal{L}(V, W)$. Then*

$$\operatorname{rank} T = \operatorname{rank} [T]_{\mathcal{B}_V,\mathcal{B}_W} \qquad \textit{and} \qquad \operatorname{null} T = \operatorname{null} [T]_{\mathcal{B}_V,\mathcal{B}_W}$$

for any bases $\mathcal{B}_V$ of V and $\mathcal{B}_W$ of W.

Proof Let $\mathcal{B}_V = (v_1, \ldots, v_n)$, and write $\mathrm{A} = [T]_{\mathcal{B}_V,\mathcal{B}_W}$. Since $(v_1, \ldots, v_n)$ spans V, range $T = \langle Tv_1, \ldots, Tv_n \rangle$. By Proposition 3.40, $\langle Tv_1, \ldots, Tv_n \rangle$ is isomorphic to

$$\left\langle [Tv_1]_{\mathcal{B}_W}, \ldots, [Tv_n]_{\mathcal{B}_W} \right\rangle.$$

The vectors $\left[Tv_j\right]_{\mathcal{B}_W}$ for $j = 1, \ldots, n$ are exactly the columns of A. Thus

$$\operatorname{rank} T = \dim \operatorname{range} T = \dim C(\mathrm{A}) = \operatorname{rank} \mathrm{A}.$$

For the second part, $Tv = 0$ if and only if

$$0 = [Tv]_{\mathcal{B}_W} = \mathrm{A}\, [v]_{\mathcal{B}_V}.$$

Thus $C_{\mathcal{B}_V}(\ker T) = \ker \mathrm{A}$, where $C_{\mathcal{B}_V}$ is as defined in Proposition 3.40. Proposition 3.40 implies that $\ker T$ and $\ker \mathrm{A}$ are therefore isomorphic, and so they have the same dimension. ▲

Corollary 3.59 *If* $\mathrm{A}, \mathrm{B} \in \mathrm{M}_n(\mathbb{F})$ *are similar, then*

$$\operatorname{rank} \mathrm{A} = \operatorname{rank} \mathrm{B} \qquad \textit{and} \qquad \operatorname{null} \mathrm{A} = \operatorname{null} \mathrm{B}.$$

Proof This follows immediately from Theorems 3.58 (in the case where $V = W$ and $\mathcal{B}_V = \mathcal{B}_W$) and Theorem 3.54. ▲

This last result is an illustration of the value in being flexible about choosing either a matrix-oriented viewpoint or an abstract linear map viewpoint on linear algebra. We've seen various examples so far in which results that were true in general about linear maps were easiest to prove when working directly with matrix representations (for example, the proof of the Rank–Nullity Theorem for matrices is quite simple). This last corollary shows the opposite phenomenon; it is easiest to show that similar matrices have the same rank and nullity by viewing them as two representations of the same map.

We next turn our attention to a more surprising kind of invariant: one i.e. defined directly in terms of the entries of the matrix, without an obvious connection to an associated linear map.

Definition Let $A \in M_n(\mathbb{F})$. The **trace** of A is

$$\operatorname{tr} A = \sum_{i=1}^{n} a_{ii}.$$

That the trace is an invariant of the matrix follows easily from the following proposition.

Proposition 3.60 *If* $A \in M_{m,n}(\mathbb{F})$ *and* $B \in M_{n,m}(\mathbb{F})$, *then* $\operatorname{tr} AB = \operatorname{tr} BA$.

Proof

$$\operatorname{tr} AB = \sum_{i=1}^{m} [AB]_{ii} = \sum_{i=1}^{m} \sum_{j=1}^{n} a_{ij} b_{ji} = \sum_{j=1}^{n} \sum_{i=1}^{m} b_{ji} a_{ij} = \sum_{j=1}^{n} [BA]_{jj} = \operatorname{tr} BA. \qquad \blacktriangle$$

Corollary 3.61 *If* A *and* B *are similar, then* $\operatorname{tr} A = \operatorname{tr} B$.

Proof Suppose that $B = SAS^{-1}$. By Proposition 3.60 and the associativity of matrix multiplication,

$$\operatorname{tr} B = \operatorname{tr}(SA)S^{-1} = \operatorname{tr} S^{-1}(SA) = \operatorname{tr}(S^{-1}S)A = \operatorname{tr} A. \qquad \blacktriangle$$

Quick Exercise #30. Give an example of two matrices of the same size which have the same trace but are not similar.

Corollary 3.62 *If* $T \in \mathcal{L}(V)$ *and* $\mathcal{B}, \mathcal{B}'$ *are any two bases of* V, *then* $\operatorname{tr}[T]_{\mathcal{B}} = \operatorname{tr}[T]_{\mathcal{B}'}$.

Proof This follows immediately from Corollaries 3.53 and 3.61. ▲

QA #30: For example, $\begin{bmatrix} 0 & 0 \\ 0 & 0 \end{bmatrix}$ and $\begin{bmatrix} 1 & 0 \\ 0 & -1 \end{bmatrix}$. We can tell that these aren't similar because they have different eigenvalues.

This last corollary means that the trace in the following definition is well-defined.

Definition Let V be a finite-dimensional vector space and $T \in \mathcal{L}(V)$. The trace of T is $\operatorname{tr} T = \operatorname{tr} [T]_{\mathcal{B}}$, where $\mathcal{B}$ is any basis of V.

KEY IDEAS

- The matrix $[I]_{\mathcal{B}_1,\mathcal{B}_2}$ is called the change of basis matrix from $\mathcal{B}_1$ to $\mathcal{B}_2$, because $[I]_{\mathcal{B}_1,\mathcal{B}_2} [v]_{\mathcal{B}_1} = [v]_{\mathcal{B}_2}$.
- If $\mathrm{S} = [I]_{\mathcal{B}_1,\mathcal{B}_2}$, then $[T]_{\mathcal{B}_2} = \mathrm{S} [T]_{\mathcal{B}_1} \mathrm{S}^{-1}$.
- $[I]_{\mathcal{B}_2,\mathcal{B}_1} = [I]^{-1}_{\mathcal{B}_1,\mathcal{B}_2}$.
- Matrices $\mathrm{A}, \mathrm{B} \in \mathrm{M}_n(\mathbb{F})$ are similar if there is $\mathrm{S} \in \mathrm{M}_n(\mathbb{F})$ such that $\mathrm{A} = \mathrm{SBS}^{-1}$. This is the same thing as saying that A and B represent the same operator in different bases.
- A matrix $\mathrm{A} \in \mathrm{M}_n(\mathbb{F})$ is diagonalizable if it is similar to a diagonal matrix. This is the same as there being a basis of $\mathbb{F}^n$ consisting of eigenvectors of A.
- Invariants are things which are the same for similar matrices. Examples include the rank, the nullity, the collection of eigenvalues, and the trace.

EXERCISES

3.6.1 Find the change of basis matrices $[I]_{\mathcal{B},\mathcal{C}}$ and $[I]_{\mathcal{C},\mathcal{B}}$ for each of the following pairs of bases of $\mathbb{R}^n$.

(a) $\mathcal{B} = \left(\begin{bmatrix} 2 \\ 5 \end{bmatrix}, \begin{bmatrix} 3 \\ 7 \end{bmatrix} \right)$ and $\mathcal{C} = \left(\begin{bmatrix} -1 \\ 2 \end{bmatrix}, \begin{bmatrix} 1 \\ -4 \end{bmatrix} \right)$

(b) $\mathcal{B} = \left(\begin{bmatrix} 1 \\ 2 \end{bmatrix}, \begin{bmatrix} 3 \\ 4 \end{bmatrix} \right)$ and $\mathcal{C} = \left(\begin{bmatrix} 1 \\ 3 \end{bmatrix}, \begin{bmatrix} 2 \\ 4 \end{bmatrix} \right)$

(c) $\mathcal{B} = \left(\begin{bmatrix} 1 \\ 0 \\ 0 \end{bmatrix}, \begin{bmatrix} 2 \\ 1 \\ 0 \end{bmatrix}, \begin{bmatrix} 0 \\ -1 \\ 1 \end{bmatrix} \right)$ and $\mathcal{C} = \left(\begin{bmatrix} 1 \\ 2 \\ 0 \end{bmatrix}, \begin{bmatrix} 0 \\ 1 \\ -2 \end{bmatrix}, \begin{bmatrix} 0 \\ 0 \\ 1 \end{bmatrix} \right)$

(d) $\mathcal{B} = \left(\begin{bmatrix} 1 \\ 1 \\ 0 \end{bmatrix}, \begin{bmatrix} 1 \\ 0 \\ 1 \end{bmatrix}, \begin{bmatrix} 0 \\ 1 \\ 1 \end{bmatrix} \right)$ and $\mathcal{C} = \left(\begin{bmatrix} 1 \\ 1 \\ 1 \end{bmatrix}, \begin{bmatrix} 1 \\ 1 \\ 2 \end{bmatrix}, \begin{bmatrix} 1 \\ 2 \\ 3 \end{bmatrix} \right)$

3.6.2 Find the change of basis matrices $[I]_{\mathcal{B},\mathcal{C}}$ and $[I]_{\mathcal{C},\mathcal{B}}$ for each of the following pairs of bases of $\mathbb{R}^n$.

(a) $\mathcal{B} = \left(\begin{bmatrix} 1 \\ 2 \end{bmatrix}, \begin{bmatrix} 2 \\ 3 \end{bmatrix} \right)$ and $\mathcal{C} = \left(\begin{bmatrix} 1 \\ 1 \end{bmatrix}, \begin{bmatrix} 1 \\ -1 \end{bmatrix} \right)$

(b) $\mathcal{B} = \left(\begin{bmatrix} 2 \\ 3 \end{bmatrix}, \begin{bmatrix} 3 \\ 5 \end{bmatrix} \right)$ and $\mathcal{C} = \left(\begin{bmatrix} 1 \\ 2 \end{bmatrix}, \begin{bmatrix} 0 \\ 1 \end{bmatrix} \right)$

(c) $\mathcal{B} = \left(\begin{bmatrix} 1 \\ 1 \\ 1 \end{bmatrix}, \begin{bmatrix} 1 \\ -1 \\ 0 \end{bmatrix}, \begin{bmatrix} 1 \\ 0 \\ -1 \end{bmatrix} \right)$ and $\mathcal{C} = \left(\begin{bmatrix} 1 \\ 1 \\ 1 \end{bmatrix}, \begin{bmatrix} 1 \\ 1 \\ -2 \end{bmatrix}, \begin{bmatrix} 1 \\ -2 \\ 1 \end{bmatrix} \right)$

(d) $\mathcal{B} = \left(\begin{bmatrix} 1 \\ 0 \\ 0 \end{bmatrix}, \begin{bmatrix} 2 \\ 1 \\ 0 \end{bmatrix}, \begin{bmatrix} 3 \\ 2 \\ 1 \end{bmatrix} \right)$ and $\mathcal{C} = \left(\begin{bmatrix} 1 \\ 2 \\ 3 \end{bmatrix}, \begin{bmatrix} 0 \\ 1 \\ 2 \end{bmatrix}, \begin{bmatrix} 0 \\ 0 \\ 1 \end{bmatrix} \right)$

3.6.3 Consider the basis $\mathcal{B} = \left(\begin{bmatrix} 2 \\ -3 \end{bmatrix}, \begin{bmatrix} -1 \\ 2 \end{bmatrix} \right)$ of $\mathbb{R}^2$.

(a) Find the change of basis matrices $[I]_{\mathcal{B},\mathcal{E}}$ and $[I]_{\mathcal{E},\mathcal{B}}$.

(b) Use these change of basis matrices to find the coordinate representations of each of the following vectors with respect to $\mathcal{B}$.

(i) $\begin{bmatrix} 1 \\ 0 \end{bmatrix}$ (ii) $\begin{bmatrix} 0 \\ 1 \end{bmatrix}$ (iii) $\begin{bmatrix} 2 \\ 3 \end{bmatrix}$ (iv) $\begin{bmatrix} 4 \\ -5 \end{bmatrix}$ (v) $\begin{bmatrix} -6 \\ 1 \end{bmatrix}$

3.6.4 Consider the basis $\mathcal{B} = \left(\begin{bmatrix} 1 \\ 2 \end{bmatrix}, \begin{bmatrix} 3 \\ 4 \end{bmatrix} \right)$ of $\mathbb{R}^2$.

(a) Find the change of basis matrices $[I]_{\mathcal{B},\mathcal{E}}$ and $[I]_{\mathcal{E},\mathcal{B}}$.

(b) Use these change of basis matrices to find the coordinate representations of each of the following vectors with respect to $\mathcal{B}$.

(i) $\begin{bmatrix} 1 \\ 0 \end{bmatrix}$ (ii) $\begin{bmatrix} 0 \\ 1 \end{bmatrix}$ (iii) $\begin{bmatrix} 2 \\ 3 \end{bmatrix}$ (iv) $\begin{bmatrix} 4 \\ -5 \end{bmatrix}$ (v) $\begin{bmatrix} -6 \\ 1 \end{bmatrix}$

3.6.5 Consider the basis $\mathcal{B} = \left(\begin{bmatrix} 1 \\ -1 \\ 0 \end{bmatrix}, \begin{bmatrix} 1 \\ 0 \\ 1 \end{bmatrix}, \begin{bmatrix} 0 \\ 1 \\ -1 \end{bmatrix} \right)$ of $\mathbb{R}^3$.

(a) Find the change of basis matrices $[I]_{\mathcal{B},\mathcal{E}}$ and $[I]_{\mathcal{E},\mathcal{B}}$.

(b) Use these change of basis matrices to find the coordinate representations of each of the following vectors with respect to $\mathcal{B}$.

(i) $\begin{bmatrix} 1 \\ 0 \\ 0 \end{bmatrix}$ (ii) $\begin{bmatrix} 0 \\ 0 \\ 1 \end{bmatrix}$ (iii) $\begin{bmatrix} -4 \\ 0 \\ 2 \end{bmatrix}$ (iv) $\begin{bmatrix} 1 \\ 2 \\ -3 \end{bmatrix}$ (v) $\begin{bmatrix} 6 \\ -4 \\ 1 \end{bmatrix}$

3.6.6 Consider the basis $\mathcal{B} = \left(\begin{bmatrix} 1 \\ -1 \\ 1 \end{bmatrix}, \begin{bmatrix} 1 \\ 2 \\ -1 \end{bmatrix}, \begin{bmatrix} 0 \\ 2 \\ -1 \end{bmatrix} \right)$ of $\mathbb{R}^3$.

(a) Find the change of basis matrices $[I]_{\mathcal{B},\mathcal{E}}$ and $[I]_{\mathcal{E},\mathcal{B}}$.

(b) Use these change of basis matrices to find the coordinate representations of each of the following vectors with respect to $\mathcal{B}$.

(i) $\begin{bmatrix} 1 \\ 0 \\ 0 \end{bmatrix}$ (ii) $\begin{bmatrix} 0 \\ 0 \\ 1 \end{bmatrix}$ (iii) $\begin{bmatrix} -4 \\ 0 \\ 2 \end{bmatrix}$ (iv) $\begin{bmatrix} 1 \\ 2 \\ -3 \end{bmatrix}$ (v) $\begin{bmatrix} 6 \\ -4 \\ 1 \end{bmatrix}$

3.6.7 Consider the bases $\mathcal{B} = \left(\begin{bmatrix} 2 \\ 1 \end{bmatrix}, \begin{bmatrix} 1 \\ 1 \end{bmatrix} \right)$ of $\mathbb{R}^2$ and

$\mathcal{C} = \left(\begin{bmatrix} 1 \\ 3 \\ 0 \end{bmatrix}, \begin{bmatrix} 0 \\ 2 \\ 3 \end{bmatrix}, \begin{bmatrix} 1 \\ 0 \\ -4 \end{bmatrix} \right)$ of $\mathbb{R}^3$, and the linear maps $S \in \mathcal{L}(\mathbb{R}^2, \mathbb{R}^3)$

and $T \in \mathcal{L}(\mathbb{R}^3, \mathbb{R}^2)$ given (with respect to the standard bases) by

$$[S]_{\mathcal{E},\mathcal{E}} = \begin{bmatrix} 0 & 1 \\ 3 & -3 \\ 4 & -8 \end{bmatrix} \quad \text{and} \quad [T]_{\mathcal{E},\mathcal{E}} = \begin{bmatrix} 4 & -1 & 1 \\ 3 & -1 & 1 \end{bmatrix}.$$

(a) Find the change of basis matrices $[I]_{\mathcal{B},\mathcal{E}}$ and $[I]_{\mathcal{E},\mathcal{B}}$ for $\mathbb{R}^2$, and $[I]_{\mathcal{C},\mathcal{E}}$ and $[I]_{\mathcal{E},\mathcal{C}}$ for $\mathbb{R}^3$.

(b) Use these change of basis matrices to find the following coordinate representations.

(i) $[S]_{\mathcal{B},\mathcal{E}}$ (ii) $[S]_{\mathcal{E},\mathcal{C}}$ (iii) $[S]_{\mathcal{B},\mathcal{C}}$ (iv) $[T]_{\mathcal{C},\mathcal{E}}$ (v) $[T]_{\mathcal{E},\mathcal{B}}$ (vi) $[T]_{\mathcal{C},\mathcal{B}}$

3.6.8 Consider the bases $\mathcal{B} = \left(\begin{bmatrix} 2 \\ 3 \end{bmatrix}, \begin{bmatrix} 3 \\ 5 \end{bmatrix} \right)$ of $\mathbb{R}^2$ and

$\mathcal{C} = \left(\begin{bmatrix} 1 \\ 1 \\ 0 \end{bmatrix}, \begin{bmatrix} 1 \\ 0 \\ 1 \end{bmatrix}, \begin{bmatrix} 0 \\ 1 \\ 1 \end{bmatrix} \right)$ of $\mathbb{R}^3$, and the linear maps $S \in \mathcal{L}(\mathbb{R}^2, \mathbb{R}^3)$

and $T \in \mathcal{L}(\mathbb{R}^3, \mathbb{R}^2)$ given (with respect to the standard bases) by

$$[S]_{\mathcal{E},\mathcal{E}} = \begin{bmatrix} 2 & -1 \\ 5 & 3 \\ -3 & 2 \end{bmatrix} \quad \text{and} \quad [T]_{\mathcal{E},\mathcal{E}} = \begin{bmatrix} 1 & -1 & 1 \\ 1 & 1 & -1 \end{bmatrix}.$$

(a) Find the change of basis matrices $[I]_{\mathcal{B},\mathcal{E}}$ and $[I]_{\mathcal{E},\mathcal{B}}$ for $\mathbb{R}^2$, and $[I]_{\mathcal{C},\mathcal{E}}$ and $[I]_{\mathcal{E},\mathcal{C}}$ for $\mathbb{R}^3$.

(b) Use these change of basis matrices to find the following coordinate representations.

(i) $[S]_{\mathcal{B},\mathcal{E}}$ (ii) $[S]_{\mathcal{E},\mathcal{C}}$ (iii) $[S]_{\mathcal{B},\mathcal{C}}$ (iv) $[T]_{\mathcal{C},\mathcal{E}}$ (v) $[T]_{\mathcal{E},\mathcal{B}}$ (vi) $[T]_{\mathcal{C},\mathcal{B}}$

3.6.9 Consider the bases

$$\mathcal{B} = \left(\begin{bmatrix} 1 \\ -1 \\ 0 \end{bmatrix}, \begin{bmatrix} 1 \\ 0 \\ -1 \end{bmatrix} \right) \quad \text{and} \quad \mathcal{C} = \left(\begin{bmatrix} 1 \\ 1 \\ -2 \end{bmatrix}, \begin{bmatrix} 0 \\ 1 \\ -1 \end{bmatrix} \right)$$

of $U = \left\{ \begin{bmatrix} x \\ y \\ z \end{bmatrix} \in \mathbb{R}^3 \;\middle|\; x + y + z = 0 \right\}$, and the linear map $T \in \mathcal{L}(U)$ given by

$$T \begin{bmatrix} x \\ y \\ z \end{bmatrix} = \begin{bmatrix} x \\ z \\ y \end{bmatrix}.$$

(a) Find the matrix $[T]_{\mathcal{B}}$.
(b) Find the change of basis matrices $[I]_{\mathcal{B},\mathcal{C}}$ and $[I]_{\mathcal{C},\mathcal{B}}$.
(c) Use your answers to the previous parts to find $[T]_{\mathcal{C}}$.

3.6.10 Consider the bases

$$\mathcal{B} = \left(\begin{bmatrix} 2 \\ -1 \\ 0 \end{bmatrix}, \begin{bmatrix} 0 \\ 3 \\ -2 \end{bmatrix} \right) \quad \text{and} \quad \mathcal{C} = \left(\begin{bmatrix} 1 \\ 1 \\ -1 \end{bmatrix}, \begin{bmatrix} 3 \\ 0 \\ -1 \end{bmatrix} \right)$$

of $U = \left\{ \begin{bmatrix} x \\ y \\ z \end{bmatrix} \in \mathbb{R}^3 \;\middle|\; x + 2y + 3z = 0 \right\}$, and the linear map $T \in \mathcal{L}(U)$ given by

$$T \begin{bmatrix} x \\ y \\ z \end{bmatrix} = \begin{bmatrix} 9z \\ 3y \\ x \end{bmatrix}.$$

(a) Find the matrix $[T]_{\mathcal{B}}$.
(b) Find the change of basis matrices $[I]_{\mathcal{B},\mathcal{C}}$ and $[I]_{\mathcal{C},\mathcal{B}}$.
(c) Use your answers to the previous parts to find $[T]_{\mathcal{C}}$.

3.6.11 Let $P \in \mathcal{L}(\mathbb{R}^2)$ be the orthogonal projection onto the line $y = \frac{1}{3}x$ (see Exercise 2.1.2).
(a) Find a basis $\mathcal{B}$ of $\mathbb{R}^2$ such that $[P]_{\mathcal{B}}$ is diagonal.
(b) Find the change of basis matrix $[I]_{\mathcal{B},\mathcal{E}}$, where $\mathcal{E}$ is the standard basis of $\mathbb{R}^2$.
(c) Find the change of basis matrix $[I]_{\mathcal{E},\mathcal{B}}$.
(d) Use the above information to find $[P]_{\mathcal{E}}$.

3.6.12 Let $R \in \mathcal{L}(\mathbb{R}^2)$ be the reflection across the line $y = \frac{1}{2}x$:

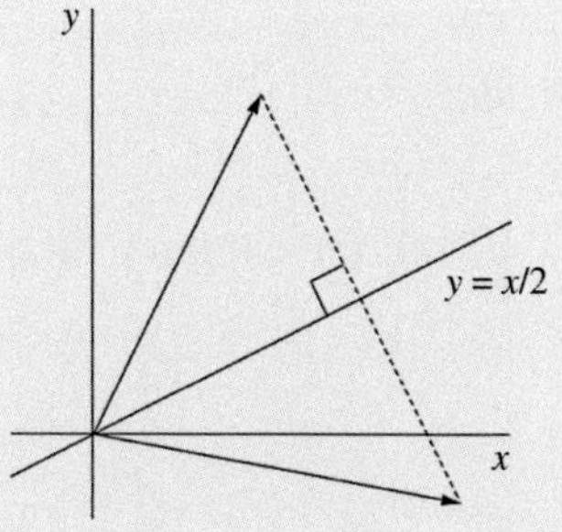

(a) Find a basis $\mathcal{B}$ of $\mathbb{R}^2$ such that $[R]_{\mathcal{B}}$ is diagonal.

(b) Find the change of basis matrix $[I]_{\mathcal{B},\mathcal{E}}$, where $\mathcal{E}$ is the standard basis of $\mathbb{R}^2$.

(c) Find the change of basis matrix $[I]_{\mathcal{E},\mathcal{B}}$.

(d) Use the above information to find $[R]_{\mathcal{E}}$.

3.6.13 Let $\mathcal{B} = (1, x, x^2)$ and let $\mathcal{B}' = \left(1, x, \frac{3}{2}x^2 - \frac{1}{2}\right)$ in $\mathcal{P}_2(\mathbb{R})$. Find the change of basis matrices $[I]_{\mathcal{B},\mathcal{B}'}$ and $[I]_{\mathcal{B}',\mathcal{B}}$.

3.6.14 Consider the bases $\mathcal{B} = (\mathbf{v}_1, \ldots, \mathbf{v}_n)$ and $\mathcal{C} = (\mathbf{w}_1, \ldots, \mathbf{w}_n)$ of $\mathbb{F}^n$ from the example on page 201.

(a) Find the change of basis matrix $[I]_{\mathcal{C},\mathcal{B}}$.

(b) Suppose $[\mathbf{v}]_{\mathcal{B}} = \begin{bmatrix} 1 \\ \vdots \\ 1 \end{bmatrix}$. Find $[\mathbf{v}]_{\mathcal{C}}$.

(c) Suppose that $T \in \mathcal{L}(\mathbb{F}^n)$ is defined by setting $T\mathbf{v}_i = \mathbf{v}_{i+1}$ for $i = 1, \ldots, n-1$ and $T\mathbf{v}_n = \mathbf{0}$, then extending by linearity. Find $[T]_{\mathcal{E}}$.

(d) Find $[T]_{\mathcal{C}}$.

3.6.15 Prove that the matrices

$$\begin{bmatrix} 5 & -2 & 3 & -7 \\ 1 & -3 & -2 & 4 \\ 6 & 0 & 1 & 3 \\ -5 & 2 & 1 & -4 \end{bmatrix} \quad \text{and} \quad \begin{bmatrix} 0 & 4 & -4 & 1 \\ 2 & 5 & -6 & 3 \\ 8 & -2 & 1 & -2 \\ 1 & -3 & 0 & 7 \end{bmatrix}$$

are not similar to each other.

3.6.16 Prove that the matrices

$$\begin{bmatrix} 2 & 0 & -1 \\ -1 & 3 & 2 \\ 1 & 3 & 1 \end{bmatrix} \quad \text{and} \quad \begin{bmatrix} 2 & 0 & -1 \\ 2 & -3 & -1 \\ 1 & 3 & 1 \end{bmatrix}$$

are not similar to each other.

3.6.17 Show that $\begin{bmatrix} -1 & -2 \\ 3 & 4 \end{bmatrix}$ is diagonalizable.

3.6.18 Show that $\begin{bmatrix} 5 & 6 \\ -1 & -2 \end{bmatrix}$ is diagonalizable.

3.6.19 Prove that the matrix $A = \begin{bmatrix} 1 & -2 \\ 1 & -1 \end{bmatrix}$ from the example on page 205 is not diagonalizable over $\mathbb{R}$.

3.6.20 Let $A = \begin{bmatrix} 1 & 1 \\ 0 & 1 \end{bmatrix}$.

(a) Find all the eigenvalues and eigenvectors of A.

(b) Prove that A is not diagonalizable (no matter what $\mathbb{F}$ is).

3.6.21 Let P be a plane through the origin in $\mathbb{R}^3$, and let $R \in \mathcal{L}(\mathbb{R}^3)$ be the reflection across P. Find $\operatorname{tr} R$.

Hint: Choose your basis wisely.

3.6.22 (a) Show that if $A, B, C \in M_n(\mathbb{F})$, then

$$\operatorname{tr} ABC = \operatorname{tr} BCA = \operatorname{tr} CAB.$$

(b) Give an example of three matrices $A, B, C \in M_n(\mathbb{F})$ such that $\operatorname{tr} ABC \neq \operatorname{tr} ACB$.

3.6.23 Suppose that $\dim V = n$ and T has n distinct eigenvalues $\lambda_1, \ldots, \lambda_n$. Show that $\operatorname{tr} T = \lambda_1 + \cdots + \lambda_n$.

3.6.24 The **geometric multiplicity** of $\lambda \in \mathbb{F}$ as an eigenvalue of $A \in M_n(\mathbb{F})$ is $m_\lambda(A) = \operatorname{null}(A - \lambda I_n)$.

(a) Show that λ is an eigenvalue of A if and only if $m_\lambda(A) > 0$.

(b) Show that for each λ, m_λ is an invariant.

3.6.25 For $A \in M_n(\mathbb{F})$, define

$$\operatorname{sum} A = \sum_{i=1}^{n} \sum_{j=1}^{n} a_{ij}.$$

Show that for $n \geq 2$, sum is *not* an invariant.

3.6.26 Prove that the top-left entry of an $n \times n$ matrix is not an invariant if $n \geq 2$.

3.6.27 (a) Show that if $A \in M_n(\mathbb{F})$, then A is similar to A.

(b) Show that if A is similar to B and B is similar to C, then A is similar to C.

Remark: Together with Quick Exercise 28, this shows that similarity is an **equivalence relation** on $M_n(\mathbb{F})$.

3.6.28 Give a second proof of the linear map half of the Rank–Nullity Theorem (Theorem 3.35), using the matrix half of Theorem 3.35 together with Theorem 3.58.

3.6.29 Use Proposition 3.45 and Theorem 3.54 to give another proof of Theorem 3.56.

3.7 Triangularization

Eigenvalues of Upper Triangular Matrices

We've seen that diagonal matrices are particularly easy to work with, and that certain types of information, like eigenvalues, can be read right off from a diagonal matrix. On the other hand, we saw in the previous section that diagonalizability is equivalent to the existence of a basis of eigenvectors, so there is no hope that we will be able to diagonalize every matrix we ever care about. It turns out that in some ways *triangular* matrices are nearly as nice as diagonal matrices, and that "triangularizability" is a more general phenomenon than diagonalizability.*

Definition An $n \times n$ matrix A is called **upper triangular** if $a_{ij} = 0$ whenever $i > j$.

The following lemma shows that we can tell at a glance if an upper triangular matrix is invertible.

Lemma 3.63 *An upper triangular matrix is invertible if and only if its diagonal entries are all nonzero.*

Proof Recall that a square matrix is invertible if and only if its row-echelon form has a pivot in every column. If A is upper triangular with all nonzero entries on the diagonal, then every column is a pivot column. If A is upper triangular with some zeroes on the diagonal, and j is the smallest index for which $a_{jj} = 0$, then the jth column of A cannot be a pivot column. ▲

More generally, we have the following:

Theorem 3.64 *Let* $A \in M_n(\mathbb{F})$ *be upper triangular. Then the eigenvalues of* A *are precisely its diagonal entries.*

---

*Some of the nice properties of upper triangular matrices have already appeared in Exercises 2.3.12, 2.4.19, 3.1.11, 3.1.12, and 3.2.18.

Proof Recall that λ is an eigenvalue of A if and only if $A - \lambda I_n$ is singular. Since $A - \lambda I_n$ is upper triangular, Lemma 3.63 says that $A - \lambda I_n$ fails to be invertible if and only if it has a 0 diagonal entry, which is true if and only if λ is one of the diagonal entries of A. ▲

Theorem 3.64 says nothing about the eigen*vectors* of an upper triangular matrix. However, there is one that we can identify.

> **Quick Exercise #31.** Show that if $A \in M_n(\mathbb{F})$ is upper triangular, then e_1 is an eigenvector of A. What is the corresponding eigenvalue?

Example Let $A = \begin{bmatrix} -1 & 1 & 2 \\ 0 & 2 & 6 \\ 0 & 0 & -1 \end{bmatrix}$. By Theorem 3.64, the eigenvalues of A are -1 and 2. Once we know those eigenvalues, we can find the eigenspaces:

$$\operatorname{Eig}_{-1}(A) = \ker \begin{bmatrix} 0 & 1 & 2 \\ 0 & 3 & 6 \\ 0 & 0 & 0 \end{bmatrix} = \left\langle \begin{bmatrix} 1 \\ 0 \\ 0 \end{bmatrix}, \begin{bmatrix} 0 \\ 2 \\ -1 \end{bmatrix} \right\rangle$$

and

$$\operatorname{Eig}_{2}(A) = \ker \begin{bmatrix} -3 & 1 & 2 \\ 0 & 0 & 6 \\ 0 & 0 & -3 \end{bmatrix} = \left\langle \begin{bmatrix} 1 \\ 3 \\ 0 \end{bmatrix} \right\rangle.$$

Since we have three linearly independent eigenvectors, they form a basis of $\mathbb{R}^3$, so A is diagonalizable. By the change of basis formula,

$$\begin{bmatrix} 1 & 0 & 1 \\ 0 & 2 & 3 \\ 0 & -1 & 0 \end{bmatrix}^{-1} A \begin{bmatrix} 1 & 0 & 1 \\ 0 & 2 & 3 \\ 0 & -1 & 0 \end{bmatrix} = \begin{bmatrix} -1 & 0 & 0 \\ 0 & -1 & 0 \\ 0 & 0 & 2 \end{bmatrix}.$$

▲

We have observed above that not all matrices can be diagonalized; Theorem 3.64 shows that not all matrices can be triangularized either, since if a matrix A is similar to an upper triangular matrix, then it must have at least one eigenvalue. The following is an example of a matrix which does not, at least from one point of view.

Example Consider the 2×2 matrix $A = \begin{bmatrix} 0 & 1 \\ -1 & 0 \end{bmatrix}$ over $\mathbb{R}$. If

$$A \begin{bmatrix} x \\ y \end{bmatrix} = \lambda \begin{bmatrix} x \\ y \end{bmatrix},$$

QA #31: $Ae_1 = a_{11}e_1$, so e_1 is an eigenvector of A with eigenvalue a_{11}.

then $-y = \lambda x$ and $x = \lambda y$. Thus $x = -\lambda^2 x$ and $y = -\lambda^2 y$. So if x and y are not both 0, then $\lambda^2 = -1$, but there is no real number λ with $\lambda^2 = -1$. That is, $\mathrm{A} \in \mathrm{M}_2(\mathbb{R})$ has no eigenvalues.

On the other hand, suppose we consider the same matrix A as an element of $\mathrm{M}_2(\mathbb{C})$. As above, if λ is an eigenvalue, then $\lambda^2 = -1$, so i and $-i$ are both possible eigenvalues. Indeed,

$$\mathrm{A}\begin{bmatrix} 1 \\ i \end{bmatrix} = i\begin{bmatrix} 1 \\ i \end{bmatrix} \quad \text{and} \quad \mathrm{A}\begin{bmatrix} 1 \\ -i \end{bmatrix} = -i\begin{bmatrix} 1 \\ -i \end{bmatrix},$$

so i and $-i$ are both eigenvalues of this map. ▲

This example indicates that what we view as the underlying field $\mathbb{F}$ is important here; it turns out that the crucial thing is whether or not $\mathbb{F}$ is *algebraically closed.*

Definition A field $\mathbb{F}$ is **algebraically closed** if every polynomial

$$p(x) = a_0 + a_1 x + \cdots + a_n x^n$$

with coefficients in $\mathbb{F}$ and $a_n \neq 0$ can be factored in the form

$$p(x) = b(x - c_1) \ldots (x - c_n)$$

for some $b, c_1, \ldots, c_n \in \mathbb{F}$.

If $\mathbb{F}$ is algebraically closed, then in particular every nonconstant polynomial over $\mathbb{F}$ has at least one root in $\mathbb{F}$. Therefore $\mathbb{R}$ is not algebraically closed, since the polynomial $x^2 + 1$ has no real roots. On the other hand, we have the following fact, which is typically proved in courses on complex analysis.

Theorem 3.65 (The Fundamental Theorem of Algebra) *The field $\mathbb{C}$ is algebraically closed.*

To make use of this property of a field $\mathbb{F}$ when working with linear maps or matrices, we need to introduce the idea of a polynomial applied to a linear map or matrix.

Definition Let $p(x) = a_0 + a_1 x + \cdots + a_n x^n$ be a polynomial with coefficients in a field $\mathbb{F}$. If V is a vector space over $\mathbb{F}$ and $T \in \mathcal{L}(V)$, then we define

$$p(T) = a_0 I + a_1 T + \cdots + a_n T^n.$$

Similarly, if $\mathrm{A} \in \mathrm{M}_n(\mathbb{F})$, then we define

$$p(\mathrm{A}) = a_0 \mathrm{I}_n + a_1 \mathrm{A} + \cdots + a_n \mathrm{A}^n.$$

For example, if $p(x) = x^2 - 3x + 1$, then

$$p\left(\begin{bmatrix} 2 & -1 \\ 3 & 1 \end{bmatrix}\right) = \begin{bmatrix} 1 & -3 \\ 9 & -2 \end{bmatrix} - 3\begin{bmatrix} 2 & -1 \\ 3 & 1 \end{bmatrix} + \begin{bmatrix} 1 & 0 \\ 0 & 1 \end{bmatrix} = \begin{bmatrix} -4 & 0 \\ 0 & -4 \end{bmatrix}.$$

Triangularization

In this section we will show that, over an algebraically closed field, every matrix or linear map can be triangularized. The following proposition is a key ingredient in the proof.

Proposition 3.66 *Suppose that V is a finite-dimensional vector space over an algebraically closed field $\mathbb{F}$. Then every $T \in \mathcal{L}(V)$ has at least one eigenvalue.*

Proof Write $n = \dim V$ and let $v \in V$ be any nonzero vector. Then the list of $n+1$ vectors

$$(v, Tv, \ldots, T^n v)$$

in V must be linearly dependent. Thus there are scalars $a_0, a_1, \ldots, a_m \in \mathbb{F}$ such that $a_m \neq 0$ and

$$a_0 v + a_1 Tv + \cdots + a_m T^m v = 0.$$

Now consider the polynomial

$$p(x) = a_0 + a_1 x + \cdots + a_m x^m$$

over $\mathbb{F}$. Since $\mathbb{F}$ is algebraically closed, we can factor this as

$$p(x) = b(x - c_1) \ldots (x - c_m)$$

for some $b, c_1, \ldots, c_m \in \mathbb{F}$. Therefore

$$a_0 I + a_1 T + \cdots + a_m T^m = b(T - c_1 I) \ldots (T - c_m I),$$

and so

$$(T - c_1 I) \ldots (T - c_m I)v = 0.$$

Now if $(T - c_m I)v = 0$, then v is an eigenvector of T with eigenvalue c_m. Otherwise, we have

$$(T - c_1 I) \ldots (T - c_{m-1} I)v' = 0,$$

for $v' = (T - c_m I)v$. If $(T - c_{m-1} I)v' = 0$, then v' is an eigenvector of T with eigenvalue c_{m-1}. Otherwise, we continue in this way, and eventually find that one of the c_j must be an eigenvalue of T. ▲

Example To illustrate what's going on in the preceding proof, we'll carry out the steps for a specific linear map. Start with the matrix $A = \begin{bmatrix} 2 & -1 \\ 5 & -2 \end{bmatrix}$ and $v = \begin{bmatrix} 1 \\ 0 \end{bmatrix}$, and look at the list of vectors

$$(v, Av, A^2v) = \left(\begin{bmatrix} 1 \\ 0 \end{bmatrix}, \begin{bmatrix} 2 \\ 5 \end{bmatrix}, \begin{bmatrix} -1 \\ 0 \end{bmatrix} \right).$$

There is a non-trivial linear dependence:

$$\begin{bmatrix} 0 \\ 0 \end{bmatrix} = \begin{bmatrix} 1 \\ 0 \end{bmatrix} + \begin{bmatrix} -1 \\ 0 \end{bmatrix} = v + A^2v,$$

so we consider the polynomial

$$p(x) = 1 + x^2 = (x + i)(x - i).$$

By the computation above,

$$(A + iI_2)(A - iI_2)v = (A^2 + I_2)v = 0.$$

Since

$$v' = (A - iI_2)v = \begin{bmatrix} 2 - i \\ 5 \end{bmatrix} \neq 0$$

and $(A + iI_2)v' = 0$, we now know that v' is an eigenvector of A with eigenvalue $-i$. (Check!) ▲

Quick Exercise #32. Find an eigenvector of A with eigenvalue i, by factoring $p(x)$ in the other order.

Theorem 3.67 *Suppose that V is a finite-dimensional vector space over an algebraically closed field $\mathbb{F}$, and $T \in \mathcal{L}(V)$. Then there is a basis $\mathcal{B}$ of V such that $[T]_{\mathcal{B}}$ is upper triangular.*

Equivalently, suppose that $A \in M_n(\mathbb{F})$ and that $\mathbb{F}$ is algebraically closed. Then A is similar to some upper triangular matrix.

The following observation is useful for the proof.

Lemma 3.68 *Suppose that $T \in \mathcal{L}(V)$ and that $\mathcal{B} = (v_1, \ldots, v_n)$ is a basis of V. Then $[T]_{\mathcal{B}}$ is upper triangular if and only if for each $j = 1, \ldots, n$,*

$$Tv_j \in \langle v_1, \ldots, v_j \rangle.$$

QA #32: $(A + iI_2)v = \begin{bmatrix} 2 + i \\ 5 \end{bmatrix}$.

Quick Exercise #33. Prove Lemma 3.68.

Proof of Theorem 3.67 We will prove just the statement about linear maps. The statement about matrices then follows by the change of basis formula.

We proceed by induction on $n = \dim V$. The theorem is obviously true for $n = 1$, since every 1×1 matrix is upper triangular.

Now suppose that $n > 1$ and that the theorem is known to be true for all vector spaces over $\mathbb{F}$ with dimension smaller than n. By Proposition 3.66, T has an eigenvalue λ and hence an eigenvector v. Write $U = \operatorname{range}(T - \lambda I)$. If $u \in U$, then since U is a subspace,

$$Tu = (T - \lambda I)u + \lambda u \in U,$$

thus T restricts to a linear map $U \to U$. Furthermore, since λ is an eigenvalue of T, $T - \lambda I$ is not injective. But this means that $T - \lambda I$ is also not surjective, and so $U \neq V$; i.e., $m = \dim U < n$.

By the induction hypothesis, there is a basis $\mathcal{B}_U = (u_1, \ldots, u_m)$ of U with respect to which T restricted to U has an upper triangular matrix. Thus, by Lemma 3.68, $Tu_j \in \langle u_1, \ldots, u_j \rangle$ for each j. We can extend $\mathcal{B}_U$ to a basis $\mathcal{B} = (u_1, \ldots, u_m, v_1, \ldots, v_k)$ of V. Now, for $j = 1, \ldots, k$,

$$Tv_j = (T - \lambda I)v_j + \lambda v_j \in \langle u_1, \ldots, u_m, v_j \rangle \subseteq \langle u_1, \ldots, u_m, v_1, \ldots, v_j \rangle,$$

which by Lemma 3.68 means that $[T]_{\mathcal{B}}$ is upper triangular. ▲

Example Suppose that you know the eigenvalues of $\mathrm{A} \in \mathrm{M}_n(\mathbb{F})$. Assuming that $\mathbb{F}$ is algebraically closed, there exist an invertible matrix $\mathrm{S} \in \mathrm{M}_n(\mathbb{F})$ and an upper triangular matrix T such that $\mathrm{A} = \mathrm{STS}^{-1}$. Then the eigenvalues of T are the same as the eigenvalues of A, and since T is upper triangular, they are the diagonal entries $t_{11}, \ldots, t_{nn}$ of T.

Now suppose you need to know about the eigenvalues of $\mathrm{A}^2 + \mathrm{A}$. Since

$$\mathrm{A}^2 + \mathrm{A} = \mathrm{STS}^{-1}\mathrm{STS}^{-1} + \mathrm{STS}^{-1} = \mathrm{S}(\mathrm{T}^2 + \mathrm{T})\mathrm{S}^{-1},$$

they are the same as the eigenvalues of $\mathrm{T}^2 + \mathrm{T}$. Since this matrix is upper triangular, its eigenvalues are its diagonal entries $t_{11}^2 + t_{11}, \ldots, t_{nn}^2 + t_{nn}$. In other words, every eigenvalue of $\mathrm{A}^2 + \mathrm{A}$ is of the form $\lambda^2 + \lambda$, where λ is an eigenvalue of A itself. ▲

KEY IDEAS

- The eigenvalues of an upper triangular matrix are exactly the diagonal entries.

QA #33: If $\mathrm{A} = [T]_{\mathcal{B}}$, then by definition $Tv_j = \sum_{i=1}^{n} a_{ij} v_i$. The matrix A is upper triangular if and only if $a_{ij} = 0$ if $i > j$, which is true if and only if $Tv_j = \sum_{i=1}^{j} a_{ij} v_i$.

- An upper triangular matrix is invertible if and only if the diagonal entries are nonzero.
- Over an algebraically closed field, every matrix is similar to an upper triangular matrix and every linear operator has a basis in which its matrix is upper triangular. This is not true if the field is not algebraically closed.

EXERCISES

3.7.1 Find all the eigenvalues and eigenspaces for each of the following matrices.

(a) $\begin{bmatrix} 1 & 2 \\ 0 & 3 \end{bmatrix}$ (b) $\begin{bmatrix} 1 & 2 & 3 \\ 0 & 4 & 5 \\ 0 & 0 & 6 \end{bmatrix}$ (c) $\begin{bmatrix} -2 & 1 & 0 \\ 0 & 3 & -1 \\ 0 & 0 & 3 \end{bmatrix}$ (d) $\begin{bmatrix} 2 & 1 & 0 & -1 \\ 0 & 1 & 3 & 0 \\ 0 & 0 & 1 & 1 \\ 0 & 0 & 0 & 2 \end{bmatrix}$

3.7.2 Find all the eigenvalues and eigenspaces for each of the following matrices.

(a) $\begin{bmatrix} 1 & -5 \\ 0 & 1 \end{bmatrix}$ (b) $\begin{bmatrix} 3 & 1 & 4 \\ 0 & 1 & 5 \\ 0 & 0 & 9 \end{bmatrix}$

(c) $\begin{bmatrix} 2 & 1 & 0 & 0 \\ 0 & 2 & 1 & 0 \\ 0 & 0 & 2 & 1 \\ 0 & 0 & 0 & 2 \end{bmatrix}$ (d) $\begin{bmatrix} 1 & 0 & -2 & 0 \\ 0 & 2 & 1 & 0 \\ 0 & 0 & 2 & 1 \\ 0 & 0 & 0 & 1 \end{bmatrix}$

3.7.3 Determine whether each of the following matrices is diagonalizable.

(a) $\begin{bmatrix} 8 & -13 \\ 0 & \sqrt{5} \end{bmatrix}$ (b) $\begin{bmatrix} 2 & -2 & 0 \\ 0 & 1 & 0 \\ 0 & 0 & 1 \end{bmatrix}$ (c) $\begin{bmatrix} 2 & -2 & 0 \\ 0 & 1 & -1 \\ 0 & 0 & 1 \end{bmatrix}$ (d) $\begin{bmatrix} 9 & 8 & 7 \\ 0 & 5 & 4 \\ 0 & 0 & 1 \end{bmatrix}$

3.7.4 Determine whether each of the following matrices is diagonalizable.

(a) $\begin{bmatrix} i & 2+\sqrt{3}i \\ 0 & \sqrt{17}-4i \end{bmatrix}$ (b) $\begin{bmatrix} 1 & -2 & 0 \\ 0 & 1 & -2 \\ 0 & 0 & 1 \end{bmatrix}$ (c) $\begin{bmatrix} 3 & 0 & 1 \\ 0 & 3 & -1 \\ 0 & 0 & 1 \end{bmatrix}$

(d) $\begin{bmatrix} 2 & 0 & 7 \\ 0 & 1 & 8 \\ 0 & 0 & 2 \end{bmatrix}$

3.7.5 Recall that a matrix $A \in M_n(\mathbb{F})$ is called **lower triangular** if $a_{ij} = 0$ whenever $i < j$. Show that every lower triangular matrix is similar to an upper triangular matrix (which means that we could have developed the theory in this section around lower triangular matrices instead of upper triangular ones).
Hint: Lemma 3.68 may be useful here.

3.7.6 Let $A \in M_{m,n}(\mathbb{F})$. Show that if $PA = LU$ is an LUP decomposition, then $\ker A = \ker U$.

3.7.7 Suppose that $A, B \in M_n(\mathbb{F})$ are upper triangular.

(a) Show that every eigenvalue of $A + B$ is the sum of an eigenvalue of A with an eigenvalue of B.

(b) Show that every eigenvalue of AB is the product of an eigenvalue of A with an eigenvalue of B.

(c) Give examples to show that both of the above statements may be false if A and B are not upper triangular.

3.7.8 Show that if $A \in M_n(\mathbb{F})$ is upper triangular with k nonzero diagonal entries, then $\operatorname{rank} A \geq k$.

3.7.9 Show that if $A \in M_n(\mathbb{F})$ is upper triangular with distinct diagonal entries, then A is diagonalizable.

3.7.10 Suppose that $x \in \mathbb{F}^n$ is an eigenvector of $A \in M_n(\mathbb{F})$ with eigenvalue λ, and let $p(x)$ be any polynomial with coefficients in $\mathbb{F}$.

(a) Prove that x is also an eigenvector of $p(A)$ with eigenvalue $p(\lambda)$.

(b) Prove that if $p(A)x = 0$, then $p(\lambda) = 0$.

(c) Prove that if $p(A) = 0$, then $p(\lambda) = 0$ for every eigenvalue λ.

3.7.11 Suppose that $\mathbb{F}$ is algebraically closed, V is finite-dimensional, $T \in \mathcal{L}(V)$, and $p(x)$ is any polynomial with coefficients in $\mathbb{F}$. Prove that every eigenvalue of $p(T)$ is of the form $p(\lambda)$, where λ is some eigenvalue of T.

3.7.12 A matrix $A \in M_n(\mathbb{F})$ is called **strictly upper triangular** if $a_{ij} = 0$ whenever $i \geq j$. That is, A is upper triangular with all diagonal entries equal to 0.

Show that if $A \in M_n(\mathbb{F})$ is strictly upper triangular, then $A^n = 0$.

3.7.13 Suppose that $\mathbb{F}$ is algebraically closed and that $A \in M_n(\mathbb{F})$ has only one eigenvalue $\lambda \in \mathbb{F}$. Show that $(A - \lambda I_n)^n = 0$.

Hint: Use Exercise 3.7.12.

3.7.14 Show that if V is finite-dimensional and $T \in \mathcal{L}(V)$, then there is a nonzero polynomial $p(x)$ with coefficients in $\mathbb{F}$ such that $p(T) = 0$.

3.7.15 Show that the field $\mathbb{F} = \{a + bi : a, b \in \mathbb{Q}\}$ (see Exercise 1.4.2) is not algebraically closed.

3.7.16 Show that the matrix $\begin{bmatrix} 1 & 1 \\ 1 & 0 \end{bmatrix}$ over $\mathbb{F}_2$ does not have an eigenvalue.

3.7.17 Prove that if $\mathbb{F}$ is a field with only finitely many elements, then $\mathbb{F}$ is not algebraically closed.

Hint: Come up with a polynomial over $\mathbb{F}$ which has every element of $\mathbb{F}$ as a root, then add 1 to it.

PERSPECTIVES: Bases

The list $(v_1, \ldots, v_n)$ of vectors in a vector space V over a field $\mathbb{F}$ is a basis of V if any of the following hold.

- $(v_1, \ldots, v_n)$ is linearly independent and spans V.
- Every $v \in V$ can be uniquely represented as $v = c_1 v_1 + \cdots + c_n v_n$, with $c_i \in \mathbb{F}$ for all i.
- $(v_1, \ldots, v_n)$ is linearly independent and $\dim(V) = n$.
- $(v_1, \ldots, v_n)$ spans V and $\dim(V) = n$.
- The RREF of the matrix

$$\begin{bmatrix} | & & | \\ \mathbf{v}_1 & \cdots & \mathbf{v}_n \\ | & & | \end{bmatrix}$$

  is the identity I_n, where $(\mathbf{v}_1, \ldots, \mathbf{v}_n)$ are the representations of $(v_1, \ldots, v_n)$ in any coordinate system.

PERSPECTIVES: Eigenvalues

A scalar $\lambda \in \mathbb{F}$ is an eigenvalue of $T \in \mathcal{L}(V)$ if either of the following holds.

- There is a nonzero $v \in V$ with $Tv = \lambda v$.
- The map $T - \lambda I$ is not invertible.

PERSPECTIVES: Isomorphisms

Let V and W be n-dimensional vector spaces over $\mathbb{F}$. A linear map $T : V \to W$ is an isomorphism if any of the following hold.

- T is bijective.
- T is invertible.
- T is injective, or equivalently, $\operatorname{null}(T) = 0$.
- T is surjective, or equivalently, $\operatorname{rank}(T) = n$.
- If $(v_1, \ldots, v_n)$ is a basis of T, then $(T(v_1), \ldots, T(v_n))$ is a basis of W.
- If A is the matrix of T with respect to bases $\mathcal{B}_V$ and $\mathcal{B}_W$, then the columns of A form a basis of $\mathbb{F}^n$.
- If A is the matrix of T with respect to bases $\mathcal{B}_V$ and $\mathcal{B}_W$, the RREF of A is the identity I_n.

4

Inner Products

4.1 Inner Products

We saw in Section 1.3 that there were various ways in which the geometry of $\mathbb{R}^n$ could shed light on linear systems of equations. We used a very limited amount of geometry, though; we only made use of general vector space operations. The geometry of $\mathbb{R}^n$ is much richer than that of an arbitrary vector space because of the concepts of length and angles; it is extensions of these ideas that we will explore in this chapter.

The Dot Product in $\mathbb{R}^n$

Recall the following definition from Euclidean geometry.

Definition Let $\mathbf{x}, \mathbf{y} \in \mathbb{R}^n$. The **dot product** or **inner product** of $\mathbf{x}$ and $\mathbf{y}$ is denoted $\langle \mathbf{x}, \mathbf{y} \rangle$ and is defined by

$$\langle \mathbf{x}, \mathbf{y} \rangle = \sum_{j=1}^{n} x_j y_j,$$

where $\mathbf{x} = \begin{bmatrix} x_1 \\ \vdots \\ x_n \end{bmatrix}$ and $\mathbf{y} = \begin{bmatrix} y_1 \\ \vdots \\ y_n \end{bmatrix}$.

Quick Exercise #1. Show that $\langle \mathbf{x}, \mathbf{y} \rangle = \mathbf{y}^T \mathbf{x}$ for $\mathbf{x}, \mathbf{y} \in \mathbb{R}^n$.

The dot product is intimately related to the ideas of length and angle: the length $\|\mathbf{x}\|$ of a vector $\mathbf{x} \in \mathbb{R}^n$ is given by

$$\|\mathbf{x}\| = \sqrt{x_1^2 + \cdots + x_n^2} = \sqrt{\langle \mathbf{x}, \mathbf{x} \rangle},$$

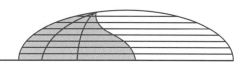

and the angle $\theta_{\mathbf{x},\mathbf{y}}$ between two vectors $\mathbf{x}$ and $\mathbf{y}$ is given by

$$\theta_{\mathbf{x},\mathbf{y}} = \cos^{-1}\left(\frac{\langle \mathbf{x}, \mathbf{y}\rangle}{\|\mathbf{x}\|\ \|\mathbf{y}\|}\right).$$

In particular, the dot product gives us a condition for perpendicularity: two vectors $\mathbf{x}, \mathbf{y} \in \mathbb{R}^n$ are **perpendicular** if they meet at a right angle, which by the formula above is equivalent to the condition $\langle \mathbf{x}, \mathbf{y}\rangle = 0$. For example, the standard basis vectors $\mathbf{e}_1, \ldots, \mathbf{e}_n \in \mathbb{R}^n$ are perpendicular to each other, since $\langle \mathbf{e}_i, \mathbf{e}_j\rangle = 0$ for $i \neq j$.

Perpendicularity is an extremely useful concept in the context of linear algebra; the following proposition gives a first hint as to why.

Proposition 4.1 *Let* $(\mathbf{e}_1, \ldots, \mathbf{e}_n)$ *denote the standard basis of* $\mathbb{R}^n$. *If* $\mathbf{v} = \begin{bmatrix} v_1 \\ \vdots \\ v_n \end{bmatrix} \in \mathbb{R}^n$, *then for each* i, $v_i = \langle \mathbf{v}, \mathbf{e}_i\rangle$.

Proof For $\mathbf{v}$ as above,

$$\langle \mathbf{v}, \mathbf{e}_j\rangle = \sum_k v_k (\mathbf{e}_j)_k = v_j,$$

since $\mathbf{e}_j$ has a 1 in the jth position and zeroes everywhere else. ▲

We will soon see that, while the computations are particularly easy with the standard basis, the crucial property that makes Proposition 4.1 work is the perpendicularity of the basis elements.

Inner Product Spaces

Motivated by the considerations above, we introduce the following extra kind of structure for vector spaces. Here and for the rest of the chapter, we will only allow the base field to be $\mathbb{R}$ or $\mathbb{C}$. Recall that for $z = a + ib \in \mathbb{C}$, the **complex conjugate** $\overline{z}$ of z is defined by $\overline{z} = a - ib$, and the **absolute value** or **modulus** $|z|$ is defined by $|z| = \sqrt{a^2 + b^2}$.

Definition Let $\mathbb{F}$ be either $\mathbb{R}$ or $\mathbb{C}$, and let V be a vector space over $\mathbb{F}$. An **inner product** is an operation on V, written $\langle v, w\rangle$, such that the following properties hold:

The inner product of two vectors is a scalar: For each $v, w \in V$, $\langle v, w\rangle \in \mathbb{F}$.
Distributive law: For each $u, v, w \in V$, $\langle u + v, w\rangle = \langle u, w\rangle + \langle v, w\rangle$.

Homogeneity: For each $v, w \in V$ and $a \in \mathbb{F}$, $\langle av, w\rangle = a\langle v, w\rangle$.
Symmetry: For each $v, w \in V$, $\langle v, w\rangle = \overline{\langle w, v\rangle}$.
Nonnegativity: For each $v \in V$, $\langle v, v\rangle \geq 0$.
Definiteness: If $\langle v, v\rangle = 0$, then $v = 0$.

A vector space together with an inner product is called an **inner product space.**

Notice that if $\mathbb{F} = \mathbb{R}$, the symmetry property just says that

$$\langle v, w\rangle = \langle w, v\rangle ,$$

since by the first requirement, inner products over a real vector space are real numbers. Even if $\mathbb{F} = \mathbb{C}$, it follows by symmetry that $\langle v, v\rangle \in \mathbb{R}$ (we are implicitly using this observation in our nonnegativity requirement).

Examples

1. $\mathbb{R}^n$ with the inner product defined on page 225 is an inner product space.
2. For $\mathrm{w}, \mathrm{z} \in \mathbb{C}^n$, the standard inner product $\langle \mathrm{w}, \mathrm{z}\rangle$ is defined by

$$\langle \mathrm{w}, \mathrm{z}\rangle := \sum_{j=1}^{n} w_j \overline{z_j},$$

where $\mathrm{w} = \begin{bmatrix} w_1 \\ \vdots \\ w_n \end{bmatrix}$ and $\mathrm{z} = \begin{bmatrix} z_1 \\ \vdots \\ z_n \end{bmatrix}$.

Quick Exercise #2. Show that if $\mathrm{w}, \mathrm{z} \in \mathbb{C}^n$, then $\langle \mathrm{w}, \mathrm{z}\rangle = \mathrm{z}^*\mathrm{w}$, where $\mathrm{z}^* = \begin{bmatrix} \overline{z_1} & \cdots & \overline{z_n} \end{bmatrix} \in \mathrm{M}_{1,n}(\mathbb{C})$ is the **conjugate transpose** of z.

3. Let ℓ^2 denote the set of square-summable sequences over $\mathbb{R}$; i.e., $a = (a_1, a_2, \ldots) \in \ell^2$ if $\sum_{j=1}^{\infty} a_j^2 < \infty$. Then

$$\langle a, b\rangle := \sum_{j=1}^{\infty} a_j b_j \tag{4.1}$$

defines an inner product on ℓ^2. Verifying all but the first property is similarly straightforward to the case of the usual dot product on $\mathbb{R}^n$. In order to verify that the definition above gives $\langle a, b\rangle \in \mathbb{R}$ for any $a, b \in \ell^2$, one has to show that the sum in equation (4.1) always converges; we will postpone confirming this until later in the section (Example 4 on page 235).

4. Let $C([0, 1])$ denote the vector space of continuous real-valued functions on $[0, 1]$. Define an inner product on $C([0, 1])$ by

$$\langle f, g\rangle := \int_0^1 f(x)g(x)\,dx.$$

The properties in the definition are easily verified.

5. If $C_{\mathbb{C}}([0, 1])$ denotes the complex vector space of continuous complex-valued functions on $[0, 1]$, then one can define an inner product on $C_{\mathbb{C}}([0, 1])$ by

$$\langle f, g\rangle := \int_0^1 f(x)\overline{g(x)}\,dx.$$

▲

For the rest of this section, V will always be an inner product space.

As in the case of fields and vector spaces, we have assumed as little as possible in the definition of an inner product; the following properties all follow easily from the definition.

Proposition 4.2 *Suppose that V is an inner product space. Then the following all hold:*

1. *For each $u, v, w \in V$, $\langle u, v + w\rangle = \langle u, v\rangle + \langle u, w\rangle$.*
2. *For each $v, w \in V$ and $a \in \mathbb{F}$, $\langle v, aw\rangle = \overline{a}\,\langle v, w\rangle$.*
3. *For each $v \in V$, $\langle 0, v\rangle = \langle v, 0\rangle = 0$.*
4. *If $v \in V$ and $\langle v, w\rangle = 0$ for every $w \in V$, then $v = 0$.*

Proof 1. By the symmetry and distributive properties,

$$\langle u, v + w\rangle = \overline{\langle v + w, u\rangle} = \overline{\langle v, u\rangle + \langle w, u\rangle} = \overline{\langle v, u\rangle} + \overline{\langle w, u\rangle} = \langle u, v\rangle + \langle u, w\rangle\,.$$

2. By the symmetry and homogeneity properties,

$$\langle v, aw\rangle = \overline{\langle aw, v\rangle} = \overline{a\,\langle w, v\rangle} = \overline{a}\overline{\langle w, v\rangle} = \overline{a}\,\langle v, w\rangle\,.$$

3. By the distributive property,

$$\langle 0, v\rangle = \langle 0 + 0, v\rangle = \langle 0, v\rangle + \langle 0, v\rangle\,.$$

Subtracting $\langle 0, v\rangle$ from both sides proves that $\langle 0, v\rangle = 0$. It then follows from symmetry that $\langle v, 0\rangle = \overline{0} = 0$.

4. Picking $w = v$, we have $\langle v, v\rangle = 0$. By the definiteness property, this implies that $v = 0$.

▲

The following definition gives an analog to length in $\mathbb{R}^n$.

> **Definition** Let $v \in V$. The **norm** of v is
>
> $$\|v\| = \sqrt{\langle v, v\rangle}.$$
>
> If $\|v\| = 1$, then v is called a **unit vector**.

Examples

1. If $\mathrm{v} = \begin{bmatrix} 2 \\ i \\ -1 \end{bmatrix}$ in $\mathbb{C}^3$ with the standard inner product, then

$$\|\mathrm{v}\| = \sqrt{|2|^2 + |i|^2 + |-1|^2} = \sqrt{2^2 + 1^2 + 1^2} = \sqrt{6}.$$

2. Consider the function $f(x) = 2x$ in the function space $C([0, 1])$ with the inner product discussed above. Then

$$\|f\| = \sqrt{\int_0^1 4x^2 \, dx} = \frac{2}{\sqrt{3}}.$$

In this example, the norm does not seem much like the length of an arrow; the object whose norm we are taking is a function. We nevertheless think of the norm as a kind of measure of size. ▲

Note that $\langle v, v\rangle \geq 0$ by the definition of an inner product, so the square root is defined and nonnegative. Note also that, by the definiteness property of the inner product, $\|v\| = 0$ if and only if $v = 0$.

If $c \in \mathbb{F}$, then

$$\|cv\| = \sqrt{\langle cv, cv\rangle} = \sqrt{|c|^2 \langle v, v\rangle} = |c| \, \|v\| \, .$$

We refer to this property as **positive homogeneity**; it means that if we multiply a vector by a positive scalar, its “length” gets multiplied by that same scalar.

Orthogonality

We observed at the beginning of the section that the idea of perpendicularity was often useful; the following is the generalization of the familiar geometric notion to arbitrary inner product spaces.

> **Definition** Two vectors $v, w \in V$ are **orthogonal** or **perpendicular** if $\langle v, w\rangle = 0$. A list of vectors $(v_1, \ldots, v_n)$ in V is called **orthogonal** if $\langle v_j, v_k\rangle = 0$ whenever $j \neq k$.

The following result is one reason for the importance of orthogonality in linear algebra.

Theorem 4.3 *An orthogonal list of nonzero vectors is linearly independent.*

Proof Let $(v_1, \ldots, v_n)$ be an orthogonal list of nonzero vectors, so that $\langle v_j, v_k \rangle = 0$ if $j \neq k$, and suppose that

$$\sum_{j=1}^{n} a_j v_j = 0$$

for some scalars $a_1, \ldots, a_n \in \mathbb{F}$. Then for each $k = 1, \ldots, n$,

$$a_k \|v_k\|^2 = a_k \langle v_k, v_k \rangle = \sum_{j=1}^{n} a_j \langle v_j, v_k \rangle = \left\langle \sum_{j=1}^{n} a_j v_j, v_k \right\rangle = \langle 0, v_k \rangle = 0.$$

Since $v_k \neq 0$, we know that $\|v_k\|^2 \neq 0$, and therefore this implies that $a_k = 0$. Since this is true for each k, we conclude that $(v_1, \ldots, v_n)$ is linearly independent. ▲

Quick Exercise #3. Use Theorem 4.3 to verify that the list of vectors

$$\left(\begin{bmatrix} 2 \\ 0 \\ -1 \\ 3 \end{bmatrix}, \begin{bmatrix} -1 \\ 1 \\ 1 \\ 1 \end{bmatrix}, \begin{bmatrix} 3 \\ 1 \\ 3 \\ -1 \end{bmatrix} \right)$$

in $\mathbb{R}^4$ is linearly independent.

The following result is sometimes called the Pythagorean Theorem for general inner product spaces.

Theorem 4.4 *If $(v_1, \ldots, v_n)$ is an orthogonal list of vectors, then*

$$\|v_1 + \cdots + v_n\|^2 = \|v_1\|^2 + \cdots + \|v_n\|^2$$

Proof By the definition of the norm and the distributive laws for the inner product,

$$\left\| \sum_{j=1}^{n} v_j \right\|^2 = \left\langle \sum_{j=1}^{n} v_j, \sum_{k=1}^{n} v_k \right\rangle = \sum_{j=1}^{n} \sum_{k=1}^{n} \langle v_j, v_k \rangle.$$

QA #3: The list is orthogonal, so it is linearly independent by Theorem 4.3.

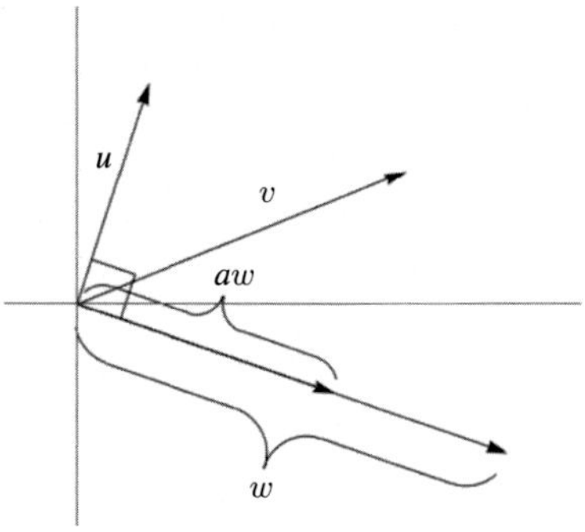

Figure 4.1 Decomposing a vector v as $v = aw + u$ where u is orthogonal to w.

By orthogonality, the only nonzero terms in this sum are those for which $j = k$, and so the sum is equal to

$$\sum_{j=1}^{n} \langle v_j, v_j \rangle = \sum_{j=1}^{n} \|v_j\|^2 .$$

▲

The next result shows us how we can make use of orthogonality even when we start off with nonorthogonal vectors.

Lemma 4.5 *Given $v, w \in V$, v can be written as*

$$v = aw + u, \tag{4.2}$$

where $a \in \mathbb{F}$ and $u \in V$ is orthogonal to w. Moreover, if $w \neq 0$, we can take

$$a = \frac{\langle v, w \rangle}{\|w\|^2} \qquad \text{and} \qquad u = v - \frac{\langle v, w \rangle}{\|w\|^2} w.$$

Proof Suppose that $w \neq 0$. (For the case that $w = 0$, see Exercise 4.1.21.) It is obvious that equation (4.2) holds for the stated values of a and u, so we could simply verify that u is orthogonal to w, but let's first see where those values come from.

The thing to notice is that if there is an $a \in \mathbb{F}$ such that $v = aw + u$ with u orthogonal to w, then by linearity,

$$\langle v, w \rangle = \langle aw + u, w \rangle = a \langle w, w \rangle + \langle u, w \rangle = a \|w\|^2 ,$$

so the only possibility is $a = \frac{\langle v,w \rangle}{\|w\|^2}$. Once a has been determined, the choice of u is determined by equation (4.2) (which says $u = v - aw$), and it is easy to check that u is in fact orthogonal to w. ▲

The following innocuous-looking inequality is of central importance in the theory of inner product spaces.

Theorem 4.6 (The Cauchy–Schwarz inequality*) *For every $v, w \in V$,*

$$|\langle v, w\rangle| \leq \|v\| \, \|w\| \, ,$$

with equality if and only if v and w are collinear.

Proof If $w = 0$, then both sides of the inequality are 0, and also $w = 0v$, so v and w are collinear.

If $w \neq 0$, then by Lemma 4.5, we can write

$$v = aw + u,$$

where $a = \frac{\langle v,w\rangle}{\|w\|^2}$ and u is orthogonal to w. By Theorem 4.4,

$$\|v\|^2 = \|aw\|^2 + \|u\|^2 = |a|^2 \, \|w\|^2 + \|u\|^2 = \frac{|\langle v, w\rangle|^2}{\|w\|^2} + \|u\|^2 \, .$$

Therefore

$$\|v\|^2 \geq \frac{|\langle v, w\rangle|^2}{\|w\|^2},$$

which is equivalent to the claimed inequality. We get equality if and only if $u = 0$. That is true if and only if $v = aw$ for some scalar a: in other words, exactly when v and w are collinear. ▲

The next result is also fundamental and is a consequence of the Cauchy–Schwarz inequality.

Theorem 4.7 (The triangle inequality) *For any $v, w \in V$,*

$$\|v + w\| \leq \|v\| + \|w\| \, .$$

Proof By linearity and the Cauchy–Schwarz inequality,

$$\begin{aligned}
\|v + w\|^2 &= \langle v + w, v + w\rangle \\
&= \langle v, v\rangle + \langle v, w\rangle + \langle w, v\rangle + \langle w, w\rangle \\
&= \|v\|^2 + 2 \operatorname{Re} \langle v, w\rangle + \|w\|^2 \\
&\leq \|v\|^2 + 2 \, |\langle v, w\rangle| + \|w\|^2 \\
&\leq \|v\|^2 + 2 \, \|v\| \, \|w\| + \|w\|^2 \\
&= \left(\|v\| + \|w\|\right)^2.
\end{aligned}$$

*This result is named after the French mathematician Augustin-Louis Cauchy and the German mathematician Hermann Schwarz, who each proved special cases of it. It is also often called *Cauchy's inequality* (especially by the French), *Schwarz's inequality* (especially by Germans), the *Cauchy–Bunyakovsky–Schwarz inequality* (especially by Russians, who have a point – Bunyakovsky proved Schwarz's special case before Schwarz did), and the *Cauchy–Schwartz inequality* (by people who don't know how to spell Schwarz properly).

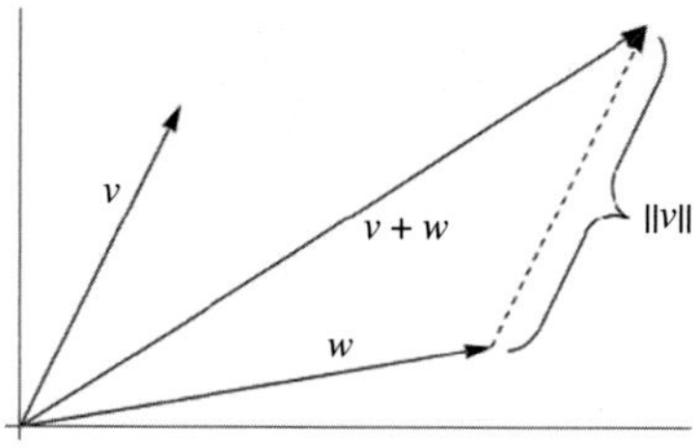

Figure 4.2 The length of the $v + w$ side of the triangle is less than or equal to the sum of the lengths of the other sides.

The result follows by taking square roots. ▲

This last proof is our first illustration of the following words to live by:

> $\|v\|^2$ is much easier to work with than $\|v\|$.
> See if you can get away with just taking square roots at the end of your proof.

The next corollary is just a slightly different way of looking at the triangle inequality, which moves the triangle in question away from the origin.

Corollary 4.8 *For any $u, v, w \in V$,*

$$\|u - v\| \leq \|u - w\| + \|w - v\| .$$

Quick Exercise #4. Prove Corollary 4.8.

More Examples of Inner Product Spaces

1. Let $c_1, \ldots, c_n > 0$ be fixed positive numbers, and define

$$\langle \mathbf{x}, \mathbf{y} \rangle = \sum_{j=1}^{n} c_j x_j y_j$$

for $\mathbf{x}, \mathbf{y} \in \mathbb{R}^n$. This defines an inner product on $\mathbb{R}^n$ which is different from the standard inner product (which corresponds to $c_j = 1$ for each j). The standard basis vectors are still orthogonal, but they are no longer unit vectors. Instead, for each j, $\frac{1}{\sqrt{c_j}}\mathbf{e}_j$ is a unit vector. This means that the collection of all unit vectors (the unit sphere for this inner product space) is an ellipsoid (or, in two dimensions, simply an ellipse):

QA #4: $u - v = (u - w) + (w - v)$. Apply the triangle inequality.

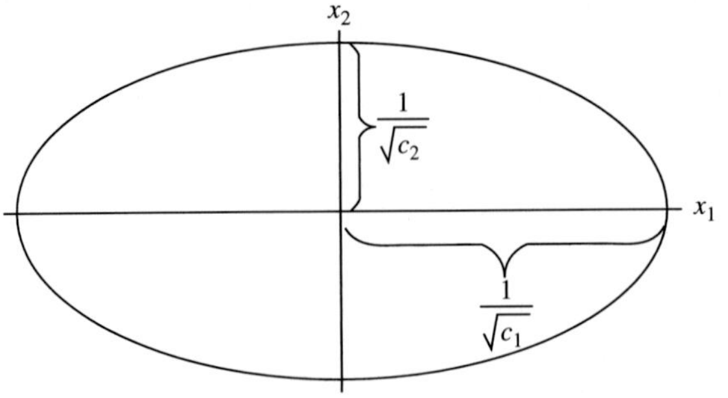

Figure 4.3 The unit circle for the inner product $\langle \mathbf{x}, \mathbf{y} \rangle = c_1 x_1 y_1 + c_2 x_2 y_2$ on $\mathbb{R}^2$.

2. The following is the function space version of the previous example. Fix a function $h \in C([a, b])$ such that $h(x) > 0$ for every $x \in [a, b]$. For $f, g \in C([a, b])$, define

$$\langle f, g \rangle = \int_a^b f(x) g(x) h(x) \, dx.$$

This defines an inner product.

3. As we've discussed before, we can think of the space $\mathrm{M}_{m,n}(\mathbb{C})$ of $m \times n$ complex matrices as being $\mathbb{C}^{mn}$ written in an unusual way. The standard inner product on $\mathbb{C}^{mn}$ then becomes

$$\langle \mathrm{A}, \mathrm{B} \rangle = \sum_{j=1}^{m} \sum_{k=1}^{n} a_{jk} \overline{b_{jk}} \tag{4.3}$$

for $\mathrm{A}, \mathrm{B} \in \mathrm{M}_{m,n}(\mathbb{C})$. Recalling that the conjugate transpose of B has (k, j) entry $\overline{b_{jk}}$, we can see that $\sum_{k=1}^{n} a_{jk} \overline{b_{jk}} = [\mathrm{AB}^*]_{jj}$, and so equation (4.3) becomes

$$\langle \mathrm{A}, \mathrm{B} \rangle = \operatorname{tr} \mathrm{AB}^*.$$

Since this involves both matrix multiplication and the trace, this definition of inner product on $\mathrm{M}_{m,n}(\mathbb{C})$ seems at least possibly sensible to work with. It is most often called the **Frobenius inner product**, and the associated norm

$$\|\mathrm{A}\|_F = \sqrt{\operatorname{tr} \mathrm{AA}^*}$$

is most often called the **Frobenius norm*** of A.

Quick Exercise #5. Suppose that $\mathrm{A}, \mathrm{B} \in \mathrm{M}_n(\mathbb{R})$ are **symmetric**, meaning that $\mathrm{A} = \mathrm{A}^{\mathrm{T}}$ and $\mathrm{B} = \mathrm{B}^{\mathrm{T}}$. Prove that

$$(\operatorname{tr} \mathrm{AB})^2 \leq (\operatorname{tr} \mathrm{A}^2)(\operatorname{tr} \mathrm{B}^2).$$

*It's also called the Hilbert–Schmidt norm or the Schur norm, among other names.

QA #5: This is just the Cauchy–Schwarz inequality applied to the Frobenius inner product of A and B.

4. Returning to the example of ℓ^2 of square-summable sequences of real numbers, we still need to verify that the formula

$$\langle a, b\rangle = \sum_{j=1}^{\infty} a_j b_j$$

does indeed define a real number when $a, b \in \ell^2$. The issue is whether the infinite sum on the right-hand side converges. The assumption $a, b \in \ell^2$ means that

$$\sum_{j=1}^{\infty} a_j^2 < \infty \qquad \text{and} \qquad \sum_{j=1}^{\infty} b_j^2 < \infty.$$

By the Cauchy–Schwarz inequality, for each $n \in \mathbb{N}$,

$$\sum_{j=1}^{n} |a_j b_j| \le \sqrt{\sum_{j=1}^{n} a_j^2}\sqrt{\sum_{j=1}^{n} b_j^2} \le \sqrt{\sum_{j=1}^{\infty} a_j^2}\sqrt{\sum_{j=1}^{\infty} b_j^2}.$$

Taking the limit as $n \to \infty$, we obtain

$$\sum_{j=1}^{\infty} |a_j b_j| \le \sqrt{\sum_{j=1}^{\infty} a_j^2}\sqrt{\sum_{j=1}^{\infty} b_j^2} < \infty.$$

So the original series $\sum_{j=1}^{\infty} a_j b_j$ converges absolutely; by a theorem from calculus, it therefore converges to some real number. This completes the demonstration that ℓ^2 is an inner product space.

KEY IDEAS

- An inner product space is a vector space with an inner product $\langle \cdot, \cdot \rangle$, which is a homogeneous, conjugate-symmetric, positive-definite scalar-valued function on pairs of vectors.
- Inner products define orthogonality (perpendicularity): v and w are orthogonal if $\langle v, w\rangle = 0$.
- Inner products define length: $\|v\| := \sqrt{\langle v, v\rangle}$.
- Examples of inner products include the usual dot product on $\mathbb{R}^n$, a modified version on $\mathbb{C}^n$, and extensions of the same to sequence spaces and function spaces. The Frobenius inner product on matrices is $\langle A, B\rangle := \mathrm{tr}(AB^*)$.
- If v and w are orthogonal, $\|v + w\| = \sqrt{\|v\|^2 + \|w\|^2}$.
- The Cauchy–Schwarz inequality: $|\langle v, w\rangle| \le \|v\|\,\|w\|$.
- The triangle inequality: $\|v + w\| \le \|v\| + \|w\|$.

EXERCISES

4.1.1 Use Theorem 4.3 to show that each of the following lists is linearly independent.

(a) $\left(\begin{bmatrix} 1 \\ 1 \\ 1 \\ 1 \end{bmatrix}, \begin{bmatrix} 1 \\ 1 \\ -1 \\ -1 \end{bmatrix}, \begin{bmatrix} 1 \\ -1 \\ 1 \\ -1 \end{bmatrix}, \begin{bmatrix} 1 \\ -1 \\ -1 \\ 1 \end{bmatrix} \right)$ in $\mathbb{R}^4$

(b) $\left(\begin{bmatrix} 1 \\ 1 \\ 1 \end{bmatrix}, \begin{bmatrix} 1+2i \\ -3+i \\ 2-3i \end{bmatrix}, \begin{bmatrix} 21-16i \\ -21-i \\ 17i \end{bmatrix} \right)$ in $\mathbb{C}^3$

(c) $\left(\begin{bmatrix} 2 & 0 & 1 \\ 0 & -3 & 4 \end{bmatrix}, \begin{bmatrix} 0 & 2 & 6 \\ -1 & 2 & 0 \end{bmatrix}, \begin{bmatrix} -2 & 3 & 0 \\ 6 & 0 & 1 \end{bmatrix} \right)$ in $\mathrm{M}_{2,3}(\mathbb{R})$

(d) $\left(5x^2 - 6x + 1, x^2 - 2x + 1, 10x^2 - 8x + 1\right)$ in $C([0, 1])$

4.1.2 Suppose that V is a real inner product space, $v, w \in V$, and that

$$\|v - w\|^2 = 3 \qquad \text{and} \qquad \|v + w\|^2 = 7.$$

Compute each of the following:

(a) $\langle v, w \rangle$ (b) $\|v\|^2 + \|w\|^2$

4.1.3 Suppose that V is a real inner product space, $v, w \in V$, and that

$$\|v + w\|^2 = 10 \qquad \text{and} \qquad \|v - w\|^2 = 16.$$

Compute each of the following:

(a) $\langle v, w \rangle$ (b) $\|v\|^2 + \|w\|^2$

4.1.4 Let $\mathrm{A}, \mathrm{B} \in \mathrm{M}_{m,n}(\mathbb{C})$. Show that

$$\langle \mathrm{A}, \mathrm{B} \rangle_F = \sum_{j=1}^{n} \langle \mathbf{a}_j, \mathbf{b}_j \rangle$$

and

$$\|\mathrm{A}\|_F^2 = \sum_{j=1}^{n} \left\| \mathbf{a}_j \right\|^2,$$

where the $\mathbf{a}_j$ and $\mathbf{b}_j$ are the columns of A and B, respectively.

4.1.5 For $\mathrm{A} \in \mathrm{M}_n(\mathbb{C})$, the **Hermitian part** of A is the matrix

$$\operatorname{Re} \mathrm{A} := \frac{1}{2}(\mathrm{A} + \mathrm{A}^*)$$

and the **anti-Hermitian part** of A is the matrix

$$\operatorname{Im} \mathrm{A} := \frac{1}{2i}(\mathrm{A} - \mathrm{A}^*).$$

(The notation is analogous to the real and imaginary parts of a complex number.)

(a) Show that if $\mathrm{A} \in \mathrm{M}_n(\mathbb{R})$, then $\operatorname{Re} \mathrm{A}$ and $\operatorname{Im} \mathrm{A}$ are orthogonal (with respect to the Frobenius inner product).

(b) Show that if $A \in M_n(\mathbb{R})$, then $\|A\|_F^2 = \|\mathrm{Re}\, A\|_F^2 + \|\mathrm{Im}\, A\|_F^2$.

4.1.6 Show that if that $A = \begin{bmatrix} a_1 & \cdots & a_n \end{bmatrix} \in M_{m,n}(\mathbb{C})$ and $B = \begin{bmatrix} b_1 & \cdots & b_p \end{bmatrix} \in M_{m,p}(\mathbb{C})$, then the (j,k) entry of A^*B is $a_j^* b_k = \langle b_k, a_j \rangle$.

4.1.7 Equip $C([0, 2\pi])$ with the inner product

$$\langle f, g \rangle = \int_0^{2\pi} f(x)g(x)\, dx.$$

Let $f(x) = \sin x$ and $g(x) = \cos x$. Compute each of the following:
(a) $\|f\|$ (b) $\|g\|$ (c) $\langle f, g \rangle$
Then find $\|af + bg\|$, where $a, b \in \mathbb{R}$ are any constants, *without* computing any more integrals.

4.1.8 Suppose that $u, v \in V$, $\|u\| = 2$ and $\|v\| = 11$. Prove that there is no vector $w \in V$ with $\|u - w\| \leq 4$ and $\|v - w\| \leq 4$.

4.1.9 Suppose that V is a vector space, W is an inner product space, and $T : V \to W$ is injective. For $v_1, v_2 \in V$, define

$$\langle v_1, v_2 \rangle_T := \langle Tv_1, Tv_2 \rangle,$$

where the right-hand side involves the given inner product on W. Prove that this defines an inner product on V.

4.1.10 Suppose that V is an inner product space, W is a vector space, and $T : V \to W$ is an isomorphism. For $w_1, w_2 \in W$, define

$$\langle w_1, w_2 \rangle = \langle T^{-1}w_1, T^{-1}w_2 \rangle,$$

where the right-hand side involves the given inner product on V. Prove that this defines an inner product on W.

4.1.11 Define

$$\left\langle \begin{bmatrix} x_1 \\ x_2 \end{bmatrix}, \begin{bmatrix} y_1 \\ y_2 \end{bmatrix} \right\rangle = x_1 y_1 + 4x_2 y_2.$$

(a) Use Exercise 4.1.9 to verify that this is an inner product (different from the standard inner product!) on $\mathbb{R}^2$.
(b) Give an example of two vectors in $\mathbb{R}^2$ which are orthogonal with respect to the standard inner product, but not with respect to this one.
(c) Given an example of two vectors in $\mathbb{R}^2$ which are orthogonal with respect to this inner product, but not with respect to the standard inner product.

4.1.12 (a) Show that $v = 0$ if and only if $\langle v, w \rangle = 0$ for every $w \in V$.
(b) Show that $v = w$ if and only if $\langle v, u \rangle = \langle w, u \rangle$ for every $u \in V$.
(c) Let $S, T \in \mathcal{L}(V, W)$. Show that $S = T$ if and only if $\langle Sv_1, v_2 \rangle = \langle Tv_1, v_2 \rangle$ for every $v_1, v_2 \in V$.

4.1.13 Show that the Frobenius norm is not an invariant on $\mathrm{M}_n(\mathbb{C})$.

4.1.14 (a) Prove that if V is a real inner product space, then

$$\langle v, w\rangle = \frac{1}{4}\left(\|v+w\|^2 - \|v-w\|^2\right)$$

for each $v, w \in V$.

(b) Prove that if V is a complex inner product space, then

$$\langle v, w\rangle = \frac{1}{4}\left(\|v+w\|^2 - \|v-w\|^2 + i\,\|v+iw\|^2 - i\,\|v-iw\|^2\right)$$

for each $v, w \in V$.

These are known as the **polarization identities.**

4.1.15 Show that if $a_1, \ldots, a_n, b_1, \ldots, b_n \in \mathbb{R}$, then

$$\left(\sum_{i=1}^{n} a_i b_i\right) \le \left(\sum_{i=1}^{n} a_i^2\right)\left(\sum_{i=1}^{n} b_i^2\right).$$

This is known as *Cauchy's inequality.*

4.1.16 Show that if $f, g : [a, b] \to \mathbb{R}$ are continuous, then

$$\left(\int_a^b f(x)g(x)\,dx\right)^2 \le \left(\int_a^b f(x)^2\,dx\right)\left(\int_a^b g(x)^2\,dx\right).$$

This is known as *Schwarz's inequality.*

4.1.17 Prove that

$$(a_1b_1+a_2b_2+\cdots+a_nb_n)^2 \le (a_1^2+2a_2^2+\cdots+na_n^2)\left(b_1^2+\frac{b_2^2}{2}+\cdots+\frac{b_n^2}{n}\right)$$

for all $a_1, \ldots, a_n, b_1, \ldots, b_n \in \mathbb{R}$.

4.1.18 Under what circumstances does equality hold in the triangle inequality?

4.1.19 Suppose that V is a complex inner product space. Define

$$\langle v, w\rangle_{\mathbb{R}} := \operatorname{Re}\langle v, w\rangle,$$

where the inner product on the right is the original inner product on V. Show that if we think of V as just a real vector space, then $\langle \cdot, \cdot\rangle_{\mathbb{R}}$ is an inner product on V.

4.1.20 Give another proof of part 3 of Proposition 4.2 using homogeneity instead of the distributive property.

4.1.21 Prove Lemma 4.5 in the case that $w = 0$.

4.1.22 Given $v, w \in V$, define the function

$$f(t) = \|v+tw\|^2$$

for $t \in \mathbb{R}$. Find the minimum value c of f, and use the fact that $c \ge 0$ to give another proof of the Cauchy–Schwarz inequality.

4.2 Orthonormal Bases

Orthonormality

Working in bases within an inner product space is often easier than it is in general vector spaces, because of the existence of bases with the following special property.

Definition A finite or countable set $\{e_i\}$ of vectors in V is called **orthonormal** if

$$\langle e_j, e_k \rangle = \begin{cases} 0 & \text{if } j \neq k, \\ 1 & \text{if } j = k. \end{cases}$$

If $\mathcal{B} = (e_1, \ldots, e_n)$ is an orthonormal list which is also a basis of V, then we call $\mathcal{B}$ an **orthonormal basis** of V.

In other words, an orthonormal set of vectors is a collection of mutually perpendicular unit vectors.

Examples

1. The standard basis $(\mathbf{e}_1, \ldots, \mathbf{e}_n)$ of $\mathbb{R}^n$ or of $\mathbb{C}^n$ is orthonormal.
2. The list $\left(\frac{1}{\sqrt{2}} \begin{bmatrix} 1 \\ 1 \end{bmatrix}, \frac{1}{\sqrt{2}} \begin{bmatrix} 1 \\ -1 \end{bmatrix} \right)$ is an orthonormal basis of either $\mathbb{R}^2$ or $\mathbb{C}^2$.

Quick Exercise #6. Show that for any fixed $\theta \in \mathbb{R}$, $\left(\begin{bmatrix} \cos\theta \\ \sin\theta \end{bmatrix}, \begin{bmatrix} -\sin\theta \\ \cos\theta \end{bmatrix} \right)$ is an orthonormal basis of $\mathbb{R}^2$ and of $\mathbb{C}^2$.

3. Consider the space $C_{2\pi}(\mathbb{R})$ of continuous 2π-periodic functions $f : \mathbb{R} \to \mathbb{C}$ with the inner product

$$\langle f, g \rangle := \int_0^{2\pi} f(\theta)\overline{g(\theta)}\, d\theta.$$

Then $\{1, \sin(n\theta), \cos(n\theta) \mid n \in \mathbb{N}\}$ is an orthogonal collection of functions:

$$\int_0^{2\pi} \sin(n\theta)\cos(m\theta)\, d\theta = \left\{ \begin{array}{ll} \left. \frac{m\sin(m\theta)\sin(n\theta) + n\cos(m\theta)\cos(n\theta)}{m^2 - n^2} \right|_0^{2\pi} & \text{if } n \neq m, \\ \left. -\frac{\cos^2(n\theta)}{2n} \right|_0^{2\pi} & \text{if } m = n \end{array} \right\} = 0,$$

and if $n \neq m$,

$$\int_0^{2\pi} \sin(n\theta)\sin(m\theta)\, d\theta = \left. \frac{n\sin(m\theta)\cos(n\theta) - m\cos(m\theta)\sin(n\theta)}{m^2 - n^2} \right|_0^{2\pi} = 0,$$

and

$$\int_0^{2\pi} \cos(n\theta)\cos(m\theta)d\theta = \left.\frac{m\sin(m\theta)\cos(n\theta) - n\cos(m\theta)\sin(n\theta)}{m^2 - n^2}\right|_0^{2\pi} = 0.$$

On the other hand,

$$\int_0^{2\pi} \sin^2(n\theta)\, d\theta = \left.\left(\frac{\theta}{2} - \frac{\sin(2n\theta)}{4n}\right)\right|_0^{2\pi} = \pi,$$

and

$$\int_0^{2\pi} \cos^2(n\theta)\, d\theta = \left.\left(\frac{\theta}{2} + \frac{\sin(2n\theta)}{4n}\right)\right|_0^{2\pi} = \pi.$$

It follows that $\mathcal{T} := \left\{ \frac{1}{\sqrt{2\pi}}, \frac{1}{\sqrt{\pi}}\sin(n\theta), \frac{1}{\sqrt{\pi}}\cos(n\theta) \,\middle|\, n \in \mathbb{N} \right\}$ is an orthonormal set, sometimes called the **trigonometric system.** It is not an orthonormal basis of $C_{2\pi}(\mathbb{R})$ because that space is infinite-dimensional.* Nevertheless, any finite subcollection of $\mathcal{T}$ is an orthonormal basis of its span.

4. The set $\left\{ \frac{1}{\sqrt{2\pi}} e^{in\theta} \,\middle|\, n \in \mathbb{Z} \right\}$ is orthonormal in $C_{2\pi}(\mathbb{R})$. Indeed, if $n = m$, then $\int_0^{2\pi} e^{in\theta} e^{-im\theta}\, d\theta = 2\pi$ trivially, and if $n \neq m$, then

$$\int_0^{2\pi} e^{in\theta} e^{-im\theta}\, d\theta = \left.\frac{e^{i(n-m)\theta}}{i(n-m)}\right|_0^{2\pi} = 0.$$

Notice that the real and imaginary parts of these functions recover the trigonometric system described above.

5. Given $1 \leq j \leq m$ and $1 \leq k \leq n$, let E_{jk} be the $m \times n$ matrix whose (j, k) entry is 1 and whose other entries are all 0. Then

$$(E_{11}, \dots, E_{1n}, E_{21}, \dots, E_{2n}, \dots, E_{m1}, \dots, E_{mn})$$

is an orthonormal basis for $M_{m,n}(\mathbb{F})$ equipped with the Frobenius inner product.

6. **Not every natural looking basis is orthonormal!** If we put the inner product

$$\langle f, g \rangle = \int_0^1 f(x)g(x)\, dx$$

on the space $\mathcal{P}_n(\mathbb{R})$, then the obvious basis $(1, x, \dots, x^n)$ is *not* orthonormal. In fact, the basis vectors are not even orthogonal:

$$\int_0^1 x^j x^k\, dx = \int_0^1 x^{j+k}\, dx = \frac{1}{j+k+1}$$

for any $0 \leq j, k \leq n$.

*There are important ways in which these functions do act like a basis of this infinite-dimensional space, but we won't get into that here.

Coordinates in Orthonormal Bases

One of the nicest things about orthonormal bases is that finding coordinate representations in them is very easy, as follows.

Theorem 4.9 *Let $\mathcal{B}_V = (e_1, \dots, e_n)$ and $\mathcal{B}_W = (f_1, \dots, f_m)$ be orthonormal bases of V and W, respectively. Then for every $v \in V$ and $T \in \mathcal{L}(V, W)$,*

$$[v]_{\mathcal{B}_V} = \begin{bmatrix} \langle v, e_1 \rangle \\ \vdots \\ \langle v, e_n \rangle \end{bmatrix} \tag{4.4}$$

and

$$\left[[T]_{\mathcal{B}_V, \mathcal{B}_W} \right]_{jk} = \left\langle Te_k, f_j \right\rangle. \tag{4.5}$$

Proof Since $\mathcal{B}_V$ is a basis, there exist $a_1, \dots, a_n \in \mathbb{F}$ such that

$$v = \sum_{j=1}^{n} a_j e_j.$$

Taking the inner product of each side with e_k and using the linearity properties of the inner product and the orthonormality of $\mathcal{B}_V$, we obtain

$$\langle v, e_k \rangle = \left\langle \sum_{j=1}^{n} a_j e_j, e_k \right\rangle = \sum_{j=1}^{n} a_j \left\langle e_j, e_k \right\rangle = a_k.$$

Putting these together, we get

$$v = \sum_{j=1}^{n} \left\langle v, e_j \right\rangle e_j,$$

which is equivalent to equation (4.4).

Applying this same fact to a vector $w \in W$, we have

$$w = \sum_{j=1}^{m} \left\langle w, f_j \right\rangle f_j,$$

and so in particular for $Te_k \in W$,

$$Te_k = \sum_{j=1}^{m} \left\langle Te_k, f_j \right\rangle f_j,$$

which is equivalent to equation (4.5). ▲

Quick Exercise #7. What is the coordinate representation of $\begin{bmatrix} 2 \\ 3 \end{bmatrix} \in \mathbb{R}^2$ with respect to the orthonormal basis $\left(\frac{1}{\sqrt{2}} \begin{bmatrix} 1 \\ 1 \end{bmatrix}, \frac{1}{\sqrt{2}} \begin{bmatrix} 1 \\ -1 \end{bmatrix} \right)$?

Example Consider the orthonormal basis

$$\mathcal{B} = (\mathbf{v}_1, \mathbf{v}_2, \mathbf{v}_3, \mathbf{v}_4) = \left(\frac{1}{2} \begin{bmatrix} 1 \\ 1 \\ 1 \\ 1 \end{bmatrix}, \frac{1}{2} \begin{bmatrix} 1 \\ 1 \\ -1 \\ -1 \end{bmatrix}, \frac{1}{2} \begin{bmatrix} 1 \\ -1 \\ 1 \\ -1 \end{bmatrix}, \frac{1}{2} \begin{bmatrix} 1 \\ -1 \\ -1 \\ 1 \end{bmatrix} \right)$$

of $\mathbb{R}^4$, and let $T \in \mathcal{L}(\mathbb{R}^4)$ be the linear map represented (with respect to the standard basis) by

$$\mathbf{A} = \begin{bmatrix} 1 & 0 & 5 & 2 \\ 0 & -1 & -2 & -5 \\ -5 & -2 & -1 & 0 \\ 2 & 5 & 0 & 1 \end{bmatrix}.$$

The (j, k) entry of the matrix $[T]_{\mathcal{B}}$ is $\langle T\mathbf{v}_k, \mathbf{v}_j \rangle = \mathbf{v}_j^* \mathbf{A} \mathbf{v}_k$. So, for instance, the $(2, 3)$ entry is

$$\frac{1}{4} \begin{bmatrix} 1 & 1 & -1 & -1 \end{bmatrix} \begin{bmatrix} 1 & 0 & 5 & 2 \\ 0 & -1 & -2 & -5 \\ -5 & -2 & -1 & 0 \\ 2 & 5 & 0 & 1 \end{bmatrix} \begin{bmatrix} 1 \\ -1 \\ 1 \\ -1 \end{bmatrix} = 4$$

and the $(3, 1)$ entry is

$$\frac{1}{4} \begin{bmatrix} 1 & -1 & 1 & -1 \end{bmatrix} \begin{bmatrix} 1 & 0 & 5 & 2 \\ 0 & -1 & -2 & -5 \\ -5 & -2 & -1 & 0 \\ 2 & 5 & 0 & 1 \end{bmatrix} \begin{bmatrix} 1 \\ 1 \\ 1 \\ 1 \end{bmatrix} = 0.$$

Doing the remaining fourteen similar computations, we obtain that

$$[T]_{\mathcal{B}} = \begin{bmatrix} 0 & 0 & 0 & -2 \\ 0 & 0 & 4 & 0 \\ 0 & -6 & 0 & 0 \\ 8 & 0 & 0 & 0 \end{bmatrix}.$$

QA #7: $\frac{1}{\sqrt{2}} \begin{bmatrix} 5 \\ -1 \end{bmatrix}$

Theorem 4.10 *Let $\mathcal{B} = (e_1, \ldots, e_n)$ be an orthonormal basis for V. Then for any $u, v \in V$,*

$$\langle u, v \rangle = \sum_{j=1}^{n} \langle u, e_j \rangle \overline{\langle v, e_j \rangle}$$

and

$$\|v\|^2 = \sum_{j=1}^{n} |\langle v, e_j \rangle|^2 .$$

The point of the theorem is that the inner product of two vectors in V is the same as the usual inner product of their coordinate representations in $\mathbb{C}^n$.

Proof By Theorem 4.9,

$$u = \sum_{j=1}^{n} \langle u, e_j \rangle e_j \qquad \text{and} \qquad v = \sum_{j=1}^{n} \langle v, e_j \rangle e_j .$$

Therefore

$$\begin{aligned} \langle u, v \rangle &= \left\langle \sum_{j=1}^{n} \langle u, e_j \rangle e_j, \sum_{k=1}^{n} \langle v, e_k \rangle e_k \right\rangle \\ &= \sum_{j=1}^{n} \sum_{k=1}^{n} \langle u, e_j \rangle \overline{\langle v, e_k \rangle} \langle e_j, e_k \rangle \\ &= \sum_{j=1}^{n} \langle u, e_j \rangle \overline{\langle v, e_j \rangle} \end{aligned}$$

by the linearity properties of the inner product and the orthonormality of $\mathcal{B}$. The formula for $\|v\|^2$ follows by considering the case $u = v$. ▲

Example Consider the space $C_{2\pi}(\mathbb{R})$ of 2π-periodic continuous functions, and the functions

$$f(\theta) = (2+i)e^{i\theta} - 3e^{-2i\theta} \quad \text{and} \quad g(\theta) = ie^{i\theta} - e^{2i\theta}.$$

Recalling that the functions $\left(\frac{1}{\sqrt{2\pi}}, \frac{1}{\sqrt{2\pi}} e^{\pm i\theta}, \frac{1}{\sqrt{2\pi}} e^{\pm 2i\theta}, \ldots \right)$ are orthonormal in $C_{2\pi}(\mathbb{R})$ (and so $\left(\frac{1}{\sqrt{2\pi}} e^{i\theta}, \frac{1}{\sqrt{2\pi}} e^{2i\theta}, \frac{1}{\sqrt{2\pi}} e^{-2i\theta} \right)$ are an orthonormal basis of their span) , if we write

$$\begin{aligned} f(\theta) &= (2+i)\sqrt{2\pi} \frac{1}{\sqrt{2\pi}} e^{i\theta} + 0 \frac{1}{\sqrt{2\pi}} e^{2i\theta} - 3\sqrt{2\pi} \frac{1}{\sqrt{2\pi}} e^{-2i\theta}, \\ g(\theta) &= i\sqrt{2\pi} \frac{1}{\sqrt{2\pi}} e^{i\theta} - \sqrt{2\pi} \frac{1}{\sqrt{2\pi}} e^{2i\theta} + 0 \frac{1}{\sqrt{2\pi}} e^{-2i\theta}, \end{aligned}$$

we can compute norms and inner products using Theorem 4.9 without evaluating any integrals:

$$\langle f, g\rangle = 2\pi\big[(2+i)(-i)+0+0\big] = 2\pi(1-2i),$$
$$\|f\|^2 = 2\pi\big[|2+i|^2+0+3^2\big] = 28\pi,$$
$$\|g\|^2 = 2\pi\big[|i|^2+(-1)^2+0\big] = 4\pi.$$ ▲

The Gram–Schmidt Process

We've said that having an orthonormal basis for an inner product space is useful because many computations become quite easy, but that won't do us much good unless we know how to find one. In some cases, for example $\mathbb{R}^n$, an orthonormal basis is staring us in the face, but in other cases this may not be true. Even in the quite explicit case of a plane in $\mathbb{R}^3$ spanned by two nonorthogonal vectors, it takes some effort to find an orthonormal basis. The following algorithm gives us a systematic way to do it.

Algorithm 4.11 (The Gram–Schmidt process) To find an orthonormal basis of a finite-dimensional inner product space V:

- Start with any basis $(v_1, \dots, v_n)$ of V.
- Define $e_1 = \frac{1}{\|v_1\|} v_1$.
- For $j = 2, \dots, n$, iteratively define e_j by

$$\tilde{e}_j = v_j - \sum_{k=1}^{j-1} \langle v_j, e_k\rangle e_k, \qquad \text{and} \qquad e_j = \frac{1}{\|\tilde{e}_j\|}\tilde{e}_j.$$

Then $(e_1, \dots, e_n)$ is an orthonormal basis of V. Moreover, for each $j = 1, \dots, n$, $\langle e_1, \dots, e_j\rangle = \langle v_1, \dots, v_j\rangle$.

In words, to perform the Gram–Schmidt process, you start with a basis and, for each vector, you remove the components in the directions of the previous basis vectors and then normalize to get a unit vector.

Proof We will prove by induction that $(e_1, \dots, e_j)$ is orthonormal for each $j = 1, \dots, n$, and that $\langle e_1, \dots, e_j\rangle = \langle v_1, \dots, v_j\rangle$.

First observe that $v_1 \neq 0$, so that $\|v_1\| \neq 0$ and thus e_1 is in fact defined and is a unit vector. Clearly $\langle e_1\rangle = \langle v_1\rangle$.

Now suppose that $j \geq 2$ and it is already known that $(e_1, \dots, e_{j-1})$ is orthonormal and $\langle e_1, \dots, e_{j-1}\rangle = \langle v_1, \dots, v_{j-1}\rangle$. Since $(v_1, \dots, v_n)$ is linearly independent,

$$v_j \notin \langle v_1, \dots, v_{j-1} \rangle = \langle e_1, \dots, e_{j-1} \rangle,$$

which implies that $\tilde{e}_j \neq 0$, so e_j is defined and is a unit vector. For each $m = 1, \dots, j-1$,

$$\begin{aligned} \langle \tilde{e}_j, e_m \rangle &= \left\langle v_j - \sum_{k=1}^{j-1} \langle v_j, e_k \rangle e_k, e_m \right\rangle \\ &= \langle v_j, e_m \rangle - \sum_{k=1}^{j-1} \langle v_j, e_k \rangle \langle e_k, e_m \rangle \\ &= \langle v_j, e_m \rangle - \langle v_j, e_m \rangle \\ &= 0 \end{aligned}$$

since $(e_1, \dots, e_{j-1})$ is orthonormal, and so $\langle e_j, e_m \rangle = 0$. Thus $(e_1, \dots, e_j)$ is orthonormal.

It follows from the definition that

$$e_j \in \langle e_1, \dots, e_{j-1}, v_j \rangle = \langle v_1, \dots, v_j \rangle,$$

and thus $\langle e_1, \dots, e_j \rangle \subseteq \langle v_1, \dots, v_j \rangle$. Finally, we know that $(v_1, \dots, v_j)$ is linearly independent (by assumption) and that $(e_1, \dots, e_j)$ is linearly independent (since it is orthonormal), and so

$$\dim \langle v_1, \dots, v_j \rangle = \dim \langle e_1, \dots, e_j \rangle = j.$$

Therefore in fact $\langle e_1, \dots, e_j \rangle = \langle v_1, \dots, v_j \rangle$.

Since the previous argument applies for each $j = 1, \dots, n$, it follows in particular that $\langle e_1, \dots, e_n \rangle = \langle v_1, \dots, v_n \rangle = V$, and so $(e_1, \dots, e_n)$ is also a basis of V. ▲

Examples

1. Consider the plane $U = \left\{ \begin{bmatrix} x \\ y \\ z \end{bmatrix} \middle| \, x + y + z = 0 \right\}$. Then the list $(\mathbf{v}_1, \mathbf{v}_2) = \left(\begin{bmatrix} 1 \\ -1 \\ 0 \end{bmatrix}, \begin{bmatrix} 1 \\ 0 \\ -1 \end{bmatrix} \right)$ is a non-orthonormal basis for U. We construct an orthonormal basis $(\mathbf{f}_1, \mathbf{f}_2)$ from it using the Gram–Schmidt process:
   - $\mathbf{f}_1 = \dfrac{1}{\|\mathbf{v}_1\|} \mathbf{v}_1 = \dfrac{1}{\sqrt{2}} \begin{bmatrix} 1 \\ -1 \\ 0 \end{bmatrix}.$
   - $\tilde{\mathbf{f}}_2 = \mathbf{v}_2 - \langle \mathbf{v}_2, \mathbf{f}_1 \rangle \mathbf{f}_1 = \begin{bmatrix} 1 \\ 0 \\ -1 \end{bmatrix} - \dfrac{1}{2} \begin{bmatrix} 1 \\ -1 \\ 0 \end{bmatrix} = \begin{bmatrix} \frac{1}{2} \\ \frac{1}{2} \\ -1 \end{bmatrix}.$

- $\mathbf{f}_2 = \frac{1}{\left\|\tilde{\mathbf{f}}_2\right\|}\tilde{\mathbf{f}}_2 = \frac{1}{\sqrt{3/2}}\begin{bmatrix} \frac{1}{2} \\ \frac{1}{2} \\ -1 \end{bmatrix} = \frac{1}{\sqrt{6}}\begin{bmatrix} 1 \\ 1 \\ -2 \end{bmatrix}.$

> **Quick Exercise #8.** How could you extend $(\mathbf{f}_1, \mathbf{f}_2)$ to an orthonormal basis of all of $\mathbb{R}^3$?

2. Suppose we want an orthonormal basis of $\mathcal{P}_2(\mathbb{R})$ equipped with the inner product

$$\langle f, g\rangle = \int_0^1 f(x)g(x)\,dx. \tag{4.6}$$

We've already seen that $(1, x, x^2)$ is not orthonormal, but the Gram–Schmidt process can be used to produce an orthonormal basis of its span:

- $e_1(x) = \frac{1}{\|1\|}1 = 1.$
- $\tilde{e}_2(x) = x - \langle x, 1\rangle\, 1 = x - \int_0^1 y\,dy = x - \frac{1}{2}.$
- $\|\tilde{e}_2(x)\| = \sqrt{\int_0^1 (x - \frac{1}{2})^2\,dx} = \sqrt{\frac{1}{12}} = \frac{1}{2\sqrt{3}}.$
- $e_2(x) = \frac{\tilde{e}_2(x)}{\|\tilde{e}_2\|} = 2\sqrt{3}x - \sqrt{3}.$
- $\tilde{e}_3(x) = x^2 - \langle x^2, 1\rangle 1 - \langle x^2, 2\sqrt{3}x - \sqrt{3}\rangle(2\sqrt{3}x - \sqrt{3}) = x^2 - x + \frac{1}{6}.$
- $\|\tilde{e}_3\| = \sqrt{\int_0^1 \left(x^2 - x + \frac{1}{6}\right)^2 dx} = \sqrt{\frac{1}{180}} = \frac{1}{6\sqrt{5}}.$
- $e_3(x) = \frac{\tilde{e}_3(x)}{\|\tilde{e}_3\|} = 6\sqrt{5}x^2 - 6\sqrt{5}x + \sqrt{5}.$

We therefore have that $(1, 2\sqrt{3}x - \sqrt{3}, 6\sqrt{5}x^2 - 6\sqrt{5}x + \sqrt{5})$ is an orthonormal basis of $\langle 1, x, x^2\rangle$, with respect to the inner product in equation (4.6). ▲

> **Quick Exercise #9.** What would happen if you tried to apply the Gram–Schmidt process to a linearly *dependent* list $(v_1, \ldots, v_n)$?

While we usually think of the Gram–Schmidt process as a computational tool, it also has important theoretical consequences, as in the following results.

Corollary 4.12 *Suppose that V is a finite-dimensional inner product space. Then there is an orthonormal basis for V.*

QA #8: Pick any vector $v_3 \notin U$ (e.g., e_1), and then apply the Gram–Schmidt process to $(\mathbf{f}_1, \mathbf{f}_2, \mathbf{v}_3)$. Alternatively, observe that $\begin{bmatrix} 1 \\ 1 \\ 1 \end{bmatrix}$ is orthogonal to every vector in U, so $\mathbf{f}_3 = \frac{1}{\sqrt{3}}\begin{bmatrix} 1 \\ 1 \\ 1 \end{bmatrix}$ works.

QA #9: By the Linear Dependence Lemma, at some point we would have $v_j \in \langle e_1, \ldots, e_{j-1}\rangle$, and so $\tilde{e}_j = 0$, making it impossible to compute e_j.

Proof By Corollary 3.12, there is a basis $(v_1, \ldots, v_n)$ for V. The Gram–Schmidt process (Algorithm 4.11) can then be used to produce an orthonormal basis for V. ▲

More generally:

Corollary 4.13 *Suppose $\mathcal{B} = (v_1, \ldots, v_k)$ is an orthonormal list in a finite-dimensional inner product space V. Then $\mathcal{B}$ can be extended to an orthonormal basis $\mathcal{B}'$ of V.*

Proof By Theorem 3.27, $\mathcal{B}$ can be extended to a basis $(v_1, \ldots, v_n)$ of V. Applying the Gram–Schmidt process (Algorithm 4.11) to this basis produces an orthonormal basis $\mathcal{B}' = (e_1, \ldots, e_n)$ for V. Furthermore, since $(v_1, \ldots, v_k)$ are already orthonormal, $e_j = v_j$ for $j = 1, \ldots, k$. Thus $\mathcal{B}'$ extends the original list $\mathcal{B}$. ▲

KEY IDEAS

- An orthonormal basis is a basis consisting of unit vectors which are all perpendicular to each other.
- With respect to an orthonormal basis, coordinates are gotten by taking inner products with the basis vectors.
- Every finite-dimensional inner product space has an orthonormal basis.
- The Gram–Schmidt process lets you start with any basis of a finite-dimensional inner product space and orthonormalize it.

EXERCISES

4.2.1 Verify that each of the following lists of vectors is an orthonormal basis.

(a) $\left(\frac{1}{\sqrt{13}}\begin{bmatrix}2\\-3\end{bmatrix}, \frac{1}{\sqrt{13}}\begin{bmatrix}3\\2\end{bmatrix}\right)$ in $\mathbb{R}^2$

(b) $\left(\frac{1}{\sqrt{30}}\begin{bmatrix}1\\2\\-5\end{bmatrix}, \frac{1}{2\sqrt{5}}\begin{bmatrix}4\\-2\\0\end{bmatrix}, \frac{1}{2\sqrt{6}}\begin{bmatrix}2\\4\\2\end{bmatrix}\right)$ in $\mathbb{R}^3$

(c) $\left(\frac{1}{\sqrt{3}}\begin{bmatrix}1\\1\\1\end{bmatrix}, \frac{1}{\sqrt{3}}\begin{bmatrix}1\\e^{2\pi i/3}\\e^{4\pi i/3}\end{bmatrix}, \frac{1}{\sqrt{3}}\begin{bmatrix}1\\e^{4\pi i/3}\\e^{2\pi i/3}\end{bmatrix}\right)$ in $\mathbb{C}^3$

(d) $\left(\frac{1}{\sqrt{3}}\begin{bmatrix}0\\1\\1\\1\end{bmatrix}, \frac{1}{\sqrt{15}}\begin{bmatrix}3\\-2\\1\\1\end{bmatrix}, \frac{1}{\sqrt{35}}\begin{bmatrix}3\\3\\-4\\1\end{bmatrix}, \frac{1}{\sqrt{7}}\begin{bmatrix}1\\1\\1\\-2\end{bmatrix}\right)$ in $\mathbb{R}^4$

4.2.2 Verify that each of the following lists of vectors is an orthonormal basis.

(a) $\left(\frac{1}{\sqrt{2}}\begin{bmatrix}1\\0\\-1\end{bmatrix}, \frac{1}{\sqrt{6}}\begin{bmatrix}1\\-2\\1\end{bmatrix}\right)$ in $\left\{\mathbf{x} \in \mathbb{R}^3 \mid x_1 + x_2 + x_3 = 0\right\}$

(b) $\left(\frac{1}{2}\begin{bmatrix}1 & 1\\1 & 1\end{bmatrix}, \frac{1}{2}\begin{bmatrix}1 & 1\\-1 & -1\end{bmatrix}, \frac{1}{2}\begin{bmatrix}1 & -1\\1 & -1\end{bmatrix}, \frac{1}{2}\begin{bmatrix}1 & -1\\-1 & 1\end{bmatrix}\right)$ in $\mathrm{M}_2(\mathbb{R})$

(c) $\left(\frac{1}{2}\begin{bmatrix}1\\1\\1\\1\end{bmatrix}, \frac{1}{2}\begin{bmatrix}1\\i\\-1\\-i\end{bmatrix}, \frac{1}{2}\begin{bmatrix}1\\-1\\1\\-1\end{bmatrix}, \frac{1}{2}\begin{bmatrix}1\\-i\\-1\\i\end{bmatrix}\right)$ in $\mathbb{C}^4$

4.2.3 Find the coordinate representation of each of the following vectors with respect to the orthonormal basis given in the corresponding part of Exercise 4.2.1.

(a) $\begin{bmatrix}-10\\3\end{bmatrix}$ (b) $\begin{bmatrix}3\\2\\1\end{bmatrix}$ (c) $\begin{bmatrix}1\\0\\1\end{bmatrix}$ (d) $\begin{bmatrix}-2\\1\\0\\4\end{bmatrix}$

4.2.4 Find the coordinate representation of each of the following vectors with respect to the orthonormal basis given in the corresponding part of Exercise 4.2.2.

(a) $\begin{bmatrix}1\\2\\-3\end{bmatrix}$ (b) $\begin{bmatrix}5 & 7\\7 & 2\end{bmatrix}$ (c) $\begin{bmatrix}-3\\0\\1\\2\end{bmatrix}$

4.2.5 Find the matrix representing each of the following linear maps with respect to the orthonormal basis given in the corresponding part of Exercise 4.2.1.

(a) Reflection across the line $y = \frac{2}{3}x$

(b) Projection onto the x–y plane

(c) $T\begin{bmatrix}x\\y\\z\end{bmatrix} = \begin{bmatrix}y\\z\\x\end{bmatrix}$ (d) $T\begin{bmatrix}x\\y\\z\\w\end{bmatrix} = \begin{bmatrix}z\\w\\x\\y\end{bmatrix}$

4.2.6 Find the matrix representing each of the following linear maps with respect to the orthonormal basis given in the corresponding part of Exercise 4.2.2.

(a) $T\begin{bmatrix} x \\ y \\ z \end{bmatrix} = \begin{bmatrix} y \\ z \\ x \end{bmatrix}$ (b) $T(A) = A^T$ (c) $T\begin{bmatrix} x \\ y \\ z \\ w \end{bmatrix} = \begin{bmatrix} y \\ x \\ w \\ z \end{bmatrix}$

4.2.7 (a) Verify that $\mathcal{B} = \left(\frac{1}{\sqrt{2}}\begin{bmatrix} 1 \\ -1 \\ 0 \end{bmatrix}, \frac{1}{\sqrt{6}}\begin{bmatrix} 1 \\ 1 \\ -2 \end{bmatrix} \right)$ is an orthonormal basis of

$$U = \left\{ \begin{bmatrix} x \\ y \\ z \end{bmatrix} \in \mathbb{R}^3 \,\middle|\, x + y + z = 0 \right\}.$$

(b) Find the coordinate representation with respect to $\mathcal{B}$ of $\begin{bmatrix} 1 \\ 2 \\ -3 \end{bmatrix}$.

(c) Find the matrix with respect to $\mathcal{B}$ of the linear map $T : U \to U$ given by

$$T\begin{bmatrix} x \\ y \\ z \end{bmatrix} = \begin{bmatrix} y \\ z \\ x \end{bmatrix}.$$

4.2.8 A function of the form

$$f(\theta) = \sum_{k=1}^{n} a_k \sin(k\theta) + b_0 + \sum_{\ell=1}^{m} b_\ell \cos(\ell\theta)$$

for $a_1, \ldots, a_n, b_0, \ldots, b_m \in \mathbb{R}$ is called a **trigonometric polynomial.** If we put the inner product

$$\langle f, g \rangle = \int_0^{2\pi} f(\theta) g(\theta)\, d\theta$$

on the space of trigonometric polynomials, express $\|f\|$ in terms of the a_k and b_ℓ without computing any integrals.

4.2.9 Recall the basis $\mathcal{B} = \left(1, 2\sqrt{3}x - \sqrt{3}, 6\sqrt{5}x^2 - 6\sqrt{5}x + \sqrt{5}\right)$ of $\mathcal{P}_2(\mathbb{R})$ from the example on page 246, and let $D \in \mathcal{L}(\mathcal{P}_2(\mathbb{R}))$ be the derivative operator. Find $[D]_\mathcal{B}$.

4.2.10 Consider the inner product $\langle f, g \rangle = \int_0^1 f(x)g(x)\, dx$ on $\mathcal{P}_2(\mathbb{R})$ and the basis $(1, x, x^2)$.

(a) Compute $\|3 - 2x + x^2\|^2$.

(b) You should have found that $\|3 - 2x + x^2\|^2 \neq 3^2 + 2^2 + 1^2$. Why does this not contradict Theorem 4.10?

4.2.11 Carry out the Gram–Schmidt process on each of the following lists of vectors.

(a) $\left(\begin{bmatrix}1\\0\\-1\end{bmatrix},\begin{bmatrix}2\\1\\0\end{bmatrix},\begin{bmatrix}0\\-1\\2\end{bmatrix}\right)$ (b) $\left(\begin{bmatrix}0\\-1\\2\end{bmatrix},\begin{bmatrix}2\\1\\0\end{bmatrix},\begin{bmatrix}1\\0\\-1\end{bmatrix}\right)$

(c) $\left(\begin{bmatrix}0\\1\\1\\1\end{bmatrix},\begin{bmatrix}1\\0\\1\\1\end{bmatrix},\begin{bmatrix}1\\1\\0\\1\end{bmatrix},\begin{bmatrix}1\\1\\1\\0\end{bmatrix}\right)$

4.2.12 Carry out the Gram–Schmidt process on each of the following lists of vectors.

(a) $\left(\begin{bmatrix}1\\0\\0\end{bmatrix},\begin{bmatrix}1\\1\\0\end{bmatrix},\begin{bmatrix}1\\1\\1\end{bmatrix}\right)$ (b) $\left(\begin{bmatrix}1\\i\\0\end{bmatrix},\begin{bmatrix}0\\1\\i\end{bmatrix},\begin{bmatrix}i\\0\\1\end{bmatrix}\right)$

(c) $\left(\begin{bmatrix}1\\0\\0\\-1\end{bmatrix},\begin{bmatrix}0\\1\\0\\-1\end{bmatrix},\begin{bmatrix}0\\0\\1\\-1\end{bmatrix}\right)$

4.2.13 Use the idea of Quick Exercise #9 to determine whether each of the following lists is linearly independent.

(a) $\left(\begin{bmatrix}1\\2\\3\end{bmatrix},\begin{bmatrix}2\\3\\1\end{bmatrix},\begin{bmatrix}3\\1\\2\end{bmatrix}\right)$ (b) $\left(\begin{bmatrix}1\\1\\1\end{bmatrix},\begin{bmatrix}1\\2\\4\end{bmatrix},\begin{bmatrix}1\\3\\9\end{bmatrix}\right)$

(c) $\left(\begin{bmatrix}-2\\-1\\1\end{bmatrix},\begin{bmatrix}1\\1\\-1\end{bmatrix},\begin{bmatrix}1\\2\\-2\end{bmatrix}\right)$

4.2.14 Use the idea of Quick Exercise #9 to determine whether each of the following lists is linearly independent.

(a) $\left(\begin{bmatrix}2\\1\\-3\end{bmatrix},\begin{bmatrix}-1\\2\\-1\end{bmatrix},\begin{bmatrix}0\\1\\-1\end{bmatrix}\right)$ (b) $\left(\begin{bmatrix}1\\1\\-1\end{bmatrix},\begin{bmatrix}1\\-1\\1\end{bmatrix},\begin{bmatrix}-1\\1\\1\end{bmatrix}\right)$

(c) $\left(\begin{bmatrix}1\\0\\1\end{bmatrix},\begin{bmatrix}1\\1\\0\end{bmatrix},\begin{bmatrix}0\\1\\-1\end{bmatrix}\right)$

4.2.15 Carry out the Gram–Schmidt process on the list of polynomials $(1, x, x^2, x^3)$ with respect to each of the following inner products.

(a) $\langle p, q \rangle = \int_0^1 p(x)q(x)\,dx$ (b) $\langle p, q \rangle = \int_{-1}^1 p(x)q(x)\,dx$

(c) $\langle p, q \rangle = \int_{-1}^1 p(x)q(x)x^2\,dx$

4.2.16 Find an orthonormal basis $(\mathbf{f}_1, \mathbf{f}_2, \mathbf{f}_3)$ of $\mathbb{R}^3$ such that $\mathbf{f}_1$ and $\mathbf{f}_2$ are in the subspace

$$U = \left\{ \begin{bmatrix} x \\ y \\ z \end{bmatrix} \middle| \, x - 2y + 3z = 0 \right\}.$$

4.2.17 Let $\mathbf{A} = \begin{bmatrix} 0 & 1 & 1 \\ 1 & 0 & 1 \\ 1 & 1 & 0 \end{bmatrix} \in \mathrm{M}_3(\mathbb{R})$, and define an inner product on $\mathbb{R}^3$ by

$$\langle \mathbf{x}, \mathbf{y} \rangle_{\mathbf{A}} = \langle \mathbf{A}\mathbf{x}, \mathbf{A}\mathbf{y} \rangle,$$

where the inner product on the right-hand side is the standard inner product on $\mathbb{R}^3$. Find an orthonormal basis of $\mathbb{R}^3$ with respect to $\langle \cdot, \cdot \rangle_{\mathbf{A}}$.

4.2.18 Let $\mathbf{A} \in \mathrm{M}_n(\mathbb{C})$ be an invertible matrix, and define a non-standard inner product $\langle \cdot, \cdot \rangle_{\mathbf{A}}$ on $\mathbb{C}^n$ by

$$\langle \mathbf{x}, \mathbf{y} \rangle_{\mathbf{A}} = \langle \mathbf{A}\mathbf{x}, \mathbf{A}\mathbf{y} \rangle,$$

where the inner product on the right-hand side is the standard inner product on $\mathbb{C}^n$.

(a) Find $\langle \mathbf{e}_j, \mathbf{e}_k \rangle_{\mathbf{A}}$ for all j and k.

(b) Under what circumstances is the standard basis $(\mathbf{e}_1, \ldots, \mathbf{e}_n)$ of $\mathbb{C}^n$ orthonormal with respect to $\langle \cdot, \cdot \rangle_{\mathbf{A}}$?

4.2.19 Prove that the space $C_{2\pi}(\mathbb{R})$ of continuous 2π-periodic functions is infinite-dimensional.

4.2.20 Suppose that the matrix of $T \in \mathcal{L}(V)$ with respect to some basis $\mathcal{B}$ is upper triangular. Show that if $\mathcal{C}$ is the orthonormal basis obtained by applying the Gram–Schmidt process to $\mathcal{B}$, then $[T]_{\mathcal{C}}$ is also upper triangular.

4.2.21 Prove that if $(e_1, \ldots, e_n)$ is the orthonormal basis constructed from $(v_1, \ldots, v_n)$ via the Gram–Schmidt process, then $\langle v_j, e_j \rangle > 0$ for each j.

4.2.22 Suppose that $\mathbf{A} \in \mathrm{M}_n(\mathbb{C})$ is upper triangular and invertible. Show that if the Gram–Schmidt process is applied to the list of columns of $\mathbf{A}$, it produces an orthonormal basis of the form $(\omega_1\mathbf{e}_1, \ldots, \omega_n\mathbf{e}_n)$, where $(\mathbf{e}_1, \ldots, \mathbf{e}_n)$ is the standard basis and $|\omega_j| = 1$ for each j.

4.3 Orthogonal Projections and Optimization

Orthogonal Complements and Direct Sums

The idea of decomposing a vector v into a component in the direction of u and a component orthogonal to u has come up several times now: for example, in the proof of the Cauchy–Schwarz inequality or in the Gram–Schmidt algorithm. We can begin to make more systematic use of this idea through the concept of orthogonal complements.

Definition Let U be a subspace of an inner product space V. The **orthogonal complement** of U is the subspace

$$U^{\perp} = \{v \in V \mid \langle u, v \rangle = 0 \text{ for every } u \in U\}.$$

That is, $U^{\perp}$ consists of all those vectors v which are orthogonal to *every* vector in U.

Quick Exercise #10. Verify that $U^{\perp}$ really is a subspace.

Examples

1. If L is a line through the origin in $\mathbb{R}^2$, then $L^{\perp}$ is the perpendicular line through the origin.
2. If P is a plane through the origin in $\mathbb{R}^3$, then $P^{\perp}$ is the line perpendicular to P through the origin. If L is a line through the origin in $\mathbb{R}^3$, then $L^{\perp}$ is the plane through the origin to which L is perpendicular. ▲

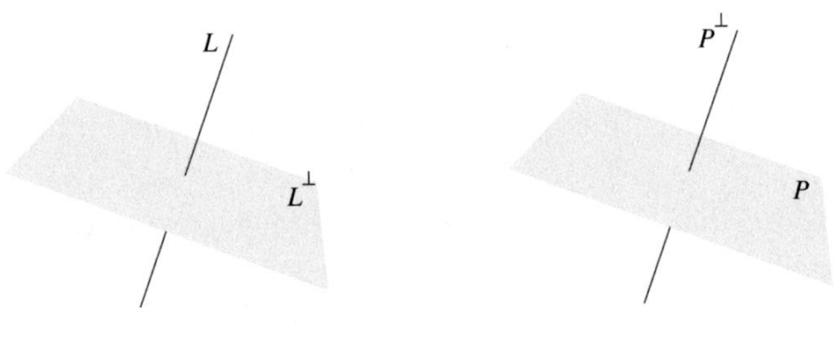

Figure 4.4 Orthogonal complements in $\mathbb{R}^3$.

QA #10: $\langle u, 0 \rangle = 0$, so $0 \in U^{\perp}$. If $\langle u, v_1 \rangle = \langle u, v_2 \rangle = 0$, then $\langle u, v_1 + v_2 \rangle = 0$ by additivity, and if $\langle u, v \rangle = 0$, then $\langle u, av \rangle = 0$ by (conjugate) homogeneity.

The following theorem allows us to uniquely decompose any vector in an inner product space V into a part in a subspace U and a part in the orthogonal complement $U^{\perp}$.

Theorem 4.14 *Let V be an inner product space. If U is a finite-dimensional subspace of V, then every vector $v \in V$ can be uniquely written in the form*

$$v = u + w,$$

where $u \in U$ and $w \in U^{\perp}$.

Proof Since U is finite-dimensional, there exists an orthonormal basis $(e_1, \ldots, e_m)$ of U. Given $v \in V$, we define

$$u = \sum_{j=1}^{m} \langle v, e_j \rangle e_j,$$

and let $w = v - u$. Then $v = u + w$ and $u \in U$ trivially. Note that, for each k,

$$\langle u, e_k \rangle = \langle v, e_k \rangle,$$

so

$$\langle w, e_k \rangle = \langle v, e_k \rangle - \langle u, e_k \rangle = 0.$$

Now if u' is any other vector in U, then write

$$u' = \sum_{j=1}^{m} a_j e_j,$$

so that

$$\langle u', w \rangle = \sum_{j=1}^{m} a_j \langle e_j, w \rangle = 0.$$

Therefore $w \in U^{\perp}$.

It remains to show that u and w are unique. Suppose that $u_1, u_2 \in U$, $w_1, w_2 \in U^{\perp}$, and that

$$u_1 + w_1 = u_2 + w_2.$$

Consider the vector

$$x = u_1 - u_2 = w_2 - w_1.$$

Since U and $U^{\perp}$ are subspaces, $x \in U$ and $x \in U^{\perp}$, so $\langle x, x \rangle = 0$, which means that $x = 0$. Therefore $u_1 = u_2$ and $w_1 = w_2$. ▲

When we can decompose every vector in an inner product space as a sum as in Theorem 4.14, we say the space itself is the **orthogonal direct sum** of the subspaces. More specifically, we have the following definition.

Definition Suppose that $U_1, \ldots, U_m$ are subspaces of an inner product space V. We say that V is the **orthogonal direct sum** of $U_1, \ldots, U_m$ if:

- every vector $v \in V$ can be written as
$$v = u_1 + \cdots + u_m \tag{4.7}$$
for some $u_1 \in U_1, \ldots, u_m \in U_m$, and
- whenever $u_j \in U_j$ and $u_k \in U_k$ for $j \neq k$, u_j and u_k are orthogonal.

In that case we* write $V = U_1 \oplus \cdots \oplus U_m$.

Examples

1. If L_1 and L_2 are any two perpendicular lines through the origin in $\mathbb{R}^2$, then $\mathbb{R}^2 = L_1 \oplus L_2$.
2. If P is a plane through the origin in $\mathbb{R}^3$ and L is the line through the origin perpendicular to P, then $\mathbb{R}^3 = P \oplus L$. ▲

Theorem 4.14 says that $V = U \oplus U^\perp$ for any subspace $U \subseteq V$, generalizing these examples. One consequence of this observation is the following intuitively appealing property of orthogonal complements.

Proposition 4.15 *If U is any subspace of a finite-dimensional inner product space V, then $(U^\perp)^\perp = U$.*

Proof Suppose that $u \in U$. Then for each $v \in U^\perp$, $\langle u, v \rangle = 0$. This implies that $u \in (U^\perp)^\perp$, and so $U \subseteq (U^\perp)^\perp$.

Now suppose that $w \in (U^\perp)^\perp$. By Theorem 4.14, we can write $w = u + v$ for some $u \in U$ and $v \in U^\perp$; we'd like to show that in fact $v = 0$. Since $v = w - u$, and $u \in U \subseteq (U^\perp)^\perp$, $v \in (U^\perp)^\perp$. So v is in both $U^\perp$ and $(U^\perp)^\perp$, which means

*Warning: Some people use this notation for a more general notion of direct sum, without meaning to imply that the subspaces U_j are orthogonal.

v is orthogonal to itself, thus $v = 0$. We therefore have that $w = u \in U$, and so $(U^\perp)^\perp \subseteq U$. ▲

Quick Exercise #11. Convince yourself geometrically that the proposition is true for all subspaces of $\mathbb{R}^2$ and $\mathbb{R}^3$.

Orthogonal Projections

Definition Let U be a subspace of V. The **orthogonal projection** onto U is the function $P_U : V \to V$ defined by $P_U(v) = u$, where $v = u + w$ for $u \in U$ and $w \in U^\perp$.

We think of P_U as picking off the part of v which lies in U.

Theorem 4.16 (Algebraic properties of orthogonal projections) *Let U be a finite-dimensional subspace of V.*

1. *P_U is a linear map.*
2. *If $(e_1, \ldots, e_m)$ is any orthonormal basis of U, then*

$$P_U v = \sum_{j=1}^{m} \langle v, e_j \rangle e_j$$

   for each $v \in V$.
3. *For each $v \in V$, $v - P_U v \in U^\perp$.*
4. *For each $v, w \in V$,*

$$\langle P_U v, w \rangle = \langle P_U v, P_U w \rangle = \langle v, P_U w \rangle .$$

5. *Suppose $\mathcal{B} = (e_1, \ldots, e_n)$ is an orthonormal basis of V such that $(e_1, \ldots, e_m)$ is an orthonormal basis of U. Then*

$$[P_U]_{\mathcal{B}} = \operatorname{diag}(1, \ldots, 1, 0, \ldots, 0),$$

   with the first m diagonal entries 1, and the remaining diagonal entries 0.
6. $\operatorname{range} P_U = U$, *and $P_U u = u$ for each $u \in U$.*
7. $\ker P_U = U^\perp$.
8. *If V is finite-dimensional, then $P_{U^\perp} = I - P_U$.*
9. $P_U^2 = P_U$.

Proof 1. Suppose that $v_1 = u_1 + w_1$ and $v_2 = u_2 + w_2$ for $u_1, u_2 \in U$ and $w_1, w_2 \in U^\perp$. Then

$$v_1 + v_2 = (u_1 + u_2) + (w_1 + w_2)$$

with $u_1 + u_2 \in U$ and $w_1 + w_2 \in U^\perp$ and so

$$P_U(v_1 + v_2) = u_1 + u_2 = P_U v_1 + P_U v_2.$$

Thus P_U is additive. Homogeneity follows similarly.

2. This follows from the proof of Theorem 4.14.
3. Suppose $v = u + w$ with $u \in U$ and $w \in U^\perp$. Then $v - P_U v = v - u = w \in U^\perp$.
4. By definition, $P_U v \in U$, and by part 3, $w - P_U w \in U^\perp$. Therefore

$$\langle P_U v, w\rangle = \langle P_U v, P_U w\rangle + \langle P_U v, w - P_U w\rangle = \langle P_U v, P_U w\rangle .$$

The other equality is proved similarly.

The remaining parts of the proof are left as exercises (see Exercise 4.3.22). ▲

The following result uses part 2 of Theorem 4.16 to give a formula for the matrix of P_U.

Proposition 4.17 *Let U be a subspace of $\mathbb{R}^n$ or $\mathbb{C}^n$, with orthonormal basis $(\mathbf{f}_1, \ldots, \mathbf{f}_m)$. Then the matrix of P_U with respect to the standard basis $\mathcal{E}$ is*

$$[P_U]_{\mathcal{E}} = \sum_{j=1}^{m} \mathbf{f}_j \mathbf{f}_j^*.$$

Proof By Quick Exercise #2 in Section 4.1 and part 2 of Theorem 4.16, for every $\mathbf{v}$,

$$\sum_{j=1}^{m} \mathbf{f}_j \mathbf{f}_j^* \mathbf{v} = \sum_{j=1}^{m} \mathbf{f}_j \langle \mathbf{v}, \mathbf{f}_j\rangle = \sum_{j=1}^{m} \langle \mathbf{v}, \mathbf{f}_j\rangle \mathbf{f}_j = P_U \mathbf{v}.$$ ▲

Example Recall that in Section 4.2 we found that if $\mathbf{f}_1 = \frac{1}{\sqrt{2}}\begin{bmatrix} 1 \\ -1 \\ 0 \end{bmatrix}$ and $\mathbf{f}_2 = \frac{1}{\sqrt{6}}\begin{bmatrix} 1 \\ 1 \\ -2 \end{bmatrix}$, then $(\mathbf{f}_1, \mathbf{f}_2)$ is an orthonormal basis of $U = \left\{ \begin{bmatrix} x \\ y \\ z \end{bmatrix} \middle| \, x + y + z = 0 \right\}$ in $\mathbb{R}^3$. By Proposition 4.17,

$$[P_U]_{\mathcal{E}} = \mathbf{f}_1\mathbf{f}_1^* + \mathbf{f}_2\mathbf{f}_2^* = \frac{1}{2}\begin{bmatrix} 1 & -1 & 0 \\ -1 & 1 & 0 \\ 0 & 0 & 0 \end{bmatrix} + \frac{1}{6}\begin{bmatrix} 1 & 1 & -2 \\ 1 & 1 & -2 \\ -2 & -2 & 4 \end{bmatrix} = \frac{1}{3}\begin{bmatrix} 2 & -1 & -1 \\ -1 & 2 & -1 \\ -1 & -1 & 2 \end{bmatrix}.$$

Quick Exercise # 12. What is the orthogonal projection of $\begin{bmatrix}1\\2\\3\end{bmatrix}$ onto the subspace U in the example above?

If a subspace U of $\mathbb{R}^n$ or $\mathbb{C}^n$ is described in terms of a basis which is not orthonormal, one option for finding the orthogonal projection onto U is to perform the Gram–Schmidt process, then apply Proposition 4.17. The following result gives an alternative approach.

Proposition 4.18 *Let U be a subspace of $\mathbb{R}^n$ or $\mathbb{C}^n$ with basis $(\mathrm{v}_1, \dots, \mathrm{v}_k)$. Let A be the $n \times k$ matrix with columns $\mathrm{v}_1, \dots, \mathrm{v}_k$. Then*

$$[P_U]_{\mathcal{E}} = \mathrm{A}(\mathrm{A}^*\mathrm{A})^{-1}\mathrm{A}^*.$$

Note the implicit claim that the matrix $\mathrm{A}^*\mathrm{A}$ is invertible.

Proof First observe that $P_U\mathrm{x}$ is necessarily an element of U, and so it is a linear combination of the columns of A. In other words, there is some $\hat{\mathrm{x}}$ such that

$$P_U\mathrm{x} = \mathrm{A}\hat{\mathrm{x}}.$$

By part 3 of Theorem 4.16,

$$\mathrm{x} - P_U\mathrm{x} = \mathrm{x} - \mathrm{A}\hat{\mathrm{x}} \in U^{\perp}.$$

In particular,

$$\langle \mathrm{x} - \mathrm{A}\hat{\mathrm{x}}, \mathrm{v}_j \rangle = \mathrm{v}_j^* \left(\mathrm{x} - \mathrm{A}\hat{\mathrm{x}}\right) = 0$$

for each j. Rewriting this system of k equations in matrix form gives

$$\mathrm{A}^* \left(\mathrm{x} - \mathrm{A}\hat{\mathrm{x}}\right) = \mathbf{0} \quad \Longleftrightarrow \quad \mathrm{A}^*\mathrm{A}\hat{\mathrm{x}} = \mathrm{A}^*\mathrm{x}.$$

Assuming that $\mathrm{A}^*\mathrm{A}$ is indeed invertible, multiplying both sides by $\mathrm{A}\left(\mathrm{A}^*\mathrm{A}\right)^{-1}$ completes the proof, since $\mathrm{A}\hat{\mathrm{x}}$ is exactly $P_U\mathrm{x}$.

To see that $\mathrm{A}^*\mathrm{A}$ is invertible, it suffices to show that the null space of $\mathrm{A}^*\mathrm{A}$ is trivial. Suppose then that $\mathrm{A}^*\mathrm{A}\mathrm{x} = \mathbf{0}$. Then

$$0 = \langle \mathrm{A}^*\mathrm{A}\mathrm{x}, \mathrm{x} \rangle = \mathrm{x}^*(\mathrm{A}^*\mathrm{A}\mathrm{x}) = (\mathrm{A}\mathrm{x})^*(\mathrm{A}\mathrm{x}) = \|\mathrm{A}\mathrm{x}\|^2.$$

But A has rank k, since its columns are assumed to be linearly independent, and so $\operatorname{null} \mathrm{A} = 0$. It follows that $\mathrm{x} = \mathbf{0}$, and so $\mathrm{A}^*\mathrm{A}$ is invertible. ▲

QA #12: $\frac{1}{3}\begin{bmatrix}2 & -1 & -1\\-1 & 2 & -1\\-1 & -1 & 2\end{bmatrix}\begin{bmatrix}1\\2\\3\end{bmatrix} = \begin{bmatrix}-1\\0\\1\end{bmatrix}.$

Theorem 4.19 (Geometric properties of orthogonal projections) *Let U be a finite-dimensional subspace of V.*

1. *For each $v \in V$, $\|P_U v\| \le \|v\|$, with equality if and only if $v \in U$.*
2. *For each $v \in V$ and $u \in U$,*

$$\|v - P_U v\| \le \|v - u\|,$$

with equality if and only if $u = P_U v$.

The first part of Theorem 4.19 says that projections are **contractions**; i.e., they can only make vectors shorter. The second part says that $P_U v$ is the closest point in U to the vector v.

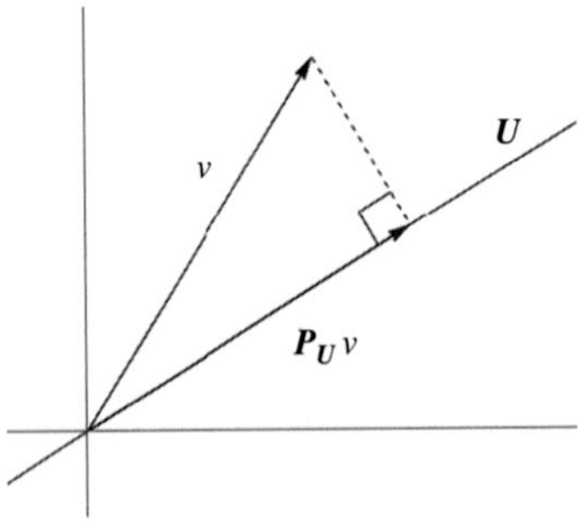

Figure 4.5 $\|P_U v\| \le \|v\|$.

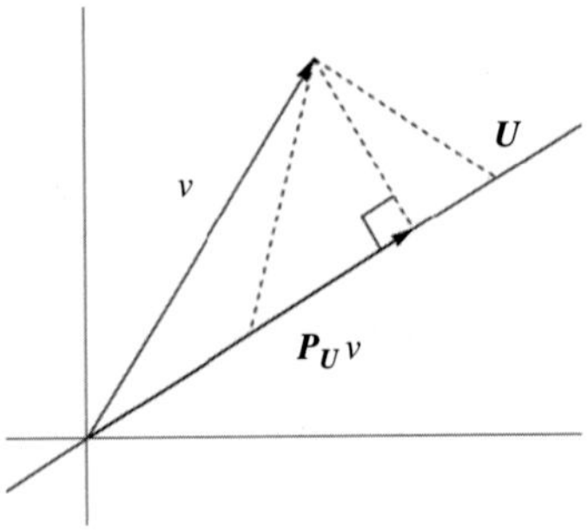

Figure 4.6 The dashed line perpendicular to U is the shortest line between points in U and v.

Proof 1. Since $v - P_U v \in U^\perp$, $v - P_U v$ is orthogonal to $P_U v$, and so

$$\|v\|^2 = \|P_U v + (v - P_U v)\|^2 = \|P_U v\|^2 + \|v - P_U v\|^2 \ge \|P_U v\|^2.$$

Equality holds in the inequality here if and only if $\|v - P_U v\|^2 = 0$, which is true if and only if $v = P_U v$, and this holds exactly when $v \in U$.

2. Since $v - P_U v \in U^\perp$ and $P_U v - u \in U$,

$$\begin{aligned}\|v - u\|^2 &= \|(v - P_U v) + (P_U v - u)\|^2 \\ &= \|v - P_U v\|^2 + \|P_U v - u\|^2 \\ &\ge \|v - P_U v\|^2.\end{aligned}$$

Equality holds in the inequality here if and only if $\|P_U v - u\|^2 = 0$, which is true if and only if $u = P_U v$. ▲

Finding the closest point to a given vector v within a specified subspace U can be seen as finding the best approximation of v by a point in U. The fact that the best possible approximation is given by orthogonal projection of v onto U (and so finding it is easy!) is crucial in many applications; we illustrate with two of the most important examples.

Linear Least Squares

Consider a set of points $\{(x_i, y_i)\}_{i=1}^n \subseteq \mathbb{R}^2$; think of them as data resulting from an experiment. A common problem in applications is to find the line $y = mx + b$ which comes "closest" to containing this set of points.

Define the subspace

$$U := \left\{ \begin{bmatrix} mx_1 + b \\ mx_2 + b \\ \vdots \\ mx_n + b \end{bmatrix} \middle| \; m, b \in \mathbb{R} \right\} = \{m\mathbf{x} + b\mathbf{1} \mid m, b \in \mathbb{R}\} \subseteq \mathbb{R}^n$$

spanned by the vector $\mathbf{x}$ of first coordinates of the data points and the vector $\mathbf{1}$ with all entries given by 1. The points (x_i, y_i) all lie on some line $y = mx + b$ if and only if the vector $\mathbf{y}$ of second coordinates lies in U. Since U is only a two-dimensional subspace of $\mathbb{R}^n$, however, this is not typically the case. By part 2 of Theorem 4.19, the closest point to $\mathbf{y}$ in U is $P_U\mathbf{y}$; taking the corresponding values of m and b gives us our best-fitting line.

This approach to fitting points to a line is called **simple linear regression**. It is also called **the method of least squares**, because minimizing the distance of $\mathbf{y}$ to U means finding the values of m and b for which

$$\sum_{i=1}^{n} (mx_i + b - y_i)^2$$

is minimal.

Example Consider the set of five points $\{(0, 1), (1, 2), (1, 3), (2, 4), (2, 3)\}$. The subspace $U \subseteq \mathbb{R}^5$ defined above is spanned by $\mathbf{x} = \begin{bmatrix} 0 \\ 1 \\ 1 \\ 2 \\ 2 \end{bmatrix}$ and $\mathbf{1} = \begin{bmatrix} 1 \\ 1 \\ 1 \\ 1 \\ 1 \end{bmatrix}$; let

$$\mathrm{A} = \begin{bmatrix} 0 & 1 \\ 1 & 1 \\ 1 & 1 \\ 2 & 1 \\ 2 & 1 \end{bmatrix}.$$

Quick Exercise #13. Show that

$$A^*A = \begin{bmatrix} 10 & 6 \\ 6 & 5 \end{bmatrix} \quad \text{and} \quad (A^*A)^{-1} = \frac{1}{14}\begin{bmatrix} 5 & -6 \\ -6 & 10 \end{bmatrix}.$$

Then by Proposition 4.18,

$$P_U \begin{bmatrix} 1 \\ 2 \\ 3 \\ 4 \\ 3 \end{bmatrix} = \begin{bmatrix} 0 & 1 \\ 1 & 1 \\ 1 & 1 \\ 2 & 1 \\ 2 & 1 \end{bmatrix} \left(\frac{1}{14} \begin{bmatrix} 5 & -6 \\ -6 & 10 \end{bmatrix} \right) \begin{bmatrix} 0 & 1 & 1 & 2 & 2 \\ 1 & 1 & 1 & 1 & 1 \end{bmatrix} \begin{bmatrix} 1 \\ 2 \\ 3 \\ 4 \\ 3 \end{bmatrix} = \begin{bmatrix} 0 & 1 \\ 1 & 1 \\ 1 & 1 \\ 2 & 1 \\ 2 & 1 \end{bmatrix} \begin{bmatrix} \frac{17}{14} \\ \frac{8}{7} \end{bmatrix}.$$

We thus take $m = \frac{17}{14}$ and $b = \frac{8}{7}$ to get the best-fitting line to the original points:

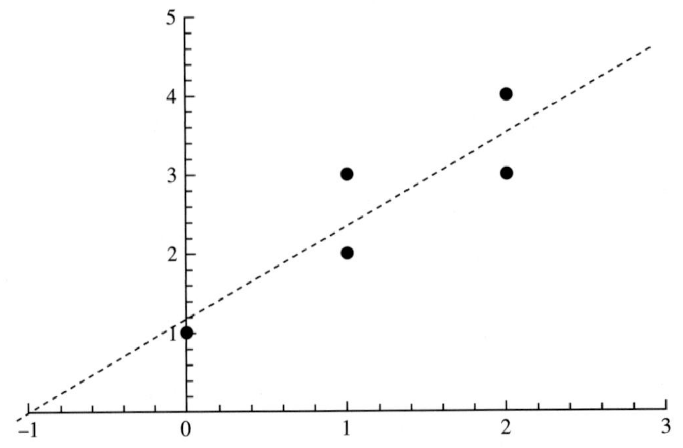

Figure 4.7 The best-fitting line to $\{(0, 1), (1, 2), (1, 3), (2, 4), (2, 3)\}$. ▲

Approximation of Functions

Another important kind of approximation problem is to approximate a complicated function by a simpler function. (Just what counts as "complicated" or "simple" depends on the context.) For example, although we know how to do lots of abstract manipulations with exponential functions, actually computing with them is much harder than with polynomials. So we may wish to find a good approximation of an exponential function by, say, a quadratic polynomial. In your calculus classes, you've probably already encountered one way to do this, namely Taylor polynomials. Taylor polynomials, however, are designed to closely approximate a function near one specific point, and may not be the best thing to approximate a function far from that point. On the other hand, the norm

$$\|f - g\| = \sqrt{\int_a^b |f(x) - g(x)|^2 \, dx}$$

is designed to measure how similar two functions are over an entire interval. Theorem 4.19 tells us exactly how to find the best approximation of a function in terms of such a norm.

Say for example we want to approximate the function $f(x) = e^x$ by a quadratic polynomial $q(x)$ on the interval $[0, 1]$ so that

$$\int_0^1 |f(x) - q(x)|^2 \, dx$$

is as small as possible. By part 2 of Theorem 4.19, the best possible approximation is $q = P_{\mathcal{P}_2(\mathbb{R})} f$, using the inner product

$$\langle f, g \rangle = \int_0^1 f(x) g(x) \, dx$$

on $C([0, 1])$. In the last section we found an orthonormal basis (e_1, e_2, e_3) of $\mathcal{P}_2(\mathbb{R})$ with this inner product:

$$e_1(x) = 1, \qquad e_2(x) = 2\sqrt{3}x - \sqrt{3}, \qquad \text{and} \qquad e_3(x) = (6\sqrt{5}x^2 - 6\sqrt{5}x + \sqrt{5}).$$

From this we can compute

$$\begin{aligned}
\langle f, e_1 \rangle &= \int_0^1 e^x \, dx = e - 1, \\
\langle f, e_2 \rangle &= \int_0^1 e^x (2\sqrt{3}x - \sqrt{3}) \, dx = -\sqrt{3}e + 3\sqrt{3}, \\
\langle f, e_3 \rangle &= \int_0^1 e^x (6\sqrt{5}x^2 - 6\sqrt{5}x + \sqrt{5}) \, dx = 7\sqrt{5}e - 19\sqrt{5},
\end{aligned}$$

and so

$$\begin{aligned}
q(x) = \left(P_{\mathcal{P}_2(\mathbb{R})} f\right)(x) &= \langle f, e_1 \rangle e_1(x) + \langle f, e_2 \rangle e_2(x) + \langle f, e_3 \rangle e_3(x) \\
&= (39e - 105) + (-216e + 588)x + (210e - 570)x^2.
\end{aligned}$$

Here are plots showing $f(x) = e^x$, the approximation $q(x)$ computed above, and, for comparison, the Taylor polynomial $1 + x + \frac{x^2}{2}$ of f about 0:

KEY IDEAS

- The orthogonal complement of a subspace is the set of all vectors perpendicular to the subspace. For example, the orthogonal complement of the x–y plane in $\mathbb{R}^3$ is the z-axis.
- If $(u_1, \ldots, u_m, w_1, \ldots, w_p)$ is an orthonormal basis of V such that $(u_1, \ldots, u_m)$ is an orthonormal basis of a subspace U, then any vector $v \in V$ can be expanded as

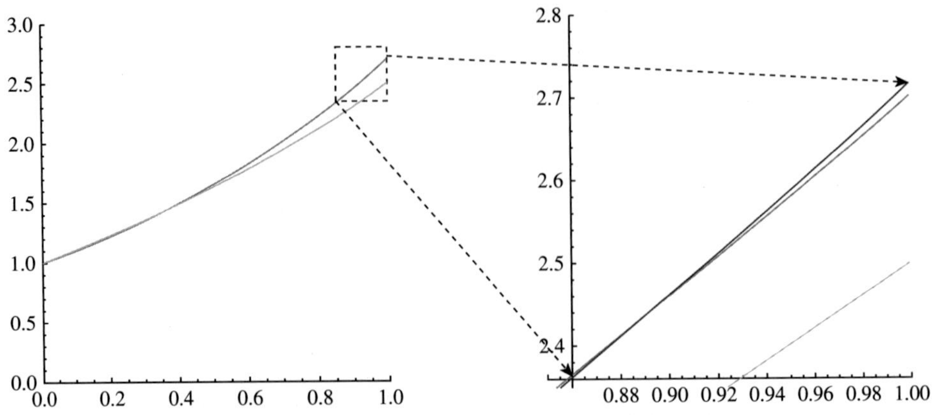

Figure 4.8 Graphs of $f(x) = e^x$, the approximation $q(x)$ and the Taylor polynomial of f about 0. The two graphs which are virtually indistinguishable in these plots are $f(x)$ and $q(x)$. The third is the Taylor polynomial, which gets to be a much worse approximation the farther you get from 0.

$$v = \sum_{j=1}^{m} \langle v, u_j \rangle u_j + \sum_{k=1}^{p} \langle v, w_k \rangle w_k,$$

and the orthogonal projection of v onto U is

$$P_U v = \sum_{j=1}^{m} \langle v, u_j \rangle u_j.$$

- Orthogonal projection of v onto U gives the closest point in U to v.
- Linear least squares and good approximations of functions over intervals can both be seen as applications of the fact that $P_U v$ is the closest point to v in U.

EXERCISES

4.3.1 Find the matrix (with respect to the standard basis) of the orthogonal projection onto the span of each of the following lists of vectors.

(a) $\left(\begin{bmatrix} 1 \\ 0 \\ -1 \\ 2 \end{bmatrix}, \begin{bmatrix} 2 \\ -1 \\ 1 \\ 0 \end{bmatrix} \right)$ in $\mathbb{R}^4$ (b) $\left(\begin{bmatrix} 1+i \\ 1 \\ i \end{bmatrix}, \begin{bmatrix} -1 \\ 2-i \\ 0 \end{bmatrix} \right)$ in $\mathbb{C}^3$

(c) $\left(\begin{bmatrix} 1 \\ 2 \\ 3 \\ -4 \\ -5 \end{bmatrix}, \begin{bmatrix} 1 \\ 1 \\ 1 \\ 1 \\ 1 \end{bmatrix} \right)$ in $\mathbb{R}^5$ (d) $\left(\begin{bmatrix} 1 \\ \vdots \\ 1 \end{bmatrix} \right)$ in $\mathbb{R}^n$

4.3.2 Find the matrix (with respect to the standard basis) of the orthogonal projection onto the span of each of the following lists of vectors.

(a) $\left(\begin{bmatrix}1\\1\\1\\1\end{bmatrix}, \begin{bmatrix}1\\2\\3\\4\end{bmatrix}\right)$ in $\mathbb{R}^4$ (b) $\left(\begin{bmatrix}1\\i\\0\end{bmatrix}, \begin{bmatrix}0\\1\\i\end{bmatrix}\right)$ in $\mathbb{C}^3$

(c) $\left(\begin{bmatrix}1\\0\\1\\0\\1\end{bmatrix}, \begin{bmatrix}0\\1\\0\\1\\0\end{bmatrix}\right)$ in $\mathbb{R}^5$ (d) $\left(\begin{bmatrix}5\\7\\7\\2\\1\end{bmatrix}\right)$ in $\mathbb{R}^5$

4.3.3 Find the matrix (with respect to the standard basis) of the orthogonal projection onto each of the following subspaces.

(a) $\left\langle \begin{bmatrix}1\\2\\3\end{bmatrix} \right\rangle^{\perp} \subseteq \mathbb{R}^3$

(b) $\left\{ \begin{bmatrix}x\\y\\z\end{bmatrix} \in \mathbb{R}^3 \;\middle|\; 3x - y - 5z = 0 \right\}$

(c) $\left\{ \begin{bmatrix}x\\y\\z\end{bmatrix} \in \mathbb{C}^3 \;\middle|\; 2x + (1-i)y + iz = 0 \right\}$

(d) $\ker \begin{bmatrix}2 & 0 & -1 & 3\\ 1 & 2 & 3 & 0\end{bmatrix} \subseteq \mathbb{R}^4$

4.3.4 Find the matrix (with respect to the standard basis) of the orthogonal projection onto each of the following subspaces.

(a) $\left\langle \begin{bmatrix}3\\1\\4\end{bmatrix} \right\rangle^{\perp} \subseteq \mathbb{R}^3$

(b) $\left\{ \begin{bmatrix}x\\y\\z\end{bmatrix} \in \mathbb{R}^3 \;\middle|\; 3x + 2y + z = 0 \right\}$

(c) $\left\{ \begin{bmatrix}x\\y\\z\end{bmatrix} \in \mathbb{C}^3 \;\middle|\; x + iy - iz = 0 \right\}$

(d) $C\left(\begin{bmatrix}1 & 0 & 2\\ -1 & 3 & -5\\ 0 & 2 & -2\\ 2 & 1 & 3\end{bmatrix}\right) \subseteq \mathbb{R}^4$

4.3.5 In each of the following, find the point in the subspace U which is closest to the point $\mathbf{x}$.

(a) $U = \left\langle \begin{bmatrix} 1 \\ 0 \\ -1 \\ 2 \end{bmatrix}, \begin{bmatrix} 2 \\ -1 \\ 1 \\ 0 \end{bmatrix} \right\rangle \subseteq \mathbb{R}^4$ and $\mathbf{x} = \begin{bmatrix} 1 \\ 2 \\ 3 \\ 4 \end{bmatrix}$

(b) $U = \left\langle \begin{bmatrix} 1+i \\ 1 \\ i \end{bmatrix}, \begin{bmatrix} -1 \\ 2-i \\ 0 \end{bmatrix} \right\rangle \subseteq \mathbb{C}^3$ and $\mathbf{x} = \begin{bmatrix} 3 \\ i \\ 1 \end{bmatrix}$

(c) $U = \left\{ \begin{bmatrix} x \\ y \\ z \end{bmatrix} \in \mathbb{R}^3 \;\middle|\; 3x - y - 5z = 0 \right\}$ and $\mathbf{x} = \begin{bmatrix} 1 \\ 1 \\ 1 \end{bmatrix}$

(d) $U = \ker \begin{bmatrix} 2 & 0 & -1 & 3 \\ 1 & 2 & 3 & 0 \end{bmatrix} \subseteq \mathbb{R}^4$ and $\mathbf{x} = \begin{bmatrix} -2 \\ 1 \\ 1 \\ 2 \end{bmatrix}$

4.3.6 In each of the following, find the point in the subspace U which is closest to the point $\mathbf{x}$.

(a) $U = \left\langle \begin{bmatrix} 1 \\ 1 \\ 1 \\ 1 \end{bmatrix}, \begin{bmatrix} 1 \\ 2 \\ 3 \\ 4 \end{bmatrix} \right\rangle \subseteq \mathbb{R}^4$ and $\mathbf{x} = \begin{bmatrix} 4 \\ 3 \\ 2 \\ 1 \end{bmatrix}$

(b) $U = \left\langle \begin{bmatrix} 1 \\ i \\ 0 \end{bmatrix}, \begin{bmatrix} 0 \\ 1 \\ i \end{bmatrix} \right\rangle \subseteq \mathbb{C}^3$ and $\mathbf{x} = \begin{bmatrix} -2 \\ 0 \\ 1 \end{bmatrix}$

(c) $U = \left\{ \begin{bmatrix} x \\ y \\ z \end{bmatrix} \in \mathbb{C}^3 \;\middle|\; x + iy - iz = 0 \right\}$ and $\mathbf{x} = \begin{bmatrix} 2 \\ 1 \\ -1 \end{bmatrix}$

(d) $U = C\left(\begin{bmatrix} 1 & 0 & 2 \\ -1 & 3 & -5 \\ 0 & 2 & -2 \\ 2 & 1 & 3 \end{bmatrix} \right) \subseteq \mathbb{R}^4$ and $\mathbf{x} = \begin{bmatrix} 1 \\ 1 \\ 1 \\ 1 \end{bmatrix}$

4.3.7 Use simple linear regression to find the line in the plane which comes closest to containing the data points $(-2, 1)$, $(-1, 2)$, $(0, 5)$, $(1, 4)$, $(2, 8)$.

4.3.8 Use simple linear regression to find the line in the plane which comes closest to containing the data points $(-2, 5)$, $(-1, 3)$, $(0, 2)$, $(1, 2)$, $(2, 3)$.

4.3.9 Show that the point on the line $y = mx$ which is closest to the point (a, b) is $\left(\frac{a+mb}{m^2+1}, m\frac{a+mb}{m^2+1} \right)$.

4.3.10 Find the quadratic polynomial $p \in \mathcal{P}_2(\mathbb{R})$ such that

$$\int_{-1}^{1} (p(x) - |x|)^2 \, dx$$

is as small as possible.

4.3.11 Find the best approximation of $f(x) = e^x$ by a linear polynomial, with respect to the inner product

$$\langle f, g \rangle = \int_0^1 f(x)g(x) \, dx$$

on $C([0, 1])$. Show that this approximation is different from the first two terms of the quadratic approximation $q(x)$ found in the section.

4.3.12 Find the best approximation of $f(x) = e^x$ by a quadratic polynomial, with respect to the inner product

$$\langle f, g \rangle = \int_{-1}^1 f(x)g(x) \, dx$$

on $C([-1, 1])$. Show that this approximation is different from the quadratic approximation $q(x)$ found in the section.

4.3.13 Find the best approximation of $f(x) = x$ by a function of the form

$$g(x) = a_1 \sin(x) + a_2 \sin(2x) + b_0 + b_1 \cos(x) + b_2 \cos(2x),$$

with respect to the inner product

$$\langle f, g \rangle = \int_0^{2\pi} f(x)g(x) \, dx$$

on $C([0, 2\pi])$.

4.3.14 (a) Show that $V = \left\{A \in M_n(\mathbb{R}) \mid A^T = A\right\}$ and $W = \left\{A \in M_n(\mathbb{R}) \mid A^T = -A\right\}$ are subspaces of $M_n(\mathbb{R})$.

(b) Show that $V^\perp = W$ (where $M_n(\mathbb{R})$ is given, as usual, the Frobenius inner product).

(c) Show that, for any $A \in M_n(\mathbb{R})$,

$$P_V(A) = \operatorname{Re} A \qquad \text{and} \qquad P_W(A) = i \operatorname{Im} A,$$

where $\operatorname{Re} A$ and $\operatorname{Im} A$ are as defined in Exercise 4.1.5.

4.3.15 Suppose that V is an inner product space and U_1, U_2 are subspaces of V with $U_1 \subseteq U_2$. Prove that $U_2^\perp \subseteq U_1^\perp$.

4.3.16 Show that if V is a finite-dimensional inner product space and U is a subspace of V, then

$$\dim U + \dim U^\perp = \dim V.$$

4.3.17 Suppose that an inner product space V is an orthogonal direct sum $U_1 \oplus \cdots \oplus U_m$. Show that if

$$v = v_1 + \cdots + v_m \qquad \text{and} \qquad w = w_1 + \cdots + w_m$$

where $v_j, w_j \in U_j$ for each j, then

$$\langle v, w \rangle = \langle v_1, w_1 \rangle + \cdots + \langle v_m, w_m \rangle$$

and

$$\|v\|^2 = \|v_1\|^2 + \cdots + \|v_m\|^2.$$

4.3.18 Show that if V is the orthogonal direct sum of subspaces $U_1, \ldots, U_m$, then for every $v \in V$, the decomposition in equation (4.7) is unique.

4.3.19 Show that if V is the orthogonal direct sum of subspaces $U_1, \ldots, U_m$, then

$$\dim V = \dim U_1 + \cdots + \dim U_m.$$

4.3.20 Suppose that U is a subspace of a finite-dimensional inner product space V, and that U is the orthogonal direct sum of subspaces $U_1, \ldots, U_m$. Show that

$$\mathbf{P}_U = \mathbf{P}_{U_1} + \cdots + \mathbf{P}_{U_m}.$$

4.3.21 Show that if U is a subspace of a finite-dimensional inner product space V, then $\operatorname{tr} \mathbf{P}_U = \operatorname{rank} \mathbf{P}_U = \dim U$.

4.3.22 (a) Prove part 5 of Theorem 4.16.
(b) Prove part 6 of Theorem 4.16.
(c) Prove part 7 of Theorem 4.16.
(d) Prove part 8 of Theorem 4.16.
(e) Prove part 9 of Theorem 4.16.

4.3.23 Use part 2 of Theorem 4.16 to give another proof of part 1 of Theorem 4.16.

4.4 Normed Spaces

In the last several sections, we have explored the implications of the fact that in inner product spaces there is a notion of orthogonality. Another important geometric feature of inner product spaces is that there is a way to define length, but actually the notion of length is more general than the notion of angles.

General Norms

Definition Let $\mathbb{F}$ be either $\mathbb{R}$ or $\mathbb{C}$, and let V be a vector space over $\mathbb{F}$. A **norm** on V is a real-valued function on V, written $\|v\|$, such that the following properties hold:

Nonnegativity: For each $v \in V$, $\|v\| \geq 0$.

Definiteness: If $\|v\| = 0$, then $v = 0$.

Positive homogeneity: For each $v \in V$ and $a \in \mathbb{F}$, $\|av\| = |a| \, \|v\|$.

Triangle inequality: For each $v, w \in V$, $\|v + w\| \leq \|v\| + \|w\|$.

A vector space together with a norm is called a **normed space**.

Quick Exercise #14. Show that in any normed space, $\|0\| = 0$.

As we saw in Section 4.1, every inner product space has a norm defined by $\|v\| = \sqrt{\langle v, v \rangle}$ which satisfies all the above properties. Thus every inner product space is a normed space.

However, not every norm comes from an inner product. Here are some examples of normed spaces which are not inner product spaces (proving that they are not inner product spaces is Exercise 4.4.1).

Examples

1. The ℓ^1 **norm** on $\mathbb{R}^n$ or $\mathbb{C}^n$ is defined by

$$\|\mathbf{x}\|_1 := \sum_{j=1}^{n} |x_j| .$$

   The fact that the ℓ^1 norm is indeed a norm follows easily from the fact that the absolute value $|\cdot|$ is a norm on $\mathbb{R}$ or $\mathbb{C}$.

2. The ℓ^∞ **norm*** on $\mathbb{R}^n$ or $\mathbb{C}^n$ is defined by

$$\|\mathbf{x}\|_\infty := \max_{1 \leq j \leq n} |x_j| .$$

*The norm $\|\mathbf{x}\| = \sqrt{\sum_{j=1}^{n} |x|_j^2}$ which comes from the standard inner product on $\mathbb{R}^n$ or $\mathbb{C}^n$ is sometimes called the ℓ^2 norm, and written $\|\mathbf{x}\|_2$. There is a whole family of ℓ^p norms given by $\|\mathbf{x}\|_p = \left(\sum_{j=1}^{n} |x_j|^p\right)^{1/p}$ for any $p \geq 1$, which we won't discuss any further here.

QA #14: By positive homogeneity, $\|0\| = \|00\| = 0 \, \|0\| = 0$. Make sure you understand which 0s are scalars and which are vectors!

The norm properties again follow from those of $|\cdot|$, together with the fact that

$$\max_{1\le j\le n}(a_j + b_j) \le \max_{1\le j\le n} a_j + \max_{1\le j\le n} b_j.$$

3. The L^1 **norm** on the space $C([0,1])$ is defined by

$$\|f\|_1 := \int_0^1 |f(x)| \; dx.$$

This is the functional analog of the ℓ^1 norm described above.
Confirming that the L^1 norm is in fact a norm on $C([0,1])$ mostly follows easily from the fact that $|\cdot|$ is a norm on $\mathbb{R}$. Definiteness takes slightly more work: suppose that $f \in C([0,1])$ such that

$$\int_0^1 |f(x)| \; dx = 0.$$

If there is an $x_0 \in [0,1]$ such that $f(x_0) \neq 0$, then by continuity of f, there is an interval $I \subseteq [0,1]$ containing x_0 such that $|f(y)| > \frac{|f(x_0)|}{2}$ for all $y \in I$. If $I = [a,b]$, then

$$\int_0^1 |f(y)| \; dy \ge \int_a^b |f(y)| \; dy \ge \frac{(b-a)\,|f(x_0)|}{2} > 0,$$

which is a contradiction.

4. The **supremum norm** on $C([0,1])$ is defined by

$$\|f\|_\infty := \max_{0\le x\le 1} |f(x)|.$$

(Recall from calculus that a continuous function on a closed interval has a maximum, either inside the interval or at one of the endpoints.) The supremum norm is the functional analog of the ℓ^∞ norm above; checking that it is in fact a norm is also analogous. ▲

The following result is a fundamental fact about norms that come from inner products. Its name derives from how it relates the lengths of the sides of a parallelogram to the lengths of its diagonals.

Proposition 4.20 (The parallelogram identity) *If V is an inner product space, then for any $v, w \in V$,*

$$\|v+w\|^2 + \|v-w\|^2 = 2\|v\|^2 + 2\|w\|^2. \tag{4.8}$$

Proof See Exercise 4.4.20. ▲

The point of Proposition 4.20 is that, although it is a theorem about inner product spaces, it directly involves only norms, not inner products or even orthogonality. That means that it can be used to check whether or not a given norm

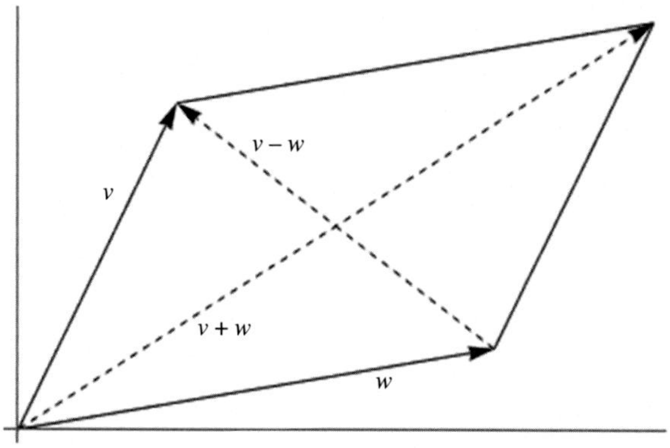

Figure 4.9 The sum of the squares of the lengths of the four sides of the parallelogram equals the sum of the squares of the lengths of the diagonals.

comes from an inner product: if you can find any two vectors v and w in a normed space which fail to satisfy equation (4.8), then the norm cannot be the norm associated to any inner product.

An easy consequence of the triangle inequality in a normed space is the following fact, which we will need below.

Proposition 4.21 *If V is a normed space, then for any $v, w \in V$,*

$$\left| \|v\| - \|w\| \right| \leq \|v - w\| .$$

Proof By the triangle inequality,

$$\|v\| = \|v - w + w\| \leq \|v - w\| + \|w\| ,$$

and so $\|v\| - \|w\| \leq \|v - w\|$. The same argument implies that $\|w\| - \|v\| \leq \|v - w\|$. ▲

The Operator Norm

There are various norms on vector spaces consisting of matrices or linear maps; we have already seen one, namely the Frobenius norm. The operator norm is one of the most important norms on matrices or linear maps. To introduce it, we need the following fact about linear maps in normed spaces.

Lemma 4.22 *Let V and W be finite-dimensional inner product spaces, and let $T \in \mathcal{L}(V, W)$. Then there is a constant $C \geq 0$ such that, for every $v_1, v_2 \in V$,*

$$\left| \|Tv_1\| - \|Tv_2\| \right| \leq \|Tv_1 - Tv_2\| \leq C \|v_1 - v_2\| . \tag{4.9}$$

In particular, the function $f : V \to \mathbb{R}$ given by $f(v) = \|Tv\|$ is continuous.

Proof The first inequality in formula (4.9) follows from Proposition 4.21.

For the second, let $(e_1, \ldots, e_n)$ be an orthonormal basis of V. Then for any $v \in V$,

$$\|Tv\| = \left\| T\left(\sum_{j=1}^{n} \langle v, e_j \rangle e_j\right) \right\| = \left\| \sum_{j=1}^{n} \langle v, e_j \rangle Te_j \right\| \leq \sum_{j=1}^{n} \left|\langle v, e_j \rangle\right| \left\|Te_j\right\|,$$

where the last inequality follows from the triangle inequality. Now, by the Cauchy–Schwarz inequality, $\left|\langle v, e_j \rangle\right| \leq \|v\| \left\|e_j\right\| = \|v\|$, so

$$\|Tv\| \leq \|v\| \sum_{j=1}^{n} \left\|Te_j\right\|. \tag{4.10}$$

The second inequality in formula (4.9) now follows by applying the inequality (4.10) with $v = v_1 - v_2$, with

$$C := \sum_{j=1}^{n} \left\|Te_j\right\|.$$

▲

Quick Exercise #15. The proof of Lemma 4.22 doesn't always give the best possible value of C (although this isn't important for how we will use the lemma below). If $T = I$ is the identity map on V, then:

(a) What is the best possible value of C?
(b) What value of C does the proof produce?

Lemma 4.23 *Let V and W be finite-dimensional inner product spaces, and let $T \in \mathcal{L}(V, W)$. Then there is a vector $u \in V$ with $\|u\| = 1$ such that*

$$\|Tv\| \leq \|Tu\|$$

whenever $v \in V$ and $\|v\| = 1$.

We summarize the situation in Lemma 4.23 by writing

$$\|Tu\| = \max_{\substack{v \in V, \\ \|v\|=1}} \|Tv\|.$$

For the proof, we will need a fundamental fact from multivariable calculus. For $S \subseteq V$, we say that S is **bounded** if there is a constant $C > 0$ such that

$$\|s\| \leq C$$

for each $s \in S$. We say that S is **closed** if S contains its own boundary, i.e., if every convergent sequence in S has its limit point in S as well (picture the difference between the open ball $\{v \in V \mid \|v\| < 1\}$ and the closed ball $\{v \in V \mid \|v\| \leq 1\}$). The

QA #15: (a) 1. (b) $n = \dim V$.

crucial calculus fact that we need is the following: if $f : V \to \mathbb{R}$ is a continuous function on a finite-dimensional inner product space, and $S \subseteq V$ is a closed, bounded subset of V, then there is a point $s_0 \in S$ such that $f(s) \leq f(s_0)$ for every $s \in S$; i.e.,

$$\max_{s \in S} f(s) = f(s_0).$$

This is a generalization of the familiar fact that if $f : [a, b] \to \mathbb{R}$ is continuous, then f achieves a maximum value at some point in the interval $[a, b]$, either in the interior or at one of the endpoints.

Proof of Lemma 4.23 Consider the set

$$S := \{v \in V \mid \|v\| = 1\};$$

S is a closed and bounded subset of V, and by Lemma 4.22, $f(v) = \|Tv\|$ is a continuous function $f : V \to \mathbb{R}$. Thus there is a vector $u \in S$ such that

$$\max_{v \in S} \|Tv\| = \|Tu\|.$$ ▲

Definition Let V and W be finite-dimensional inner product spaces, and let $T \in \mathcal{L}(V, W)$. The **operator norm*** of T is

$$\|T\|_{op} := \max_{\substack{v \in V, \\ \|v\|=1}} \|Tv\|.$$

An equivalent formulation of the definition (see Exercise 4.4.3) is that $\|T\|_{op}$ is the smallest number C such that

$$\|Tv\| \leq C \|v\| \tag{4.11}$$

for every $v \in V$. Lemma 4.23 says that there is at least one unit vector v for which equality holds in the inequality (4.11) for $C = \|T\|_{op}$.

We have once again given a definition containing an implicit claim, namely that the quantity $\|T\|_{op}$ defined above does define a norm on $\mathcal{L}(V, W)$. The next theorem verifies this claim.

Theorem 4.24 *Let V and W be finite-dimensional inner product spaces. Then the operator norm is a norm on $\mathcal{L}(V, W)$.*

*Also called the **spectral norm** (among other things), for reasons which will be explained later (see Section 5.4).

Proof It is obvious that $\|T\|_{op} \geq 0$ for every $T \in \mathcal{L}(V, W)$.

If $\|T\|_{op} = 0$, then for every $v \in V$, $\|Tv\| = 0$ by the inequality (4.11). Therefore $Tv = 0$ for every $v \in V$, and thus $T = \mathbf{0}$, the zero operator.

If $a \in \mathbb{F}$ and $T \in \mathcal{L}(V, W)$, then for every $v \in V$,

$$\|(aT)v\| = \|a(Tv)\| = |a| \, \|Tv\| \,,$$

which implies that

$$\|aT\|_{op} = \max_{\substack{v \in V, \\ \|v\|=1}} \|(aT)v\| = \max_{\substack{v \in V, \\ \|v\|=1}} |a| \, \|Tv\| = |a| \max_{\substack{v \in V, \\ \|v\|=1}} \|Tv\| = |a| \, \|T\|_{op} \,.$$

Finally, if $S, T \in \mathcal{L}(V, W)$ and $v \in V$, then

$$\|(S+T)v\| = \|Sv + Tv\| \leq \|Sv\| + \|Tv\| \leq \|S\|_{op} \|v\| + \|T\|_{op} \|v\|$$

by the triangle inequality in V and the inequality (4.11). Therefore

$$\|S+T\|_{op} = \max_{\substack{v \in V, \\ \|v\|=1}} \|(S+T)v\| \leq \max_{\substack{v \in V, \\ \|v\|=1}} \left\{ (\|S\|_{op} + \|T\|_{op}) \|v\| \right\} = \|S\|_{op} + \|T\|_{op} \,.$$

▲

Quick Exercise #16. What is the operator norm of the identity map I on a finite-dimensional inner product space?

As usual, anything that can be done for linear maps can be done for matrices.

Definition If $\mathrm{A} \in \mathrm{M}_{m,n}(\mathbb{F})$, then the **operator norm*** $\|\mathrm{A}\|_{op}$ of A is the norm of the linear map in $\mathcal{L}(\mathbb{F}^n, \mathbb{F}^m)$ whose matrix is A. That is,

$$\|\mathrm{A}\|_{op} = \max_{\substack{\mathrm{v} \in \mathbb{F}^n, \\ \|\mathrm{v}\|=1}} \|\mathrm{Av}\| \,.$$

*Also called (as before) the **spectral norm**, and also the **induced norm**. This is often denoted $\|\mathrm{A}\|_2$, because of its relationship to the ℓ^2 norms used for vectors in the definition, but the same notation is sometimes also used for other things.

QA #16: $Iv = v$, and hence $\|Iv\| = \|v\|$ for every v, so $\|I\|_{op} = 1$.

Quick Exercise #17. What is $\|\mathrm{I}_n\|_{op}$? What is the Frobenius norm $\|\mathrm{I}_n\|_F$?

One of the most important uses of the operator norm is to understand the effect of error on solutions of linear systems.* Say you want to solve an $n \times n$ linear system $\mathrm{Ax} = \mathbf{b}$ over $\mathbb{R}$. If A is invertible, then this can be solved by matrix multiplication:

$$\mathbf{x} = \mathrm{A}^{-1}\mathrm{Ax} = \mathrm{A}^{-1}\mathbf{b}.$$

Now suppose that you don't know the entries of $\mathbf{b}$ exactly; this could be because they come from real-world measurements, or were calculated by a computer and are subject to round-off error, or some other reason. Intuitively, if the error in $\mathbf{b}$ is small, then the error in the computed solution $\mathbf{x}$ should also be small. But how small?

Suppose that $\mathbf{h} \in \mathbb{R}^n$ is the vector of errors in $\mathbf{b}$. That is, instead of the true vector $\mathbf{b}$, you actually have $\hat{\mathbf{b}} = \mathbf{b} + \mathbf{h}$ to work with. In that case, instead of the true vector $\mathbf{x} = \mathrm{A}^{-1}\mathbf{b}$, the solution you compute will be

$$\mathrm{A}^{-1}(\hat{\mathbf{b}}) = \mathrm{A}^{-1}(\mathbf{b} + \mathbf{h}) = \mathrm{A}^{-1}\mathbf{b} + \mathrm{A}^{-1}\mathbf{h} = \mathbf{x} + \mathrm{A}^{-1}\mathbf{h}.$$

Thus the error in your computed solution is $\mathrm{A}^{-1}\mathbf{h}$. We can bound the size of this error using the operator norm of A^{-1}:

$$\left\|\mathrm{A}^{-1}\mathbf{h}\right\| \leq \left\|\mathrm{A}^{-1}\right\|_{op} \|\mathbf{h}\|,$$

where the first and third norms here are the standard norm on $\mathbb{R}^n$. That is, the operator norm $\left\|\mathrm{A}^{-1}\right\|_{op}$ tells us how an error of a given size in the vector $\mathbf{b}$ propagates to an error in the solution of the system $\mathrm{Ax} = \mathbf{b}$.

KEY IDEAS

- A normed space is a vector space together with a notion of length: a norm is a positively homogeneous, definite, real-valued function which satisfies the triangle inequality.
- Inner products define norms, but not all norms come from inner products. The parallelogram identity lets you check whether a norm is coming from an inner product.
- The vector space of linear maps between normed spaces V and W has a norm called the operator norm: $\|T\|_{op} := \max_{\|v\|=1} \|Tv\|$.
- The operator norm of A^{-1} tells you how much an error in $\mathbf{b}$ propagates to an error in the solution of $\mathrm{Ax} = \mathbf{b}$.

*Here we're using the word *error* in the sense of small changes being made in data, not in the sense of making mistakes in your work. The latter issue is much harder to deal with!

QA #17: The last quick exercise shows that $\|\mathrm{I}_n\|_{op} = 1$, since the identity matrix represents the identity map. On the other hand, $\|\mathrm{I}_n\|_F = \sqrt{n}$.

EXERCISES

4.4.1 (a) Show that the ℓ^1 norm is not the norm associated to any inner product on $\mathbb{R}^n$ or $\mathbb{C}^n$.

(b) Show that the ℓ^∞ norm is not the norm associated to any inner product on $\mathbb{R}^n$ or $\mathbb{C}^n$.

4.4.2 Show that the operator norm is not the norm associated to any inner product on $\mathrm{M}_{m,n}(\mathbb{C})$ when $m, n \geq 2$.

4.4.3 Show that if V and W are finite-dimensional inner product spaces and $T \in \mathcal{L}(V, W)$, then $\|T\|_{op}$ is the smallest constant C such that

$$\|Tv\| \leq C\,\|v\|$$

for all $v \in V$.

4.4.4 Show that if $\lambda \in \mathbb{C}$ is an eigenvalue of $\mathrm{A} \in \mathrm{M}_n(\mathbb{C})$, then $|\lambda| \leq \|\mathrm{A}\|_{op}$.

4.4.5 Suppose you are solving the $n \times n$ linear system $\mathrm{Ax} = \mathbf{b}$ using A^{-1}, but there are errors in $\mathbf{b}$. If you know that at most m of the entries of $\mathbf{b}$ have errors, and that each entry has an error of size at most $\varepsilon > 0$, what can you say about the size of the error in the $\mathbf{x}$ you compute?

4.4.6 Show that

$$\|\mathbf{diag}(d_1, \ldots, d_n)\|_{op} = \left\| \begin{bmatrix} d_1 \\ \vdots \\ d_n \end{bmatrix} \right\|_\infty .$$

4.4.7 Show that if $\mathrm{A} \in \mathrm{M}_{m,n}(\mathbb{C})$, then $\|\mathrm{A}\|_{op} \geq \|\mathbf{a}_j\|$ for each j, where $\mathbf{a}_j$ is the jth column of A.

4.4.8 Show that if $\mathrm{A} \in \mathrm{M}_{m,n}(\mathbb{C})$ and $\mathrm{B} \in \mathrm{M}_{n,p}(\mathbb{C})$, then $\|\mathrm{AB}\|_F \leq \|\mathrm{A}\|_{op}\,\|\mathrm{B}\|_F$.

4.4.9 Suppose that U, V, and W are finite-dimensional inner product spaces, and that $S \in \mathcal{L}(U, V)$ and $T \in \mathcal{L}(V, W)$. Show that $\|TS\|_{op} \leq \|T\|_{op}\,\|S\|_{op}$.

4.4.10 Suppose that $\mathrm{A} \in \mathrm{M}_n(\mathbb{C})$ is invertible.

(a) Show that $\|\mathbf{x}\| \leq \left\|\mathrm{A}^{-1}\right\|_{op} \|\mathrm{A}\mathbf{x}\|$ for every $\mathbf{x} \in \mathbb{C}^n$.

(b) Show that if $\|\mathrm{A} - \mathrm{B}\|_{op} < \left\|\mathrm{A}^{-1}\right\|_{op}^{-1}$, then B is invertible.

Hint: First show that if $\mathbf{x} \in \ker \mathrm{B}$ and $\mathbf{x} \neq \mathbf{0}$, then $\|\mathrm{A}\|_{op}^{-1} \|\mathrm{A}\mathbf{x}\| < \|\mathbf{x}\|$.

4.4.11 The **condition number** of an invertible matrix $\mathrm{A} \in \mathrm{M}_n(\mathbb{C})$ is $\kappa(\mathrm{A}) = \|\mathrm{A}\|_{op}\left\|\mathrm{A}^{-1}\right\|_{op}$.

(a) Show that $\kappa(\mathrm{A}) \geq 1$ for every invertible $\mathrm{A} \in \mathrm{M}_n(\mathbb{C})$.

(b) Suppose that you know A^{-1} and are using it to solve the $n \times n$ system $\mathrm{A}\mathbf{x} = \mathbf{b}$. Suppose that there is some error $\mathbf{h} \in \mathbb{C}^n$ in $\mathbf{b}$; i.e., the system you are actually solving is $\mathrm{A}\mathbf{y} = \mathbf{b} + \mathbf{h}$. Show that

$$\frac{\|A^{-1}h\|}{\|x\|} \leq \kappa(A)\frac{\|h\|}{\|b\|}.$$

That is, the condition number of A can be used to bound the *relative* error in the solution you compute in terms of the relative error in the input b.

4.4.12 Show that if $A \in M_{m,n}(\mathbb{R})$, then

$$\max_{\substack{v\in\mathbb{R}^n,\\ \|v\|=1}} \|Av\| = \max_{\substack{v\in\mathbb{C}^n,\\ \|v\|=1}} \|Av\|.$$

This means that the operator norm of A is the same whether we think of it as a real matrix or a complex matrix.

4.4.13 Prove that if $A \in M_{m,n}(\mathbb{R})$ has rank 1, then $\|A\|_{op} = \|A\|_F$.
Hint: By Quick Exercise #16 in Section 3.4, $A = vw^T$ for some nonzero vectors $v \in \mathbb{R}^m$ and $w \in \mathbb{R}^n$. You can compute Ax and $\|A\|_F$ explicitly in terms of v and w. Use the formulas you get to show that $\|A\|_{op} \geq \|A\|_F$ by finding a specific unit vector $x \in \mathbb{R}^n$ with $\|Ax\| \geq \|A\|_F$. The opposite inequality follows from Exercise 4.4.17.

4.4.14 Show that the operator norm is not an invariant on $M_n(\mathbb{C})$.
Hint: Consider the matrices $\begin{bmatrix} 1 & 0 \\ 0 & 2 \end{bmatrix}$ and $\begin{bmatrix} 1 & 1 \\ 0 & 2 \end{bmatrix}$, and use Exercises 4.4.6 and 4.4.7.

4.4.15 Show that if $x \in \mathbb{C}^n$, then $\|x\|_\infty \leq \|x\|_2 \leq \sqrt{n}\,\|x\|_\infty$.

4.4.16 Show that if $x \in \mathbb{C}^n$, then $\|x\|_2 \leq \|x\|_1 \leq \sqrt{n}\,\|x\|_2$.
Hint: For the first inequality, use the triangle inequality for the ℓ^2 norm. For the second inequality, use the Cauchy–Schwarz inequality.

4.4.17 Show that if $A \in M_{m,n}(\mathbb{C})$, then $\|A\|_{op} \leq \|A\|_F \leq \sqrt{n}\,\|A\|_{op}$.

4.4.18 For $f \in C([0,1])$, let $\|f\| = \sqrt{\int_0^1 |f(x)|^2\, dx}$.

(a) Show that if $f \in C([0,1])$, then

$$\|f\|_1 \leq \|f\| \leq \|f\|_\infty.$$

Hint: Use the Cauchy–Schwarz inequality.

(b) Show that there is no constant $C > 0$ such that

$$\|f\| \leq C\,\|f\|_1$$

for every $f \in C([0,1])$.
Hint: Consider the functions

$$f_n(x) = \begin{cases} 1 - nx & \text{if } 0 \leq x \leq \frac{1}{n}, \\ 0 & \text{if } \frac{1}{n} < x \leq 1. \end{cases}$$

(c) Show that there is no constant $C > 0$ such that

$$\|f\|_\infty \leq C \|f\|$$

for every $f \in C([0, 1])$.

Hint: Consider the same functions f_n as in the previous part.

4.4.19 A norm on a real vector space V is called **strictly convex** if, for every $v, w \in V$ with $v \neq w$ and $\|v\| = \|w\| = 1$, we have $\left\| \frac{1}{2}(v + w) \right\| < 1$.

(a) Draw a picture illustrating what this means geometrically.

(b) Use the parallelogram identity to show that the standard norm on $\mathbb{R}^n$ is strictly convex.

(c) Show that the ℓ^1 norm on $\mathbb{R}^n$ is not strictly convex if $n \geq 2$.

4.4.20 Prove Proposition 4.20 (the parallelogram identity).

4.5 Isometries

Preserving Lengths and Angles

Recall the discussion in Section 2.1 in which we introduced linear maps: in mathematics, once you have defined a new structure, you also need to formalize the notion of *sameness* of that structure. For general vector spaces, we observed that the right way to do this was via linear maps: i.e., maps between vector spaces which respected the vector space operations. When working with normed spaces, we need to consider linear maps which not only respect the linear structure, but respect the norm as well:

Definition Let V and W be normed spaces. A linear map $T : V \to W$ is called an **isometry** if T is surjective and

$$\|Tv\| = \|v\|$$

for every $v \in V$. If there is an isometry $T : V \to W$, we say that V and W are **isometric.**

Quick Exercise #18. Suppose $V = \mathbb{R}^2$ with the ℓ^1 norm and $W = \mathbb{R}^2$ with the ℓ^∞ norm. Show that the identity map $I : V \to W$ is not an isometry. Does this mean that V and W are not isometric?

QA #18: For example, $\left\| \begin{bmatrix} 1 \\ 1 \end{bmatrix} \right\|_1 = 2$ but $\left\| I \begin{bmatrix} 1 \\ 1 \end{bmatrix} \right\|_\infty = \left\| \begin{bmatrix} 1 \\ 1 \end{bmatrix} \right\|_\infty = 1$. This does *not* mean that V and W are not isometric: in fact, the linear map represented by $\begin{bmatrix} 1 & 1 \\ 1 & -1 \end{bmatrix}$ is an isometry (see Exercise 4.5.17).

Lemma 4.25 *Let V and W be normed spaces. If $T \in \mathcal{L}(V, W)$ is an isometry, then T is an isomorphism.*

Proof Since an isometry is surjective by definition, we only need to show that T is injective.

Suppose that $Tv = 0$. Then

$$\|v\| = \|Tv\| = \|0\| = 0,$$

so $v = 0$. Thus $\ker T = \{0\}$, and so T is injective. ▲

A natural extension of the discussion above is that, after defining the structure of inner product spaces, we should consider linear maps which preserve that structure as well. In fact, we already have: it turns out that the linear maps between inner product spaces which respect the inner product structure are exactly the isometries.

Theorem 4.26 *Let V and W be inner product spaces. A surjective linear map $T : V \to W$ is an isometry if and only if*

$$\langle Tv_1, Tv_2 \rangle = \langle v_1, v_2 \rangle \tag{4.12}$$

for every $v_1, v_2 \in V$.

Proof Suppose first that equation (4.12) holds for every $v_1, v_2 \in V$. Then for each $v \in V$,

$$\|Tv\| = \sqrt{\langle Tv, Tv \rangle} = \sqrt{\langle v, v \rangle} = \|v\| .$$

For the other implication, recall the following identities, called the **polarization identities**, from Exercise 4.1.14: for $\mathbb{F} = \mathbb{R}$,

$$\langle v_1, v_2 \rangle = \frac{1}{4}\left(\|v_1 + v_2\|^2 - \|v_1 - v_2\|^2\right)$$

and for $\mathbb{F} = \mathbb{C}$,

$$\langle v_1, v_2 \rangle = \frac{1}{4}\left(\|v_1 + v_2\|^2 - \|v_1 - v_2\|^2 + i\,\|v_1 + iv_2\|^2 - i\,\|v_1 - iv_2\|^2\right).$$

Suppose that $T : V \to W$ is an isometry. If $\mathbb{F} = \mathbb{R}$, then

$$\begin{aligned}
\langle Tv_1, Tv_2 \rangle &= \frac{1}{4}\left(\|Tv_1 + Tv_2)\|^2 - \|Tv_1 - Tv_2)\|^2\right) \\
&= \frac{1}{4}\left(\|T(v_1 + v_2)\|^2 - \|T(v_1 - v_2)\|^2\right) \\
&= \frac{1}{4}\left(\|v_1 + v_2\|^2 - \|v_1 - v_2\|^2\right) \\
&= \langle v_1, v_2 \rangle .
\end{aligned}$$

The proof in the case $\mathbb{F} = \mathbb{C}$ is similar (see Exercise 4.5.23). ▲

Geometrically, this means that a linear map between inner product spaces which preserves lengths of vectors must also preserve the angles between vectors. The following result gives a related perspective.

Theorem 4.27 *An invertible linear map $T \in \mathcal{L}(V, W)$ between inner product spaces is an isometry if and only if, for each $v \in V$ and $w \in W$,*

$$\langle Tv, w \rangle = \langle v, T^{-1}w \rangle . \tag{4.13}$$

Proof Suppose first that T is an isometry. Since T is surjective, $w = Tu$ for some $u \in V$, and so

$$\langle Tv, w \rangle = \langle Tv, Tu \rangle = \langle v, u \rangle = \langle v, T^{-1}w \rangle .$$

Now suppose that T is invertible and that equation (4.13) holds for each $v \in V$ and $w \in W$. Then T is surjective, and for each $v_1, v_2 \in V$,

$$\langle Tv_1, Tv_2 \rangle = \langle v_1, T^{-1}Tv_2 \rangle = \langle v_1, v_2 \rangle .$$

Thus T is an isometry. ▲

Example Let $R_\theta : \mathbb{R}^2 \to \mathbb{R}^2$ be the counterclockwise rotation of the plane by an angle of θ radians. Then $R_\theta^{-1} = R_{-\theta}$, and the theorem above says that, for any $v \in \mathbb{R}^2$,

$$\langle R_\theta v, w \rangle = \langle v, R_{-\theta} w \rangle ;$$

i.e., rotating v by θ and then measuring the angle with w is the same as measuring the angle of v with the rotation of w in the opposite direction. ▲

Theorem 4.28 *Suppose V and W are inner product spaces and $(e_1, \ldots, e_n)$ is an orthonormal basis of V. Then $T \in \mathcal{L}(V, W)$ is an isometry if and only if $(Te_1, \ldots, Te_n)$ is an orthonormal basis of W.*

Proof Suppose first that T is an isometry. Then for each $1 \le j, k \le n$,

$$\langle Te_j, Te_k \rangle = \langle e_j, e_k \rangle ,$$

and so $(Te_1, \ldots, Te_n)$ is orthonormal because $(e_1, \ldots, e_n)$ is. Furthermore, since T is an isometry, it is an isomorphism (Lemma 4.25), so by Theorem 3.15, $(Te_1, \ldots, Te_n)$ is a basis of W.

Now suppose that $(Te_1, \dots, Te_n)$ is an orthonormal basis for W. For each $v \in V$,

$$v = \sum_{j=1}^{n} \langle v, e_j \rangle e_j,$$

and so

$$Tv = \sum_{j=1}^{n} \langle v, e_j \rangle Te_j.$$

Since $(Te_1, \dots, Te_n)$ is orthonormal, this means that

$$\|Tv\|^2 = \sum_{j=1}^{n} \left|\langle v, e_j \rangle\right|^2 = \|v\|^2.$$

Furthermore, since $(Te_1, \dots, Te_n)$ spans W, T is surjective.

Quick Exercise #19. Fill in the details of the preceding claim.

We have thus shown that $\|Tv\| = \|v\|$ for all $v \in V$ and T is surjective; i.e., T is an isometry.

▲

Examples

1. Let $\theta \in [0, 2\pi)$, and let $R_\theta : \mathbb{R}^2 \to \mathbb{R}^2$ be the counterclockwise rotation by θ radians. Geometrically, it is obvious that R_θ is an isometry: rotating a vector does not change its length.
   To verify this explicitly, one can simply note that
   $$(R_\theta \mathbf{e}_1, R_\theta \mathbf{e}_2) = \left(\begin{bmatrix} \cos(\theta) \\ \sin(\theta) \end{bmatrix}, \begin{bmatrix} -\sin(\theta) \\ \cos(\theta) \end{bmatrix} \right)$$
   is an orthonormal basis of $\mathbb{R}^2$, so R_θ is an isometry by Theorem 4.28.
2. Let $T : \mathbb{R}^2 \to \mathbb{R}^2$ be given by reflection across the line $y = 2x$. Again, it is geometrically clear that T preserves lengths.
   Recall that we found in the example on page 189 that
   $$(T\mathbf{e}_1, T\mathbf{e}_2) = \left(\begin{bmatrix} -\frac{3}{5} \\ \frac{4}{5} \end{bmatrix}, \begin{bmatrix} \frac{4}{5} \\ \frac{3}{5} \end{bmatrix} \right),$$
   which is an orthonormal basis of $\mathbb{R}^2$.
3. Consider the inner product space $C[0, 1]$ with
   $$\langle f, g \rangle = \int_0^1 f(x)g(x)dx.$$

QA #19: If $w \in W$, then there are $c_i \in \mathbb{F}$ such that $w = \sum_i c_i Te_i = T\left(\sum_i c_i e_i\right)$.

Define the map

$$Tf(x) := f(1 - x).$$

Then T is a linear map, which is its own inverse and is therefore surjective. Since

$$\|Tf\| = \sqrt{\int_0^1 (Tf(x))^2 dx} = \sqrt{\int_0^1 (f(1-x))^2 dx} = \sqrt{\int_0^1 (f(u))^2 du} = \|f\|,$$

T is an isometry.

4. Let V be an inner product space with orthonormal basis $(e_1, \ldots, e_n)$. Let π be a **permutation** of $\{1, \ldots, n\}$; i.e., $\pi : \{1, \ldots, n\} \to \{1, \ldots, n\}$ is a bijective function. We can define a linear map $T_\pi : V \to V$ by letting $T_\pi(e_j) = e_{\pi(j)}$ and extending by linearity. Then T_π is clearly an isometry, since it takes an orthonormal basis to an orthonormal basis. In the case of, say, $\mathbb{R}^2$ and the standard basis, this is another example along the lines of the previous one, since the map which swaps $\mathbf{e}_1$ and $\mathbf{e}_2$ is exactly reflection in the line $y = x$. ▲

Recall that two finite-dimensional vector spaces are isomorphic if and only if they have the same dimension; the following result extends this to inner product spaces.

Corollary 4.29 *Let V and W be finite-dimensional inner product spaces. Then V and W are isometric if and only if* $\dim V = \dim W$.

Proof Since an isometry is an isomorphism, Theorem 3.23 implies that if V and W are isometric then $\dim V = \dim W$.

Now suppose that $\dim V = \dim W = n$. By Corollary 4.12, V and W have orthonormal bases $(e_1, \ldots, e_n)$ and $(f_1, \ldots, f_n)$, respectively. By Theorem 3.14, there is a linear map $T : V \to W$ such that $Te_j = f_j$ for each j, which by Theorem 4.28 is an isometry. ▲

Corollary 4.30 *Suppose that $\mathcal{B}_V$ and $\mathcal{B}_W$ are orthonormal bases of V and W, respectively. Then $T \in \mathcal{L}(V, W)$ is an isometry if and only if the columns of the matrix $[T]_{\mathcal{B}_V,\mathcal{B}_W}$ form an orthonormal basis of $\mathbb{F}^n$.*

Proof The columns of $[T]_{\mathcal{B}_V,\mathcal{B}_W}$ are the column vectors $[Te_k]_{\mathcal{B}_W}$, where $\mathcal{B}_V = (e_1, \ldots, e_n)$. According to Theorem 4.28, T is an isometry if and only if the vectors Te_k form an orthonormal basis of W, and by Theorem 4.10, this happens if and only if their coordinate representations $[Te_k]_{\mathcal{B}_W}$ form an orthonormal basis of $\mathbb{F}^n$. ▲

Orthogonal and Unitary Matrices

The following proposition gives a convenient characterization of orthonormality of the columns of a square matrix.

Proposition 4.31 *The columns of a matrix* $\mathrm{A} \in \mathrm{M}_n(\mathbb{F})$ *are orthonormal if and only if* $\mathrm{A}^*\mathrm{A} = \mathrm{I}_n$.

Proof Write $\mathrm{A} = \begin{bmatrix} | & & | \\ \mathbf{a}_1 & \cdots & \mathbf{a}_n \\ | & & | \end{bmatrix}$. By Lemma 2.14, the (j, k) entry of $\mathrm{A}^*\mathrm{A}$ is $\mathbf{a}_j{}^*\mathbf{a}_k = \langle \mathbf{a}_k, \mathbf{a}_j \rangle$. Thus $(\mathbf{a}_1, \ldots, \mathbf{a}_n)$ is orthonormal if and only if $\mathrm{A}^*\mathrm{A}$ has diagonal entries equal to 1, and all other entries 0. ▲

Square matrices with orthonormal columns are sufficiently important that they get a name:

Definition A matrix $\mathrm{A} \in \mathrm{M}_n(\mathbb{C})$ is called **unitary** if $\mathrm{A}^*\mathrm{A} = \mathrm{I}_n$. A matrix $\mathrm{A} \in \mathrm{M}_n(\mathbb{R})$ is called **orthogonal** if $\mathrm{A}^{\mathrm{T}}\mathrm{A} = \mathrm{I}_n$.

Note that if A has only real entries then $\mathrm{A}^* = \mathrm{A}^{\mathrm{T}}$. Therefore an orthogonal matrix is the same thing as a unitary matrix with real entries.

Thus Corollary 4.30 says that isometries between inner product spaces are represented (with respect to *orthonormal* bases) by unitary matrices in the complex case, and by orthogonal matrices in the real case. This is one of the many reasons that inner product spaces are especially nice to work with: it is trivial to compute the inverses of the corresponding structure-preserving maps, since if A is unitary, then $\mathrm{A}^{-1} = \mathrm{A}^*$.

Quick Exercise #20. Verify directly that the matrix $\begin{bmatrix} -\frac{3}{5} & \frac{4}{5} \\ \frac{4}{5} & \frac{3}{5} \end{bmatrix}$ which represents reflection in the line $y = 2x$ (see page 189) is orthogonal.

The discussion above tells us that certain maps of Euclidean space which are interesting geometrically (those that preserve lengths and angles) can be simply described algebraically. This simple algebraic description actually lets us go back and understand the geometry better. The situation is simplest in $\mathbb{R}^2$: consider an orthogonal matrix $\mathrm{U} \in \mathrm{M}_2(\mathbb{R})$. Now, the first column $\mathbf{u}_1$ of U has to be a unit vector, so we can write it as $\begin{bmatrix}\cos(\theta)\\ \sin(\theta)\end{bmatrix}$ for some $\theta \in [0, 2\pi)$. The second column $\mathbf{u}_2$ is also a unit vector which is perpendicular to $\mathbf{u}_1$, so there are only two possibilities:

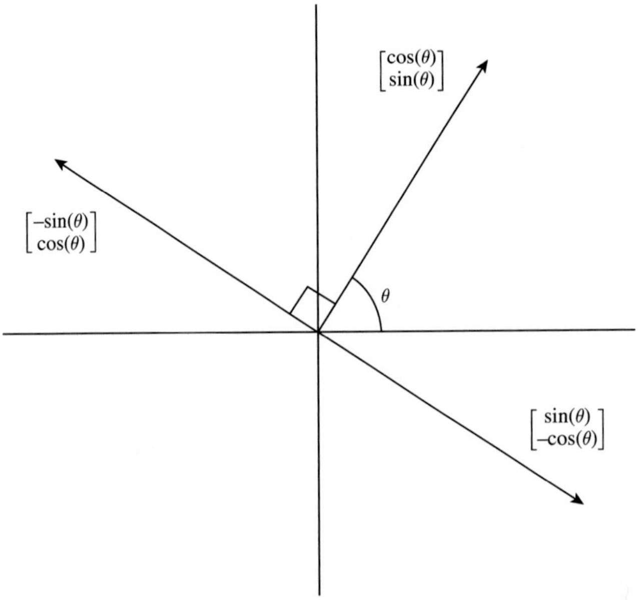

Figure 4.10 $\mathbf{u}_2 = \pm\begin{bmatrix}-\sin(\theta)\\ \cos(\theta)\end{bmatrix}$.

That is, we must have either

$$\mathrm{U} = \begin{bmatrix}\cos(\theta) & -\sin(\theta)\\ \sin(\theta) & \cos(\theta)\end{bmatrix} \quad \text{or} \quad \mathrm{U} = \begin{bmatrix}\cos(\theta) & \sin(\theta)\\ \sin(\theta) & -\cos(\theta)\end{bmatrix}.$$

The first case is a counterclockwise rotation by θ. In the second case, we can factor U as

$$\mathrm{U} = \begin{bmatrix}\cos(\theta) & -\sin(\theta)\\ \sin(\theta) & \cos(\theta)\end{bmatrix}\begin{bmatrix}1 & 0\\ 0 & -1\end{bmatrix},$$

and so U corresponds to a reflection across the x-axis, followed by a rotation by θ. This proves the following well-known fact:

> All length-preserving linear transformations of $\mathbb{R}^2$ are combinations of rotations and reflections.

Really, we have shown more, namely that every length-preserving map is either a single rotation or a reflection across the x-axis followed by a rotation. Exercise 4.5.5 asks you to show that these last maps are actually reflections of $\mathbb{R}^2$ (across a line determined by θ).

The QR Decomposition

The following result is known as the QR decomposition and is particularly useful in numerical linear algebra.

Theorem 4.32 *If* $\mathrm{A} \in \mathrm{M}_n(\mathbb{F})$ *is invertible, then there exist a matrix* $\mathrm{Q} \in \mathrm{M}_n(\mathbb{F})$ *with orthonormal columns and an upper triangular matrix* $\mathrm{R} \in \mathrm{M}_n(\mathbb{F})$ *such that*

$$\mathrm{A} = \mathrm{QR}.$$

Proof Since A is invertible, its list of columns $(\mathbf{a}_1, \ldots, \mathbf{a}_n)$ forms a basis of $\mathbb{F}^n$. Applying the Gram–Schmidt process to this basis produces an orthonormal basis $(\mathbf{q}_1, \ldots, \mathbf{q}_n)$ of $\mathbb{F}^n$ such that

$$\langle \mathbf{a}_1, \ldots, \mathbf{a}_j \rangle = \langle \mathbf{q}_1, \ldots, \mathbf{q}_j \rangle$$

for each j.

The matrix Q with columns $\mathbf{q}_1, \ldots, \mathbf{q}_n$ is unitary, and so

$$\mathrm{R} := \mathrm{Q}^{-1}\mathrm{A} = \mathrm{Q}^*\mathrm{A}$$

has (j, k) entry

$$r_{jk} = \mathbf{q}_j^* \mathbf{a}_k = \langle \mathbf{a}_k, \mathbf{q}_j \rangle .$$

Since $\mathbf{a}_k \in \langle \mathbf{q}_1, \ldots, \mathbf{q}_k \rangle$ and the $\mathbf{q}_j$ are orthonormal, if $j > k$ then $r_{jk} = 0$. That is, R is upper triangular. Finally, $\mathrm{A} = \mathrm{QR}$ by the definition of R. ▲

Quick Exercise #21. Suppose that the columns of $\mathrm{A} \in \mathrm{M}_n(\mathbb{C})$ are orthogonal and nonzero. What is the QR decomposition of A?

The QR decomposition is useful in solving systems of equations: given $\mathrm{A}\mathbf{x} = \mathbf{b}$, if $\mathrm{A} = \mathrm{QR}$ as above, then

$$\mathrm{A}\mathbf{x} = \mathbf{b} \quad \Longleftrightarrow \quad \mathrm{R}\mathbf{x} = \mathrm{Q}^*\mathbf{b}.$$

Since Q is known *a priori* to be unitary, its inverse $\mathrm{Q}^{-1} = \mathrm{Q}^*$ is trivial to compute and, since R is upper triangular, the system on the right is easy to solve via back-substitution. More significantly for real-life applications, the fact that Q^* is an isometry means it doesn't tend to magnify errors in the vector $\mathbf{b}$. Solving an upper triangular system by back-substitution is also less prone to round-off and other sources of error than going through the full Gaussian elimination algorithm.

QA #21: In this case the Gram–Schmidt process just normalizes the columns $\mathbf{a}_1, \ldots, \mathbf{a}_n$ of A, so the jth column of Q is $\frac{1}{\|\mathbf{a}_j\|}\mathbf{a}_j$, and $\mathrm{R} = \mathrm{diag}(\|\mathbf{a}_1\|, \ldots, \|\mathbf{a}_n\|)$.

Example To solve

$$\begin{bmatrix} 3 & 5 \\ 4 & 6 \end{bmatrix} \begin{bmatrix} x \\ y \end{bmatrix} = \begin{bmatrix} -1 \\ 1 \end{bmatrix}, \tag{4.14}$$

first find the QR decomposition of the coefficient matrix via the Gram–Schmidt process:

$$\mathbf{q}_1 = \frac{1}{\|\mathbf{a}_1\|}\mathbf{a}_1 = \frac{1}{5}\begin{bmatrix} 3 \\ 4 \end{bmatrix},$$

so

$$\tilde{\mathbf{q}}_2 = \mathbf{a}_2 - \langle \mathbf{a}_2, \mathbf{q}_1 \rangle \mathbf{q}_1 = \begin{bmatrix} 5 \\ 6 \end{bmatrix} - \frac{39}{25}\begin{bmatrix} 3 \\ 4 \end{bmatrix} = \frac{1}{25}\begin{bmatrix} 8 \\ -6 \end{bmatrix},$$

and

$$\mathbf{q}_2 = \frac{1}{\|\tilde{\mathbf{q}}_2\|}\tilde{\mathbf{q}}_2 = \frac{1}{5}\begin{bmatrix} 4 \\ -3 \end{bmatrix}.$$

Now,

$$\mathbf{Q}^{-1} = \mathbf{Q}^{\mathrm{T}} = \frac{1}{5}\begin{bmatrix} 3 & 4 \\ 4 & -3, \end{bmatrix}$$

and multiplying the system in equation (4.14) through by $\mathbf{Q}^{-1}$ yields the upper triangular system

$$\frac{1}{5}\begin{bmatrix} 25 & 39 \\ 0 & 2 \end{bmatrix}\begin{bmatrix} x \\ y \end{bmatrix} = \frac{1}{5}\begin{bmatrix} 1 \\ -7 \end{bmatrix}.$$

From this we can read off that $y = -\frac{7}{2}$, and then

$$x = \frac{1}{25}(1 - 39y) = \frac{11}{2}.$$

▲

KEY IDEAS

- Isometries are linear maps of normed spaces which preserve lengths. In the case of inner product spaces, they also preserve the inner product.
- A matrix $\mathbf{U} \in \mathrm{M}_n(\mathbb{R})$ is orthogonal if $\mathbf{U}\mathbf{U}^{\mathrm{T}} = \mathbf{I}_n$.
- A matrix $\mathbf{U} \in \mathrm{M}_n(\mathbb{C})$ is unitary if $\mathbf{U}\mathbf{U}^* = \mathbf{I}_n$.
- With respect to an orthonormal basis, the matrix of an isometry is orthogonal (in the real case) or unitary (in the complex case).
- The QR decomposition factors any invertible matrix $\mathbf{A} = \mathbf{QR}$, where $\mathbf{Q}$ is orthogonal/unitary and $\mathbf{R}$ is upper triangular. The matrix $\mathbf{Q}$ is gotten by performing the Gram–Schmidt process on the columns of $\mathbf{A}$; then $\mathbf{R} = \mathbf{Q}^*\mathbf{A}$.

EXERCISES

4.5.1 Let U be a finite-dimensional subspace of an inner product space V. Define the map

$$R_U := 2P_U - I,$$

where P_U is orthogonal projection onto U.

(a) Describe the action of R_U geometrically.

(b) Show that R_U is an isometry.

4.5.2 A **Householder matrix** is a matrix of the form $\mathrm{I}_n - 2\mathrm{x}\mathrm{x}^*$, where $\mathrm{x} \in \mathbb{C}^n$ is a unit vector. Show that every Householder matrix is unitary.

4.5.3 The $n \times n$ **DFT matrix** (for **discrete Fourier transform**) is the matrix $\mathrm{F} \in \mathrm{M}_n(\mathbb{C})$ with entries

$$f_{jk} = \frac{1}{\sqrt{n}}\omega^{jk},$$

where $\omega = e^{2\pi i/n} = \cos(2\pi/n) + i\sin(2\pi/n)$. Prove that F is unitary.
Hint: The necessary facts about complex exponentials are summarized in the appendix. The following identity is also useful here:

$$\sum_{k=0}^{n-1} z^k = \frac{1 - z^n}{1 - z}$$

for any $z \in \mathbb{C}$ with $z \neq 1$. (This can be proved by simply multiplying both sides by $1 - z$.)

4.5.4 Let $\mathrm{R} \in \mathcal{L}(\mathbb{R}^2)$ be the reflection across the x-axis.

(a) Find the matrix A of R with respect to the basis $\left(\begin{bmatrix}1\\0\end{bmatrix}, \begin{bmatrix}1\\1\end{bmatrix}\right)$ of $\mathbb{R}^2$.

(b) Show that A is *not* orthogonal.

(c) Explain why this fact does not contradict Corollary 4.30.

4.5.5 Show that the orthogonal matrix

$$\mathrm{U} = \begin{bmatrix}\cos(\theta) & \sin(\theta)\\ \sin(\theta) & -\cos(\theta)\end{bmatrix},$$

which we have seen describes reflection across the x-axis followed by rotation by θ, also describes reflection across the line making an angle $\frac{\theta}{2}$ with the x-axis.

4.5.6 Let $C_{2\pi}(\mathbb{R})$ denote the space of continuous 2π-periodic functions $f : \mathbb{R} \to \mathbb{C}$, equipped with the inner product

$$\langle f, g\rangle = \int_0^{2\pi} f(x)\overline{g(x)}\, dx,$$

and fix $t \in \mathbb{R}$. Define $T \in \mathcal{L}(C_{2\pi}(\mathbb{R}))$ by $(Tf)(x) = f(x + t)$. Show that T is an isometry.

4.5.7 Find a QR decomposition of each of the following matrices.

(a) $\begin{bmatrix} 1 & 2 \\ 3 & 4 \end{bmatrix}$ (b) $\begin{bmatrix} 0 & 1 & 1 \\ 1 & 0 & 1 \\ 1 & 1 & 0 \end{bmatrix}$ (c) $\begin{bmatrix} 1 & -1 & 2 \\ 0 & 2 & -1 \\ -1 & 1 & 0 \end{bmatrix}$

(d) $\begin{bmatrix} 1 & 1 & 1 & 1 \\ 1 & 1 & -1 & -1 \\ 1 & -1 & 1 & -1 \\ 1 & -1 & -1 & 1 \end{bmatrix}$

4.5.8 Find a QR decomposition of each of the following matrices.

(a) $\begin{bmatrix} 2 & 0 \\ -1 & 3 \end{bmatrix}$ (b) $\begin{bmatrix} 1 & -2 & 0 \\ 0 & 1 & 2 \\ 2 & 1 & -1 \end{bmatrix}$ (c) $\begin{bmatrix} 1 & 1 & 0 \\ 1 & 0 & 1 \\ 1 & -1 & -1 \end{bmatrix}$

(d) $\begin{bmatrix} 3 & 1 & 4 & 1 \\ 0 & 5 & 9 & 2 \\ 0 & 0 & 6 & 5 \\ 0 & 0 & 0 & 3 \end{bmatrix}$

4.5.9 Solve each of the following linear systems by the method of the example on page 284.

(a) $\begin{aligned} -3x + 4y &= -2 \\ 4x - 3y &= 5 \end{aligned}$ (b) $\begin{aligned} x - y + z &= 2 \\ y - z &= -1 \\ x + 2z &= 1 \end{aligned}$

4.5.10 Solve each of the following linear systems by the method of the example on page 284.

(a) $\begin{aligned} 3x - y &= 4 \\ x + 5y &= 9 \end{aligned}$ (b) $\begin{aligned} x - 2y + z &= 1 \\ y + z &= 0 \\ -x + y + z &= -3 \end{aligned}$

4.5.11 Use the method of the proof of Theorem 4.32 to find a QR decomposition of $\mathrm{A} = \begin{bmatrix} 2 & 3 \\ 0 & -4 \end{bmatrix}$. Conclude that the QR decomposition of a matrix is not unique.

4.5.12 (a) Show that if V is any normed space, $T \in \mathcal{L}(V)$ is an isometry, and λ is an eigenvalue of T, then $|\lambda| = 1$.

(b) Show that if $\mathrm{U} \in \mathrm{M}_n(\mathbb{C})$ is unitary and λ is an eigenvalue of U, then $|\lambda| = 1$.

4.5.13 Suppose that V and W are normed spaces with $\dim V = \dim W$, and that $T \in \mathcal{L}(V, W)$ satisfies $\|Tv\| = \|v\|$ for every $v \in V$. Show that T is an isometry.

4.5.14 Suppose that $U \in M_n(\mathbb{C})$ is unitary.

(a) Compute $\|U\|_{op}$.

(b) Compute $\|U\|_F$.

4.5.15 Suppose that $U \in M_m(\mathbb{C})$ and $V \in M_n(\mathbb{C})$ are unitary, and $A \in M_{m,n}(\mathbb{C})$. Prove that

$$\|UAV\|_{op} = \|A\|_{op} \quad \text{and} \quad \|UAV\|_F = \|A\|_F.$$

Norms with this property are called **unitarily invariant.**

4.5.16 Let π be a permutation of $\{1, \ldots, n\}$, i.e., a bijective function $\pi : \{1, \ldots, n\} \to \{1, \ldots, n\}$, and define $T_\pi \in \mathcal{L}(\mathbb{R}^n)$ by setting $T_\pi e_j = e_{\pi(j)}$.

(a) Show that T_π is an isometry when $\mathbb{R}^n$ is equipped with the ℓ^1 norm.

(b) Show that T_π is an isometry when $\mathbb{R}^n$ is equipped with the ℓ^∞ norm.

4.5.17 Let V denote $\mathbb{R}^2$ equipped with the ℓ^1 norm and let W denote $\mathbb{R}^2$ equipped with the ℓ^∞ norm. Show that the linear map in $\mathcal{L}(V, W)$ represented by $\begin{bmatrix} 1 & 1 \\ 1 & -1 \end{bmatrix}$ is an isometry.

Remark: On the other hand, if $n \geq 3$, then $\mathbb{R}^n$ with the ℓ^1 norm and $\mathbb{R}^n$ with the ℓ^∞ norm are *not* isometric.

4.5.18 Let $\kappa(A)$ be the condition number of $A \in M_n(\mathbb{C})$ (see Exercise 4.4.11). Show that if $\kappa(A) = 1$, then A is a scalar multiple of a unitary matrix.

4.5.19 Let $\kappa(A)$ be the condition number of $A \in M_n(\mathbb{C})$ (see Exercise 4.4.11). Show that if $A = QR$ is a QR decomposition of A, then $\kappa(A) = \kappa(R)$.

4.5.20 Show that if we find a QR decomposition of $A \in M_n(\mathbb{C})$ as in the proof of Theorem 4.32, then $r_{jj} > 0$ for every j.

4.5.21 Suppose that $A = QR$ is a QR decomposition of $A \in M_n(\mathbb{C})$. Show that for each j, $|r_{jj}| \leq \|a_j\|$.

4.5.22 Show that Theorem 4.32 is true even if A is singular.

Hint: First perform the Gram–Schmidt process on just the columns of A which do not lie in the span of the previous columns.

4.5.23 Complete the proof of Theorem 4.26 in the case $\mathbb{F} = \mathbb{C}$.

PERSPECTIVES:
Isometries

A linear map T between inner product spaces V and W is an isometry if any of the following hold.

- T is surjective and $\|Tv\| = \|v\|$ for all $v \in V$.
- T is surjective and $\langle Tv, Tw \rangle = \langle v, w \rangle$ for all $v, w \in V$.
- If $(e_1, \ldots, e_n)$ is an orthonormal basis of V, then $(Te_1, \ldots, Te_n)$ is an orthonormal basis of W.
- If A is the matrix of T with respect to orthonormal bases $\mathcal{B}_V$ and $\mathcal{B}_W$, then the columns of A are an orthonormal basis of $\mathbb{F}^n$.
- If A is the matrix of T with respect to orthonormal bases $\mathcal{B}_V$ and $\mathcal{B}_W$, then A is unitary (in the complex case) or orthogonal (in the real case).

5

Singular Value Decomposition and the Spectral Theorem

5.1 Singular Value Decomposition of Linear Maps

Singular Value Decomposition

Back in Chapter 3, Theorem 3.44 said that given finite-dimensional vector spaces V and W and $T \in \mathcal{L}(V, W)$, there are bases of V and W such that the matrix of T with respect to those bases has 1s on all or part of the diagonal and zeroes everywhere else. We commented at the time that this theorem is actually pretty pointless, because we don't really know anything about these bases other than they make the matrix of T look simple. The following result is much more powerful, because it gives us *orthonormal* bases in which the matrix of T is very simple.

Theorem 5.1 (Singular value decomposition (SVD)) *Let V and W be finite-dimensional inner product spaces and let $T \in \mathcal{L}(V, W)$ have rank r. Then there exist orthonormal bases $(e_1, \dots, e_n)$ of V and $(f_1, \dots, f_m)$ of W and numbers $\sigma_1 \geq \cdots \geq \sigma_r > 0$ such that $Te_j = \sigma_j f_j$ for $j = 1, \dots, r$ and $Te_j = 0$ for $j = r + 1, \dots, n$.*

The numbers $\sigma_1, \dots, \sigma_r$ are called **singular values**. The vectors $e_1, \dots, e_n$ are called **right singular vectors** and $f_1, \dots, f_m$ are called **left singular vectors**. (The left/right terminology will become clearer in the next section.)

Example Let $T : \mathbb{R}^3 \to \mathbb{R}^3$ be the map which first projects onto the x–y plane, then rotates space counterclockwise through an angle of θ radians about the z-axis. Then

$$T\mathbf{e}_1 = \begin{bmatrix} \cos\theta \\ \sin\theta \\ 0 \end{bmatrix}, \quad T\mathbf{e}_2 = \begin{bmatrix} -\sin\theta \\ \cos\theta \\ 0 \end{bmatrix}, \quad T\mathbf{e}_3 = \mathbf{0}.$$

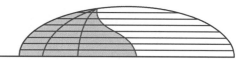

If we define $\mathbf{f}_1 = T\mathbf{e}_1$, $\mathbf{f}_2 = T\mathbf{e}_2$, and $\mathbf{f}_3 = \mathbf{e}_3$, then $(\mathbf{f}_1, \mathbf{f}_2, \mathbf{f}_3)$ is an orthonormal basis of $\mathbb{R}^3$. We therefore have a singular value decomposition of T with right singular vectors $(\mathbf{e}_1, \mathbf{e}_2, \mathbf{e}_3)$, left singular vectors $(\mathbf{f}_1, \mathbf{f}_2, \mathbf{f}_3)$, and singular values $\sigma_1 = 1$ and $\sigma_2 = 1$. ▲

In some ways, a singular value decomposition of a linear map can serve as a substitute for eigenvalues and eigenvectors. An advantage of the singular value decomposition is that, thanks to Theorem 5.1, it always exists.

Quick Exercise #1. Show that there is no basis of $\mathbb{R}^3$ consisting of eigenvectors of the map T in the example above.

Example Let $T \in \mathcal{L}(\mathbb{R}^3, \mathbb{R}^2)$ be the linear map with matrix (with respect to the standard bases) $\begin{bmatrix} 1 & 0 & 1 \\ 0 & 1 & 0 \end{bmatrix}$. In this case the images of the standard basis vectors are not all orthogonal, but if you are lucky (or have an exceptionally good eye for such things), you may notice that

$$T\begin{bmatrix} 1 \\ 0 \\ 1 \end{bmatrix} = \begin{bmatrix} 2 \\ 0 \end{bmatrix}, \quad T\begin{bmatrix} 0 \\ 1 \\ 0 \end{bmatrix} = \begin{bmatrix} 0 \\ 1 \end{bmatrix}, \quad T\begin{bmatrix} 1 \\ 0 \\ -1 \end{bmatrix} = \mathbf{0}.$$

The key insight is that, with these choices, both the input and the nonzero output vectors above are orthogonal. Renormalizing in order to have ortho*normal* singular vectors, we get that T has a singular value decomposition with right singular vectors $\left(\frac{1}{\sqrt{2}}\begin{bmatrix} 1 \\ 0 \\ 1 \end{bmatrix}, \begin{bmatrix} 0 \\ 1 \\ 0 \end{bmatrix}, \frac{1}{\sqrt{2}}\begin{bmatrix} 1 \\ 0 \\ -1 \end{bmatrix} \right)$, left singular vectors $\left(\begin{bmatrix} 1 \\ 0 \end{bmatrix}, \begin{bmatrix} 0 \\ 1 \end{bmatrix} \right)$, and singular values $\sigma_1 = \sqrt{2}$ and $\sigma_2 = 1$. ▲

If this choice of input/output vectors didn't jump out at you, don't worry; we will see how to find singular value decompositions systematically in Section 5.3.

The proof of Theorem 5.1 takes some work; we begin with a preliminary lemma.

Lemma 5.2 *Let V and W be finite-dimensional inner product spaces, and let $T \in \mathcal{L}(V, W)$. Let $e \in V$ be a unit vector such that*

$$\|T\|_{op} = \|Te\|. \tag{5.1}$$

Then for any vector $u \in V$ with $\langle u, e \rangle = 0$, we have

$$\langle Tu, Te \rangle = 0.$$

QA #1: The 0-eigenspace is one-dimensional, and $T\mathbf{x}$ never points in the same direction as $\mathbf{x}$.

Proof To simplify notation, write $\sigma = \|T\|_{op}$. Observe first that if $\sigma = 0$, then $T = 0$ and the result is trivial.

Next, recall that

$$\|Tv\| \leq \sigma \|v\|$$

for every $v \in V$. Thus for any $a \in \mathbb{F}$ and $u \in V$ with $\langle u, e\rangle = 0$,

$$\|T(e + au)\|^2 \leq \sigma^2 \|e + au\|^2 = \sigma^2 \left(\|e\|^2 + \|au\|^2\right) = \sigma^2 \left(1 + |a|^2 \|u\|^2\right), \quad (5.2)$$

where the first equality follows from Theorem 4.4. Expanding the left-hand side of inequality (5.2),

$$\begin{aligned} \|T(e + au)\|^2 &= \langle Te + aTu, Te + aTu\rangle \\ &= \|Te\|^2 + 2\,\mathrm{Re}\,(a\,\langle Tu, Te\rangle) + \|aTu\|^2 \\ &\geq \sigma^2 + 2\,\mathrm{Re}\,(a\,\langle Tu, Te\rangle)\,. \end{aligned} \quad (5.3)$$

Combining inequalities (5.2) and (5.3), we get that

$$2\,\mathrm{Re}\,(a\,\langle Tu, Te\rangle) \leq \sigma^2 \|u\|^2 |a|^2$$

for every scalar $a \in \mathbb{F}$. Letting $a = \langle Te, Tu\rangle\, \varepsilon$ for $\varepsilon > 0$, we get

$$2\,|\langle Tu, Te\rangle|^2\, \varepsilon \leq \sigma^2 \|u\|^2 |\langle Tu, Te\rangle|^2\, \varepsilon^2 \quad (5.4)$$

for every $\varepsilon > 0$. But if $\langle Tu, Te\rangle \neq 0$, then inequality (5.4) can only be true for

$$\varepsilon \geq \frac{2}{\sigma^2 \|u\|^2}.$$

Thus it must be that $\langle Tu, Te\rangle = 0$. ▲

Proof of Theorem 5.1 Let

$$S_1 = \{v \in V \mid \|v\| = 1\},$$

and let $e_1 \in S_1$ be a unit vector such that $\|Te_1\| = \|T\|_{op}$. (Recall that Lemma 4.23 guarantees that such a vector exists.) Write

$$\sigma_1 := \|T\|_{op} = \|Te_1\|\,.$$

If $\sigma_1 = 0$, then $Tv = 0$ for every $v \in V$, so the theorem holds trivially. If $\sigma_1 > 0$, then define

$$f_1 = \frac{1}{\sigma_1} Te_1 \in W,$$

so that $Te_1 = \sigma_1 f_1$ and $\|f_1\| = \frac{1}{\sigma_1}\|Te_1\| = 1$.

Now we repeat the argument in the subspace $V_2 := \langle e_1\rangle^\perp$. Let T_2 denote the restriction of T to V_2. Let

$$S_2 = \{v \in V_2 \mid \|v\| = 1\},$$

let $e_2 \in S_2$ be a unit vector such that

$$\|Te_2\| = \|T_2\|_{op} = \max_{v \in S_2} \|Tv\| ;$$

note that $\langle e_2, e_1 \rangle = 0$ by construction. Define

$$\sigma_2 := \|T_2\|_{op} = \|Te_2\| .$$

Quick Exercise #2. Show that $\sigma_2 \leq \sigma_1$.

If $\sigma_2 = 0$, then $Tv = 0$ for every $v \in V_2$. We can thus extend (e_1) and (f_1) to orthonormal bases any way we like and have $Te_j = 0$ for $j \geq 2$. If $\sigma_2 \neq 0$, then define a unit vector

$$f_2 = \frac{1}{\sigma_2} Te_2 \in W.$$

By Lemma 5.2, since $\langle e_1, e_2 \rangle = 0$ and $\|Te_1\| = \|T\|_{op}$, we have that $\langle Te_1, Te_2 \rangle = 0$ and so

$$\langle f_1, f_2 \rangle = \frac{1}{\sigma_1 \sigma_2} \langle Te_1, Te_2 \rangle = 0.$$

We continue in this way: after constructing orthonormal vectors $e_1, \ldots, e_k \in V$ such that $f_1 = Te_1, \ldots, f_k = Te_k$ are orthonormal, we define

$$S_{k+1} = \left\{ v \in \langle e_1, \ldots, e_k \rangle^{\perp} \;\middle|\; \|v\| = 1 \right\}$$

and pick e_{k+1} such that $\|Tv\|$ is maximized on S_{k+1} by $\sigma_{k+1} = \|Te_{k+1}\|$. If at any point $\sigma_{k+1} = 0$, then we are finished; we can fill out the list $(e_1, \ldots, e_k)$ to an orthonormal basis of V, and since all of the vectors e_j with $j \geq k+1$ lie in S_{k+1}, $Te_j = 0$ for $j \geq k+1$. Since V is finite-dimensional, at some point the process must terminate, either because we have $Te_{k+1} = 0$ or because we've constructed a full basis of V. At this point, the list $(f_1, \ldots, f_k)$ can be extended to an orthonormal basis of W (if necessary), and this completes the construction of the singular values and vectors.

It remains to show that if k is the largest index for which $\sigma_k > 0$, then $k = r = \operatorname{rank} T$. We will do this by showing that $(f_1, \ldots, f_k)$ is an orthonormal basis of range T.

Since the f_j were constructed to be orthonormal (in particular, linearly independent), they form a basis for their span. Now, for each $j \in \{1, \ldots, k\}$, $f_j = \frac{1}{\sigma_j} Te_j$, and so

$$\langle f_1, \ldots, f_k \rangle \subseteq \operatorname{range} T.$$

QA #2: $S_2 \subseteq S_1$, so $\max_{v \in S_2} \|Tv\| \leq \max_{v \in S_1} \|Tv\|$.

Conversely, if $w \in \operatorname{range} T$, then $w = Tv$ for some $v \in V$. Expanding v with respect to the orthonormal basis $(e_1, \ldots, e_n)$,

$$w = Tv = T\left(\sum_{j=1}^{n} \langle v, e_j \rangle e_j\right) = \sum_{j=1}^{n} \langle v, e_j \rangle Te_j = \sum_{j=1}^{k} \langle v, e_j \rangle \sigma_j f_j,$$

and so $\operatorname{range} T \subseteq \langle f_1, \ldots, f_k \rangle$. ▲

Uniqueness of Singular Values

Singular value decompositions are not unique. As a trivial example, consider the identity map $I : V \to V$. *Any* orthonormal basis $(e_1, \ldots, e_n)$ has the property that $I(e_j) = e_j$ for all j. Notice, however, that even though we are free to take any basis we like, the values σ_j do not change; in this example, $\sigma_j = 1$ for each j. This is true in general: while the singular *vectors* in Theorem 5.1 are not unique, the singular *values* are.

Theorem 5.3 *Let V and W be finite-dimensional inner product spaces, and let $T \in \mathcal{L}(V, W)$ with $\operatorname{rank}(T) = r$. Suppose there are orthonormal bases $(e_1, \ldots, e_n)$ and $(\tilde{e}_1, \ldots, \tilde{e}_n)$ of V and $(f_1, \ldots, f_m)$ and $(\tilde{f}_1, \ldots, \tilde{f}_m)$ of W, and real scalars $\sigma_1 \geq \cdots \geq \sigma_r > 0$ and $\tilde{\sigma}_1 \geq \cdots \geq \tilde{\sigma}_r > 0$ such that*

$$Te_j = \begin{cases} \sigma_j f_j & \text{if } 1 \leq j \leq r, \\ 0 & \text{otherwise;} \end{cases} \qquad T\tilde{e}_j = \begin{cases} \tilde{\sigma}_j \tilde{f}_j & \text{if } 1 \leq j \leq r, \\ 0 & \text{otherwise.} \end{cases}$$

Then $\sigma_j = \tilde{\sigma}_j$ for all $j \in \{1, \ldots, r\}$.

Proof We first show that $\sigma_1 = \tilde{\sigma}_1 = \|T\|_{op}$, and that

$$\#\left\{j \mid \sigma_j = \sigma_1\right\} = \#\left\{j \mid \tilde{\sigma}_j = \tilde{\sigma}_1\right\}.$$

Let

$$U := \left\{v \in V \;\middle|\; \|Tv\| = \|T\|_{op} \|v\|\right\}.$$

For $v \in V$, write $v = \sum_{j=1}^{n} \langle v, e_j \rangle e_j$. Then

$$Tv = T\left(\sum_{j=1}^{n} \langle v, e_j \rangle e_j\right) = \sum_{j=1}^{n} \langle v, e_j \rangle Te_j = \sum_{j=1}^{r} \langle v, e_j \rangle \sigma_j f_j,$$

and so since the f_j are orthonormal and $\sigma_1 \geq \cdots \geq \sigma_r > 0$,

$$\|Tv\|^2 = \sum_{j=1}^{r} \left|\langle v, e_j \rangle\right|^2 \sigma_j^2 \leq \sigma_1^2 \sum_{j=1}^{r} \left|\langle v, e_j \rangle\right|^2 \leq \sigma_1^2 \sum_{j=1}^{n} \left|\langle v, e_j \rangle\right|^2 = \sigma_1^2 \|v\|^2, \quad (5.5)$$

with equality throughout if $v = e_1$. We thus have that

$$\|T\|_{op} = \max_{\|u\|=1} \|Tu\| = \sigma_1,$$

and in the same way, $\|T\|_{op} = \tilde{\sigma}_1$.

Moreover, the fact that $\sigma_1 = \|T\|_{op}$ means that, for $v \in U$, we must have equality throughout equation (5.5). For equality to hold in the first inequality, it must be that whenever $j \in \{1, \ldots, r\}$ with $\langle v, e_j \rangle \neq 0$, then $\sigma_j = \sigma_1$. For the second inequality to be an equality, we must have $v \in \langle e_1, \ldots, e_r \rangle$. It follows that

$$U \subseteq \langle e_1, \ldots, e_{r_1} \rangle,$$

where r_1 is the largest index such that $\sigma_{r_1} = \sigma_1$.

On the other hand, if $v \in \langle e_1, \ldots, e_{r_1} \rangle$, then writing $v = \sum_{j=1}^{r_1} \langle v, e_j \rangle e_j$ leads to

$$\|Tv\|^2 = \left\| \sum_{j=1}^{r_1} \langle v, e_j \rangle \sigma_j f_j \right\|^2 = \sigma_1^2 \sum_{j=1}^{r_1} |\langle v, e_j \rangle|^2 = \sigma_1^2 \|v\|^2,$$

and so

$$\langle e_1, \ldots, e_{r_1} \rangle \subseteq U.$$

That is, we have shown that $\langle e_1, \ldots, e_{r_1} \rangle = U$; the same argument shows that $\langle \tilde{e}_1, \ldots, \tilde{e}_{\tilde{r}_1} \rangle = U$, where $\tilde{r}_1$ is the largest index such that $\tilde{\sigma}_{\tilde{r}_1}| = \tilde{\sigma}_1 = \|T\|_{op}$. We thus have that

$$\# \{ j \mid \sigma_j = \sigma_1 \} = \# \{ j \mid \tilde{\sigma}_j = \tilde{\sigma}_1 \} = \dim(U).$$

To continue, we apply the same argument as above to the restriction $T_2 = T|_{U^\perp}$. If

$$U_2 := \left\{ v \in U^\perp \;\middle|\; \|T_2 v\| = \|T_2\|_{op} \|v\| \right\},$$

then for $k_2 = \dim(U_2)$, it follows as above that

$$\sigma_{r_1+1} = \cdots = \sigma_{r_1+k_2} = \tilde{\sigma}_{r_1+1} = \cdots = \tilde{\sigma}_{r_1+k_2} = \|T_2\|_{op},$$

and that both $\sigma_{r_1+k_2+1} < \sigma_{r_1+1}$ and $\tilde{\sigma}_{r_1+k_2+1} < \sigma_{r_1+1}$. Continuing in this fashion, V is decomposed into subspaces depending only on T, and the (common) values of the σ_j and $\tilde{\sigma}_j$ are the operator norms of restrictions of T to those subspaces, with the number of times they occur equal to the dimensions of the corresponding subspaces. ▲

Now that we have verified uniqueness, we can legitimately define *the* singular values of a map.

Definition Let V and W be finite-dimensional inner product spaces, and let $T \in \mathcal{L}(V, W)$. The **singular values** of T are the numbers $\sigma_1 \geq \cdots \geq \sigma_p \geq 0$, where $p = \min\{m, n\}$, $\sigma_1, \ldots, \sigma_r$ are given in the statement of Theorem 5.1, and $\sigma_j = 0$ for $r + 1 \leq j \leq p$.

For instance, in the example on page 289, we saw a map $T \in \mathcal{L}(\mathbb{R}^3)$ with a singular value decomposition in which $\sigma_1 = 1$ and $\sigma_2 = 1$. We therefore say that its singular values are 1, 1, and 0.

Quick Exercise #3. Suppose that $T \in \mathcal{L}(V, W)$ is an isometry. What are the singular values of T?

KEY IDEAS

- SVD: For any $T \in \mathcal{L}(V, W)$, there are orthonormal bases $(e_1, \ldots, e_n)$ and $(f_1, \ldots, f_m)$ of V and W, and real numbers $\sigma_1 \geq \cdots \geq \sigma_r > 0$ (with $r = \operatorname{rank} T$), such that

$$Te_j = \begin{cases} \sigma_j f_j & \text{if } 1 \leq j \leq r, \\ 0 & \text{otherwise.} \end{cases}$$

- The singular values $\sigma_1 \geq \cdots \geq \sigma_r > 0 = \sigma_{r+1} = \cdots = \sigma_p$ are unique, but the bases are not. ($r = \operatorname{rank} T$ and $p = \min\{\dim(V), \dim(W)\} \geq r$.)
- The largest singular value is the operator norm of the map.

EXERCISES

5.1.1 Find a singular value decomposition for the linear map $T \in \mathcal{L}(\mathbb{R}^4, \mathbb{R}^2)$ given by $T\begin{bmatrix} x \\ y \\ z \\ w \end{bmatrix} = \begin{bmatrix} x + y \\ z + w \end{bmatrix}$.

5.1.2 Find a singular value decomposition for the linear map $T \in \mathcal{L}(\mathbb{R}^2, \mathbb{R}^3)$ with matrix (with respect to the standard bases) $\begin{bmatrix} 1 & 0 \\ 0 & 1 \\ -1 & 1 \end{bmatrix}$.

5.1.3 Let $T \in \mathcal{L}(\mathbb{R}^2)$ be the linear map which first rotates counterclockwise through an angle of $\pi/3$ radians, then projects orthogonally onto the line $y = x$. Find a singular value decomposition for T.

5.1.4 Let $T \in \mathcal{L}(\mathbb{R}^3)$ be the linear map which first rotates space counterclockwise through an angle of $\pi/2$ radians about the z-axis, then projects

QA #3: Since $\|Tv\| = \|v\|$ for every v, the proof of Theorem 5.1 shows that $\sigma_j = 1$ for each $j = 1, \ldots, n$.

orthogonally onto the y-z plane. Find a singular value decomposition for T.

5.1.5 Consider the space $C^\infty([0, 2\pi])$ of infinitely differentiable functions $f : [0, 2\pi] \to \mathbb{R}$ with the inner product

$$\langle f, g \rangle = \int_0^{2\pi} f(x)g(x)\, dx.$$

Fix $n \in \mathbb{N}$, and let $V \subseteq C^\infty([0, 2\pi])$ be the subspace spanned by the functions

$$1, \sin(x), \sin(2x), \ldots, \sin(nx), \cos(x), \cos(2x), \ldots, \cos(nx).$$

Find a singular value decomposition for the derivative operator $D \in \mathcal{L}(V)$.

5.1.6 Suppose that $P \in \mathcal{L}(V)$ is an orthogonal projection on a finite-dimensional inner product space V. Show that:

(a) the singular values of P are all 1 or 0,

(b) P has a singular value decomposition in which the left singular vectors are the same as the right singular vectors.

5.1.7 Suppose that $T \in \mathcal{L}(V)$ has singular values $\sigma_1 \geq \cdots \geq \sigma_n \geq 0$. Show that if λ is an eigenvalue of T, then $\sigma_n \leq |\lambda| \leq \sigma_1$.

5.1.8 Show that $T \in \mathcal{L}(V, W)$ is invertible if and only if $\dim V = \dim W$ and all the singular values of T are nonzero.

5.1.9 Suppose that $T \in \mathcal{L}(V, W)$ is invertible. Given a singular value decomposition of T, find a singular value decomposition of T^{-1}.

5.1.10 Suppose that $T \in \mathcal{L}(V, W)$ is invertible with singular values $\sigma_1 \geq \cdots \geq \sigma_n$. Show that

$$\sigma_n = \min_{\|v\|=1} \|Tv\| = \left\|T^{-1}\right\|_{op}^{-1}.$$

5.1.11 Show that if V is a finite-dimensional inner product space and all the singular values of $T \in \mathcal{L}(V)$ are 1, then T is an isometry.

5.1.12 Let V and W be finite-dimensional inner product spaces and suppose that $T \in \mathcal{L}(V, W)$ has singular values $\sigma_1 \geq \cdots \geq \sigma_p$. Show that, given any $s \in \mathbb{R}$ with $\sigma_p \leq s \leq \sigma_1$, there is a unit vector $v \in V$ with $\|Tv\| = s$.

5.1.13 Let V and W be finite-dimensional inner product spaces and $T \in \mathcal{L}(V, W)$. Show that there exist:

- subspaces $V_0, V_1 \ldots, V_k$ of V such that V is the orthogonal direct sum $V_0 \oplus V_1 \oplus \cdots \oplus V_k$,
- subspaces $W_0, W_1, \ldots, W_k$ of W such that W is the orthogonal direct sum $W_0 \oplus W_1 \oplus \cdots \oplus W_k$,

- isometries $T_j \in \mathcal{L}(V_j, W_j)$ for $j = 1, \ldots, k$,
- distinct real scalars $\tau_1, \ldots, \tau_k > 0$,

such that

$$T = \sum_{j=1}^{k} \tau_j T_j P_{V_j}.$$

5.1.14 Suppose that $T \in \mathcal{L}(V, W)$ has a singular value decomposition with right singular vectors $(e_1, \ldots, e_n)$ and singular values $\sigma_1 \geq \cdots \geq \sigma_p$.

(a) For $j = 1, \ldots, p$, let $U_j = \langle e_j, \ldots, e_n \rangle$. Show that, for every $v \in U_j$, we have $\|Tv\| \leq \sigma_j \|v\|$.

(b) For $j = 1, \ldots, p$, let $V_j = \langle e_1, \ldots, e_j \rangle$. Show that, for every $v \in V_j$, we have $\|Tv\| \geq \sigma_j \|v\|$.

(c) Show that if U is a subspace of V with $\dim U = n - j + 1$, then there exists a nonzero $v \in U$ such that $\|Tv\| \geq \sigma_j \|v\|$.
Hint: Use Lemma 3.22.

(d) Use the above results to show that

$$\sigma_j = \min_{\dim U = n-j+1} \max_{v \in U,\ \|v\|=1} \|Tv\| .$$

Remark: This gives another proof of Theorem 5.3.

5.1.15 In Theorem 5.3, we assumed the number of nonzero singular values in each singular value decomposition is equal to the rank of T. Show that this assumption is unnecessary: Suppose that there are orthonormal bases $(e_1, \ldots, e_n)$ of V and $(f_1, \ldots, f_m)$ of W, and scalars $\sigma_1 \geq \cdots \geq \sigma_k > 0$ such that $Te_j = \sigma_j f_j$ for $j = 1, \ldots, k$, and $Te_j = 0$ for $j > k$. Prove that $\operatorname{rank} T = k$.

5.2 Singular Value Decomposition of Matrices

Matrix Version of SVD

In this section we explore the consequences of Theorem 5.1 for matrices. We begin by re-expressing the theorem in matrix language.

Theorem 5.4 (Singular value decomposition (SVD)) *Let* $\mathrm{A} \in \mathrm{M}_{m,n}(\mathbb{F})$ *have rank* r. *Then there exist matrices* $\mathrm{U} \in \mathrm{M}_m(\mathbb{F})$ *and* $\mathrm{V} \in \mathrm{M}_n(\mathbb{F})$ *with* U *and* V *unitary (if* $\mathbb{F} = \mathbb{C}$*) or orthogonal (if* $\mathbb{F} = \mathbb{R}$*) and unique real numbers* $\sigma_1 \geq \cdots \geq \sigma_r > 0$ *such that*

$$\mathrm{A} = \mathrm{U}\Sigma\mathrm{V}^*,$$

where $\Sigma \in \mathrm{M}_{m,n}(\mathbb{R})$ *has* (j, j) *entry* σ_j *for* $1 \leq j \leq r$*, and all other entries* 0.

Proof Let $T \in \mathcal{L}(\mathbb{F}^n, \mathbb{F}^m)$ be the linear map given by $T\mathbf{v} = \mathbf{A}\mathbf{v}$. By Theorem 5.1, there are orthonormal bases $\mathcal{B} = (\mathbf{u}_1, \ldots, \mathbf{u}_m)$ of $\mathbb{F}^m$ and $\mathcal{B}' = (\mathbf{v}_1, \ldots, \mathbf{v}_n)$ of $\mathbb{F}^n$ such that

$$[T]_{\mathcal{B}',\mathcal{B}} = \Sigma,$$

with $\sigma_1, \ldots, \sigma_r$ the singular values of T (which are unique by Theorem 5.3). By the change of basis formula,

$$\Sigma = [T]_{\mathcal{B}',\mathcal{B}} = [I]_{\mathcal{E},\mathcal{B}} [T]_{\mathcal{E}} [I]_{\mathcal{B}',\mathcal{E}} .$$

Now $[T]_{\mathcal{E}} = \mathbf{A}$ by definition, and

$$\mathbf{V} := [I]_{\mathcal{B}',\mathcal{E}} = \begin{bmatrix} | & & | \\ \mathbf{v}_1 & \cdots & \mathbf{v}_n \\ | & & | \end{bmatrix}$$

is unitary (or orthogonal, if $\mathbb{F} = \mathbb{R}$) since its columns are an orthonormal basis of $\mathbb{F}^n$. Similarly,

$$\mathbf{U} := \begin{bmatrix} | & & | \\ \mathbf{u}_1 & \cdots & \mathbf{u}_m \\ | & & | \end{bmatrix}$$

is unitary (orthogonal), and

$$[I]_{\mathcal{E},\mathcal{B}} = [I]_{\mathcal{B},\mathcal{E}}^{-1} = \mathbf{U}^{-1}.$$

Therefore

$$\Sigma = \mathbf{U}^{-1}\mathbf{A}\mathbf{V},$$

and so

$$\mathbf{A} = \mathbf{U}\Sigma\mathbf{V}^{-1} = \mathbf{U}\Sigma\mathbf{V}^*.$$

▲

Example Consider the matrix

$$\mathbf{A} = \begin{bmatrix} -1 & 1 & 3 & 5 & 6 \\ 3 & -1 & 3 & -1 & 6 \\ -1 & 3 & -3 & 1 & -6 \end{bmatrix}.$$

Then

$$(\mathbf{v}_1, \mathbf{v}_2, \mathbf{v}_3, \mathbf{v}_4, \mathbf{v}_5) = \left(\frac{1}{4\sqrt{3}} \begin{bmatrix} 1 \\ -1 \\ 3 \\ 1 \\ 6 \end{bmatrix}, \frac{1}{\sqrt{6}} \begin{bmatrix} -1 \\ 1 \\ 0 \\ 2 \\ 0 \end{bmatrix}, \frac{1}{\sqrt{2}} \begin{bmatrix} 1 \\ 1 \\ 0 \\ 0 \\ 0 \end{bmatrix}, \frac{1}{4} \begin{bmatrix} 1 \\ -1 \\ 3 \\ 1 \\ -2 \end{bmatrix}, \frac{1}{2} \begin{bmatrix} 1 \\ -1 \\ -1 \\ -1 \\ 0 \end{bmatrix} \right)$$

is an orthonormal basis of $\mathbb{R}^5$,

$$(\mathbf{u}_1, \mathbf{u}_2, \mathbf{u}_3) = \left(\frac{1}{\sqrt{3}} \begin{bmatrix} 1 \\ 1 \\ -1 \end{bmatrix}, \frac{1}{\sqrt{6}} \begin{bmatrix} 2 \\ -1 \\ 1 \end{bmatrix}, \frac{1}{\sqrt{2}} \begin{bmatrix} 0 \\ 1 \\ 1 \end{bmatrix} \right)$$

is an orthonormal basis of $\mathbb{R}^3$, and

$$\mathrm{A}\mathbf{v}_1 = 12\mathbf{u}_1, \quad \mathrm{A}\mathbf{v}_2 = 6\mathbf{u}_2, \quad \mathrm{A}\mathbf{v}_3 = 2\mathbf{u}_3, \quad \mathrm{A}\mathbf{v}_4 = \mathbf{0}, \quad \text{and} \quad \mathrm{A}\mathbf{v}_5 = \mathbf{0}.$$

So, as in the proof above, we get the factorization

$$\mathrm{A} = \begin{bmatrix} \frac{1}{\sqrt{3}} & \frac{2}{\sqrt{6}} & 0 \\ \frac{1}{\sqrt{3}} & \frac{-1}{\sqrt{6}} & \frac{1}{\sqrt{2}} \\ \frac{-1}{\sqrt{3}} & \frac{1}{\sqrt{6}} & \frac{1}{\sqrt{2}} \end{bmatrix} \begin{bmatrix} 12 & 0 & 0 & 0 & 0 \\ 0 & 6 & 0 & 0 & 0 \\ 0 & 0 & 2 & 0 & 0 \end{bmatrix} \begin{bmatrix} \frac{1}{4\sqrt{3}} & \frac{-1}{4\sqrt{3}} & \frac{3}{4\sqrt{3}} & \frac{1}{4\sqrt{3}} & \frac{6}{4\sqrt{3}} \\ \frac{-1}{\sqrt{6}} & \frac{1}{\sqrt{6}} & 0 & \frac{2}{\sqrt{6}} & 0 \\ \frac{1}{\sqrt{2}} & \frac{1}{\sqrt{2}} & 0 & 0 & 0 \\ \frac{1}{4} & \frac{-1}{4} & \frac{3}{4} & \frac{1}{4} & \frac{-2}{4} \\ \frac{1}{2} & \frac{-1}{2} & \frac{-1}{2} & \frac{-1}{2} & 0 \end{bmatrix}. \quad \blacktriangle$$

Definition Let $\mathrm{A} \in \mathrm{M}_{m,n}(\mathbb{C})$. The **singular values** of A are the entries $[\Sigma]_{jj}$ of the matrix Σ in Theorem 5.4; that is, if

$$\mathrm{A} = \mathrm{U}\Sigma\mathrm{V}^*$$

and $p = \min\{m, n\}$, then the singular values of A are σ_j for $1 \le j \le r$ and $\sigma_j := 0$ for $r + 1 \le j \le p$.

The columns of U are called the **left singular vectors** of A and the columns of V are the **right singular vectors** of A.

Quick Exercise #4. Suppose that $\mathrm{A} \in \mathrm{M}_{m,n}(\mathbb{C})$, and that $\mathrm{W} \in \mathrm{M}_m(\mathbb{C})$ and $\mathrm{Y} \in \mathrm{M}_n(\mathbb{C})$ are both unitary. Show that WAY has the same singular values as A.

In the example above, we can just read off the singular values

$$\sigma_1 = 12, \qquad \sigma_2 = 6, \qquad \sigma_3 = 2,$$

because we were handed the singular vectors $(\mathbf{v}_1, \mathbf{v}_2, \mathbf{v}_3, \mathbf{v}_4, \mathbf{v}_5)$ and $(\mathbf{u}_1, \mathbf{u}_2, \mathbf{u}_3)$. The following proposition gives a way to actually find the singular values of A when all we have is A itself.

Proposition 5.5 *Let $\mathrm{A} \in \mathrm{M}_{m,n}(\mathbb{F})$ have singular values $\sigma_1 \ge \cdots \ge \sigma_p$, where $p = \min\{m, n\}$. Then for each $j = 1, \ldots, p$, σ_j^2 is an eigenvalue of $\mathrm{A}^*\mathrm{A}$ and of AA^*. Moreover, whichever of $\mathrm{A}^*\mathrm{A}$ and AA^* is a $p \times p$ matrix has exactly*

QA #4: If $\mathrm{A} = \mathrm{U}\Sigma\mathrm{V}^*$ is the singular value decomposition of A, then $\mathrm{WAY} = \mathrm{WU}\Sigma\mathrm{V}^*\mathrm{Y} = (\mathrm{WU})\Sigma(\mathrm{Y}^*\mathrm{V})^*$. Since WU and $\mathrm{Y}^*\mathrm{V}$ are both unitary, this is also a singular value decomposition, with the same Σ.

$\sigma_1^2, \ldots, \sigma_p^2$ as eigenvalues; if the other is larger, it has eigenvalues $\sigma_1^2, \ldots, \sigma_p^2$ together with zero.

That is,

> The singular values of A are the square roots of the eigenvalues of $\mathrm{A^*A}$ and of $\mathrm{AA^*}$.

Proof Let $\mathrm{A} = \mathrm{U}\Sigma\mathrm{V}^*$ be the singular value decomposition of A. Then

$$\mathrm{A^*A} = \mathrm{V}\Sigma^*\mathrm{U}^*\mathrm{U}\Sigma\mathrm{V}^* = \mathrm{V}\Sigma^*\Sigma\mathrm{V}^*.$$

Since V is unitary, $\mathrm{V}^* = \mathrm{V}^{-1}$, so $\mathrm{A^*A}$ is similar to $\Sigma^*\Sigma$, and therefore has the same eigenvalues.

Now $\Sigma^*\Sigma = \mathrm{diag}(\sigma_1^2, \ldots, \sigma_p^2)$ if $p = n$ (i.e., if $n \leq m$), and if $p = m$ (i.e., if $n > m$), then $\Sigma^*\Sigma = \mathrm{diag}(\sigma_1^2, \ldots, \sigma_p^2, 0, \ldots, 0)$. Since this matrix is diagonal, its eigenvalues are its diagonal entries.

The argument for $\mathrm{AA^*}$ is analogous. ▲

Example If $\mathrm{A} = \frac{1}{5}\begin{bmatrix} 6 & -2 \\ 2 & -9 \end{bmatrix}$, then $\mathrm{AA^*} = \frac{1}{5}\begin{bmatrix} 8 & 6 \\ 6 & 17 \end{bmatrix}$. To simplify computations, we will first find the eigenvalues of $\mathrm{B} = 5\mathrm{AA^*}$; λ is an eigenvalue of B if and only if

$$\mathrm{B} - \lambda \mathrm{I}_2 = \begin{bmatrix} 8-\lambda & 6 \\ 6 & 17-\lambda \end{bmatrix}$$

has rank smaller than 2. Row-reducing $\mathrm{B} - \lambda\mathrm{I}_2$ leads to

$$\begin{bmatrix} 8-\lambda & 6 \\ 6 & 17-\lambda \end{bmatrix} \xrightarrow{\text{R3}} \begin{bmatrix} 6 & 17-\lambda \\ 8-\lambda & 6 \end{bmatrix} \xrightarrow{\text{R1}} \begin{bmatrix} 6 & 17-\lambda \\ 0 & -\frac{1}{6}(\lambda^2 - 25\lambda + 100) \end{bmatrix},$$

which has rank 1 when $\lambda = 5, 20$. So the eigenvalues of B are 5 and 20, which means that the eigenvalues of $\mathrm{AA^*}$ are 4 and 1, and so the singular values of A are 2 and 1. ▲

We have shown how to find the singular values of an arbitrary matrix, but at this point you may be wondering how to go about finding a full SVD; how do we get our hands on the singular vectors? The full algorithm is given in the next section (Algorithm 5.17); for now, we discuss a few more implications of the existence of SVDs.

SVD and Geometry

The decomposition $A = U\Sigma V^*$ can be interpreted geometrically as a sequence of transformations of $\mathbb{F}^n$:

$$\mathbf{x} \longmapsto V^*\mathbf{x} \longmapsto \Sigma(V^*\mathbf{x}) \longmapsto U(\Sigma(V^*\mathbf{x})).$$

Since V^* is unitary, V^* acts as an isometry and so $V^*\mathbf{x}$ has the same length as $\mathbf{x}$. In fact, we saw that in $\mathbb{R}^2$, an isometry like V^* must act as either a rotation or a reflection; something like this is true in higher dimensions as well. From a more algebraic perspective, we have that

$$V^* = [I]_{\mathcal{E},\mathcal{B}'},$$

where $\mathcal{B}' = (\mathbf{v}_1, \ldots, \mathbf{v}_n)$ and $\mathcal{E} = (\mathbf{e}_1, \ldots, \mathbf{e}_n)$ is the standard basis. In other words, V^* is the isometry of $\mathbb{F}^n$ which sends the orthonormal basis $\mathcal{B}'$ to $\mathcal{E}$.

Now, as a map in $\mathcal{L}(\mathbb{F}^n, \mathbb{F}^m)$, Σ stretches the standard basis vectors according to the singular values: if $j \in \{1, \ldots, r\}$, then $\mathbf{e}_j$ is sent to $\sigma_j \mathbf{e}_j$; if $j > r$, then $\mathbf{e}_j$ is collapsed to $\mathbf{0}$. A nice way to visualize this different stretching in different directions is to notice that when the standard basis vectors are stretched in this way, the unit sphere $\{\mathbf{x} \in \mathbb{F}^n \mid \|\mathbf{x}\| = 1\}$ gets mapped to an ellipsoid, with the lengths of the semi-axes determined by the singular values.

Finally, $U = [I]_{\mathcal{B},\mathcal{E}}$ is the isometry of $\mathbb{F}^m$ that sends the standard basis $\mathcal{E} = (\mathbf{e}_1, \ldots, \mathbf{e}_m)$ to the orthonormal basis $\mathcal{B} = (\mathbf{u}_1, \ldots, \mathbf{u}_m)$; it's some combination of rotations and reflections.

Here is a sequence of pictures showing the geometry of $A = U\Sigma V^*$ in action:

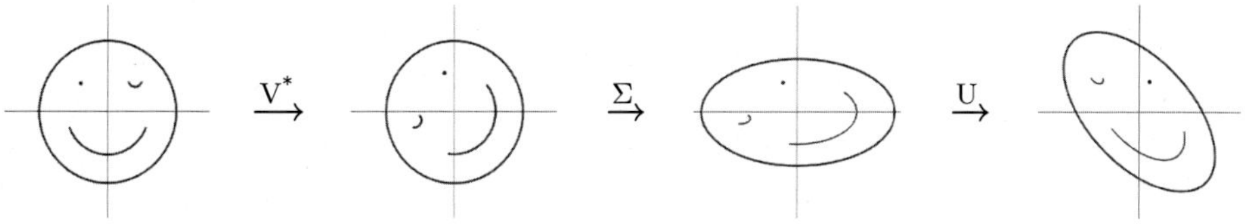

Figure 5.1 The effect of multiplication by $A = U\Sigma V^*$ on $\mathbb{R}^2$.

Example Consider the matrix

$$A = \begin{bmatrix} 2 & 3 \\ -2 & 3 \end{bmatrix} = \begin{bmatrix} \frac{1}{\sqrt{2}} & \frac{1}{\sqrt{2}} \\ \frac{1}{\sqrt{2}} & -\frac{1}{\sqrt{2}} \end{bmatrix} \begin{bmatrix} 3\sqrt{2} & 0 \\ 0 & 2\sqrt{2} \end{bmatrix} \begin{bmatrix} 0 & 1 \\ 1 & 0 \end{bmatrix}.$$

Consider the effect of A on the unit circle in $\mathbb{R}^2$. First, we multiply by $V^* = \begin{bmatrix} 0 & 1 \\ 1 & 0 \end{bmatrix}$, but since this is an isometry, the unit circle is unchanged. Then we apply

$\Sigma = \begin{bmatrix} 3\sqrt{2} & 0 \\ 0 & 2\sqrt{2} \end{bmatrix}$, which stretches the x-axis by a factor of $3\sqrt{2}$ and the y-axis by a factor of $2\sqrt{2}$:

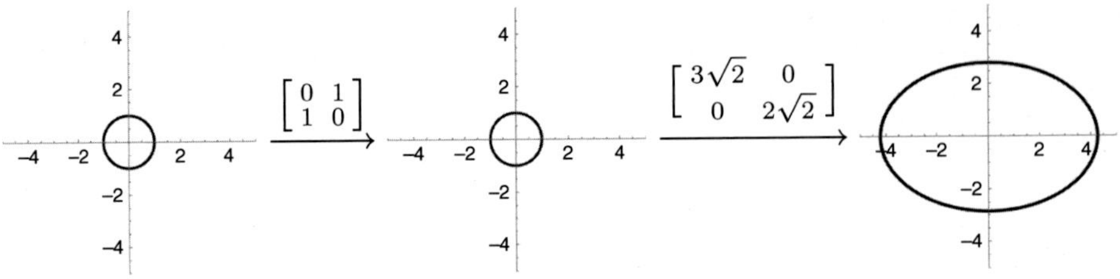

Figure 5.2 The effect on the unit circle of multiplication by V^* and Σ.

Finally, we apply $U = \begin{bmatrix} \frac{1}{\sqrt{2}} & \frac{1}{\sqrt{2}} \\ \frac{1}{\sqrt{2}} & -\frac{1}{\sqrt{2}} \end{bmatrix}$, which is a rotation by 45° in the counter-clockwise direction. The image of the unit circle under multiplication by A is thus the ellipse with major axis of length $6\sqrt{2}$ along the line $y = x$, and minor axis of length $4\sqrt{2}$ along the line $y = -x$:

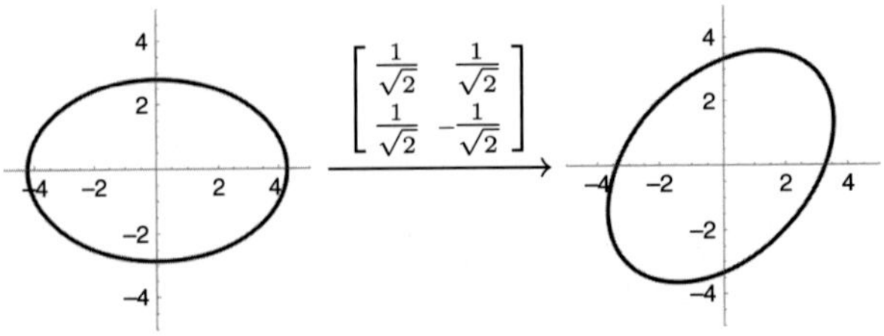

Figure 5.3 The effect of multiplication by U.

Quick Exercise #5. Add the (correctly transformed) winking smiley faces, as in Figure 5.1, to Figures 5.2 and 5.3.

▲

We have seen that $\|A\|_{op} = \sigma_1$, the length of the longest semi-axis of the ellipse described above; there is also a simple expression for the Frobenius norm $\|A\|_F$ in terms of the singular values of A.

Proposition 5.6 *If $\sigma_1, \dots, \sigma_r$ are the positive singular values of $A \in M_{m,n}(\mathbb{C})$, then*

$$\|A\|_F = \sqrt{\sum_{j=1}^{r} \sigma_j^2}.$$

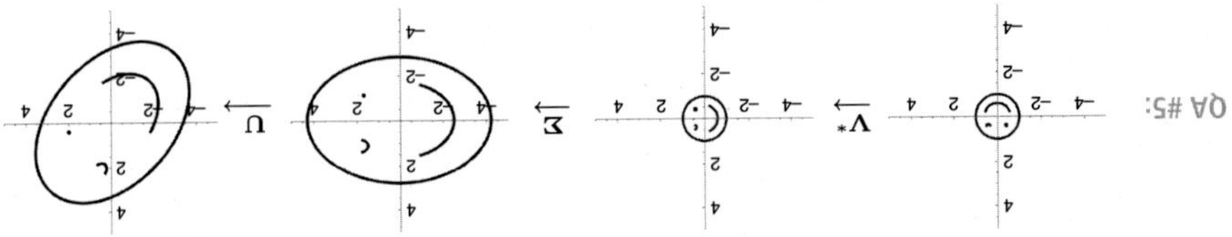

That is, the Frobenius norm measures a kind of average distortion in the third picture in Figure 5.1, whereas the operator norm measures the maximum distortion.

Proof Let $A = U\Sigma V^*$ be the singular value decomposition of A. Then

$$AA^* = U\Sigma V^* V\Sigma^* U^* = U\Sigma\Sigma^* U^*$$

since V is unitary, and so by Proposition 3.60,

$$\|A\|_F^2 = \operatorname{tr} U\Sigma\Sigma^* U^* = \operatorname{tr} U^* U\Sigma\Sigma^* = \operatorname{tr}\Sigma\Sigma^* = \|\Sigma\|_F^2 = \sum_{j=1}^{r}\sigma_j^2.$$

▲

Quick Exercise #6. In the example on page 298, we saw that

$$A = \begin{bmatrix} -1 & 1 & 3 & 5 & 6 \\ 3 & -1 & 3 & -1 & 6 \\ -1 & 3 & -3 & 1 & -6 \end{bmatrix}$$

has singular values 12, 6, and 2. Verify Proposition 5.6 for this matrix by computing $\|A\|_F$ both by definition and in terms of the singular values.

Low-rank Approximation

An important consequence of SVD for applications is that it gives a formula for the best approximation of A by a lower-rank matrix. Approximating by low-rank matrices is important to make big computations feasible; in general, it takes mn numbers to describe an $m \times n$ matrix, but as the following alternate form of SVD demonstrates, a rank r matrix in $M_{m,n}(\mathbb{F})$ can be specified using just $(m+n)r$ numbers (the entries of $\sigma_1\mathbf{u}_1, \ldots, \sigma_r\mathbf{u}_r$ and of $\mathbf{v}_1, \ldots, \mathbf{v}_r$).

Theorem 5.7 (Singular value decomposition (SVD)) *Let $A \in M_{m,n}(\mathbb{F})$. Then there exist orthonormal bases $(\mathbf{u}_1, \ldots, \mathbf{u}_m)$ of $\mathbb{F}^m$ and $(\mathbf{v}_1, \ldots, \mathbf{v}_n)$ of $\mathbb{F}^n$ such that*

$$A = \sum_{j=1}^{r}\sigma_j \mathbf{u}_j \mathbf{v}_j^*,$$

where $\sigma_1 \geq \cdots \geq \sigma_r > 0$ are the nonzero singular values of A.

The proof of Theorem 5.7 in general is left as an exercise (see Exercise 5.2.20), but we will see how it works in a particular case.

QA #6: $12^2 + 6^2 + 2^2 = 184 = 3 \cdot 6^2 + 5^2 + 5 \cdot 3^2 + 6 \cdot 1^2$.

Example In the example on page 298, we found that for

$$A = \begin{bmatrix} -1 & 1 & 3 & 5 & 6 \\ 3 & -1 & 3 & -1 & 6 \\ -1 & 3 & -3 & 1 & -6 \end{bmatrix},$$

there were particular orthonormal bases $(\mathbf{u}_1, \mathbf{u}_2, \mathbf{u}_3)$ of $\mathbb{R}^3$ and $(\mathbf{v}_1, \mathbf{v}_2, \mathbf{v}_3, \mathbf{v}_4, \mathbf{v}_5)$ of $\mathbb{R}^5$ such that

$$A\mathbf{v}_1 = 12\mathbf{u}_1, \quad A\mathbf{v}_2 = 6\mathbf{u}_2, \quad A\mathbf{v}_3 = 2\mathbf{u}_3, \quad A\mathbf{v}_4 = \mathbf{0}, \quad \text{and} \quad A\mathbf{v}_5 = \mathbf{0}.$$

Consider the matrix

$$B = 12\mathbf{u}_1\mathbf{v}_1^* + 6\mathbf{u}_2\mathbf{v}_2^* + 2\mathbf{u}_3\mathbf{v}_3^*.$$

If you're in the mood to waste paper, you can of course just work out all the entries of B in order to show that $A = B$. But since $(\mathbf{v}_1, \mathbf{v}_2, \mathbf{v}_3, \mathbf{v}_4, \mathbf{v}_5)$ is orthonormal, it is easy to compute that

$$B\mathbf{v}_1 = 12\mathbf{u}_1, \quad B\mathbf{v}_2 = 6\mathbf{u}_2, \quad B\mathbf{v}_3 = 2\mathbf{u}_3, \quad B\mathbf{v}_4 = \mathbf{0}, \quad \text{and} \quad B\mathbf{v}_5 = \mathbf{0}.$$

Since $(\mathbf{v}_1, \mathbf{v}_2, \mathbf{v}_3, \mathbf{v}_4, \mathbf{v}_5)$ is a basis of $\mathbb{R}^5$, this is enough to show that $A = B$. ▲

We now show that the best rank k approximation of A is gotten by truncating the singular value decomposition of A, in the form given in Theorem 5.7.

Theorem 5.8 *Let* $A \in M_{m,n}(\mathbb{C})$ *be fixed with positive singular values* $\sigma_1 \geq \cdots \geq \sigma_r > 0$. *Then for any* $B \in M_{m,n}(\mathbb{C})$ *with rank* $k < r$,

$$\|A - B\|_{op} \geq \sigma_{k+1},$$

and equality is achieved for

$$B = \sum_{j=1}^{k} \sigma_j \mathbf{u}_j \mathbf{v}_j^*,$$

where the notation is as in the statement of Theorem 5.7.

Example Returning again to the example from page 298, the best approximation of A by a matrix of rank 1 is

$$B = 12\mathbf{u}_1\mathbf{v}_1^* = \begin{bmatrix} 1 \\ 1 \\ -1 \end{bmatrix} \begin{bmatrix} 1 & -1 & 3 & 1 & 6 \end{bmatrix} = \begin{bmatrix} 1 & -1 & 3 & 1 & 6 \\ 1 & -1 & 3 & 1 & 6 \\ -1 & 1 & -3 & -1 & -6 \end{bmatrix},$$

and the best approximation by A by a matrix of rank 2 is

$$\mathrm{B} = 12\mathbf{u}_1\mathbf{v}_1^* + 6\mathbf{u}_2\mathbf{v}_2^* = \begin{bmatrix} -1 & 1 & 3 & 5 & 6 \\ 2 & -2 & 3 & -1 & 6 \\ -2 & 2 & -3 & 1 & -6 \end{bmatrix}. \qquad \blacktriangle$$

Quick Exercise #7. What is $\|\mathrm{A} - \mathrm{B}\|_{op}$ for each of the approximations above? (Recall that the singular values of A are 12, 6, and 2.)

Proof of Theorem 5.8 Let

$$\mathrm{A} = \sum_{j=1}^{r} \sigma_j \mathbf{u}_j \mathbf{v}_j^*$$

be the singular value decomposition of A, and let B have rank k. By the Rank–Nullity Theorem, $\dim \ker \mathrm{B} = n - k$. So by Lemma 3.22,

$$\langle \mathbf{v}_1, \ldots, \mathbf{v}_{k+1} \rangle \cap \ker \mathrm{B} \neq \{\mathbf{0}\}.$$

That is, there exists a nonzero vector $\mathbf{x} \in \langle \mathbf{v}_1, \ldots, \mathbf{v}_{k+1} \rangle$ such that $\mathrm{B}\mathbf{x} = \mathbf{0}$. Since $(\mathbf{v}_1, \ldots, \mathbf{v}_n)$ is orthonormal, $\mathbf{x}$ is orthogonal to $\mathbf{v}_j$ for $j > k + 1$. Therefore

$$\mathrm{A}\mathbf{x} = \sum_{j=1}^{r} \sigma_j \mathbf{u}_j \mathbf{v}_j^* \mathbf{x} = \sum_{j=1}^{r} \sigma_j \langle \mathbf{x}, \mathbf{v}_j \rangle \mathbf{u}_j = \sum_{j=1}^{k+1} \sigma_j \langle \mathbf{x}, \mathbf{v}_j \rangle \mathbf{u}_j.$$

Then

$$\|(\mathrm{A} - \mathrm{B})\mathbf{x}\|^2 = \|\mathrm{A}\mathbf{x}\|^2 = \left\| \sum_{j=1}^{k+1} \sigma_j \langle \mathbf{x}, \mathbf{v}_j \rangle \mathbf{u}_j \right\|^2 = \sum_{j=1}^{k+1} \sigma_j^2 \left| \langle \mathbf{x}, \mathbf{v}_j \rangle \right|^2,$$

since $(\mathbf{u}_1, \ldots, \mathbf{u}_m)$ is orthonormal, and so

$$\|(\mathrm{A} - \mathrm{B})\mathbf{x}\|^2 \geq \sigma_{k+1}^2 \sum_{j=1}^{k+1} \left| \langle \mathbf{x}, \mathbf{v}_j \rangle \right|^2 = \sigma_{k+1}^2 \|\mathbf{x}\|^2.$$

By inequality (4.11), this implies that $\|\mathrm{A} - \mathrm{B}\|_{op} \geq \sigma_{k+1}$.

For the proof that equality is achieved for the stated matrix B, see Exercise 5.2.21. $\blacktriangle$

Using the methods of Section 4.3, we can also prove a version of Theorem 5.8 for the Frobenius norm.

QA #7: For the rank 1 approximation, $\|\mathrm{A} - \mathrm{B}\|_{op} = 6$. For the rank 2 approximation, $\|\mathrm{A} - \mathrm{B}\|_{op} = 2$.

Theorem 5.9 *Let* $A \in M_{m,n}(\mathbb{C})$ *be fixed with positive singular values* $\sigma_1 \geq \cdots \geq \sigma_r > 0$. *Then for any* $B \in M_{m,n}(\mathbb{C})$ *with rank* $k < r$,

$$\|A - B\|_F \geq \sqrt{\sum_{j=k+1}^{r} \sigma_j^2}$$

and equality is achieved for

$$B = \sum_{j=1}^{k} \sigma_j \mathbf{u}_j \mathbf{v}_j^*,$$

where the notation is as in the statement of Theorem 5.7.

Together, Theorem 5.9 and Theorem 5.8 imply the surprising fact that the best rank k approximation to a given matrix is the same regardless of whether we use the operator norm or Frobenius norm to judge "best approximation." (Generally, the "best" way of doing something depends a lot on how you define what is "best"; see Exercise 5.2.17.)

Quick Exercise #8. Let A be the 3×5 matrix from the example on page 298.

(a) What are the best rank 1 and rank 2 approximations of A with respect to the Frobenius norm?

(b) For each of those approximations B, what is $\|A - B\|_F$?

To prove Theorem 5.9, we will need the following lemma.

Lemma 5.10 *Let* $W \subseteq \mathbb{C}^m$ *be a fixed k-dimensional subspace, and define*

$$V_W := \left\{ B \in M_{m,n}(\mathbb{C}) \;\middle|\; C(B) \subseteq W \right\}.$$

Then V_W *is a subspace of* $M_{m,n}(\mathbb{C})$.

Moreover, if $(\mathbf{w}_1, \ldots, \mathbf{w}_k)$ *is an orthonormal basis of* W, *then the matrices*

$$W_{j,\ell} = \begin{bmatrix} | & & | & & | \\ 0 & \cdots & \mathbf{w}_j & \cdots & 0 \\ | & & | & & | \end{bmatrix}$$

(with $\mathbf{w}_j$ *in the* ℓ*th column) form an orthonormal basis of* V_W.

Proof See Exercise 5.2.22. ▲

QA #8: (a) The same as the two matrices in the example on pages 304–305. (b) $\sqrt{6^2 + 2^2} = 2\sqrt{5}$ and 2.

Proof of Theorem 5.9 To prove that, for any B of rank k,

$$\|A - B\|_F \geq \sqrt{\sum_{j=k+1}^{r} \sigma_j^2}, \tag{5.6}$$

observe first that if $A = U\Sigma V^*$ with U and V unitary, then for any $B \in M_{m,n}(\mathbb{C})$,

$$\|A - B\|_F = \left\|U\Sigma V^* - B\right\|_F = \left\|U(\Sigma - U^*BV)V^*\right\|_F = \left\|\Sigma - U^*BV\right\|_F.$$

Furthermore, $\operatorname{rank} U^*BV = \operatorname{rank} B$, since U and V are unitary, hence isomorphisms. It thus suffices to prove the lower bound of Theorem 5.9 when

$$A = \Sigma = \begin{bmatrix} | & & | & | & & | \\ \sigma_1 e_1 & \cdots & \sigma_r e_r & 0 & \cdots & 0 \\ | & & | & | & & | \end{bmatrix}.$$

(The final columns of zeroes are not there if $r = n$.)

Now, for each fixed k-dimensional subspace $W \subset \mathbb{C}^m$, $\|A - B\|_F$ is minimized among $B \in V_W$ by $P_{V_W}A$. Since the $W_{j,\ell}$ are an orthonormal basis of V_W,

$$\begin{aligned} P_{V_W}A &= \sum_{j=1}^{k}\sum_{\ell=1}^{n} \langle A, W_{j,\ell}\rangle W_{j,\ell} \\ &= \sum_{j=1}^{k}\sum_{\ell=1}^{n} \langle a_\ell, w_j\rangle W_{j,\ell} \\ &= \begin{bmatrix} | & & | \\ \sum_{j=1}^{k} \langle a_1, w_j\rangle w_j & \cdots & \sum_{j=1}^{k} \langle a_n, w_j\rangle w_j \\ | & & | \end{bmatrix} \\ &= \begin{bmatrix} | & & | \\ P_W a_1 & \cdots & P_W a_n \\ | & & | \end{bmatrix}. \end{aligned}$$

Since $A - P_{V_W}A$ is orthogonal to $P_{V_W}A$ with respect to the Frobenius inner product,

$$\|A - P_{V_W}A\|_F^2 = \|A\|_F^2 - \|P_{V_W}A\|_F^2.$$

Moreover, since $a_\ell = \sigma_\ell e_\ell$ for $1 \leq \ell \leq r$ and $a_\ell = 0$ for $\ell > r$, the formula above for $P_{V_W}A$ gives that

$$\left\|P_{V_W}A\right\|_F^2 = \sum_{\ell=1}^{n} \|P_W a_\ell\|^2 = \sum_{\ell=1}^{r} \sigma_\ell^2 \|P_W e_\ell\|^2,$$

and so

$$\|A - P_{V_W}A\|_F^2 = \sum_{\ell=1}^{r} \sigma_\ell^2 - \sum_{\ell=1}^{r} \sigma_\ell^2 \|P_W \mathbf{e}_\ell\|^2 .$$

Now, for each ℓ, $\|P_W \mathbf{e}_\ell\|^2 \le \|\mathbf{e}_\ell\|^2 = 1$, and

$$\sum_{\ell=1}^{m} \|P_W \mathbf{e}_\ell\|^2 = \sum_{\ell=1}^{m}\sum_{j=1}^{k} \left|\langle \mathbf{e}_\ell, \mathbf{w}_j\rangle\right|^2 = \sum_{j=1}^{k} \left\|\mathbf{w}_j\right\|^2 = k.$$

Since $\sigma_1 \ge \cdots \ge \sigma_r > 0$, this means that the expression $\sum_{\ell=1}^{r} \sigma_\ell^2 \|P_W \mathbf{e}_\ell\|^2$ is largest if W is chosen so that $\|P_W \mathbf{e}_\ell\|^2 = 1$ for $1 \le \ell \le k$ and is zero otherwise, that is, if $W = \langle \mathbf{e}_1, \dots, \mathbf{e}_k\rangle$. It follows that

$$\left\|A - P_{V_W}A\right\|_F^2 = \sum_{\ell=1}^{r} \sigma_\ell^2 - \sum_{\ell=1}^{r} \sigma_\ell^2 \|P_W \mathbf{e}_\ell\|^2 \ge \sum_{\ell=1}^{r} \sigma_\ell^2 - \sum_{\ell=1}^{k} \sigma_\ell^2 = \sum_{j=k+1}^{r} \sigma_j^2.$$

Since every B of rank k is in some V_W, this completes the proof of inequality (5.6).

To prove that if $B = \sum_{j=1}^{k} \sigma_j \mathbf{u}_j \mathbf{v}_j^*$, we have equality in (5.6), it suffices to compute $\|A - B\|_F$ for this choice of B:

$$\begin{aligned}
\left\| A - \sum_{j=1}^{k} \sigma_j \mathbf{u}_j \mathbf{v}_j^* \right\|_F^2 &= \left\| \sum_{j=k+1}^{r} \sigma_j \mathbf{u}_j \mathbf{v}_j^* \right\|_F^2 \\
&= \left\langle \sum_{j=k+1}^{r} \sigma_j \mathbf{u}_j \mathbf{v}_j^*, \sum_{\ell=k+1}^{r} \sigma_\ell \mathbf{u}_\ell \mathbf{v}_\ell^* \right\rangle \\
&= \sum_{j=k+1}^{r} \sum_{\ell=k+1}^{r} \sigma_j \sigma_\ell \operatorname{tr}(\mathbf{u}_j \mathbf{v}_j^* \mathbf{v}_\ell \mathbf{u}_\ell^*) \\
&= \sum_{j=k+1}^{r} \sum_{\ell=k+1}^{r} \sigma_j \sigma_\ell \left\langle \mathbf{v}_\ell, \mathbf{v}_j\right\rangle \left\langle \mathbf{u}_j, \mathbf{u}_\ell\right\rangle = \sum_{j=k+1}^{r} \sigma_j^2.
\end{aligned}$$

▲

KEY IDEAS

- SVD: given $A \in M_{m,n}(\mathbb{F})$, there are orthogonal/unitary U, V, and $\Sigma \in M_{m,n}(\mathbb{F})$ with σ_j in the (j, j) entry for $1 \le j \le p$ and zeroes otherwise, such that

$$A = U\Sigma V^*.$$

- The best approximation of A by a rank k matrix, either in the operator norm or the Frobenius norm, is gotten by truncating the SVD $A = \sum_{j=1}^{r} \sigma_j \mathbf{u}_j \mathbf{v}_j^*$ at the kth singular value.

EXERCISES

5.2.1 Find the singular values of the following matrices.

(a) $\begin{bmatrix} 2 & 1 \\ 1 & 2 \end{bmatrix}$ (b) $\begin{bmatrix} -1+3i & 1+3i \\ 1+3i & -1+3i \end{bmatrix}$ (c) $\begin{bmatrix} 1 & -1 & 2 \\ 1 & -1 & -2 \end{bmatrix}$

(d) $\begin{bmatrix} 1 & -1 & 2 \\ 1 & -1 & -2 \\ 0 & 0 & 0 \end{bmatrix}$ (e) $\begin{bmatrix} 0 & 0 & -1 \\ 0 & 2 & 0 \\ 1 & 0 & 0 \end{bmatrix}$

5.2.2 Find the singular values of the following matrices.

(a) $\begin{bmatrix} 1 & 2 \\ -2 & 1 \end{bmatrix}$ (b) $\begin{bmatrix} 4 & -3 \\ 6 & 8 \end{bmatrix}$ (c) $\begin{bmatrix} i & 0 & 0 \\ 0 & \sqrt{2} & 0 \end{bmatrix}$ (d) $\begin{bmatrix} 1 & 2 & 1 \\ 1 & -1 & 1 \end{bmatrix}$

(e) $\begin{bmatrix} 0 & 1 & 0 \\ 0 & 0 & 2 \\ 3 & 0 & 0 \end{bmatrix}$

5.2.3 For each $z \in \mathbb{C}$, let $A_z = \begin{bmatrix} 1 & z \\ 0 & 2 \end{bmatrix}$. Show that the eigenvalues of A_z don't depend on z, but the singular values of A_z do depend on z.
Hint: You don't need to calculate the singular values of A_z. Use something related to the singular values which is simpler to compute.

5.2.4 Let A be the matrix of $T \in \mathcal{L}(V, W)$ with respect to some orthonormal bases of V and W. Show that the singular values of A are the same as the singular values of T.

5.2.5 Show that the singular values of A^* are the same as the singular values of A.

5.2.6 Show that $\operatorname{rank} A^* = \operatorname{rank} A$.

5.2.7 Show that $\operatorname{rank}(AA^*) = \operatorname{rank}(A^*A) = \operatorname{rank} A$.

5.2.8 Show that if $A = \mathbf{diag}(\lambda_1, \ldots, \lambda_n)$, then the singular values of A are $|\lambda_1|, \ldots, |\lambda_n|$ (though not necessarily in the same order).

5.2.9 Let

$$A = \frac{1}{5}\begin{bmatrix} 8 & 9 \\ -6 & 12 \end{bmatrix} = \begin{bmatrix} \frac{3}{5} & \frac{4}{5} \\ \frac{4}{5} & -\frac{3}{5} \end{bmatrix}\begin{bmatrix} 3 & 0 \\ 0 & 2 \end{bmatrix}\begin{bmatrix} 0 & 1 \\ 1 & 0 \end{bmatrix}.$$

Draw a sequence of pictures analogous to Figure 5.1 (page 301) demonstrating the effect on the unit square

$$S = \left\{(x, y) \in \mathbb{R}^2 \mid 0 \le x, y \le 1\right\}$$

of the map on $\mathbb{R}^2$ with matrix A.

5.2.10 Prove that if $A \in M_n(\mathbb{C})$ has singular values $\sigma_1, \ldots, \sigma_n$, then

$$|\operatorname{tr} A| \le \sum_{j=1}^{n} \sigma_j.$$

Hint: Use Theorem 5.7 and the Cauchy–Schwarz inequality.

5.2.11 Let $\kappa(A)$ be the condition number of an invertible matrix $A \in M_n(\mathbb{C})$ (see Exercise 4.4.11), and let $\sigma_1 \geq \cdots \geq \sigma_n > 0$ be the singular values of A. Show that $\kappa(A) = \frac{\sigma_1}{\sigma_n}$.

5.2.12 Let $A \in M_{m,n}(\mathbb{R})$, and suppose that $B \in M_{m,n}(\mathbb{C})$ is the best rank k approximation of A. Show that all the entries of B are real numbers.

5.2.13 Let $A \in M_n(\mathbb{F})$ be invertible. Show that the minimum value of $\|A - B\|_{op}$ for B a singular matrix is σ_n. Show that the same holds if $\|A - B\|_{op}$ is replaced with $\|A - B\|_F$.

5.2.14 If $A \in M_{m,n}(\mathbb{C})$ is nonzero, then the **stable rank** of A is defined as

$$\operatorname{srank} A := \frac{\|A\|_F^2}{\|A\|_{op}^2}.$$

(a) Show that $\operatorname{srank} A \leq \operatorname{rank} A$.

(b) Suppose that $k \geq \alpha \operatorname{srank} A$ for some $\alpha > 0$, and B is the best rank k approximation to A. Show that

$$\frac{\|A - B\|_{op}}{\|A\|_{op}} \leq \frac{1}{\sqrt{\alpha}}.$$

5.2.15 Let $A = U\Sigma V^*$ be a singular value decomposition of $A \in M_{m,n}(\mathbb{C})$. Define the **pseudoinverse** $A^\dagger \in M_{n,m}(\mathbb{C})$ of A by

$$A^\dagger = V\Sigma^\dagger U^*,$$

where $\Sigma^\dagger$ is an $n \times m$ matrix with $\frac{1}{\sigma_j}$ in the (j,j) entry for $1 \leq j \leq r$ and zeroes otherwise.

(a) Show that if A is invertible, then $A^\dagger = A^{-1}$.

(b) Compute $\Sigma\Sigma^\dagger$ and $\Sigma^\dagger\Sigma$.

(c) Show that $(AA^\dagger)^* = AA^\dagger$ and $(A^\dagger A)^* = A^\dagger A$.

(d) Show that $AA^\dagger A = A$ and that $A^\dagger AA^\dagger = A^\dagger$.

5.2.16 Let $A^\dagger$ be the pseudoinverse of $A \in M_{m,n}(\mathbb{C})$, as defined in Exercise 5.2.15.

(a) Show that the linear system $Ax = b$ has solutions if and only if $AA^\dagger b = b$.

(b) Show that if the system $Ax = b$ is consistent, and $x \in \mathbb{C}^n$ is any solution, then $\|x\| \geq \|A^\dagger b\|$.
(For this reason, $A^\dagger b$ is called the **least-squares** solution of the system.)

(c) Show that if the system $Ax = b$ is consistent, then its set of solutions is given by

$$\left\{ A^\dagger b + [I_n - A^\dagger A]w \,\middle|\, w \in \mathbb{C}^n \right\}.$$

Hint: First show that $\ker A = C(I_n - A^{\dagger}A)$; one inclusion is trivial using Exercise 5.2.15 and the other is easiest by dimension counting.

5.2.17 Usually, the notion of "best approximation" depends on how we measure. Show that the closest point on the line $y = \frac{1}{2}x$ to the point $(1,1)$ is:

(a) $(6/5, 3/5)$, when distance is measured with the standard norm on $\mathbb{R}^2$;

(b) $(1, 1/2)$, when distance is measured with the ℓ^1 norm;

(c) $(4/3, 2/3)$, when distance is measured with the ℓ^∞ norm.

5.2.18 Let V_W be as in Lemma 5.10, and let $A \in M_{m,n}(\mathbb{C})$.

(a) Show that for $x \in \mathbb{C}^n$, $\left[P_{V_W}A\right]x = P_W(Ax)$.

(b) Find a k-dimensional subspace W of $\mathbb{C}^m$ such that $P_{V_W}A = \sum_{j=1}^{k} \sigma_j u_j v_j^*$, where $A = \sum_{j=1}^{r} \sigma_j u_j v_j^*$ is the SVD of A.

5.2.19 Use Theorem 5.7 to give another proof of Proposition 5.6.

5.2.20 Prove Theorem 5.7.

5.2.21 Prove the stated equality case in Theorem 5.8. That is, suppose that $A = \sum_{j=1}^{r} \sigma_j u_j v_j^*$ is the SVD of A (in the form of Theorem 5.7) and that $B = \sum_{j=1}^{k} \sigma_j u_j v_j^*$ for some $1 \le k < r$, and prove that

$$\|A - B\|_{op} = \sigma_{k+1}.$$

5.2.22 Prove Lemma 5.10.

5.3 Adjoint Maps

The Adjoint of a Linear Map

Recall the representation of the standard dot product in $\mathbb{C}^n$ by matrix-multiplication: for $x, y \in \mathbb{C}^n$,

$$\langle x, y \rangle = y^* x.$$

If we consider an $n \times n$ matrix A as an operator on $\mathbb{C}^n$, then since $(By)^* = y^* B^*$,

$$\langle Ax, y \rangle = y^* A x = (A^* y)^* x = \left\langle x, A^* y \right\rangle.$$

We take this observation as motivation for the following definition.

Definition Let V and W be inner product spaces, and let $T \in \mathcal{L}(V, W)$. A linear map $S \in \mathcal{L}(W, V)$ is called an **adjoint** of T if

$$\langle Tv, w \rangle = \langle v, Sw \rangle$$

for every $v \in V$ and $w \in W$. We then write $S = T^*$.

Examples

1. By the discussion above, if $T : \mathbb{C}^n \to \mathbb{C}^n$ is given by multiplication by A, then T^* is represented by multiplication by A^*.
2. Part 4 of Theorem 4.16 says that if $P_U \in \mathcal{L}(V, V)$ is the orthogonal projection onto a finite-dimensional subspace U of V, then $P_U^* = P_U$.
3. Theorem 4.27 says that if $T \in \mathcal{L}(V, W)$ is an isometry between inner product spaces, then $T^* = T^{-1}$. ▲

Lemma 5.11 *Let $T \in \mathcal{L}(V, W)$ be a linear map between inner product spaces. If T has an adjoint, then its adjoint is unique.*

Proof Suppose that $S_1, S_2 \in \mathcal{L}(W, V)$ are both adjoints of T. Then for each $v \in V$ and $w \in W$,

$$\langle v, (S_1 - S_2)w \rangle = 0.$$

Setting $v = (S_1 - S_2)w$, we have

$$\langle (S_1 - S_2)w, (S_1 - S_2)w \rangle = 0$$

for every $w \in W$, and so $S_1 = S_2$. ▲

Thus T^* (if it exists at all) is the one and only operator with the property that

$$\langle Tv, w \rangle = \langle v, T^* w \rangle$$

for every $v \in V$ and $w \in W$.

Quick Exercise #9. Show that if $S, T \in \mathcal{L}(V, W)$ have adjoints and $a \in \mathbb{F}$,

$$(aT)^* = \overline{a}T^* \qquad \text{and} \qquad (S + T)^* = S^* + T^*.$$

Theorem 5.12 *If V is a finite-dimensional inner product space and W is any inner product space, then every $T \in \mathcal{L}(V, W)$ has an adjoint.*

Proof Let $(e_1, \ldots, e_n)$ be an orthonormal basis of V, and define

$$Sw = \sum_{j=1}^{n} \langle w, Te_j \rangle e_j$$

QA #9: For any $v \in V$ and $w \in W$, $\langle aTv, w \rangle = a \langle Tv, w \rangle = a \langle v, T^* w \rangle = \langle v, \overline{a}T^* w \rangle$. The additivity property is shown similarly.

for $w \in W$. Using the linearity properties of the inner product, it is easy to check that S is linear. Moreover, for $v \in V$ and $w \in W$,

$$\begin{aligned}\langle Tv, w\rangle &= \left\langle T\sum_{j=1}^{n}\langle v, e_j\rangle e_j, w\right\rangle\\ &= \sum_{j=1}^{n}\langle v, e_j\rangle\langle Te_j, w\rangle\\ &= \sum_{j=1}^{n}\langle v, e_j\rangle\overline{\langle w, Te_j\rangle}\\ &= \left\langle v, \sum_{j=1}^{n}\langle w, Te_j\rangle e_j\right\rangle\\ &= \langle v, Sw\rangle .\end{aligned}$$

▲

Motivated by our observations in $\mathbb{C}^n$, we've been using the same symbol for the adjoint of a linear map that we previously used for the conjugate transpose of a matrix. We know this is reasonable for representations of linear maps on $\mathbb{C}^n$ with respect to the standard basis; the following result says that this works more generally, as long as we are working in coordinates associated to an *orthonormal* basis.

Theorem 5.13 *Suppose that $\mathcal{B}_V$ and $\mathcal{B}_W$ are orthonormal bases of V and W, respectively. Then for $T \in \mathcal{L}(V, W)$,*

$$[T^*]_{\mathcal{B}_W,\mathcal{B}_V} = \left([T]_{\mathcal{B}_V,\mathcal{B}_W}\right)^*.$$

WARNING! This only works if both the bases $\mathcal{B}_V$ and $\mathcal{B}_W$ are orthonormal!

Proof If we write $\mathcal{B}_V = (e_1, \ldots, e_n)$ and $\mathcal{B}_W = (f_1, \ldots, f_m)$, then by Theorem 4.9, the (j, k) entry of $[T^*]_{\mathcal{B}_W,\mathcal{B}_V}$ is

$$\langle T^*f_k, e_j\rangle = \overline{\langle e_j, T^*f_k\rangle} = \overline{\langle Te_j, f_k\rangle},$$

which is the complex conjugate of the (k, j) entry of $[T]_{\mathcal{B}_V,\mathcal{B}_W}$. ▲

The following basic properties of the adjoint follow easily from the definition.

Proposition 5.14 *Let U, V, and W be finite-dimensional inner product spaces, $S, T \in \mathcal{L}(U, V)$, and let $R \in \mathcal{L}(V, W)$. Then*

1. $(T^*)^* = T$,
2. $(S + T)^* = S^* + T^*$,

3. $(aT)^* = \overline{a}T^*$,
4. $(RT)^* = T^*R^*$,
5. *if T is invertible, then* $\left(T^{-1}\right)^* = (T^*)^{-1}$.

Proof Parts 2 and 3 were proved in Quick Exercise #9. We will prove part 1, and leave the remaining parts as exercises (see Exercise 5.3.21).

For any $u \in U$ and $v \in V$,

$$\langle v, Tu\rangle = \overline{\langle Tu, v\rangle} = \overline{\langle u, T^*v\rangle} = \left\langle T^*v, u\right\rangle.$$

Therefore T is the adjoint of T^*. ▲

Note the appearance of the Rat Poison Principle: we have shown that $(T^*)^* = T$ by showing that T does the job that defines $(T^*)^*$.

Self-adjoint Maps and Matrices

It turns out that operators that are their own adjoints have some very special properties.

Definition A linear map $T \in \mathcal{L}(V)$ is **self-adjoint** if $T^* = T$.

A matrix $A \in M_n(\mathbb{C})$ is **Hermitian** if $A^* = A$. A matrix $A \in M_n(\mathbb{R})$ is **symmetric** if $A^T = A$.

Theorem 5.13 implies that a linear map T on a finite inner product space V is self-adjoint if and only if its matrix with respect to an orthonormal basis is Hermitian. If V is a real inner product space, then T is self-adjoint if and only if its matrix with respect to an orthonormal basis is symmetric.

Quick Exercise #10. Show that if $T \in \mathcal{L}(V)$ is self-adjoint, then every eigenvalue of T is real.

Hint: If $v \in V$ is an eigenvector of T, consider $\langle Tv, v\rangle$.

Recall that if v_1, v_2 are both eigenvectors of a linear map T, with distinct eigenvalues, then v_1 and v_2 are linearly independent. If T is self-adjoint, then more is true:

Proposition 5.15 *Let* $T \in \mathcal{L}(V)$ *be self-adjoint. If* $Tv_1 = \lambda_1 v_1$ *and* $Tv_2 = \lambda_2 v_2$, *and* $\lambda_1 \neq \lambda_2$, *then* $\langle v_1, v_2\rangle = 0$.

QA #10: $\langle Tv, v\rangle = \lambda \|v\|^2$, but also $\langle Tv, v\rangle = \left\langle v, T^*v\right\rangle = \langle v, Tv\rangle = \overline{\lambda}\|v\|^2$, so $\lambda = \overline{\lambda}$.

Proof Since

$$\lambda_1 \langle v_1, v_2 \rangle = \langle \lambda_1 v_1, v_2 \rangle = \langle Tv_1, v_2 \rangle = \langle v_1, Tv_2 \rangle = \langle v_1, \lambda_2 v_2 \rangle = \lambda_2 \langle v_1, v_2 \rangle,$$

and $\lambda_1 \neq \lambda_2$, it must be that $\langle v_1, v_2 \rangle = 0$. ▲

Example The second derivative is a self-adjoint operator on the space of 2π-periodic smooth functions on $\mathbb{R}$ (see Exercise 5.3.7). It then follows from Proposition 5.15 that, for example, $\sin(nx)$ and $\sin(mx)$ are orthogonal if $n \neq m$, since they are both eigenvectors of the second derivative operator, with distinct eigenvalues $-n^2$ and $-m^2$. ▲

The Four Subspaces

The following fundamental result gives the geometric relationship between the range and kernel of an operator and of its adjoint.

Proposition 5.16 *Let V and W be finite-dimensional inner product spaces and let $T \in \mathcal{L}(V, W)$. Then*

1. $\ker T^* = (\operatorname{range} T)^\perp$,
2. $\operatorname{range} T^* = (\ker T)^\perp$.

This result gives a new proof of the Rank–Nullity Theorem in inner product spaces: we can orthogonally decompose V as

$$V = \ker T \oplus (\ker T)^\perp = \ker T \oplus \operatorname{range} T^*. \tag{5.7}$$

The dimension of the right-hand side is $\operatorname{null} T + \operatorname{rank} T^* = \operatorname{null} T + \operatorname{rank} T$ (by Exercise 5.3.20), and so we recover the Rank–Nullity Theorem. We get much more, though, from equation (5.7) itself: every vector in V breaks into two orthogonal pieces, one in the kernel of T and one in the range of T^*. The subspaces range T, ker T, range T^*, and ker T^* are sometimes called **the four fundamental subspaces** associated to the operator T.

Proof of Proposition 5.16 Suppose that $w \in \ker T^*$; we'd like to show that $\langle w, w' \rangle = 0$ for every $w' \in \operatorname{range} T$. If $w' \in \operatorname{range} T$, then there is some $v' \in V$ such that $w' = Tv'$. We then have

$$\langle w, w' \rangle = \langle w, Tv' \rangle = \langle T^* w, v' \rangle = 0,$$

since $T^* w = 0$ by assumption.

Conversely, if $w \in (\operatorname{range} T)^\perp$, then for every $v \in V$,

$$0 = \langle w, Tv \rangle = \langle T^* w, v \rangle.$$

By part 4 of Proposition 4.2, this means that $T^* w = 0$.

To get the second statement, we can just apply the first statement to T^* and then take orthogonal complements:

$$\ker T = \ker(T^*)^* = (\operatorname{range} T^*)^\perp,$$

and so

$$(\ker T)^\perp = ((\operatorname{range} T^*)^\perp)^\perp = \operatorname{range} T^*. \qquad \blacktriangle$$

The four fundamental subspaces are intimately connected with the singular value decomposition; see Exercise 5.3.20.

Computing SVD

In Section 5.2, we gave a method for finding the singular values of a matrix; we are now in a position to complete the picture by giving the full algorithm for finding the SVD of an arbitrary matrix.

Algorithm 5.17 To find an SVD for the matrix $A \in M_{m,n}(\mathbb{F})$:

- Find the eigenvalues of A^*A. The square roots of the largest $p = \min\{m, n\}$ of them are the singular values $\sigma_1 \geq \cdots \geq \sigma_p$ of A.
- Find an orthonormal basis for each eigenspace of A^*A; put the resulting collection of vectors $(\mathbf{v}_1, \dots, \mathbf{v}_n)$ in a list so that the corresponding eigenvalues are in decreasing order. Take

$$V := \begin{bmatrix} | & & | \\ \mathbf{v}_1 & \cdots & \mathbf{v}_n \\ | & & | \end{bmatrix}.$$

- Let $r = \operatorname{rank} A$. For $1 \leq j \leq r$, define $\mathbf{u}_j := \frac{1}{\sigma_j} A\mathbf{v}_j$, and extend $(\mathbf{u}_1, \dots, \mathbf{u}_r)$ to an orthonormal basis $(\mathbf{u}_1, \dots, \mathbf{u}_m)$ of $\mathbb{F}^m$. Take

$$U := \begin{bmatrix} | & & | \\ \mathbf{u}_1 & \cdots & \mathbf{u}_m \\ | & & | \end{bmatrix}.$$

Then if $\Sigma \in M_{m,n}(\mathbb{F})$ has σ_j in the (j, j) position for $1 \leq j \leq p$ and 0 otherwise,

$$A = U\Sigma V^*$$

is an SVD of A.

Proof We have already seen (Proposition 5.5) that the singular values are exactly the square roots of the p largest eigenvalues of A^*A. Defining the matrices U, Σ, and V as above makes it automatic that $A = U\Sigma V^*$; it remains only to check that U and V are in fact unitary.

The vectors $(\mathbf{v}_1, \dots, \mathbf{v}_n)$ constructed as above are chosen to be orthonormal within eigenspaces, and Proposition 5.15 shows that vectors from distinct eigenspaces of A^*A are orthogonal. Thus $(\mathbf{v}_1, \dots, \mathbf{v}_n)$ is orthonormal, and so V is unitary.

Now for $1 \le j, k \le r$,

$$\langle \mathbf{u}_j, \mathbf{u}_k \rangle = \frac{1}{\sigma_j \sigma_k} \langle A\mathbf{v}_j, A\mathbf{v}_k \rangle = \frac{1}{\sigma_j \sigma_k} \langle A^*A\mathbf{v}_j, \mathbf{v}_k \rangle = \frac{\sigma_j}{\sigma_k} \langle \mathbf{v}_j, \mathbf{v}_k \rangle = \begin{cases} 1 & \text{if } j = k, \\ 0 & \text{otherwise.} \end{cases}$$

Thus $(\mathbf{u}_1, \dots, \mathbf{u}_r)$ is orthonormal, and can indeed be extended to an orthonormal basis of $\mathbb{F}^m$; U is then unitary by construction. ▲

Example Let $A = \begin{bmatrix} 1 & -2 & 3 \\ 2 & 1 & 1 \\ 1 & -2 & -3 \\ 2 & 1 & -1 \end{bmatrix}$; then $A^*A = \begin{bmatrix} 10 & 0 & 0 \\ 0 & 10 & 0 \\ 0 & 0 & 20 \end{bmatrix}$, and so we see immediately that $\sigma_1 = \sqrt{20}$ and $\sigma_2 = \sigma_3 = \sqrt{10}$. Moreover, we can take $\mathbf{v}_1 = \mathbf{e}_3$, $\mathbf{v}_2 = \mathbf{e}_1$, and $\mathbf{v}_3 = \mathbf{e}_2$. (Since the 10-eigenspace is two-dimensional, we might have needed to work a bit harder to find an orthonormal basis, but in this case, it was obvious.) Now we can find most of the matrix U:

$$\mathbf{u}_1 = \frac{1}{\sqrt{20}} A\mathbf{v}_1 = \frac{1}{\sqrt{20}} \begin{bmatrix} 3 \\ 1 \\ -3 \\ -1 \end{bmatrix},$$

$$\mathbf{u}_2 = \frac{1}{\sqrt{10}} A\mathbf{v}_2 = \frac{1}{\sqrt{10}} \begin{bmatrix} 1 \\ 2 \\ 1 \\ 2 \end{bmatrix},$$

$$\mathbf{u}_3 = \frac{1}{\sqrt{10}} A\mathbf{v}_3 = \frac{1}{\sqrt{10}} \begin{bmatrix} -2 \\ 1 \\ -2 \\ 1 \end{bmatrix}.$$

To fill out the matrix U, we need to find $\mathbf{u}_4$ such that $(\mathbf{u}_1, \mathbf{u}_2, \mathbf{u}_3, \mathbf{u}_4)$ is an orthonormal basis of $\mathbb{R}^4$. We can take any vector that's not in $\langle \mathbf{u}_1, \mathbf{u}_2, \mathbf{u}_3 \rangle$, say $\mathbf{e}_1$, and use the Gram-Schmidt process to produce $\mathbf{u}_4$:

$$\tilde{\mathbf{u}}_4 = \mathbf{e}_1 - \langle \mathbf{e}_1, \mathbf{u}_1 \rangle \mathbf{u}_1 - \langle \mathbf{e}_1, \mathbf{u}_2 \rangle \mathbf{u}_2 - \langle \mathbf{e}_1, \mathbf{u}_3 \rangle \mathbf{u}_3 = \frac{1}{20} \begin{bmatrix} 1 \\ -3 \\ -1 \\ 3 \end{bmatrix},$$

and so $\mathbf{u}_4 = \frac{1}{\sqrt{20}}\begin{bmatrix} 1 \\ -3 \\ -1 \\ 3 \end{bmatrix}$. So finally,

$$A = U\Sigma V^* = \frac{1}{\sqrt{20}}\begin{bmatrix} 3 & \sqrt{2} & -2\sqrt{2} & 1 \\ 1 & 2\sqrt{2} & \sqrt{2} & -3 \\ -3 & \sqrt{2} & -2\sqrt{2} & -1 \\ -1 & 2\sqrt{2} & \sqrt{2} & 3 \end{bmatrix}\begin{bmatrix} \sqrt{20} & 0 & 0 \\ 0 & \sqrt{10} & 0 \\ 0 & 0 & \sqrt{10} \\ 0 & 0 & 0 \end{bmatrix}\begin{bmatrix} 0 & 0 & 1 \\ 1 & 0 & 0 \\ 0 & 1 & 0 \end{bmatrix}.$$

▲

KEY IDEAS

- Every $T \in \mathcal{L}(V, W)$ has a unique adjoint $T^* \in \mathcal{L}(W, V)$, defined by the fact that, for every $v \in V$ and $w \in W$,

$$\langle Tv, w\rangle = \langle v, T^*w\rangle.$$

- If $\mathcal{B}_V$ and $\mathcal{B}_W$ are orthonormal bases, then $[T^*]_{\mathcal{B}_W,\mathcal{B}_V} = \left([T]_{\mathcal{B}_V,\mathcal{B}_W}\right)^*$.
- T is self-adjoint if $T = T^*$. For matrices, this is called symmetric if $\mathbb{F} = \mathbb{R}$, or Hermitian if $\mathbb{F} = \mathbb{C}$.
- To find the SVD of A, first find the eigenvalues of A^*A; their square roots are the singular values. Then find the corresponding eigenvectors; that gives you V. Then most of U is determined by the definition of an SVD, and the rest can be filled out arbitrarily to make U orthogonal/unitary.

EXERCISES

5.3.1 Find an SVD for each of the following matrices.

(a) $\begin{bmatrix} 2 & 1 \\ 1 & 2 \end{bmatrix}$ (b) $\begin{bmatrix} -1+3i & 1+3i \\ 1+3i & -1+3i \end{bmatrix}$ (c) $\begin{bmatrix} 1 & -1 & 2 \\ 1 & -1 & -2 \end{bmatrix}$

(d) $\begin{bmatrix} 1 & -1 & 2 \\ 1 & -1 & -2 \\ 0 & 0 & 0 \end{bmatrix}$ (e) $\begin{bmatrix} 0 & 0 & -1 \\ 0 & 2 & 0 \\ 1 & 0 & 0 \end{bmatrix}$

5.3.2 Find an SVD for each of the following matrices.

(a) $\begin{bmatrix} 1 & 2 \\ -2 & 1 \end{bmatrix}$ (b) $\begin{bmatrix} 4 & -3 \\ 6 & 8 \end{bmatrix}$ (c) $\begin{bmatrix} i & 0 & 0 \\ 0 & \sqrt{2} & 0 \end{bmatrix}$ (d) $\begin{bmatrix} 1 & 2 & 1 \\ 1 & -1 & 1 \end{bmatrix}$

(e) $\begin{bmatrix} 0 & 1 & 0 \\ 0 & 0 & 2 \\ 3 & 0 & 0 \end{bmatrix}$

5.3.3 Compute the pseudoinverse of $A = \begin{bmatrix} 1 & -2 \\ 2 & -4 \end{bmatrix}$ (see Exercise 5.2.15).

5.3.4 Compute the pseudoinverse of $\mathrm{A} = \begin{bmatrix} 1 & -1 & 2 \\ 1 & -1 & -2 \end{bmatrix}$ (see Exercise 5.2.15).

5.3.5 Let L be a line through the origin in $\mathbb{R}^2$, and let $\boldsymbol{R} \in \mathcal{L}(\mathbb{R}^2)$ be the reflection across L. Show that $\boldsymbol{R}$ is self-adjoint.

5.3.6 Let $\mathrm{A} \in \mathrm{M}_n(\mathbb{C})$ be invertible, and define an inner product on $\mathbb{C}^n$ by

$$\langle \mathrm{x}, \mathrm{y} \rangle_{\mathrm{A}} := \langle \mathrm{Ax}, \mathrm{Ay} \rangle,$$

where the inner product on the right-hand side is the standard inner product on $\mathbb{C}^n$.

Suppose that $T \in \mathcal{L}(\mathbb{C}^n)$ is represented with respect to the standard basis by $\mathrm{B} \in \mathrm{M}_n(\mathbb{C})$. What is the matrix of T^*, where the adjoint is defined in terms of this inner product?

5.3.7 Let $C^\infty_{2\pi}(\mathbb{R})$ be the space of 2π-periodic infinitely differentiable functions $f : \mathbb{R} \to \mathbb{R}$ with the inner product

$$\langle f, g \rangle = \int_0^{2\pi} f(x)g(x)\, dx,$$

and let $\boldsymbol{D} \in \mathcal{L}(C^\infty_{2\pi}(\mathbb{R}))$ be the differentiation operator.

(a) Show that $\boldsymbol{D}^* = -\boldsymbol{D}$.

(b) Show that 0 is the only real eigenvalue of $\boldsymbol{D}$.

(c) Show that $\boldsymbol{D}^2$ is self-adjoint.

5.3.8 Let $\mathcal{P}(\mathbb{R})$ denote the space of polynomials over $\mathbb{R}$, with the inner product defined by

$$\langle p, q \rangle := \int_{-\infty}^{\infty} p(x)q(x)e^{-x^2/2}dx.$$

(a) Find the adjoint of the derivative operator $\boldsymbol{D} \in \mathcal{L}(\mathcal{P}(\mathbb{R}))$.

(b) Now consider the restriction of the derivative operator $\boldsymbol{D} : \mathcal{P}_n(\mathbb{R}) \to \mathcal{P}_{n-1}(\mathbb{R})$. Find the matrix of $\boldsymbol{D}$ with respect to the bases $(1, x, \ldots, x^n)$ and $(1, x, \ldots, x^{n-1})$.

(c) Find the matrix of $\boldsymbol{D}^* : \mathcal{P}_{n-1}(\mathbb{R}) \to \mathcal{P}_n(\mathbb{R})$ with respect to the bases $(1, x, \ldots, x^{n-1})$ and $(1, x, \ldots, x^n)$.

(d) Why does what you found not contradict Theorem 5.13?

5.3.9 Let U be a subspace of a finite-dimensional inner product space V, and consider the orthogonal projection $\boldsymbol{P}_U$ as an element of $\mathcal{L}(V, U)$. Define the **inclusion map** $\boldsymbol{J} \in \mathcal{L}(U, V)$ by $\boldsymbol{J}(u) = u$ for all $u \in U$. Show that $\boldsymbol{P}_U$ and $\boldsymbol{J}$ are adjoint maps.

5.3.10 Let V be a finite-dimensional inner product space, and let $T \in \mathcal{L}(V)$.

(a) Show that $\operatorname{tr} T^* = \overline{(\operatorname{tr} T)}$.

(b) Show that if T is self-adjoint, then $\operatorname{tr} T \in \mathbb{R}$.

5.3.11 Let $A \in M_n(\mathbb{C})$. Show that the Hermitian and anti-Hermitian parts of A (defined in Exercise 4.1.5) are both Hermitian.
Remark: This may seem rather strange at first glance, but remember that the imaginary part of a complex number is in fact a real number!

5.3.12 Let V be a finite-dimensional inner product space. Suppose that $T \in \mathcal{L}(V)$ is self-adjoint and $T^2 = T$. Prove that T is the orthogonal projection onto $U = \operatorname{range} T$.

5.3.13 Let V and W be finite-dimensional inner product spaces and let $T \in \mathcal{L}(V, W)$. Show that $T^*T = 0$ if and only if $T = 0$.

5.3.14 Let V and W be finite-dimensional inner product spaces and let $T \in \mathcal{L}(V, W)$. Show that $\|T^*\|_{op} = \|T\|_{op}$.

5.3.15 Let V and W be finite-dimensional inner product spaces and let $T \in \mathcal{L}(V, W)$. Show that $\|T^*T\|_{op} = \|TT^*\|_{op} = \|T\|_{op}^2$.

5.3.16 Let V be a finite-dimensional inner product space. Show that the set W of all self-adjoint linear maps on V is a real vector space.

5.3.17 Let V and W be finite-dimensional inner product spaces. Show that

$$\langle S, T\rangle_F := \operatorname{tr}(ST^*)$$

defines an inner product on $\mathcal{L}(V, W)$.

5.3.18 Suppose that V is a complex inner product space, $T \in \mathcal{L}(V)$, and $T^* = -T$. Prove that every eigenvalue of T is purely imaginary (i.e., of the form ia for some $a \in \mathbb{R}$).

5.3.19 Suppose that V is a complex inner product space, $T \in \mathcal{L}(V)$, and $T^* = -T$. Prove that eigenvectors of T with distinct eigenvalues are orthogonal.

5.3.20 Suppose that $T \in \mathcal{L}(V, W)$ has an SVD with right singular vectors $e_1, \ldots, e_n \in V$, left singular vectors $f_1, \ldots, f_m \in W$, and singular values $\sigma_1 \geq \cdots \geq \sigma_r > 0$ (where $r = \operatorname{rank} T$). Show that:
(a) $(f_1, \ldots, f_r)$ is an orthonormal basis of range T.
(b) $(e_{r+1}, \ldots, e_n)$ is an orthonormal basis of ker T.
(c) $(f_{r+1}, \ldots, f_m)$ is an orthonormal basis of ker T^*.
(d) $(e_1, \ldots, e_r)$ is an orthonormal basis of range T^*.

5.3.21 Prove parts 4 and 5 of Proposition 5.14.

5.3.22 State and prove a version of Algorithm 5.17 that starts with “Find the eigenvalues of AA^*.”

5.4 The Spectral Theorems

In this section we will definitively answer the question “which linear maps can be diagonalized in an orthonormal basis?” In the context of matrices, we say that such a matrix is **unitarily diagonalizable**.

Eigenvectors of Self-adjoint Maps and Matrices

In order to construct an orthonormal basis of eigenvectors, we will for starters need the existence of at least one eigenvector.

Lemma 5.18 *If* $\mathrm{A} \in \mathrm{M}_n(\mathbb{F})$ *is Hermitian, then* A *has an eigenvector in* $\mathbb{F}^n$.

If V *is a nonzero finite-dimensional inner product space and* $T \in \mathcal{L}(V)$ *is self-adjoint, then* T *has an eigenvector.*

Proof We will prove the matrix-theoretic statement; the version about linear maps then follows.

Let $\mathrm{A} = \mathrm{U}\Sigma\mathrm{V}^*$ be a singular value decomposition of A. Then

$$\mathrm{A}^2 = \mathrm{A}^*\mathrm{A} = \mathrm{V}\Sigma^*\Sigma\mathrm{V}^* = \mathrm{V}\Sigma^2\mathrm{V}^{-1},$$

since Σ is diagonal with real entries when $m = n$, and V is unitary. By Theorem 3.56, each column $\mathbf{v}_j$ of V is an eigenvector of A^2 with corresponding eigenvalue σ_j^2. Therefore

$$\mathbf{0} = (\mathrm{A}^2 - \sigma_j^2\mathrm{I}_n)\mathbf{v}_j = (\mathrm{A} + \sigma_j\mathrm{I}_n)(\mathrm{A} - \sigma_j\mathrm{I}_n)\mathbf{v}_j.$$

Let $\mathbf{w} = (\mathrm{A} - \sigma_1\mathrm{I}_n)\mathbf{v}_1$. If $\mathbf{w} = \mathbf{0}$, then $\mathbf{v}_1$ is an eigenvector of A with eigenvalue σ_1. On the other hand, if $\mathbf{w} \neq \mathbf{0}$, then since $(\mathrm{A} + \sigma_1\mathrm{I}_n)\mathbf{w} = \mathbf{0}$, $\mathbf{w}$ is an eigenvector of A with eigenvalue $-\sigma_1$. ▲

The following result gives an easily checked *sufficient* condition for unitary diagonalizability; it is a necessary and sufficient condition for unitary diagonalizability with real eigenvalues (see Exercise 5.4.17). The name derives from the fact that the set of eigenvalues of a linear map is sometimes called its **spectrum**.

Theorem 5.19 (The Spectral Theorem for self-adjoint maps and Hermitian matrices) *If* V *is a finite-dimensional inner product space and* $T \in \mathcal{L}(V)$ *is self-adjoint, then there is an orthonormal basis of* V *consisting of eigenvectors of* T.

If $\mathrm{A} \in \mathrm{M}_n(\mathbb{F})$ *is Hermitian (if* $\mathbb{F} = \mathbb{C}$*) or symmetric (if* $\mathbb{F} = \mathbb{R}$*), then there exist a unitary (if* $\mathbb{F} = \mathbb{C}$*) or orthogonal (if* $\mathbb{F} = \mathbb{R}$*) matrix* $\mathrm{U} \in \mathrm{M}_n(\mathbb{F})$ *and real numbers* $\lambda_1, \ldots, \lambda_n \in \mathbb{R}$ *such that*

$$\mathrm{A} = \mathrm{U}\,\mathrm{diag}(\lambda_1, \ldots, \lambda_n)\mathrm{U}^*.$$

Proof In this case we will prove the statement about linear maps; the statement about matrices then follows from that, together with the fact (see Quick Exercise #10 in Section 5.3) that eigenvalues of self-adjoint maps are real.

By Lemma 5.18, T has an eigenvector. By rescaling we can assume this eigenvector is a unit vector; call it e_1, and let λ_1 be the corresponding eigenvalue.

Now if $\langle v, e_1 \rangle = 0$, then

$$\langle Tv, e_1 \rangle = \langle v, Te_1 \rangle = \lambda_1 \langle v, e_1 \rangle = 0,$$

so T maps the subspace $\langle e_1 \rangle^\perp$ to itself.

Quick Exercise #11. Prove that T is self-adjoint on $\langle e_1 \rangle^\perp$.

By Lemma 5.18, T has an eigenvector in $\langle e_1 \rangle^\perp$, which we may assume is a unit vector e_2. Thus e_1 and e_2 are orthonormal eigenvectors of T.

We continue in this way: after constructing orthonormal eigenvectors $e_1, \ldots, e_k$, T restricts to a self-adjoint map on $\langle e_1, \ldots, e_k \rangle^\perp$. If $\dim V > k$, then $\langle e_1, \ldots, e_k \rangle^\perp \neq \{0\}$, and so by Lemma 5.18, T has a unit eigenvector $e_{k+1} \in \langle e_1, \ldots, e_k \rangle^\perp$. Since V is finite-dimensional, this process will terminate once $k = \dim V$, at which point $(e_1, \ldots, e_k)$ will be an orthonormal basis consisting of eigenvectors of T. ▲

The factorization $\mathrm{A} = \mathrm{U}\,\mathrm{diag}(\lambda_1, \ldots, \lambda_n)\mathrm{U}^*$ in Theorem 5.19 is called a **spectral decomposition** of A.

It should be kept in mind that Theorem 5.19 gives a necessary and sufficient condition for *unitary (orthogonal)* diagonalizability in the case of all real eigenvalues; there are linear maps which can be diagonalized only by a non-orthonormal basis.

Example Let $\mathrm{A} = \begin{bmatrix} 1 & 2 \\ 0 & 3 \end{bmatrix}$. Clearly A is not Hermitian, but since A has both 1 and 3 as eigenvalues, there must be (necessarily linearly independent) corresponding eigenvectors, and so A is diagonalizable. Indeed,

$$\mathrm{A}\begin{bmatrix} 1 \\ 0 \end{bmatrix} = \begin{bmatrix} 1 \\ 0 \end{bmatrix} \quad \text{and} \quad \mathrm{A}\begin{bmatrix} 1 \\ 1 \end{bmatrix} = 3\begin{bmatrix} 1 \\ 1 \end{bmatrix},$$

so A is diagonal in the nonorthogonal basis $\left(\begin{bmatrix} 1 \\ 0 \end{bmatrix}, \begin{bmatrix} 1 \\ 1 \end{bmatrix} \right)$, but A cannot be diagonalized by any orthonormal basis; see Exercise 5.4.17. ▲

The Spectral Theorem is a powerful tool for working with self-adjoint maps and Hermitian matrices. We present the following result as one important sample application.

QA #11: Given $v, w \in \langle e_1 \rangle^\perp$, $\langle Tv, w \rangle = \langle v, Tw \rangle$, so $T^* = T$ whether we think of T acting on all of V or just this subspace.

Theorem 5.20 *Let $A \in M_n(\mathbb{C})$ be Hermitian. The following are equivalent:*

1. *Every eigenvalue of A is positive.*
2. *For every $0 \neq x \in \mathbb{C}^n$, we have $\langle Ax, x\rangle > 0$.*

Definition A matrix satisfying the equivalent properties in Theorem 5.20 is called **positive definite.**

Proof of Theorem 5.20 Suppose first that every eigenvalue of A is positive. By the Spectral Theorem, we can write

$$A = U \operatorname{diag}(\lambda_1, \ldots, \lambda_n) U^*,$$

where U is unitary and $\lambda_1, \ldots, \lambda_n$ are the eigenvalues of A. Therefore

$$\langle Ax, x\rangle = \left\langle U \operatorname{diag}(\lambda_1, \ldots, \lambda_n) U^* x, x\right\rangle = \left\langle \operatorname{diag}(\lambda_1, \ldots, \lambda_n) U^* x, U^* x\right\rangle.$$

Let $y = U^* x$; since U^* acts as an isometry, $y \neq 0$ if $x \neq 0$. Then

$$\langle Ax, x\rangle = \langle \operatorname{diag}(\lambda_1, \ldots, \lambda_n) y, y\rangle = \sum_{j=1}^{n} \lambda_j \left|y_j\right|^2 > 0$$

since $\lambda_j > 0$ for each j.

Conversely, suppose that $\langle Ax, x\rangle > 0$ for every $x \neq 0$. Let λ be an eigenvalue of A with eigenvector x. Then

$$0 < \langle Ax, x\rangle = \langle \lambda x, x\rangle = \lambda \|x\|^2,$$

so $\lambda > 0$. ▲

There are some computations which are possible in principle to do directly but are greatly facilitated by the Spectral Theorem. For example, suppose that A is a Hermitian matrix, and let $k \in \mathbb{N}$. If $A = UDU^*$ is a spectral decomposition of A with $D = \operatorname{diag}(\lambda_1, \ldots, \lambda_n)$, then

$$A^k = (UDU^*)^k = \underbrace{(UDU^*) \cdots (UDU^*)}_{k \text{ times}} = UD^kU^* = U \operatorname{diag}(\lambda_1^k, \ldots, \lambda_n^k) U^*.$$

This observation makes it easy to calculate arbitrarily high powers of a Hermitian matrix. It also suggests how to go about defining more general functions of a matrix. Consider, for example, the exponential function $f(x) = e^x$. By Taylor's Theorem,

$$f(x) = \sum_{k=0}^{\infty} \frac{x^k}{k!},$$

which suggests defining the **matrix exponential** for $A \in M_n(\mathbb{C})$ by

$$e^A := \sum_{k=0}^{\infty} \frac{1}{k!} A^k.$$

Now, if A is Hermitian and has a spectral decomposition as above, then

$$e^A = \sum_{k=0}^{\infty} \frac{1}{k!} UD^kU^* = U\left(\sum_{k=0}^{\infty} \frac{1}{k!} D^k\right)U^* = U\,\mathrm{diag}(e^{\lambda_1}, \dots, e^{\lambda_n})U^*.$$

This leads us to a general definition for a function of a matrix: if $f : \mathbb{R} \to \mathbb{R}$, we define

$$f(A) := U\,\mathrm{diag}(f(\lambda_1), \dots, f(\lambda_n))U^*.$$

This route to defining $f(A)$ is sometimes called the **functional calculus**. The resulting functions sometimes, but not always, behave as you'd expect.

Examples

1. Suppose that A is Hermitian with nonnegative eigenvalues. Then $\sqrt{A}$ can be defined via the functional calculus as described above, and

$$(\sqrt{A})^2 = (U\,\mathrm{diag}(\sqrt{\lambda_1}, \dots, \sqrt{\lambda_n})U^*)^2 = U\,\mathrm{diag}(\lambda_1, \dots, \lambda_n)U^* = A.$$

2. If A and B are Hermitian and commute, then $e^{A+B} = e^A e^B$, but if A and B do not commute, this need not be the case (see Exercise 5.4.7). ▲

Normal Maps and Matrices

Besides the potential nonorthogonality of a basis of eigenvectors, another issue with self-adjoint maps and Hermitian matrices is that they necessarily have real eigenvalues. Theorem 5.19 thus has nothing to say about diagonalizability for any maps or matrices with nonreal eigenvalues.

Let's consider under what circumstances we could hope to prove a result like Theorem 5.19 for nonreal eigenvalues. Working in the matrix-theoretic framework, suppose that $A = U\,\mathrm{diag}(\lambda_1, \dots, \lambda_n)U^*$ for a unitary matrix $U \in M_n(\mathbb{C})$ and complex numbers $\lambda_1, \dots, \lambda_n \in \mathbb{C}$. Then

$$A^* = U\,\mathrm{diag}(\lambda_1, \dots, \lambda_n)^*U^* = U\,\mathrm{diag}(\overline{\lambda_1}, \dots, \overline{\lambda_n})U^*,$$

and so

$$\begin{aligned} A^*A &= U\,\mathrm{diag}(\overline{\lambda_1}, \dots, \overline{\lambda_n})U^*U\,\mathrm{diag}(\lambda_1, \dots, \lambda_n)U^* \\ &= U\,\mathrm{diag}(|\lambda_1|^2, \dots, |\lambda_n|^2)U^* \\ &= AA^*. \end{aligned}$$

So for a start, we need at least to have $A^*A = AA^*$.

Definition A matrix $A \in M_n(\mathbb{C})$ is **normal** if $A^*A = AA^*$.
A linear map $T \in \mathcal{L}(V)$ is **normal** if $T^*T = TT^*$.

By Theorem 5.13, a linear map is normal if and only if its matrix with respect to an orthonormal basis is normal.

Examples

1. Any self-adjoint map or Hermitian matrix is normal.
2. Any isometry or unitary matrix is normal: if T is an isometry, then $T^* = T^{-1}$, so $TT^* = I = T^*T$. ▲

The observations above mean that for T to be diagonalized by an orthonormal basis, T must be normal; it turns out that this condition is *necessary and sufficient*; that is, a linear map can be diagonalized by an orthonormal basis if and only if it is normal. The following lemma gives an alternative characterization of normality which will be used in the proof.

Lemma 5.21 *Let V be a complex inner product space. Given $T \in \mathcal{L}(V)$, define*

$$T_r := \frac{1}{2}(T + T^*) \qquad \text{and} \qquad T_i := \frac{1}{2i}(T - T^*).$$

Then T_r and T_i are both self-adjoint, and T is normal if and only if

$$T_r T_i = T_i T_r.$$

Proof The self-adjointness of T_r follows immediately from Proposition 5.14; for T_i,

$$T_i^* = \overline{\left(\frac{1}{2i}\right)}(T^* - T) = -\frac{1}{2i}(T^* - T) = T_i.$$

For the second claim, note that

$$T_r T_i - T_i T_r = \frac{1}{2i}(T^*T - TT^*),$$

and so $T_r T_i - T_i T_r = 0$ if and only if $T^*T - TT^* = 0$, that is, if and only if T is normal. ▲

Lemma 5.22 *Suppose that $S, T \in \mathcal{L}(V)$ and that $ST = TS$. If λ is an eigenvalue of T and $v \in \operatorname{Eig}_\lambda(T)$, then $Sv \in \operatorname{Eig}_\lambda(T)$.*

Quick Exercise #12. Prove Lemma 5.22.

Theorem 5.23 (The Spectral Theorem for normal maps and normal matrices) *If V is a finite-dimensional inner product space over $\mathbb{C}$ and $T \in \mathcal{L}(V)$ is normal, then there is an orthonormal basis of V consisting of eigenvectors of T.*

If $\mathrm{A} \in \mathrm{M}_n(\mathbb{C})$ is normal, then there exist a unitary matrix $\mathrm{U} \in \mathrm{M}_n(\mathbb{C})$ and complex numbers $\lambda_1, \ldots, \lambda_n \in \mathbb{C}$ such that

$$\mathrm{A} = \mathrm{U}\,\mathbf{diag}(\lambda_1, \ldots, \lambda_n)\mathrm{U}^*.$$

Proof Again we prove the result for linear maps; the version for matrices follows.

Define T_r and T_i as in Lemma 5.21. For each eigenvalue λ of T_r, it follows from Lemma 5.22 that T_i maps $\mathrm{Eig}_\lambda(T_r)$ into itself, and is a self-adjoint map on that subspace (see Quick Exercise #11). So by Theorem 5.19, there is an orthonormal basis of $\mathrm{Eig}_\lambda(T_r)$ consisting of eigenvectors of T_i; these vectors are then eigenvectors of both T_r and T_i.

Putting all these bases together, we get an orthonormal basis $(e_1, \ldots, e_n)$ of V, each member of which is an eigenvector of both T_r and T_i, though with different eigenvalues in general. So for each j, there are constants $\lambda_j, \mu_j \in \mathbb{R}$ such that

$$T_r e_j = \lambda_j e_j \qquad \text{and} \qquad T_i e_j = \mu_j e_j,$$

and so

$$Te_j = (T_r + iT_i)e_j = (\lambda_j + i\mu_j)e_j.$$

Thus each e_j is an eigenvector of T as well. ▲

Example Let $\mathrm{A} := \begin{bmatrix} 0 & 1 \\ -1 & 0 \end{bmatrix}$. As we've seen, multiplication by A corresponds to a rotation of $\mathbb{R}^2$ clockwise by 90°, and in particular, A has no real eigenvalues. However, A is certainly normal (it is an isometry!), and so it follows from Theorem 5.23 that A can be unitarily diagonalized over $\mathbb{C}$. Indeed,

$$\mathrm{A} - \lambda\mathrm{I} = \begin{bmatrix} -\lambda & 1 \\ -1 & -\lambda \end{bmatrix} \xrightarrow{\text{R3,R1}} \begin{bmatrix} 1 & \lambda \\ 0 & 1+\lambda^2 \end{bmatrix},$$

and so λ is an eigenvalue of A if and only if $1 + \lambda^2 = 0$, that is, if and only if $\lambda = \pm i$.

Quick Exercise #13. Check that $(i, -1)$ is an eigenvector for $\lambda = i$ and that $(i, 1)$ is an eigenvector for $\lambda = -i$.

QA #12: If $v \in \mathrm{Eig}_\lambda(T)$, then $T(Sv) = STv = \lambda Sv$.

We thus take

$$U = \frac{1}{\sqrt{2}} \begin{bmatrix} i & i \\ -1 & 1 \end{bmatrix},$$

which is indeed unitary (check!), and we have that

$$A = \left(\frac{1}{\sqrt{2}} \begin{bmatrix} i & i \\ -1 & 1 \end{bmatrix} \right) \begin{bmatrix} i & 0 \\ 0 & -i \end{bmatrix} \left(\frac{1}{\sqrt{2}} \begin{bmatrix} -i & -1 \\ -i & 1 \end{bmatrix} \right).$$ ▲

Schur Decomposition

Theorem 5.23 is mathematically very satisfying, because it gives us necessary and sufficient conditions for finding an orthonormal basis of eigenvectors, that is, being able to "unitarily diagonalize" a map or a matrix. The downside is that not all matrices and maps can be unitarily diagonalized; not all matrices and maps are normal. However, we do have the next best thing: all matrices and maps over $\mathbb{C}$ can be unitarily *triangularized*. The fact that this only works for complex inner product spaces and matrices is to be expected; recall that we showed in Section 3.7 that in order to be able to triangularize *all* matrices, the base field needs to be algebraically closed.

Corollary 5.24 (The Schur decomposition) *Suppose that V is a finite-dimensional complex inner product space, and $T \in \mathcal{L}(V)$. Then there is an orthonormal basis $\mathcal{B}$ of V such that $[T]_{\mathcal{B}}$ is upper triangular.*

Equivalently, if $A \in M_n(\mathbb{C})$, then there exist a unitary matrix $U \in M_n(\mathbb{C})$ and an upper triangular matrix $T \in M_n(\mathbb{C})$ such that $A = UTU^*$.

Proof We start with the statement for matrices. Recall that every matrix over $\mathbb{C}$ can be triangularized: by Theorem 3.67, there exist an invertible matrix $S \in M_n(\mathbb{C})$ and an upper triangular matrix $B \in M_n(\mathbb{C})$ such that $A = SBS^{-1}$. Let $S = QR$ be the QR decomposition of S, so that Q is unitary and R is upper triangular. Then

$$A = QRB(QR)^{-1} = QRBR^{-1}Q^{-1} = Q(RBR^{-1})Q^*.$$

We've seen (see Exercises 2.3.12 and 2.4.19) that inverses and products of upper triangular matrices are upper triangular, so RBR^{-1} is upper triangular. Corollary 5.24 thus follows, with $U = Q$ and $T = RBR^{-1}$.

The statement for maps follows from the statement for matrices: let $T \in \mathcal{L}(V)$ as above, and let $A = [T]_{\mathcal{B}_0}$, for some orthonormal basis $\mathcal{B}_0$ of V. Then there is a unitary matrix $U \in M_n(\mathbb{C})$ and an upper triangular matrix $R \in M_n(\mathbb{C})$ such that $A = URU^*$; that is,

$$U^* [T]_{\mathcal{B}_0} U = R.$$

Define a basis $\mathcal{B} = (v_1, \dots, v_n)$ of V so that $\left[v_j\right]_{\mathcal{B}_0} = \mathbf{u}_j$, the jth column of U. Then $\mathcal{B}$ is indeed a basis of V, because U is invertible, and moreover, since $\mathcal{B}_0$ is orthonormal,

$$\langle v_i, v_j \rangle = \langle \mathbf{u}_i, \mathbf{u}_j \rangle = \begin{cases} 1 & \text{if } i = j, \\ 0 & \text{otherwise.} \end{cases}$$

By construction, $[I]_{\mathcal{B},\mathcal{B}_0} = \mathrm{U}$, and so we have an upper triangular R such that

$$\mathrm{R} = \mathrm{U}^* \, [T]_{\mathcal{B}_0} \, \mathrm{U} = [I]_{\mathcal{B}_0,\mathcal{B}} \, [T]_{\mathcal{B}_0} \, [I]_{\mathcal{B},\mathcal{B}_0} = [T]_{\mathcal{B}} \, .$$ ▲

Quick Exercise #14. (a) Show that an upper triangular Hermitian matrix is diagonal.
(b) Use this fact and the Schur decomposition to prove the Spectral Theorem for complex Hermitian matrices.

Example Let $\mathrm{A} = \begin{bmatrix} 6 & -2 \\ 2 & 2 \end{bmatrix}$. A quick check shows that A is not normal, and so it cannot be unitarily diagonalized. To find a Schur decomposition of A, we first look for an eigenvector (which will be the first column of U):

$$\mathrm{A} - \lambda \mathrm{I} = \begin{bmatrix} 6-\lambda & -2 \\ 2 & 2-\lambda \end{bmatrix} \xrightarrow{\text{R3,R1}} \begin{bmatrix} 1 & 1 - \frac{\lambda}{2} \\ 0 & -\frac{1}{2}\left(\lambda^2 - 8\lambda + 16\right) \end{bmatrix},$$

and so A has a single eigenvalue $\lambda = 4$. An eigenvector is an element of the null space of

$$\mathrm{A} - 4\mathrm{I} \longrightarrow \begin{bmatrix} 1 & -1 \\ 0 & 0 \end{bmatrix},$$

so we can choose the first column of U to be $\mathbf{u}_1 = \frac{1}{\sqrt{2}} \begin{bmatrix} 1 \\ 1 \end{bmatrix}$. Since we only need a 2×2 unitary matrix U which triangularizes A, we can take $\mathbf{u}_2$ to be any unit vector orthogonal to $\mathbf{u}_1$; for example, $\mathbf{u}_2 = \frac{1}{\sqrt{2}} \begin{bmatrix} 1 \\ -1 \end{bmatrix}$. We can now solve for T:

$$\mathrm{A} = \mathrm{UTU}^* \quad \Longleftrightarrow \quad \mathrm{T} = \mathrm{U}^*\mathrm{AU}$$

$$= \frac{1}{2} \begin{bmatrix} 1 & 1 \\ 1 & -1 \end{bmatrix} \begin{bmatrix} 6 & -2 \\ 2 & 2 \end{bmatrix} \begin{bmatrix} 1 & 1 \\ 1 & -1 \end{bmatrix} = \begin{bmatrix} 4 & 4 \\ 0 & 4 \end{bmatrix}.$$

▲

QA #14: (a) Obvious. (b) If A is Hermitian and A = UTU* is its Schur decomposition, then T = U*AU = (U*AU)* = T*, so T is diagonal. Notice that this *doesn't* prove that U can be taken to be real orthogonal if $\mathrm{A} \in \mathrm{M}_n(\mathbb{R})$ is symmetric. That takes more work to prove.

This was of course a rather *ad hoc* way to unitarily triangularize A. In practice, the Schur decomposition of a matrix is found by an iterative version of the QR algorithm, but we will not discuss it here. Our main interest is in the purely theoretical fact that it is always possible to unitarily triangularize a matrix over $\mathbb{C}$.

KEY IDEAS

- A map or matrix is normal if it commutes with its adjoint.
- The Spectral Theorem for maps: If $T \in \mathcal{L}(V)$ is normal and V is complex, then V has an orthonormal basis of eigenvectors of T. If T is self-adjoint, then V has an orthonormal basis of eigenvectors of T and the eigenvalues are real.
- The Spectral Theorem for matrices: If $A \in M_n(\mathbb{C})$ is normal, there is a unitary U such that $A = UDU^*$, where D is diagonal with the eigenvalues of A on the diagonal. If A is Hermitian, then the entries of D are real. If A is real symmetric, then U is orthogonal.
- The Schur decomposition: If $A \in M_n(\mathbb{C})$, there is a unitary U and upper triangular T such that $A = UTU^*$.

EXERCISES

5.4.1 Find a spectral decomposition of each of the following matrices.

(a) $\begin{bmatrix} 1 & 2 \\ 2 & 4 \end{bmatrix}$ (b) $\begin{bmatrix} 1 & -1 \\ 1 & 1 \end{bmatrix}$ (c) $\begin{bmatrix} 2 & 0 & 1 \\ 0 & -1 & 0 \\ 1 & 0 & 2 \end{bmatrix}$ (d) $\begin{bmatrix} 1 & 0 & 1 \\ 0 & 1 & 1 \\ 1 & 1 & 0 \end{bmatrix}$

5.4.2 Find a spectral decomposition of each of the following matrices.

(a) $\begin{bmatrix} 7 & 6 \\ 6 & -2 \end{bmatrix}$ (b) $\begin{bmatrix} 1 & i \\ i & 1 \end{bmatrix}$ (c) $\begin{bmatrix} 1 & 0 & 0 \\ 0 & 0 & 2 \\ 0 & 2 & 0 \end{bmatrix}$ (d) $\begin{bmatrix} 1 & 0 & -1 & 0 \\ 0 & 1 & 0 & 1 \\ -1 & 0 & 1 & 0 \\ 0 & 1 & 0 & 1 \end{bmatrix}$

5.4.3 Find a Schur decomposition of each of the following matrices.

(a) $\begin{bmatrix} 1 & 0 \\ 2 & 3 \end{bmatrix}$ (b) $\begin{bmatrix} 1 & -1 \\ -2 & 0 \end{bmatrix}$

5.4.4 Find a Schur decomposition of each of the following matrices.

(a) $\begin{bmatrix} 3 & 0 \\ 1 & 4 \end{bmatrix}$ (b) $\begin{bmatrix} 3 & -1 \\ 1 & 1 \end{bmatrix}$

5.4.5 Prove that if $A \in M_n(\mathbb{R})$ is symmetric, then there is a matrix $B \in M_n(\mathbb{R})$ such that $B^3 = A$.

5.4.6 Suppose that A is an invertible Hermitian matrix. Show that defining $\frac{1}{A}$ via the functional calculus does indeed produce a matrix inverse to A.

5.4.7 Let A and B be Hermitian matrices. Show that if A and B commute, then $e^{A+B} = e^A e^B$. Show that if A and B do not commute, then this need not be true.

5.4.8 Let A be a Hermitian matrix. Show that e^A is positive definite.

5.4.9 Let $A \in M_n(\mathbb{C})$ be Hermitian. Show that the following are equivalent:

(a) Every eigenvalue of A is nonnegative.

(b) For every $x \in \mathbb{C}^n$, we have $\langle Ax, x \rangle \geq 0$.

A matrix satisfying these properties is called **positive semidefinite.**

5.4.10 Let $A \in M_{m,n}(\mathbb{C})$

(a) Show that A^*A is positive semidefinite (see Exercise 5.4.9).

(b) Show that if rank $A = n$, then A^*A is positive definite.

5.4.11 Prove that if $A \in M_n(\mathbb{F})$ is positive definite, then there is an upper triangular matrix $X \in M_n(\mathbb{F})$ such that $A = X^*X$. (This is known as a **Cholesky decomposition** of A.)

5.4.12 Let V be a finite-dimensional inner product space, and suppose that $S, T \in \mathcal{L}(V)$ are both self-adjoint and all their eigenvalues are nonnegative. Show that all the eigenvalues of $S + T$ are nonnegative.

5.4.13 Show that if $A \in M_n(\mathbb{C})$ is positive definite, then $\langle x, y \rangle = y^*Ax$ defines an inner product on $\mathbb{C}^n$.

5.4.14 Suppose that V is a finite-dimensional complex inner product space. Prove that if $T \in \mathcal{L}(V)$ is normal, then there is a map $S \in \mathcal{L}(V)$ such that $S^2 = T$.

5.4.15 A **circulant matrix** is a matrix of the form

$$\begin{bmatrix} a_1 & a_2 & a_3 & \cdots & a_{n-1} & a_n \\ a_n & a_1 & a_2 & \cdots & a_{n-2} & a_{n-1} \\ a_{n-1} & a_n & a_1 & \ddots & & \vdots \\ \vdots & \vdots & \ddots & \ddots & \ddots & \vdots \\ a_3 & a_4 & & \ddots & a_1 & a_2 \\ a_2 & a_3 & \cdots & \cdots & a_n & a_1 \end{bmatrix}$$

where $a_1, \ldots, a_n \in \mathbb{C}$.

(a) Prove that if A and B are $n \times n$ circulant matrices, then $AB = BA$.

(b) Prove that every circulant matrix is normal.

5.4.16 Suppose that $C \in M_n(\mathbb{C})$ is a circulant matrix (see Exercise 5.4.15) and that $F \in M_n(\mathbb{C})$ is the DFT matrix (see Exercise 4.5.3). Show that $C = FDF^*$ for some diagonal matrix D.

5.4.17 (a) Let $A \in M_n(\mathbb{C})$ have only real eigenvalues. Show that if there is an orthonormal basis of $\mathbb{C}^n$ consisting of eigenvectors of A, then A is Hermitian.

(b) Let $A \in M_n(\mathbb{R})$ have only real eigenvalues. Show that if there is an orthonormal basis of $\mathbb{R}^n$ consisting of eigenvectors of A, then A is symmetric.

5.4.18 Let $U \in M_n(\mathbb{C})$ be unitary. Show that there is a unitary matrix $V \in M_n(\mathbb{C})$ such that V^*UV is diagonal, with diagonal entries with absolute value 1.

If U is assumed to be real and orthogonal, can V be taken to be orthogonal?

5.4.19 Let V be a finite-dimensional inner product space and $T \in \mathcal{L}(V)$.

(a) Suppose that $\mathbb{F} = \mathbb{R}$. Show that T is self-adjoint if and only if V is the orthogonal direct sum of the eigenspaces of T.

(b) Suppose that $\mathbb{F} = \mathbb{C}$. Show that T is normal if and only if V is the orthogonal direct sum of the eigenspaces of T.

(c) Suppose that $\mathbb{F} = \mathbb{C}$. Show that T is self-adjoint if and only if all the eigenvalues of T are real and V is the orthogonal direct sum of the eigenspaces of T.

5.4.20 Prove that if $A \in M_n(\mathbb{F})$ is Hermitian, then there are an orthonormal basis $(v_1, \ldots, v_n)$ of $\mathbb{F}^n$ and numbers $\lambda_1, \ldots, \lambda_n \in \mathbb{R}$ such that

$$A = \sum_{j=1}^{n} \lambda_j v_j v_j^*.$$

5.4.21 Suppose that V is a finite-dimensional inner product space and $T \in \mathcal{L}(V)$ is self-adjoint. Prove that there are subspaces $U_1, \ldots, U_m$ of V and numbers $\lambda_1, \ldots, \lambda_m \in \mathbb{R}$ such that V is the orthogonal direct sum of $U_1, \ldots, U_m$ and

$$T = \sum_{j=1}^{m} \lambda_j P_{U_j}.$$

5.4.22 Show that if $A \in M_n(\mathbb{C})$ is normal, then eigenvectors of A with distinct eigenvalues are orthogonal.

5.4.23 Let V be a finite-dimensional inner product space, and suppose that $T \in \mathcal{L}(V)$ is self-adjoint. By reordering if necessary, let $(e_1, \ldots, e_n)$ be an orthonormal basis of V consisting of eigenvectors of T, such that the corresponding eigenvalues $\lambda_1, \ldots, \lambda_n$ are nonincreasing (i.e., $\lambda_1 \geq \cdots \geq \lambda_n$).

(a) For $j = 1, \ldots, n$, let $U_j = \langle e_j, \ldots, e_n \rangle$. Show that, for every $v \in U_j$, we have $\langle Tv, v \rangle \leq \lambda_j \|v\|^2$.

(b) For $j = 1, \ldots, n$, let $V_j = \langle e_1, \ldots, e_j \rangle$. Show that, for every $v \in V_j$, we have $\langle Tv, v \rangle \geq \lambda_j \|v\|^2$.

(c) Show that if U is a subspace of V with $\dim U = n - j + 1$, then there exists a $v \in U$ such that $\langle Tv, v \rangle \geq \lambda_j \|v\|^2$.

(d) Use the above results to show that

$$\lambda_j = \min_{\substack{U \subseteq V \\ \dim U = n-j+1}} \max_{\substack{v \in U \\ \|v\|=1}} \langle Tv, v \rangle .$$

Remark: This is known as the **Courant–Fischer min–max principle.**

5.4.24 Let $A \in M_n(\mathbb{C})$ be normal, and let λ be an eigenvalue of A. Show that, in any spectral decomposition of A, the number of times that λ appears in the diagonal factor is

$$\dim \ker(A - \lambda I_n).$$

5.4.25 Suppose that $A \in M_n(\mathbb{C})$ has distinct eigenvalues $\lambda_1, \ldots, \lambda_n$. Prove that

$$\sqrt{\sum_{j=1}^{n} |\lambda_j|^2} \leq \|A\|_F.$$

Hint: Prove this first for upper triangular matrices, then use Schur decomposition.

5.4.26 Let $A \in M_n(\mathbb{C})$ and let $\varepsilon > 0$. Show that there is a $B \in M_n(\mathbb{C})$ with n distinct eigenvalues such that $\|A - B\|_F \leq \varepsilon$.
Hint: First consider the case where A is upper triangular, then use the Schur decomposition.

5.4.27 (a) Prove that if $A \in M_n(\mathbb{C})$ is upper triangular and normal, then A is diagonal.
(b) Use this fact and the Schur decomposition to give another proof of the Spectral Theorem for normal matrices.

6

Determinants

After spending the last two chapters working exclusively over the fields $\mathbb{R}$ and $\mathbb{C}$, in this chapter we will once again allow the scalar field $\mathbb{F}$ to be any field at all.*

6.1 Determinants

Multilinear Functions

In this section we will meet a matrix invariant which, among other things, encodes invertibility. Of course we already have good algorithms based on Gaussian elimination to determine whether a matrix is invertible. But, for some theoretical purposes, it can be useful to have a formula, because a formula is often better than an algorithm as a starting point for future analysis.

We begin by noting that in the case of 1×1 matrices, there's a simple invariant which indicates whether the matrix is invertible: if $D([a]) = a$, then the matrix $[a]$ is invertible if and only if $D([a]) \neq 0$.

Quick Exercise #1. Show that $D([a]) = a$ really is an invariant of 1×1 matrices.

More generally, we've seen that an $n \times n$ matrix A fails to be invertible if and only if there is a linear dependence among the columns. The most obvious way for this to happen is if two columns are actually identical; we take this as motivation for the following definition.

Definition A function $f : \mathrm{M}_n(\mathbb{F}) \to \mathbb{F}$ is called **isoscopic**† if $f(\mathrm{A}) = 0$ whenever A has two identical columns.

*Although as usual, if you prefer to just stick with $\mathbb{R}$ and $\mathbb{C}$, go ahead.
†*Isoscopic* is a nonce word used here for convenience of exposition.

QA #1: If $[a]$ is similar to $[b]$, then there is $s \in \mathbb{F}$, $s \neq 0$ such that $[a] = [s][b][s]^{-1}$; i.e., $a = sbs^{-1} = b$.

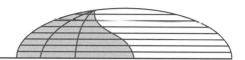

Of course, we want our function to detect *any* linear dependence among the columns, not just this very simple type. One way to do this is to require the function to be isoscopic and also to have the following property.

Definition A function $D : \mathrm{M}_n(\mathbb{F}) \to \mathbb{F}$ is called **multilinear** if, for every $\mathbf{a}_1, \dots, \mathbf{a}_n$ and $\mathbf{b}_1, \dots, \mathbf{b}_n \in \mathbb{F}^n$ and $c \in \mathbb{F}$, for each $j = 1, \dots, n$,

$$D\left(\begin{bmatrix} | & & | & & | \\ \mathbf{a}_1 & \cdots & \mathbf{a}_j + \mathbf{b}_j & \cdots & \mathbf{a}_n \\ | & & | & & | \end{bmatrix}\right)$$

$$= D\left(\begin{bmatrix} | & & | & & | \\ \mathbf{a}_1 & \cdots & \mathbf{a}_j & \cdots & \mathbf{a}_n \\ | & & | & & | \end{bmatrix}\right) + D\left(\begin{bmatrix} | & & | & & | \\ \mathbf{a}_1 & \cdots & \mathbf{b}_j & \cdots & \mathbf{a}_n \\ | & & | & & | \end{bmatrix}\right)$$

and

$$D\left(\begin{bmatrix} | & & | & & | \\ \mathbf{a}_1 & \cdots & c\mathbf{a}_j & \cdots & \mathbf{a}_n \\ | & & | & & | \end{bmatrix}\right) = cD\left(\begin{bmatrix} | & & | & & | \\ \mathbf{a}_1 & \cdots & \mathbf{a}_j & \cdots & \mathbf{a}_n \\ | & & | & & | \end{bmatrix}\right).$$

That is, $D(\mathrm{A})$ is linear when thought of as a function of one of the columns of A, with the other columns held fixed.

Lemma 6.1 *Suppose* $D : \mathrm{M}_n(\mathbb{F}) \to \mathbb{F}$ *is an isoscopic multilinear function. If* $\mathrm{A} \in \mathrm{M}_n(\mathbb{F})$ *is singular, then* $D(\mathrm{A}) = 0$.

Proof If A is singular, then rank $\mathrm{A} < n$, so some column of A is a linear combination of the other columns. Suppose that $\mathbf{a}_j = \sum_{k \neq j} c_k \mathbf{a}_k$ for some scalars $\{c_k\}_{k \neq j}$. Then by multilinearity,

$$\begin{aligned} D(\mathrm{A}) &= D\left(\begin{bmatrix} | & & | & | & | & & | \\ \mathbf{a}_1 & \cdots & \mathbf{a}_{j-1} & \sum_{k \neq j} c_k \mathbf{a}_k & \mathbf{a}_{j+1} & \cdots & \mathbf{a}_n \\ | & & | & | & | & & | \end{bmatrix}\right) \\ &= \sum_{k \neq j} c_k D\left(\begin{bmatrix} | & & | & | & | & & | \\ \mathbf{a}_1 & \cdots & \mathbf{a}_{j-1} & \mathbf{a}_k & \mathbf{a}_{j+1} & \cdots & \mathbf{a}_n \\ | & & | & | & | & & | \end{bmatrix}\right). \end{aligned}$$

Since D is isoscopic, each term of this sum is 0, and so $D(\mathrm{A}) = 0$. ▲

Examples

1. The function $D : \mathrm{M}_n(\mathbb{F}) \to \mathbb{F}$ given by $D(\mathrm{A}) = a_{11}a_{22}\cdots a_{nn}$ is multilinear, but not isoscopic. However, if D is restricted to the set of upper triangular matrices, then D is isoscopic: if two columns of an upper triangular matrix are identical, then the one farther to the right has a zero in the diagonal entry.
2. More generally, pick any $i_1, \ldots, i_n \in \{1, \ldots, n\}$. The function $D : \mathrm{M}_n(\mathbb{F}) \to \mathbb{F}$ given by $D(\mathrm{A}) = a_{i_1 1}a_{i_2 2}\cdots a_{i_n n}$ is multilinear (but not isoscopic). ▲

Quick Exercise #2. Suppose that $D : \mathrm{M}_n(\mathbb{F}) \to \mathbb{F}$ is multilinear, $\mathrm{A} \in \mathrm{M}_n(\mathbb{F})$, and $c \in \mathbb{F}$. What is $D(c\mathrm{A})$?

In the context of multilinear functions, there is a different term which is normally used in place of what we have called isoscopic.

Definition A multilinear function $D : \mathrm{M}_n(\mathbb{F}) \to \mathbb{F}$ is called **alternating** if

$$D\left(\begin{bmatrix} | & & | \\ \mathbf{a}_1 & \cdots & \mathbf{a}_n \\ | & & | \end{bmatrix}\right) = 0$$

whenever there are $i \neq j$ such that $\mathbf{a}_i = \mathbf{a}_j$.

The reason for this terminology is given by the following lemma.

Lemma 6.2 *Suppose $D : \mathrm{M}_n(\mathbb{F}) \to \mathbb{F}$ is an alternating multilinear function. Given $\mathrm{A} \in \mathrm{M}_n(\mathbb{F})$ and $1 \leq i < j \leq n$, let $\mathrm{B} \in \mathrm{M}_n(\mathbb{F})$ be the matrix obtained from* A *by exchanging the ith and jth columns. Then $D(\mathrm{B}) = -D(\mathrm{A})$.*

Proof Write $\mathrm{A} = \begin{bmatrix} | & & | \\ \mathbf{a}_1 & \cdots & \mathbf{a}_n \\ | & & | \end{bmatrix}$. Assume without loss of generality that $i < j$, so that

$$\mathrm{B} = \begin{bmatrix} | & & | & & | & & | \\ \mathbf{a}_1 & \cdots & \mathbf{a}_j & \cdots & \mathbf{a}_i & \cdots & \mathbf{a}_n \\ | & & | & & | & & | \end{bmatrix}.$$

QA #2: $c^n D(\mathrm{A})$

Since D is alternating,

$$D\left(\begin{bmatrix} | & & | & & | & & | \\ \mathrm{a}_1 & \cdots & \mathrm{a}_i + \mathrm{a}_j & \cdots & \mathrm{a}_i + \mathrm{a}_j & \cdots & \mathrm{a}_n \\ | & & | & & | & & | \end{bmatrix}\right) = 0.$$

On the other hand, by multilinearity the left-hand side is equal to

$$D(\mathrm{A}) + D(\mathrm{B}) + D\left(\begin{bmatrix} | & & | & & | & & | \\ \mathrm{a}_1 & \cdots & \mathrm{a}_i & \cdots & \mathrm{a}_i & \cdots & \mathrm{a}_n \\ | & & | & & | & & | \end{bmatrix}\right)$$
$$+ D\left(\begin{bmatrix} | & & | & & | & & | \\ \mathrm{a}_1 & \cdots & \mathrm{a}_j & \cdots & \mathrm{a}_j & \cdots & \mathrm{a}_n \\ | & & | & & | & & | \end{bmatrix}\right).$$

Since D is alternating, the last two terms are both 0, and so $D(\mathrm{A}) + D(\mathrm{B}) = 0$. ▲

The Determinant

Theorem 6.3 *For each n, there is a unique alternating multilinear function $D : \mathrm{M}_n(\mathbb{F}) \to \mathbb{F}$ such that $D(\mathrm{I}_n) = 1$.*

This function is called the **determinant**; *the determinant of a matrix* A *is denoted* $\det(\mathrm{A})$.

We will prove Theorem 6.3 later in this section, but first we will explore some of its consequences. In particular, the uniqueness asserted in Theorem 6.3 is a powerful thing. We will see that it implies in particular that the determinant is not only a good detector of singularity, but also a matrix invariant.

Examples

1. We can easily find the determinant of a 1×1 matrix from the definition alone: by multilinearity,

$$\det[a] = a \det[1] = a \det \mathrm{I}_1 = a1 = a.$$

   In particular, det is exactly the invariant which we observed detecting singularity for 1×1 matrices at the beginning of this section.
2. With somewhat more work we could deduce the formula for the determinant of a 2×2 matrix from the definition, but we'll save that for when we prove Theorem 6.3. For now, we'll simply observe that

$$\det \begin{bmatrix} a & b \\ c & d \end{bmatrix} = ad - bc \qquad (6.1)$$

defines an alternating multilinear function on $M_2(\mathbb{F})$ with $\det I_2 = 1$. So by uniqueness, this is indeed the determinant for a 2×2 matrix.

Quick Exercise #3. Verify that equation (6.1) really does define an alternating multilinear function on $M_2(\mathbb{F})$ with $\det I_2 = 1$.

3. If A is diagonal, then by multilinearity,

$$\det A = (a_{11} \dots a_{nn}) \det I_n = a_{11} \dots a_{nn}.$$

That is, the determinant of a diagonal matrix is the product of its diagonal entries. (But remember: we already saw that the product of the diagonal entries is not isoscopic in general!) ▲

The following alternative form of Theorem 6.3 is frequently useful.

Corollary 6.4 *If $D : M_n(\mathbb{F}) \to \mathbb{F}$ is an alternating multilinear function, then*

$$D(A) = D(I_n) \det A$$

for every $A \in M_n(\mathbb{F})$.

Proof Suppose first that $D(I_n) \neq 0$, and define $f : M_n(\mathbb{F}) \to \mathbb{F}$ by

$$f(A) := \frac{D(A)}{D(I_n)}.$$

Then f is alternating and multilinear, and $f(I_n) = \frac{D(I_n)}{D(I_n)} = 1$, so by uniqueness, $f(A) = \det(A)$. Multiplying through by $D(I_n)$ completes the proof.

On the other hand, if $D(I_n) = 0$, then $D(A) = 0$ for all A. Indeed, by multilinearity,

$$D(A) = D\left(\begin{bmatrix} | & & | \\ \sum_{i_1=1}^n a_{i_1 1} e_{i_1} & \cdots & \sum_{i_n=1}^n a_{i_n n} e_{i_n} \\ | & & | \end{bmatrix}\right)$$

$$= \sum_{i_1=1}^n \cdots \sum_{i_n=1}^n a_{i_1 1} \cdots a_{i_n n} D\left(\begin{bmatrix} | & & | \\ e_{i_1} & \cdots & e_{i_n} \\ | & & | \end{bmatrix}\right).$$

QA #3: Multilinearity: $\det \begin{bmatrix} a_1 + a_2 & b \\ c_1 + c_2 & d \end{bmatrix} = (a_1 + a_2)d - b(c_1 + c_2) = a_1 d - bc_1 + a_2 d - bc_2$; homogeneity and linearity in the second column work similarly. Alternating: $\det \begin{bmatrix} a & a \\ c & c \end{bmatrix} = ac - ac = 0$, $\det I_2 = 1 \cdot 1 - 0 \cdot 0 = 1$.

If any of the indices i_k are the same, then

$$D\left(\begin{bmatrix} | & & | \\ \mathbf{e}_{i_1} & \cdots & \mathbf{e}_{i_n} \\ | & & | \end{bmatrix}\right) = 0$$

because D is alternating. If the i_k are all distinct, then by swapping columns as needed, it follows from Lemma 6.2 that

$$D\left(\begin{bmatrix} | & & | \\ \mathbf{e}_{i_1} & \cdots & \mathbf{e}_{i_n} \\ | & & | \end{bmatrix}\right) = \pm D(\mathrm{I}_n) = 0.$$ ▲

The following important property of the determinant is our first impressive consequence of the uniqueness stated in Theorem 6.3.

Theorem 6.5 *If* $\mathrm{A}, \mathrm{B} \in \mathrm{M}_n(\mathbb{F})$, *then* $\det(\mathrm{AB}) = (\det \mathrm{A})(\det \mathrm{B})$.

Proof For a given $\mathrm{A} \in \mathrm{M}_n(\mathbb{F})$, we define a function $D_\mathrm{A} : \mathrm{M}_n(\mathbb{F}) \to \mathbb{F}$ by

$$D_\mathrm{A}(\mathrm{B}) = \det(\mathrm{AB}).$$

We claim that D_A is an alternating multilinear function. Once this is proved, Corollary 6.4 will imply that, for every $\mathrm{B} \in \mathrm{M}_n(\mathbb{F})$,

$$\det(\mathrm{AB}) = D_\mathrm{A}(\mathrm{B}) = D_\mathrm{A}(\mathrm{I}_n) \det \mathrm{B} = (\det \mathrm{A})(\det \mathrm{B}).$$

Write $\mathrm{B} = \begin{bmatrix} | & & | \\ \mathbf{b}_1 & \cdots & \mathbf{b}_n \\ | & & | \end{bmatrix}$, so that

$$D_\mathrm{A}(\mathrm{B}) = \det(\mathrm{AB}) = \det \begin{bmatrix} | & & | \\ \mathrm{A}\mathbf{b}_1 & \cdots & \mathrm{A}\mathbf{b}_n \\ | & & | \end{bmatrix}.$$

The linearity of matrix multiplication and the multilinearity of det imply that D_A is multilinear. Moreover, if $\mathbf{b}_i = \mathbf{b}_j$ for $i \neq j$, then $\mathrm{A}\mathbf{b}_i = \mathrm{A}\mathbf{b}_j$ and so $D_\mathrm{A}(\mathrm{B}) = \det(\mathrm{AB}) = 0$ because the determinant itself is alternating. Thus D_A is alternating.

Since $\mathrm{A} \in \mathrm{M}_n(\mathbb{F})$ was arbitrary, we now know that, for every $\mathrm{A}, \mathrm{B} \in \mathrm{M}_n(\mathbb{F})$,

$$\det(\mathrm{AB}) = D_\mathrm{A}(\mathrm{B}) = (\det \mathrm{A})(\det \mathrm{B}).$$ ▲

A first consequence of Theorem 6.5 is that the determinant is a *good* detector of invertibility, since it returns zero if and only if a matrix is singular.

Corollary 6.6 *A matrix* $\mathrm{A} \in \mathrm{M}_n(\mathbb{F})$ *is invertible iff* $\det \mathrm{A} \neq 0$. *In that case,* $\det \mathrm{A}^{-1} = (\det \mathrm{A})^{-1}$.

Proof If A is singular, then $\det A = 0$ by Lemma 6.1.
If A is invertible, then by Theorem 6.5,

$$1 = \det I_n = \det\left(AA^{-1}\right) = (\det A)\left(\det A^{-1}\right),$$

which implies that $\det A \neq 0$ and $\det A^{-1} = (\det A)^{-1}$. ▲

Theorem 6.5 also implies that the determinant is a matrix invariant.

Corollary 6.7 *If* $A, B \in M_n(\mathbb{F})$ *are similar, then* $\det A = \det B$.

Proof If $B = SAS^{-1}$, then by Theorem 6.5,

$$\det B = (\det S)(\det A)(\det S^{-1}).$$

By Corollary 6.6, $\det S^{-1} = (\det S)^{-1}$, and so $\det B = \det A$. ▲

In particular, we can define the notion of the determinant of a linear map.

Definition Let V be finite-dimensional. The **determinant** of $T \in \mathcal{L}(V)$ is $\det T = \det [T]_{\mathcal{B}}$, where $\mathcal{B}$ is any basis of V.

Quick Exercise #4. Suppose that V is a finite-dimensional vector space. What is $\det(cI)$?

Existence and Uniqueness of the Determinant

We conclude this section with the proof of Theorem 6.3. We will need the following piece of notation.

Definition Let $A \in M_n(\mathbb{F})$ for $n \geq 2$. For each $1 \leq i, j \leq n$, $A_{ij} \in M_{n-1}(\mathbb{F})$ denotes the $(n-1) \times (n-1)$ matrix obtained by removing the ith row and jth column from A.

We begin the proof of Theorem 6.3 with existence. It will be convenient to refer to an alternating multilinear function $D_n : M_n(\mathbb{F}) \to \mathbb{F}$ such that $D_n(I_n) = 1$ as a **determinant function** on $M_n(F)$ (so that Theorem 6.3 says that there exists a unique determinant function on $M_n(\mathbb{F})$).

QA #4: The matrix of cI with respect to any basis of V is cI_n, where $n = \dim V$, so $\det(cI) = \det(cI_n) = c^n = c^{\dim V}$.

Lemma 6.8 *For each n, there exists a determinant function on $\mathrm{M}_n(\mathbb{F})$.*

Proof We will prove this by induction on n. First of all, it is trivial to see that D_1 defined by $D_1([a]) = a$ works.

Now suppose that $n > 1$ and we already have a determinant function D_{n-1} on $\mathrm{M}_{n-1}(\mathbb{F})$. For any fixed i we define $D_n : \mathrm{M}_n(\mathbb{F}) \to \mathbb{F}$ by

$$D_n(\mathrm{A}) := \sum_{j=1}^{n} (-1)^{i+j} a_{ij} D_{n-1}(\mathrm{A}_{ij}). \tag{6.2}$$

We claim that D_n is a determinant function on $\mathrm{M}_n(\mathbb{F})$, and that $D_n(\mathrm{I}_n) = 1$.

Write $\mathrm{A} = \begin{bmatrix} | & & | \\ \mathbf{a}_1 & \cdots & \mathbf{a}_n \\ | & & | \end{bmatrix}$, fix $1 \le k \le n$ and $\mathbf{b}_k \in \mathbb{F}^n$, and let

$$\mathrm{B} = \begin{bmatrix} | & & | & & | \\ \mathbf{a}_1 & \cdots & \mathbf{b}_k & \cdots & \mathbf{a}_n \\ | & & | & & | \end{bmatrix} \quad \text{and} \quad \mathrm{C} = \begin{bmatrix} | & & | & & | \\ \mathbf{a}_1 & \cdots & \mathbf{a}_k + \mathbf{b}_k & \cdots & \mathbf{a}_n \\ | & & | & & | \end{bmatrix}.$$

As usual, we denote the entries of A, B, and C by a_{ij}, b_{ij}, and c_{ij}, respectively. We need to show that $D_n(\mathrm{C}) = D_n(\mathrm{A}) + D_n(\mathrm{B})$.

Using the notation introduced above, notice that

$$\mathrm{A}_{ik} = \mathrm{B}_{ik} = \mathrm{C}_{ik},$$

since A, B, and C only differ in the kth column. Furthermore, if $j \neq k$, then

$$D_{n-1}(\mathrm{C}_{ij}) = D_{n-1}(\mathrm{A}_{ij}) + D_{n-1}(\mathrm{B}_{ij}),$$

by the multilinearity of D_{n-1}. Therefore

$$\begin{aligned}
D_n(\mathrm{C}) &= \sum_{j=1}^{n} (-1)^{i+j} c_{ij} D_{n-1}(\mathrm{C}_{ij}) \\
&= \sum_{j \neq k} (-1)^{i+j} c_{ij} D_{n-1}(\mathrm{C}_{ij}) + (-1)^{i+k} c_{ik} D_{n-1}(\mathrm{C}_{ik}) \\
&= \sum_{j \neq k} (-1)^{i+j} a_{ij} \big(D_{n-1}(\mathrm{A}_{ij}) + D_{n-1}(\mathrm{B}_{ij}) \big) + (-1)^{i+k} (a_{ik} + b_{ik}) D_{n-1}(\mathrm{A}_{ik}) \\
&= \sum_{j=1}^{n} (-1)^{i+j} a_{ij} D_{n-1}(\mathrm{A}_{ij}) + \sum_{j=1}^{n} (-1)^{i+j} b_{ij} D_{n-1}(\mathrm{B}_{ij}) \\
&= D_n(\mathrm{A}) + D_n(\mathrm{B}),
\end{aligned}$$

where in the second-last line we've rearranged terms, using that $b_{ij} = a_{ij}$ for $j \neq k$.

Next, for $c \in \mathbb{F}$, let $A' = \begin{bmatrix} | & & | & & | \\ a_1 & \cdots & ca_k & \cdots & a_n \\ | & & | & & | \end{bmatrix}$. Then

$$a'_{ij} = \begin{cases} a_{ij} & \text{if } j \neq k, \\ ca_{ij} & \text{if } j = k, \end{cases} \qquad \text{and} \qquad D_{n-1}(A'_{ij}) = \begin{cases} cD_{n-1}(A_{ij}) & \text{if } j \neq k, \\ D_{n-1}(A_{ij}) & \text{if } j = k, \end{cases}$$

by the multilinearity of D_{n-1}, and so

$$D_n(A') = \sum_{j=1}^{n} (-1)^{i+j} a'_{ij} D_{n-1}(A'_{ij}) = c \sum_{j=1}^{n} (-1)^{i+j} a_{ij} D_{n-1}(A_{ij}) = cD_n(A).$$

Thus D_n is multilinear.

Next, suppose that $a_k = a_\ell$ for some $k < \ell$. We need to show that then $D_n(A) = 0$.

Note first that if $j \notin \{k, \ell\}$, then two columns of A_{ij} are equal, and so $D_{n-1}(A_{ij}) = 0$ since D_{n-1} is alternating. Therefore

$$\begin{aligned} D_n(A) &= (-1)^{i+k} a_{ik} D_{n-1}(A_{ik}) + (-1)^{i+\ell} a_{i\ell} D_{n-1}(A_{i\ell}) \\ &= (-1)^{i+k} a_{ik} \Big[D_{n-1}(A_{ik}) + (-1)^{\ell-k} D_{n-1}(A_{i\ell}) \Big], \end{aligned}$$

since $a_{ik} = a_{i\ell}$. Now notice that A_{ik} and $A_{i\ell}$ have the same columns in a different order (since in both cases one of the two identical columns of A was deleted). We can obtain $A_{i\ell}$ from A_{ik} by swapping consecutive columns $\ell - k - 1$ times (to move the $a_\ell = a_k$ column left from the $(\ell - 1)$th position to the kth position). By Lemma 6.2, this means that $D_{n-1}(A_{ik}) = (-1)^{\ell-k-1} D_{n-1}(A_{i\ell})$, and so $D_n(A) = 0$. Thus D_n is alternating.

Finally, observe that if $A = I_n$, then a_{ij} is only nonzero when $j = i$, and that $A_{ii} = I_{n-1}$. Thus

$$D_n(I_n) = (-1)^{i+i} 1 D_{n-1}(I_{n-1}) = 1. \qquad \blacktriangle$$

We now complete the proof of Theorem 6.3 by showing the uniqueness of the determinant.

Proof of Theorem 6.3 We will again proceed by induction on n. For $n = 1$, we have

$$D_1([a]) = aD_1([1]) = a$$

by linearity.

Assume now that $n \geq 2$, that D_{n-1} is known to be the unique determinant function on $M_{n-1}(\mathbb{F})$, and that D_n is *a* determinant function on $M_n(\mathbb{F})$. We will show that in fact D_n is uniquely determined by D_{n-1}.

Fix $A \in M_n(\mathbb{F})$. For each $j = 1, \dots, n$, we define a function $d_j : M_{n-1}(\mathbb{F}) \to \mathbb{F}$ by

$$d_j(B) = D_n\left(\begin{bmatrix} b_{11} & \cdots & b_{1,n-1} & a_{1j} \\ \vdots & \ddots & \vdots & \vdots \\ b_{n-1,1} & \cdots & b_{n-1,n-1} & a_{n-1,j} \\ 0 & \cdots & 0 & a_{nj} \end{bmatrix}\right). \tag{6.3}$$

Then d_j is an alternating multilinear function on $M_{n-1}(\mathbb{F})$ because D_n is alternating and multilinear. By the same argument as in the proof of Corollary 6.4, it now follows from the induction hypothesis that $d_j(B) = d_j(I_{n-1})D_{n-1}(B)$ for every $B \in M_{n-1}(\mathbb{F})$.

By multilinearity,

$$\begin{aligned} d_j(I_{n-1}) &= D_n\left(\begin{bmatrix} 1 & \cdots & 0 & a_{1j} \\ \vdots & \ddots & \vdots & \vdots \\ 0 & \cdots & 1 & a_{n-1,j} \\ 0 & \cdots & 0 & a_{nj} \end{bmatrix}\right) \\ &= D_n\left(\begin{bmatrix} | & & | & | \\ \mathbf{e}_1 & \cdots & \mathbf{e}_{n-1} & \sum_{i=1}^n a_{ij}\mathbf{e}_i \\ | & & | & | \end{bmatrix}\right) \\ &= \sum_{i=1}^n a_{ij} D_n\left(\begin{bmatrix} | & & | & | \\ \mathbf{e}_1 & \cdots & \mathbf{e}_{n-1} & \mathbf{e}_i \\ | & & | & | \end{bmatrix}\right) \\ &= a_{nj}D_n(I_n) + \sum_{i=1}^{n-1} a_{ij} D_n\left(\begin{bmatrix} | & & | & | \\ \mathbf{e}_1 & \cdots & \mathbf{e}_{n-1} & \mathbf{e}_i \\ | & & | & | \end{bmatrix}\right). \end{aligned}$$

Since D_n is alternating, all the terms in the latter sum are zero, so

$$d_j(I_{n-1}) = a_{nj}D_n(I_n) = a_{nj} \tag{6.4}$$

since D_n is a determinant function on $M_n(\mathbb{F})$. This means that

$$D_n\left(\begin{bmatrix} b_{11} & \cdots & b_{1,n-1} & a_{1j} \\ \vdots & \ddots & \vdots & \vdots \\ b_{n-1,1} & \cdots & b_{n-1,n-1} & a_{n-1,j} \\ 0 & \cdots & 0 & a_{nj} \end{bmatrix}\right) = a_{nj}D_{n-1}(B);$$

i.e., D_n is uniquely determined for matrices of this special form.

Finally, recall (see Lemma 6.1) that if A does not have full rank, then $D_n(A) = 0$, and so $D_n(A)$ is uniquely determined. If A does have full rank, then there is at least one column, say the jth, with the last entry nonzero. It then follows from the fact that D_n is alternating and multilinear that

$$D_n(\mathrm{A}) = D_n\left(\begin{bmatrix} | & & | & & | \\ \mathrm{a}_1 & \cdots & \mathrm{a}_j & \cdots & \mathrm{a}_n \\ | & & | & & | \end{bmatrix}\right)$$

$$= D_n\left(\begin{bmatrix} | & & | & & | \\ \left(\mathrm{a}_1 - \frac{a_{n1}}{a_{nj}}\mathrm{a}_j\right) & \cdots & \mathrm{a}_j & \cdots & \left(\mathrm{a}_n - \frac{a_{nn}}{a_{nj}}\mathrm{a}_j\right) \\ | & & | & & | \end{bmatrix}\right)$$

$$= -D_n\left(\begin{bmatrix} | & & | & & | \\ \left(\mathrm{a}_1 - \frac{a_{n1}}{a_{nj}}\mathrm{a}_j\right) & \cdots & \left(\mathrm{a}_n - \frac{a_{nn}}{a_{nj}}\mathrm{a}_j\right) & \cdots & \mathrm{a}_j \\ | & & | & & | \end{bmatrix}\right),$$

where the sign change in the last equality results from switching the jth and nth columns as in Lemma 6.2 (and does not take place if $j = n$). This final expression for $D_n(\mathrm{A})$ above is $D_n(\tilde{\mathrm{A}})$, where $\tilde{\mathrm{A}}$ has zeroes in all but the last entry of the bottom row as in (6.3), and so D_n has already been shown to be uniquely determined. ▲

It's worth noticing that in equation (6.2) in the proof of Lemma 6.8, we built an alternating multilinear function on $\mathrm{M}_n(\mathbb{F})$ from one on $\mathrm{M}_{n-1}(\mathbb{F})$. Now that we've finished proving the uniqueness of the determinant, equation (6.2) gives a formula for the determinant of an $n \times n$ matrix in terms of determinants of $(n-1) \times (n-1)$ matrices:

$$\det \mathrm{A} = \sum_{j=1}^{n} (-1)^{i+j} a_{ij} \det(\mathrm{A}_{ij}) \tag{6.5}$$

for each i.

Examples

1. Formula (6.5) can be used to derive formula (6.1) for the determinant of a 2×2 matrix: using $i = 1$,

$$\det \begin{bmatrix} a & b \\ c & d \end{bmatrix} = a \det \begin{bmatrix} d \end{bmatrix} - b \det \begin{bmatrix} c \end{bmatrix} = ad - bc.$$

Quick Exercise #5. Check that you'd get the same result here using $i = 2$ instead.

2. Once we know how to compute determinants of 2×2 matrices, formula (6.5) can be used to compute determinants of 3×3 matrices:

QA #5: $\det \begin{bmatrix} a & b \\ c & d \end{bmatrix} = -c \det [b] + d \det [a] = ad - bc.$

$$\det\begin{bmatrix}3 & -1 & 4\\ -1 & 5 & -9\\ 2 & -6 & 5\end{bmatrix} = 3\det\begin{bmatrix}5 & -9\\ -6 & 5\end{bmatrix} - (-1)\det\begin{bmatrix}-1 & -9\\ 2 & 5\end{bmatrix} + 4\det\begin{bmatrix}-1 & 5\\ 2 & -6\end{bmatrix}$$
$$= 3(5\cdot 5-(-9)(-6))-(-1)((-1)5-(-9)2)+4((-1)(-6)-5\cdot 2)$$
$$= 3\cdot(-29)-(-1)\cdot 13+4(-4)$$
$$= -90.$$

Of course there's no need to stop with 3×3 matrices, but we'll wait until the next section to tackle computing determinants of larger matrices. ▲

KEY IDEAS

- A function on matrices is multilinear if it is a linear function of each column (keeping the other columns fixed). A multilinear function is alternating if it gives zero whenever the argument has two identical columns.
- The determinant is the unique alternating multilinear function on matrices taking the value 1 at the identity.
- $\det(A) = 0$ if and only if A is singular (i.e., not invertible).
- $\det(AB) = \det(A)\det(B)$.
- The determinant is a matrix invariant.

EXERCISES

6.1.1 Compute the determinant of each of the following matrices (using only the techniques of this section).

(a) $\begin{bmatrix}1 & 2\\ 3 & 4\end{bmatrix}$ (b) $\begin{bmatrix}\sqrt{5}+i & 2i\\ -3 & \sqrt{5}-i\end{bmatrix}$ (c) $\begin{bmatrix}0 & 2 & -1\\ 3 & 0 & 4\\ -2 & 1 & 2\end{bmatrix}$

(d) $\begin{bmatrix}1 & 1 & 1\\ 1 & 2 & 2\\ 1 & 2 & 3\end{bmatrix}$

6.1.2 Compute the determinant of each of the following matrices (using only the techniques of this section).

(a) $\begin{bmatrix}-3 & 1\\ 4 & -2\end{bmatrix}$ (b) $\begin{bmatrix}-3i & 2+i\\ -2i & 1-i\end{bmatrix}$ (c) $\begin{bmatrix}-1 & 2 & 3\\ 4 & -5 & 6\\ 7 & -8 & 9\end{bmatrix}$

(d) $\begin{bmatrix}1 & 0 & -2\\ -3 & 0 & 2\\ 0 & 4 & -1\end{bmatrix}$

6.1.3 Suppose that $D : M_2(\mathbb{F}) \to \mathbb{F}$ is an alternating multilinear function such that $D(I_2) = 7$. Compute each of the following.

(a) $D\left(\begin{bmatrix}1 & 2\\ 3 & 4\end{bmatrix}\right)$ (b) $D\left(\begin{bmatrix}-2 & 3\\ 5 & -4\end{bmatrix}\right)$

6.1.4 Suppose that $D : M_3(\mathbb{F}) \to \mathbb{F}$ is an alternating multilinear function such that $D(I_3) = 6$. Compute each of the following.

(a) $D\left(\begin{bmatrix} 1 & 2 & 3 \\ 0 & 4 & 5 \\ 0 & 0 & 6 \end{bmatrix}\right)$ (b) $D\left(\begin{bmatrix} 3 & 2 & 4 \\ 2 & 1 & 2 \\ 2 & 0 & 1 \end{bmatrix}\right)$

6.1.5 (a) Suppose that $D : M_2(\mathbb{F}) \to \mathbb{F}$ is multilinear, and $A, B \in M_2(\mathbb{F})$. What is $D(A + B)$?

(b) Suppose that $D : M_3(\mathbb{F}) \to \mathbb{F}$ is multilinear, and $A, B \in M_3(\mathbb{F})$. What is $D(A + B)$?

6.1.6 Show that

$$\begin{bmatrix} 2 & 1 & -1 \\ 0 & -1 & 2 \\ 3 & 6 & -1 \end{bmatrix} \quad \text{and} \quad \begin{bmatrix} 3 & 5 & 2 \\ 1 & -2 & 0 \\ 0 & 6 & -1 \end{bmatrix}$$

are not similar.

6.1.7 Suppose that U is a proper subspace of a finite-dimensional inner product space V, and let $P_U \in \mathcal{L}(V)$ denote the orthogonal projection onto U. What is $\det P_U$?

6.1.8 Let $R : \mathbb{R}^3 \to \mathbb{R}^3$ denote reflection across the plane $x + 2y + 3z = 0$. What is $\det R$?

6.1.9 Suppose that $\dim V = n$ and $T \in \mathcal{L}(V)$ has n distinct eigenvalues $\lambda_1, \ldots, \lambda_n$. Prove that

$$\det T = \lambda_1 \ldots \lambda_n.$$

6.1.10 Suppose that $A \in M_n(\mathbb{R})$ factors as $A = BC$, where $B \in M_{n,m}(\mathbb{R})$, $C \in M_{m,n}(\mathbb{R})$, and $m < n$. Show that $\det A = 0$.

6.1.11 Show that if $A \in M_n(\mathbb{C})$ is Hermitian, then $\det A \in \mathbb{R}$.

6.1.12 Use formula (6.5) to find a general formula for

$$\det \begin{bmatrix} a_{11} & a_{12} & a_{13} \\ a_{21} & a_{22} & a_{23} \\ a_{31} & a_{32} & a_{33} \end{bmatrix}.$$

6.1.13 Show that the function

$$f(A) = \prod_{i \neq j} \left(\sum_{k=1}^{n} |a_{ki} - a_{kj}| \right)$$

is an isoscopic function on $M_n(\mathbb{R})$ which is not multilinear.

6.1.14 Show that the function

$$g(A) = \prod_{j=1}^{n} \left(\sum_{i=1}^{n} a_{ij} \right)$$

is a multilinear function on $M_n(\mathbb{R})$ which is not isoscopic.

6.1.15 Show that if $D : M_n(\mathbb{F}) \to \mathbb{F}$ is an alternating multilinear function, then you can add to any one column of a matrix A any linear combination of the remaining columns without changing the value of $D(A)$.

6.1.16 Let V be the set of multilinear functions $f : M_n(\mathbb{F}) \to \mathbb{F}$.

(a) Show that V is a vector space over $\mathbb{F}$, with the usual notions of addition of functions and multiplication of functions by scalars.

(b) Show that $\dim V = n^n$.

6.1.17 Let W be the set of alternating multilinear functions $f : M_n(\mathbb{F}) \to \mathbb{F}$.

(a) Show that W is a vector space over $\mathbb{F}$, with the usual notions of addition of functions and multiplication of functions by scalars.

(b) Show that $\dim W = 1$.

6.2 Computing Determinants

Basic Properties

We first recall the recursive formula for the determinant established in the previous section.

Proposition 6.9 (Laplace expansion along a row) *If* $A \in M_n(\mathbb{F})$, *then for any* $i = 1, \ldots, n$,

$$\det A = \sum_{j=1}^{n} (-1)^{i+j} a_{ij} \det(A_{ij}).$$

Laplace expansion is particularly useful for computing determinants of matrices that are *sparse*, i.e., which have lots of entries equal to 0.

Example

$$\det \begin{bmatrix} -2 & 0 & 1 & 3 & 4 \\ 0 & -1 & 0 & 1 & 0 \\ 0 & 0 & 0 & -3 & 0 \\ 1 & 0 & 2 & -2 & -5 \\ 4 & -2 & 0 & -4 & 1 \end{bmatrix} = -(-3)\det \begin{bmatrix} -2 & 0 & 1 & 4 \\ 0 & -1 & 0 & 0 \\ 1 & 0 & 2 & -5 \\ 4 & -2 & 0 & 1 \end{bmatrix}$$

$$= -3 \det \begin{bmatrix} -2 & 1 & 4 \\ 1 & 2 & -5 \\ 4 & 0 & 1 \end{bmatrix}$$

$$= -3 \left(4 \det \begin{bmatrix} 1 & 4 \\ 2 & -5 \end{bmatrix} + 1 \det \begin{bmatrix} -2 & 1 \\ 1 & 2 \end{bmatrix} \right)$$

$$= -3(4(-13) - 5)$$
$$= 171. \qquad \blacktriangle$$

> **Quick Exercise #6.** Which rows were used in the Laplace expansions in the example above?

The Laplace expansion gives a simple formula for the determinant of an upper triangular matrix, which generalizes the observation we made earlier about diagonal matrices.

Corollary 6.10 *If* $A \in M_n(\mathbb{F})$ *is upper triangular, then*

$$\det A = a_{11} \cdots a_{nn}.$$

Proof We will prove this by induction on n. The statement is trivial if $n = 1$.

Suppose that $n \geq 2$ and the theorem is known to be true for $(n-1) \times (n-1)$ matrices. Notice that if A is upper triangular, $a_{in} = 0$ for $i \leq n-1$, and A_{nn} is upper triangular with diagonal entries $a_1, \ldots, a_{n-1,n-1}$. Then by Proposition 6.9 with $i = n$,

$$\det A = \sum_{j=1}^{n} (-1)^{i+n} a_{in} \det A_{in} = (-1)^{n+n} a_{nn} \det A_{nn} = a_{11} \ldots a_{n-1,n-1} a_{nn}.$$

▲

The following convenient symmetry of the determinant can also be proved via the Laplace expansion.

Theorem 6.11 *If* $A \in M_n(\mathbb{F})$, *then* $\det A^T = \det A$.

Proof We will prove by induction on n that the function $f(A) = \det A^T$ is a determinant function on $M_n(\mathbb{F})$, which by the uniqueness of the determinant implies that $f(A) = \det A$.

Slightly unusually, this inductive proof needs two base steps (the reason for this will become clear in the inductive step of the proof). If $A \in M_1(\mathbb{F})$, then $A^T = A$, so $\det A = \det A^T$. If $A \in M_2(\mathbb{F})$, then we can show that $\det A^T = \det A$ using equation (6.1).

QA #6: Third row of the 5 × 5 matrix, second row of the 4 × 4 matrix, and third row of the 3 × 3 matrix.

Quick Exercise #7. Show that $\det A = \det A^{\mathrm{T}}$ for $A \in M_2(\mathbb{F})$.

Now suppose that $n \geq 3$, and that we already know that $\det B^{\mathrm{T}} = \det B$ for every $B \in M_{n-1}(\mathbb{F})$. By Proposition 6.9, for any $A \in M_n(\mathbb{F})$ and any $i = 1, \dots, n$,

$$f(A) = \sum_{j=1}^{n} (-1)^{i+j} a_{ji} \det(A^{\mathrm{T}})_{ij}, \tag{6.6}$$

where $(A^{\mathrm{T}})_{ij}$ means the $(n-1) \times (n-1)$ matrix obtained by removing the ith row and jth column from A^{T}. Now, for each $k = 1, \dots, n$, by using the Laplace expansion along the kth row, we see that

$$\begin{aligned}
f\left(\begin{bmatrix} | & & | & & | \\ \mathbf{a}_1 & \cdots & \mathbf{a}_k + \mathbf{b}_k & \cdots & \mathbf{a}_n \\ | & & | & & | \end{bmatrix}\right)
&= \sum_{j=1}^{n} (-1)^{k+j} (a_{jk} + b_{jk}) \det(A^{\mathrm{T}})_{kj} \\
&= \sum_{j=1}^{n} (-1)^{k+j} a_{jk} \det(A^{\mathrm{T}})_{kj} + \sum_{j=1}^{n} (-1)^{k+j} b_{jk} \det(A^{\mathrm{T}})_{kj} \\
&= f\left(\begin{bmatrix} | & & | \\ \mathbf{a}_1 & \cdots & \mathbf{a}_n \\ | & & | \end{bmatrix}\right) + f\left(\begin{bmatrix} | & & | & & | \\ \mathbf{a}_1 & \cdots & \mathbf{b}_k & \cdots & \mathbf{a}_n \\ | & & | & & | \end{bmatrix}\right),
\end{aligned}$$

since $(A^{\mathrm{T}})_{kj}$ is unaffected by changing the kth column of A. Therefore f is additive in each column. Homogeneity follows similarly, and so f is a multilinear function. (Note that we haven't yet used the induction hypothesis, that the theorem is true for $(n-1) \times (n-1)$ matrices.)

Now suppose that for some $k \neq \ell$, $\mathbf{a}_k = \mathbf{a}_\ell$, and choose $i \notin \{k, \ell\}$ (this is why we need $n \geq 3$ for this part of the proof). Then by equation (6.6) and the induction hypothesis,

$$f(A) = \sum_{j=1}^{n} (-1)^{i+j} a_{ji} \det(A_{ji})^{\mathrm{T}} = \sum_{j=1}^{n} (-1)^{i+j} a_{ji} \det A_{ji}.$$

Since $i \notin \{k, \ell\}$, A_{ji} has two columns which are equal, and so $\det A_{ji} = 0$. Therefore $f(A) = 0$ in this situation, and so f is alternating.

Finally $f(I_n) = \det I_n^{\mathrm{T}} = \det I_n = 1$, and so f is indeed a determinant function. ▲

Theorem 6.11 implies in particular that the determinant is an alternating multilinear function of the *rows* of a matrix.

QA #7: $\det \begin{bmatrix} a & c \\ b & d \end{bmatrix} = ad - bc = ad - cb = \det \begin{bmatrix} a & b \\ c & d \end{bmatrix}$.

Corollary 6.12 (Laplace expansion along a column) *If* $A \in M_n(\mathbb{F})$, *then for any* $j = 1, \ldots, n$,

$$\det A = \sum_{i=1}^{n} (-1)^{i+j} a_{ij} \det(A_{ij}).$$

Proof This follows by using Proposition 6.9 to express $\det A^T$, which by Theorem 6.11 is equal to $\det A$. ▲

Corollary 6.13 *If* $A \in M_n(\mathbb{C})$, *then* $\det A^* = \overline{\det A}$.

If V *is a finite-dimensional inner product space and* $T \in \mathcal{L}(V)$, *then* $\det T^* = \overline{\det T}$.

Proof Let $\overline{A}$ denote the matrix with entries $\overline{a_{jk}}$, i.e., the entry-wise complex conjugate of A. Since $A^* = \overline{A}^T$, by Theorem 6.11 we need to show that $\det \overline{A} = \overline{\det A}$.

This is trivial for 1×1 matrices, and then follows easily by induction on n using Proposition 6.9.

The statement for operators now follows from the definition of the determinant of a linear operator and Theorem 5.13. ▲

Determinants and Row Operations

Now that we know that the determinant is an alternating multilinear function of the *rows* of a matrix, we can perform row operations on an arbitrary matrix to bring it into upper triangular form (in fact, we need only operations R1 and R3 to do this). The effect of those row operations on the determinant of the matrix is directly determined by the alternating and multilinear properties. Viewing the row operations as multiplication by elementary matrices is a useful bookkeeping device in this context; recall that the following matrices are those that correspond to operations R1 and R3:

$$P_{c,i,j} := \begin{bmatrix} 1 & 0 & & 0 \\ 0 & \ddots & c & \\ & & \ddots & 0 \\ 0 & & 0 & 1 \end{bmatrix}, \qquad R_{i,j} := \begin{bmatrix} 1 & & & & & \\ & \ddots & & & & \\ & & 0 & & 1 & \\ & & & \ddots & & \\ & & 1 & & 0 & \\ & & & & & \ddots & \\ & & & & & & 1 \end{bmatrix}.$$

$$(i \neq j) \qquad\qquad (i \neq j)$$

Lemma 6.14 *Let $1 \leq i,j \leq n$ with $i \neq j$ and $c \in \mathbb{F}$, and let $\mathrm{P}_{c,i,j}$ and $\mathrm{R}_{i,j}$ be the elementary $n \times n$ matrices above. Then:*

- $\det \mathrm{P}_{c,i,j} = 1$,
- $\det \mathrm{R}_{i,j} = -1$.

Proof The matrices $\mathrm{P}_{c,i,j}$ are upper triangular, so by Theorem 6.10, $\det \mathrm{P}_{c,i,j}$ is the product of the diagonal entries of $\mathrm{P}_{c,i,j}$, which are all 1.

The matrix $\mathrm{R}_{i,j}$ can be obtained by exchanging the ith and jth columns of I_n, so by Lemma 6.2, $\det \mathrm{R}_{i,j} = -\det \mathrm{I}_n = -1$. ▲

This lemma lets us keep track of the effect of row operations on the determinant of a matrix, which leads to the following algorithm for computing determinants. For large matrices, this gives a much more efficient way to compute most determinants than Laplace expansion, or the sum over permutations formula we will see later in this section.

Algorithm 6.15 To compute the determinant of $\mathrm{A} \in \mathrm{M}_n(\mathbb{F})$:

- convert A into an upper triangular matrix B via row operations **R1** and **R3** only,
- let k be the number of times two rows are switched,
- $\det \mathrm{A} = (-1)^k b_{11} \cdots b_{nn}$.

Proof If B is the result of performing row operations **R1** and **R3** starting from A, then by Theorem 2.21,

$$\mathrm{B} = \mathrm{E}_1 \cdots \mathrm{E}_p \mathrm{A}$$

for some elementary matrices $\mathrm{E}_1, \ldots, \mathrm{E}_p$ of type $\mathrm{P}_{c,i,j}$ or $\mathrm{R}_{i,j}$. By Theorems 6.10 and 6.5,

$$\det \mathrm{B} = b_{11} \cdots b_{nn} = (\det \mathrm{E}_1) \cdots (\det \mathrm{E}_p) \det \mathrm{A},$$

and by Lemma 6.14,

$$(\det \mathrm{E}_1) \cdots (\det \mathrm{E}_p) = (-1)^k.$$ ▲

Example

$$\det \begin{bmatrix} 3 & 0 & 1 \\ -2 & -2 & 1 \\ 1 & 0 & 1 \end{bmatrix} = -\det \begin{bmatrix} 1 & 0 & 1 \\ -2 & -2 & 1 \\ 3 & 0 & 1 \end{bmatrix}$$

$$= -\det \begin{bmatrix} 1 & 0 & 1 \\ 0 & -2 & 3 \\ 3 & 0 & 1 \end{bmatrix}$$

$$= -\det \begin{bmatrix} 1 & 0 & 1 \\ 0 & -2 & 3 \\ 0 & 0 & -2 \end{bmatrix}$$

$$= -(1)(-2)(-2) = -4. \qquad \blacktriangle$$

Quick Exercise #8. What were the row operations used in the example above?

Permutations

In this section we will develop an alternative formula for the determinant. We first need to introduce the concept of a permutation.

Definition A **permutation** of $\{1, \dots, n\}$ is a bijective function

$$\sigma : \{1, \dots, n\} \to \{1, \dots, n\}.$$

Equivalently, $(\sigma(1), \dots, \sigma(n))$ is a list of all the numbers in $\{1, \dots, n\}$ in some order.

The set of all permutations of $\{1, \dots, n\}$ is denoted S_n and is called the **symmetric group on n letters.**

The **identity** permutation is denoted ι.

The symmetric group S_n has a natural representation as a group of $n \times n$ matrices: to a permutation $\sigma \in S_n$ we associate a **permutation matrix A_σ** with entries

$$a_{ij} = \begin{cases} 1 & \text{if } \sigma(i) = j, \\ 0 & \text{otherwise.} \end{cases}$$

Example The matrix corresponding to the permutation of $\{1, 2, 3, 4\}$ which exchanges 1 with 2 and 3 with 4 is

$$\begin{bmatrix} 0 & 1 & 0 & 0 \\ 1 & 0 & 0 & 0 \\ 0 & 0 & 0 & 1 \\ 0 & 0 & 1 & 0 \end{bmatrix}.$$

QA #8: Swap the first and third rows; add 2 times the first row to the second row; add −3 times the first row to the third row. Notice we didn't strictly follow the Gaussian elimination algorithm, but we don't need to.

The matrix corresponding to the permutation of $\{1, 2, 3, 4\}$ which fixes 2 and cyclicly permutes $1 \to 3 \to 4 \to 1$ is

$$\begin{bmatrix} 0 & 0 & 1 & 0 \\ 0 & 1 & 0 & 0 \\ 0 & 0 & 0 & 1 \\ 1 & 0 & 0 & 0 \end{bmatrix}.$$

▲

Quick Exercise #9. Show that

$$A_\sigma \begin{bmatrix} x_1 \\ x_2 \\ \vdots \\ x_n \end{bmatrix} = \begin{bmatrix} x_{\sigma(1)} \\ x_{\sigma(2)} \\ \vdots \\ x_{\sigma(n)} \end{bmatrix}.$$

That is, A_σ is exactly the matrix of the linear map on $\mathbb{F}^n$ which permutes the coordinates of a vector according to σ. This has the important consequence that

$$A_{\sigma_1 \circ \sigma_2} = A_{\sigma_1} A_{\sigma_2}. \tag{6.7}$$

Every permutation has an associated **sign**, which is a kind of parity. While the sign can be defined in terms of the permutation itself, it is most easily defined in terms of the representation of the permutation as a matrix, as follows.

Definition Let $\sigma \in S_n$ with permutation matrix A_σ. The **sign** of σ is

$$\operatorname{sgn}(\sigma) := \det A_\sigma.$$

Lemma 6.16 *Let $\sigma, \rho \in S_n$. Then*

- $\operatorname{sgn}(\sigma) \in \{\pm 1\}$,
- $\operatorname{sgn}(\sigma \circ \rho) = \operatorname{sgn}(\sigma) \operatorname{sgn}(\rho)$,
- $\operatorname{sgn}(\sigma) = \operatorname{sgn}(\sigma^{-1})$.

Proof It follows from equation (6.7) that $A_\sigma^{-1} = A_{\sigma^{-1}}$, so that A_σ is in fact invertible. The RREF of A_σ is therefore I_n; moreover, it is clear that when row-reducing A_σ, only row operation R3 is used. It thus follows from Algorithm 6.15 that $\operatorname{sgn}(\sigma) = \det A_\sigma = (-1)^k$, where k is the number of times R3 is applied.

The fact that $\operatorname{sgn}(\sigma \circ \rho) = \operatorname{sgn}(\sigma) \operatorname{sgn}(\rho)$ is immediate from equation (6.7) and the fact that $\det(AB) = (\det A)(\det B)$. In particular, this implies that

$$\operatorname{sgn}(\sigma) \operatorname{sgn}(\sigma^{-1}) = \operatorname{sgn}(\sigma\sigma^{-1}) = \operatorname{sgn}(\iota) = \det I_n = 1.$$

▲

We are now in a position to state the classical "sum over permutations" formula for the determinant.

Theorem 6.17 *For* $A \in M_n(\mathbb{F})$,

$$\det A = \sum_{\sigma \in S_n} (\operatorname{sgn}\sigma) a_{1,\sigma(1)} \dots a_{n,\sigma(n)}$$
$$= \sum_{\sigma \in S_n} (\operatorname{sgn}\sigma) a_{\sigma(1),1} \dots a_{\sigma(n),n}.$$

Proof By multilinearity,

$$\begin{aligned}\det(A) &= \det\left(\begin{bmatrix} | & & | \\ \sum_{i_1=1}^n a_{i_1 1}\mathbf{e}_{i_1} & \cdots & \sum_{i_n=1}^n a_{i_n n}\mathbf{e}_{i_n} \\ | & & | \end{bmatrix}\right) \\ &= \sum_{i_1=1}^n \cdots \sum_{i_n=1}^n a_{i_1 1}\cdots a_{i_n n} \det\left(\begin{bmatrix} | & & | \\ \mathbf{e}_{i_1} & \cdots & \mathbf{e}_{i_n} \\ | & & | \end{bmatrix}\right).\end{aligned} \tag{6.8}$$

Now, if any two of the indices $i_1, \dots, i_n$ are equal, then

$$\det\left(\begin{bmatrix} | & & | \\ \mathbf{e}_{i_1} & \cdots & \mathbf{e}_{i_n} \\ | & & | \end{bmatrix}\right) = 0$$

because the determinant is alternating. So the sum in (6.8) can be taken to be the sum over sets of *distinct* indices $\{i_1, \dots, i_n\}$, i.e., the sum over permutations

$$\sum_{\sigma \in S_n} a_{\sigma(1)1} \cdots a_{\sigma(n)n} \det\left(\begin{bmatrix} | & & | \\ \mathbf{e}_{\sigma(1)} & \cdots & \mathbf{e}_{\sigma(n)} \\ | & & | \end{bmatrix}\right). \tag{6.9}$$

Observe that the (i,j) entry of the matrix

$$\begin{bmatrix} | & & | \\ \mathbf{e}_{\sigma(1)} & \cdots & \mathbf{e}_{\sigma(n)} \\ | & & | \end{bmatrix}$$

is 1 if $i = \sigma(j)$ and 0 otherwise. That is, this matrix is exactly $A_{\sigma^{-1}}$, and so its determinant is $\operatorname{sgn}(\sigma^{-1}) = \operatorname{sgn}(\sigma)$. This completes the proof of the second version of the formula.

Alternatively, observe that

$$a_{\sigma(1)1} \cdots a_{\sigma(n)n} = a_{1\sigma^{-1}(1)} \cdots a_{n\sigma^{-1}(n)};$$

all of the same factors appear, just in a different order. We can therefore rewrite expression (6.9) as

$$\sum_{\sigma \in S_n} a_{1\sigma^{-1}(1)} \cdots a_{n\sigma^{-1}(n)} \operatorname{sgn}(\sigma^{-1}).$$

Making the change of variables $\rho = \sigma^{-1}$ gives the first version of the formula. ▲

KEY IDEAS

- Laplace expansion:

$$\det(A) = \sum_{j=1}^{n} (-1)^{i+j} a_{ij} \det(A_{ij}) = \sum_{i=1}^{n} (-1)^{i+j} a_{ij} \det(A_{ij}).$$

- $\det(A^T) = \det(A)$ and $\det(A^*) = \overline{\det(A)}$.
- To compute $\det A$ via row operations: row-reduce A to an upper triangular matrix B using only row operations R1 and R3. Then

$$\det A = (-1)^k b_{11} \cdots b_{nn},$$

where k is the number of times R3 is used.
- A permutation is a bijection from $\{1, \ldots, n\}$ to itself. Permutations can be encoded as matrices.
- The sum over permutations formula:

$$\det A = \sum_{\sigma \in S_n} (\operatorname{sgn} \sigma) a_{1,\sigma(1)} \ldots a_{n,\sigma(n)} = \sum_{\sigma \in S_n} (\operatorname{sgn} \sigma) a_{\sigma(1),1} \ldots a_{\sigma(n),n}.$$

EXERCISES

6.2.1 Find the determinant of each of the following matrices.

(a) $\begin{bmatrix} 1 & 3 & -1 \\ 0 & 2 & 2 \\ 4 & 1 & 3 \end{bmatrix}$ (b) $\begin{bmatrix} 1 & -1 & 0 & 2 \\ 3 & 3 & -1 & 1 \\ 2 & 4 & -1 & -1 \\ 1 & 1 & 1 & 1 \end{bmatrix}$

(c) $\begin{bmatrix} 1 & 0 & -1 & 3 \\ 2 & -3 & -2 & 5 \\ 3 & 0 & -1 & 9 \\ 2 & -3 & -2 & 6 \end{bmatrix}$ (d) $\begin{bmatrix} 1 & 0 & 1 & 0 & -1 \\ 1 & 2 & 1 & -1 & 1 \\ 1 & 2 & 4 & 0 & 0 \\ 1 & 2 & 4 & -1 & 0 \\ 1 & 2 & 4 & -1 & 1 \end{bmatrix}$

(e) $\begin{bmatrix} 1 & 1 & 0 \\ 1 & 0 & 1 \\ 0 & 1 & 1 \end{bmatrix}$, over $\mathbb{R}$ (f) $\begin{bmatrix} 1 & 1 & 0 \\ 1 & 0 & 1 \\ 0 & 1 & 1 \end{bmatrix}$, over $\mathbb{F}_2$

6.2.2 Find the determinant of each of the following matrices.

(a) $\begin{bmatrix} 2 & -7 & 1 \\ 8 & 2 & 8 \\ 1 & 8 & 2 \end{bmatrix}$ (b) $\begin{bmatrix} -1 & 4 & 1 \\ 2 & -1 & -2 \\ 3 & 1 & -1 \end{bmatrix}$ (c) $\begin{bmatrix} 1 & -2 & 3 & 0 \\ -2 & 1 & 0 & -1 \\ 3 & -4 & 5 & -2 \\ 0 & -1 & 2 & -3 \end{bmatrix}$

(d) $\begin{bmatrix} 2 & 0 & -1 & 1 & 3 \\ 0 & -1 & 0 & -1 & 0 \\ 2 & 1 & 4 & 0 & 3 \\ 2 & -1 & -1 & -3 & 3 \\ 0 & 2 & 0 & 2 & 1 \end{bmatrix}$ (e) $\begin{bmatrix} 1 & 0 & 0 & 0 & 3 & -2 \\ 0 & 2 & 0 & -1 & 0 & 0 \\ 0 & 0 & 1 & 0 & 2 & 0 \\ -1 & 0 & 0 & 4 & 0 & 1 \\ 1 & 0 & -2 & 0 & 0 & -1 \\ 0 & 1 & 0 & -1 & 2 & 0 \end{bmatrix}$

6.2.3 (a) Show that if $U \in M_n(\mathbb{C})$ is unitary, then $|\det U| = 1$.
(b) Show that if $\sigma_1, \ldots, \sigma_n$ are the singular values of $A \in M_n(\mathbb{C})$, then

$$|\det A| = \sigma_1 \cdots \sigma_n.$$

6.2.4 Let $A \in M_n(\mathbb{F})$.
(a) Show that if $A = LU$ is an LU decomposition, then $\det A = u_{11} \cdots u_{nn}$.
(b) Suppose that $PA = LU$ is an LUP decomposition, and that P is the permutation matrix A_σ for some $\sigma \in S_n$. Show that $\det A = (\operatorname{sgn} \sigma) u_{11} \cdots u_{nn}$.

6.2.5 Show that if $A = LDU$ is an LDU decomposition of $A \in M_n(\mathbb{F})$ (see Exercise 2.4.17), then $\det A = d_{11} \cdots d_{nn}$.

6.2.6 Show that if $A = QR$ is a QR decomposition of $A \in M_n(\mathbb{C})$, then $|\det A| = |r_{11} \cdots r_{nn}|$.

6.2.7 Show that if $A \in M_n(\mathbb{C})$, then $|\det A| \leq \prod_{j=1}^{n} \|a_j\|$. (This is called **Hadamard's inequality.**)
Hint: Use Exercise 4.5.21.

6.2.8 Define $A \in M_n(\mathbb{R})$ by $a_{ij} = \min\{i, j\}$. Show that $\det A = 1$.
Hint: Proceed by induction. For the inductive step, use row and column operations.

6.2.9 Suppose that V is a finite-dimensional complex vector space, and $T \in \mathcal{L}(V)$ has only real eigenvalues. Prove that $\det T \in \mathbb{R}$.

6.2.10 Suppose that $A \in M_{m+n}(\mathbb{F})$ has the form

$$A = \begin{bmatrix} B & C \\ 0 & D \end{bmatrix}$$

for some $B \in M_m(\mathbb{F})$, $D \in M_n(\mathbb{F})$, and $C \in M_{m,n}(\mathbb{F})$. Show that $\det A = \det B \det D$.

Hint: First show that $f(\mathrm{B}) = \det \begin{bmatrix} \mathrm{B} & \mathrm{C} \\ 0 & \mathrm{D} \end{bmatrix}$ is an alternating multilinear function on $\mathrm{M}_m(\mathbb{F})$. Then show that $g(\mathrm{D}) = \det \begin{bmatrix} \mathrm{I}_m & \mathrm{C} \\ 0 & \mathrm{D} \end{bmatrix}$ is an alternating multilinear function of the *rows* of D.

6.2.11 Suppose that $\mathrm{A} \in \mathrm{M}_n(\mathbb{C})$ is a normal matrix. Give an expression for $\det \mathrm{A}$ in terms of the eigenvalues of A. (Remember that the eigenvalues may not be distinct.)

6.2.12 Let V be a real vector space with basis $(v_1, \ldots, v_n)$, and define $T \in \mathcal{L}(V)$ by setting

$$Tv_k = \sum_{i=1}^{k} iv_i = v_1 + 2v_2 + \cdots + kv_k$$

for each $k = 1, \ldots, n$ and extending by linearity.

(a) Find $\det T$.

(b) Use this to show that $(v_1, v_1 + 2v_2, \ldots, v_1 + 2v_2 + \cdots + nv_n)$ is a basis of V.

6.2.13 Fix $\mathrm{A} \in \mathrm{M}_n(\mathbb{R})$ and for $t \in \mathbb{R}$ let

$$f(t) := \det(\mathrm{I}_n + t\mathrm{A}).$$

Show that $f'(0) = \operatorname{tr} \mathrm{A}$.

Hint: Either proceed from the sum over permutations formula, or consider the case that A is upper triangular first, then show that the general case follows.

6.2.14 The **permanent** of a matrix $\mathrm{A} \in \mathrm{M}_n(\mathbb{F})$ is defined by

$$\operatorname{per} \mathrm{A} := \sum_{\sigma \in S_n} a_{1\sigma(1)} \cdots a_{n\sigma(n)}.$$

That is, it is the same as the determinant but without the factors of $\operatorname{sgn}(\sigma)$ in Theorem 6.17.

(a) Show that the permanent is a multilinear function, but that it is not alternating.

(b) Give an example of matrices $\mathrm{A}, \mathrm{B} \in \mathrm{M}_2(\mathbb{F})$ such that $\operatorname{per}(\mathrm{AB}) \neq (\operatorname{per} \mathrm{A})(\operatorname{per} \mathrm{B})$.

6.2.15 Show that a matrix $\mathrm{A} \in \mathrm{M}_n(\mathbb{F})$ is a permutation matrix if and only if:

- each row of A contains exactly one entry equal to 1,
- each column of A contains exactly one entry equal to 1,
- all the other entries of A are 0.

6.2.16 Let $\sigma \in S_n$ for $n \geq 2$. Use the representation of permutations as permutation matrices together with linear algebra to prove the classical fact that there are transpositions (i.e., permutations which move only two elements) $\tau_1, \dots, \tau_k \in S_n$ such that

$$\sigma = \tau_1 \circ \cdots \circ \tau_k.$$

6.2.17 Suppose that $\tau_1, \dots, \tau_k$ and $\tilde{\tau}_1, \dots, \tilde{\tau}_\ell \in S_n$ are transpositions (permutations which move only two elements), and that

$$\tau_1 \circ \cdots \circ \tau_k = \tilde{\tau}_1 \circ \cdots \circ \tilde{\tau}_\ell.$$

Prove that k and ℓ are either both even or both odd.

6.2.18 Let $x_0, \dots, x_n \in \mathbb{F}$, and define

$$\mathrm{V} := \begin{bmatrix} 1 & x_0 & x_0^2 & \cdots & x_0^n \\ 1 & x_1 & x_1^2 & \cdots & x_1^n \\ \vdots & \vdots & \vdots & & \vdots \\ 1 & x_n & x_n^2 & \cdots & x_n^n \end{bmatrix} \in \mathrm{M}_{n+1}(\mathbb{F});$$

i.e., $v_{ij} = x_{i-1}^{j-1}$. Prove that

$$\det \mathrm{V} = \prod_{0 \le i < j \le n} (x_j - x_i).$$

This expression is called a **Vandermonde determinant**. Conclude that V is invertible iff the scalars $x_0, \dots, x_n$ are distinct.

Hints: Proceed by induction. For the inductive step, first use row operation **R1** to make the top-left entry a pivot. Then use column operations: subtract x_0 times column $n-1$ from column n, then subtract x_0 times column $n-2$ from column $n-1$, and so on, until the top-left entry is the only nonzero entry in the top row. Now look for common factors in the rows, and use multilinearity.

6.2.19 Use Exercise 6.2.18 to show the following.

(a) For any $x_1, \dots, x_n \in \mathbb{F}$,

$$\prod_{1 \le i < j \le n} (x_j - x_i) = \sum_{\sigma \in S_n} \operatorname{sgn}(\sigma) \prod_{j=1}^{n} x_j^{\sigma(j)-1}.$$

(b) For $z_1, \dots, z_n \in \mathbb{C}$ with $|z_j| = 1$,

$$\prod_{1 \le j < k \le n} |z_j - z_k|^2 = \sum_{\sigma, \rho \in S_n} \operatorname{sgn}(\sigma\rho^{-1}) z_1^{\sigma(1)-\rho(1)} \cdots z_n^{\sigma(n)-\rho(n)}.$$

6.3 Characteristic Polynomials

In this section, we introduce another way of looking at the eigenvalues of a matrix, namely, as the roots of its **characteristic polynomial**, which we define below.

The Characteristic Polynomial of a Matrix

Recall that λ is an eigenvalue of A if and only if $A - \lambda I_n$ is singular. In Section 6.1, we saw that a matrix is singular if and only if its determinant is zero. Together, these observations mean that λ is an eigenvalue of A if and only if

$$\det(A - \lambda I_n) = 0.$$

This observation leads us to the following definition.

Definition Let $A \in M_n(\mathbb{F})$. The polynomial $p_A(x) = \det(A - xI_n)$ is called the **characteristic polynomial** of A.

The discussion above about eigenvalues can be summarized as follows.

Proposition 6.18 *Let $A \in M_n(\mathbb{F})$. Then $\lambda \in \mathbb{F}$ is an eigenvalue of A if and only if $p_A(\lambda) = 0$.*

That is,

The eigenvalues of A are the roots of the characteristic polynomial $p_A(x)$.

Note that our definition of the characteristic polynomial is another instance of a definition which makes an implicit claim that needs to be verified, in this case, that $p_A(x)$ is in fact a polynomial.

Proposition 6.19 *Let $A \in M_n(\mathbb{F})$, and let $p_A(x) = \det(A - xI_n)$. Then $p_A(x)$ is a polynomial of degree n, with leading term $(-x)^n$.*

Proof This is easiest to check using the sum over permutations formula for the determinant. Let $B = A - xI_n$;

$$p_A(x) = \det B = \sum_{\sigma \in S_n} \operatorname{sgn}(\sigma) b_{1\sigma(1)} \cdots b_{n\sigma(n)}.$$

Given $\sigma \in S_n$, let $\mathcal{F}_\sigma \subseteq \{1, \ldots, n\}$ be the set of fixed points of σ; i.e., $\sigma(i) = i$ if and only if $i \in \mathcal{F}_\sigma$. Then

$$b_{1\sigma(1)} \cdots b_{n\sigma(n)} = \prod_{i \in \mathcal{F}_\sigma} (a_{ii} - x) \prod_{i \notin \mathcal{F}_\sigma} a_{i\sigma(i)},$$

so that each term of the sum above is a polynomial in x. Moreover, the highest possible power of x is n, and x^n can appear if and only if $\mathcal{F}_\sigma = \{1, \ldots, n\}$, i.e., if and only if σ is the identity permutation ι. In that case the order n term is exactly $\operatorname{sgn}(\iota)(-x)^n = (-x)^n$. ▲

Quick Exercise #10. Show that the characteristic polynomial is a matrix invariant.

In the case of distinct eigenvalues, we get the following simple formula for the characteristic polynomial.

Proposition 6.20 *Let* $A \in M_n(\mathbb{F})$ *have distinct eigenvalues* $\lambda_1, \ldots, \lambda_n$. *Then*

$$p_A(x) = \prod_{j=1}^{n} (\lambda_j - x).$$

Proof Since λ_j is a root of $p_A(x)$ for each j, $p_A(x)$ is divisible by $(x - \lambda_j)$ for each j. Since $p_A(x)$ is a polynomial of degree n, this means it can be completely factored:

$$p_A(x) = c \prod_{j=1}^{n} (x - \lambda_j),$$

for some $c \in \mathbb{F}$. As discussed above, $p_A(x)$ has leading term $(-x)^n$, and so $c = (-1)^n$. ▲

Many students, having vast experience solving quadratic equations, are easily misled into thinking that Proposition 6.18 is a great tool for finding the eigenvalues of a matrix: just write down the characteristic polynomial from the definition, then find its roots. The trouble is that when $n > 2$, finding roots of polynomials gets a whole lot harder. For degrees 3 and 4 there are a cubic formula and a quartic formula which are analogous to the familiar quadratic formula, but they're drastically more complicated.* When $n \geq 5$, no such formula exists.†

In fact, some computer algorithms for finding the roots of polynomials actually use Proposition 6.18 the other way around: to find the roots of a polynomial $p(x)$, they first construct a matrix A whose characteristic polynomial is $p(x)$ (one way to do this is given in Exercise 6.3.18) and then compute the eigenvalues of A by other means.

Nevertheless, the characteristic polynomial is handy for finding the eigenvalues of 2×2 matrices (which come up more in linear algebra homework than in real-world problems), and is an important theoretical tool.

*They're much too messy to include here – try googling "cubic formula" and "quartic formula."
† It's not just that no one has been able to find a formula so far – it was actually proved by the Norwegian mathematician Niels Henrik Abel that there *isn't* one.

QA #10: $p_{SAS^{-1}}(x) = \det(SAS^{-1} - xI_n) = \det(S(A - xI_n)S^{-1}) = \det(A - xI_n) = p_A(x)$.

Quick Exercise #11. Show that if A is a 2×2 matrix, then $p_A(x) = x^2 - Tx + D$, where $T = \operatorname{tr} A$ and $D = \det A$.

For the remainder of the section, it will be helpful to be working over an algebraically closed field. The following result (which we state without proof) often justifies restricting to that setting.

Proposition 6.21 *If $\mathbb{F}$ is any field, then there is an algebraically closed field $\mathbb{K}$ such that $\mathbb{F} \subseteq \mathbb{K}$, and such that the operations on $\mathbb{F}$ are the same as those on $\mathbb{K}$.*

For example, the field $\mathbb{R}$ of real numbers is not algebraically closed, but it is contained in the algebraically closed field $\mathbb{C}$ of complex numbers. Saying that the operations are the same means, for example, that $\mathbb{R}$ sits inside $\mathbb{C}$ because $+$ and $\cdot$ mean the same thing in both places, but $\mathbb{F}_2$ does not sit inside $\mathbb{C}$, since $+$ and $\cdot$ mean different things in those two fields. Proposition 6.21 says that there is an algebraically closed field which contains $\mathbb{F}_2$ with the same operations, but it is more complicated to describe.

The relevance of Proposition 6.21 in this section is that if $A \in M_n(\mathbb{F})$ and $\mathbb{F}$ is not algebraically closed, then the characteristic polynomial $p_A(x)$ may not factor into linear factors.

Example The characteristic polynomial of the matrix $A = \begin{bmatrix} 0 & 1 \\ -1 & 0 \end{bmatrix}$ is $p_A(x) = x^2 + 1$, which does not factor over $\mathbb{R}$. This corresponds to the fact, which we have seen before, that the matrix A has no eigenvalues or eigenvectors over $\mathbb{R}$.

On the other hand, if $p_A(x) = x^2 + 1$ is viewed as a polynomial over $\mathbb{C}$, then $p_A(x)$ does factor:

$$p_A(x) = (x + i)(x - i).$$

This means that over $\mathbb{C}$, the matrix A does have eigenvalues: i and $-i$. ▲

Multiplicities of Eigenvalues

Over an algebraically closed field, the characteristic polynomial of every matrix factors completely, which allows us to make the following definition.

QA #11: $\det\left(\begin{bmatrix} 1 & 0 \\ 0 & 1 \end{bmatrix} x - \begin{bmatrix} a & b \\ c & d \end{bmatrix}\right) = (x - a)(x - d) - bc = x^2 - (a + d)x + ad - bc.$

Definition Let $A \in M_n(\mathbb{F})$, where $\mathbb{F}$ is an algebraically closed field, and suppose that λ is an eigenvalue of A. Then the characteristic polynomial $p_A(x)$ of A factors as

$$p_A(x) = (-1)^n(x-\lambda)^k(x-c_1)^{k_1}\cdots(x-c_m)^{k_m},$$

where the c_j are distinct and different from λ. The power k to which the factor $(x-\lambda)$ appears in $p_A(x)$ is called the **multiplicity** of λ as an eigenvalue of A.

If $p(x)$ has degree n and $\mathbb{F}$ is algebraically closed, then although $p(x)$ may have fewer than n *distinct* roots, it has exactly n roots if we count each root **with multiplicity**. For example, the polynomial

$$p(x) = x^6 - 5x^5 + 6x^4 + 4x^3 - 8x^2 = x^2(x+1)(x-2)^3$$

has the six roots -1, 0, 0, 2, 2, and 2, counted with multiplicity.

The following result summarizes our observations.

Proposition 6.22 *If $\mathbb{F}$ is algebraically closed, every $A \in M_n(\mathbb{F})$ has n eigenvalues, counted with multiplicity.*

If $\lambda_1, \ldots, \lambda_m$ are the distinct eigenvalues of A, with respective multiplicities $k_1, \ldots, k_m$, then

$$p_A(x) = \prod_{j=1}^{m}(\lambda_j - x)^{k_j}.$$

Since the characteristic polynomial of A is an invariant, it follows that not only the eigenvalues of a matrix but also their multiplicities are invariants.

In the upper triangular case, the multiplicities of the eigenvalues are easy to see:

Lemma 6.23 *Suppose that $A \in M_n(\mathbb{F})$ is upper triangular and that λ is an eigenvalue of A. Then the multiplicity of λ as an eigenvalue of A is the number of times λ appears on the diagonal of A.*

Quick Exercise #12. Prove Lemma 6.23.

Counting eigenvalues of a matrix with multiplicity gives us the following nice expressions for the trace and determinant in terms of the eigenvalues.

QA #12: The matrix $A - xI_n$ is triangular, so $p_A(x) = (a_{11} - x)\cdots(a_{nn} - x) = (-1)^n(x - a_{11})\cdots(x - a_{nn})$.

Corollary 6.24 *Suppose that $\mathbb{F}$ is algebraically closed, and that $\lambda_1, \ldots, \lambda_n$ are the eigenvalues of $A \in M_n(\mathbb{F})$, counted with multiplicity. Then*

$$\operatorname{tr}(A) = \lambda_1 + \cdots + \lambda_n \qquad \textit{and} \qquad \det(A) = \lambda_1 \cdots \lambda_n.$$

Proof This follows immediately from Lemma 6.23 if A is upper triangular. Otherwise, it follows from the fact that A is similar to an upper triangular matrix (see Theorem 3.67) and the fact that the trace, the determinant, and the list of eigenvalues with multiplicity are all invariants. ▲

The following result is a good example of the value of viewing a field $\mathbb{F}$ as lying in a larger, algebraically closed field: even though the conclusion is true in any field, the proof works by working in the larger field.

Proposition 6.25 *If $A \in M_n(\mathbb{F})$, then the coefficient of x^{n-1} in $p_A(x)$ is $(-1)^{n-1}\operatorname{tr}(A)$ and the constant term in $p_A(x)$ is $\det A$.*

Proof First suppose that $\mathbb{F}$ is algebraically closed and that $\lambda_1, \ldots, \lambda_n$ are the eigenvalues of A, counted with multiplicity. Then

$$p_A(x) = (\lambda_1 - x) \cdots (\lambda_n - x).$$

Expanding the product gives that the terms of order $(n-1)$ are

$$\lambda_1(-x)^{n-1} + \cdots + \lambda_n(-x)^{n-1} = (-1)^{n-1}(\lambda_1 + \cdots + \lambda_n)x^{n-1} = (-1)^{n-1}(\operatorname{tr} A)x^{n-1}$$

and that the constant term is $\lambda_1 \cdots \lambda_n = \det A$.

If $\mathbb{F}$ is not algebraically closed, we may view A as a matrix over $\mathbb{K}$, where $\mathbb{K}$ is an algebraically closed field containing $\mathbb{F}$. Then the above argument applies to show that, when viewed as a polynomial over $\mathbb{K}$, the coefficient of x^{n-1} in $p_A(x)$ is $(-1)^{n-1}\operatorname{tr}(A)$ and the constant term is $\det(A)$. But, in fact, it doesn't matter how we view p_A; we have identified the order $n-1$ and order 0 terms of $p_A(x)$, and we are done. (As a sanity check, note that even though we are using expressions for tr(A) and det(A) in terms of the eigenvalues of A, which may not lie in $\mathbb{F}$, the trace and determinant themselves are indeed elements of $\mathbb{F}$, by definition.) ▲

The Cayley–Hamilton Theorem

We conclude this section with the following famous theorem.

Theorem 6.26 (The Cayley–Hamilton Theorem) *If $A \in M_n(\mathbb{F})$, then $p_A(A) = 0$.*

Before proving Theorem 6.26, let's consider an appealing but *wrong* way to try to prove it: since $p_A(x) = \det(A - xI_n)$, substituting $x = A$ gives

$$p_A(A) = \det(A - AI_n) = \det \mathbf{0} = 0. \tag{6.10}$$

The typographical conventions used in this book make it a bit easier to see that there's something fishy here – $p_A(A)$ should be a matrix, but $\det \mathbf{0}$ is the *scalar* 0! In fact the error creeps in already in the first equality in equation (6.10). Since the things on opposite sides of the equals sign are of different types (a matrix on the left, a scalar on the right), they can't possibly be equal.

Proof of Theorem 6.26 First of all, we will assume that $\mathbb{F}$ is algebraically closed. If not, then by Proposition 6.21 we could instead work over an algebraically closed field containing $\mathbb{F}$, as in the proof of Proposition 6.25.

Now since $\mathbb{F}$ is algebraically closed, by Theorem 3.67, there exist an invertible matrix $S \in M_n(\mathbb{F})$ and an upper triangular matrix $T \in M_n(\mathbb{F})$ such that $A = STS^{-1}$. Then for any polynomial $p(x)$, $p(A) = Sp(T)S^{-1}$.

Quick Exercise #13. Prove the preceding statement.

Furthermore, $p_A(x) = p_T(x)$ since the characteristic polynomial is an invariant. Therefore

$$p_A(A) = Sp_A(T)S^{-1} = Sp_T(T)S^{-1},$$

so we need to prove that $p_T(T) = \mathbf{0}$ for an upper triangular matrix T.

Now, by Proposition 6.23,

$$p_T(x) = (t_{11} - x) \cdots (t_{nn} - x),$$

so

$$p_T(T) = (t_{11}I_n - T) \cdots (t_{nn}I_n - T).$$

Note that the jth column of $t_{jj}I_n - T$ can only have nonzero entries in the first $(j-1)$ positions; in particular, the first column of $t_{11}I_n - T$ is $\mathbf{0}$. Thus,

$$(t_{jj}I_n - T)e_j \in \langle e_1, \ldots, e_{j-1} \rangle$$

for $j \geq 2$, and $(t_{11}I_n - T)e_1 = \mathbf{0}$. This implies that

$$(t_{11}I_n - T) \cdots (t_{jj}I_n - T)e_j = \mathbf{0},$$

and so

$$p_T(T)e_j = (t_{j+1\, j+1}I_n - T) \cdots (t_{nn}I_n - T)(t_{11}I_n - T) \cdots (t_{jj}I_n - T)e_j = \mathbf{0}.$$

Therefore each column of $p_T(T)$ is $\mathbf{0}$. ▲

QA #13: For each $k \geq 1$, $A^k = (STS^{-1}) \cdots (STS^{-1}) = ST^kS^{-1}$. So if $p(x) = a_0 + a_1x + \cdots + a_mx^m$, then $p(A) = a_0I_n + a_1STS^{-1} + \cdots + a_mST^mS^{-1} = Sp(T)S^{-1}$.

KEY IDEAS

- The characteristic polynomial of A is $p_A(\lambda) = \det(A - \lambda I_n)$. Its roots are the eigenvalues of A.
- The multiplicity of an eigenvalue of A is its multiplicity as a root of p_A. For an upper triangular matrix, this is the number of times the eigenvalue appears on the diagonal.
- The Cayley–Hamilton Theorem: $p_A(A) = \mathbf{0}$.

EXERCISES

6.3.1 For each of the following matrices, find the characteristic polynomial and all the eigenvalues and eigenvectors.

(a) $\begin{bmatrix} 1 & 1 \\ 2 & 0 \end{bmatrix}$ (b) $\begin{bmatrix} \cos(\theta) & -\sin(\theta) \\ \sin(\theta) & \cos(\theta) \end{bmatrix}$ (c) $\begin{bmatrix} -1 & -3 & 1 \\ 3 & 3 & 1 \\ 3 & 0 & 4 \end{bmatrix}$

(d) $\begin{bmatrix} 1 & 0 & 3 \\ 0 & -2 & 0 \\ 3 & 0 & 1 \end{bmatrix}$

6.3.2 For each of the following matrices, find the characteristic polynomial and all the eigenvalues and eigenvectors.

(a) $\begin{bmatrix} 1 & 2 \\ 3 & 2 \end{bmatrix}$ (b) $\begin{bmatrix} 2 & 2i \\ i & -1 \end{bmatrix}$

(c) $\begin{bmatrix} 2 & 1 & 1 \\ -4 & -3 & 0 \\ -2 & -2 & 1 \end{bmatrix}$ (d) $\begin{bmatrix} 0 & 2 & 1 \\ -1 & -2 & 0 \\ 1 & 1 & -1 \end{bmatrix}$

6.3.3 Let $A = \begin{bmatrix} 1 & 1 \\ 0 & 1 \end{bmatrix}$.

(a) Show that A and I_2 have the same trace, determinant, and characteristic polynomial.

(b) Show that A and I_2 are *not* similar.

6.3.4 (a) Prove that the matrices

$$\begin{bmatrix} 1 & -5,387,621.4 \\ 0 & 2 \end{bmatrix} \quad \text{and} \quad \begin{bmatrix} 5 & 3 \\ -4 & -2 \end{bmatrix}$$

are similar.

(b) Prove that the matrices

$$\begin{bmatrix} 1.000000001 & 0 \\ 0 & 2 \end{bmatrix} \quad \text{and} \quad \begin{bmatrix} 1 & 0 \\ 0 & 2 \end{bmatrix}$$

are *not* similar.

6.3.5 Suppose that the matrix $A \in M_3(\mathbb{R})$ has $\operatorname{tr} A = -1$ and $\det A = 6$. If $\lambda = 2$ is an eigenvalue of A, what are the other two eigenvalues?

6.3.6 Suppose that the matrix $A \in M_3(\mathbb{R})$ has $\operatorname{tr} A = -1$ and $\det A = -6$. If $\lambda = 2$ is an eigenvalue of A, what are the other two eigenvalues?

6.3.7 Let $A = \begin{bmatrix} 1 & 2 \\ 3 & 4 \end{bmatrix}$.

(a) Find the characteristic polynomial $p_A(x)$.

(b) Use the Cayley–Hamilton Theorem to find A^3 *without* ever multiplying two matrices together.

6.3.8 Let $A = \begin{bmatrix} 2 & 7 \\ 1 & 8 \end{bmatrix}$.

(a) Find the characteristic polynomial $p_A(x)$.

(b) Use the Cayley–Hamilton Theorem to find A^4 *without* ever multiplying two matrices together.

6.3.9 Let $A \in M_n(\mathbb{F})$. Show that $\operatorname{rank} A = n$ if and only if $p_A(0) \neq 0$.

6.3.10 Suppose λ is an eigenvalue of $A \in M_n(\mathbb{F})$. Prove that the geometric multiplicity of λ (see Exercise 3.6.24) is less than or equal to the multiplicity of λ.

6.3.11 Show that if A is a normal matrix, then the geometric multiplicities (see Exercise 3.6.24) of its eigenvalues are equal to their multiplicities.

6.3.12 Show that if $A \in M_n(\mathbb{C})$ has only real eigenvalues, then the coefficients of $p_A(x)$ are all real.

6.3.13 Suppose that $A \in M_{m,n}(\mathbb{C})$ with $m \leq n$, and that $\sigma_1 \geq \cdots \geq \sigma_m \geq 0$ are the singular values of A. Show that $\sigma_1^2, \ldots, \sigma_m^2$ are the eigenvalues of AA^*, counted with multiplicity.

6.3.14 Suppose that $\mathbb{F}$ is algebraically closed, $A \in M_n(\mathbb{F})$ is invertible, and λ is an eigenvalue of A. Show that λ^{-1} has the same multiplicity as an eigenvalue of A^{-1} that λ has as an eigenvalue of A.

6.3.15 Let $p(x)$ be any polynomial over $\mathbb{F}$. Prove that $p(A)$ lies in the span of $(I_n, A, A^2, \ldots, A^{n-1})$.

6.3.16 (a) Show that if $A \in M_n(\mathbb{F})$ is invertible, then A^{-1} lies in the span of $(I_n, A, A^2, \ldots, A^{n-1})$.

(b) Use this fact and Exercise 2.3.12 to prove that if A is invertible and upper triangular, then A^{-1} is upper triangular.

6.3.17 Suppose that $A \in M_n(\mathbb{C})$ has eigenvalues $\lambda_1, \ldots, \lambda_n$, counted with multiplicity. Prove that

$$\sqrt{\sum_{j=1}^{n} |\lambda_j|^2} \leq \|A\|_F.$$

Hint: Prove this first for upper triangular matrices, then use Schur decomposition.

6.3.18 Let $p(x) = c_0 + c_1 x + \cdots + c_{n-1} x^{n-1} + x^n$ be a given polynomial with coefficients in $\mathbb{F}$. Define

$$A = \begin{bmatrix} 0 & 0 & \cdots & 0 & -c_0 \\ 1 & 0 & \cdots & 0 & -c_1 \\ 0 & 1 & \cdots & 0 & -c_2 \\ \vdots & \vdots & \ddots & \vdots & \vdots \\ 0 & 0 & \cdots & 1 & -c_{n-1} \end{bmatrix}.$$

Show that $p_A(x) = (-1)^n p(x)$.

Remark: The matrix A is called the **companion matrix** of $p(x)$.

6.3.19 Here's another way to convince yourself that the appealing-but-wrong way to prove the Cayley–Hamilton Theorem is indeed wrong. Given $A \in M_n(\mathbb{F})$, define the **trace polynomial** $t_A(x) = \operatorname{tr}(A - xI_n)$. The same kind of argument suggests that

$$t_A(A) = \operatorname{tr}(A - AI_n) = 0.$$

Show that, in fact, if $\mathbb{F} = \mathbb{R}$ or $\mathbb{C}$ then $t_A(A) = \mathbf{0}$ if and only if $A = cI_n$ for some $c \in \mathbb{F}$.

6.4 Applications of Determinants

Volume

The connection to volume is one of the most important applications of determinants outside of linear algebra. Since much of this topic is necessarily beyond the scope of this book, we will only sketch the details.

Given real numbers $a_i \le b_i$ for each $i = 1, \ldots, n$, the set

$$\mathcal{R} = [a_1, b_1] \times \cdots \times [a_n, b_n] := \{(x_1, \ldots, x_n) \mid a_i \le x_i \le b_i \text{ for each } i = 1, \ldots, n\}$$

is called a **rectangle**. The **volume** of the rectangle $\mathcal{R}$ is defined to be

$$\operatorname{vol}(\mathcal{R}) = (b_1 - a_1) \cdots (b_n - a_n).$$

When $n = 2$, what we're calling "volume" is usually known as "area."

Note that $\operatorname{vol}(\mathcal{R}) = 0$ iff $a_i = b_i$ for some i. We say that two rectangles are **essentially disjoint** if their intersection has volume 0. (So two essentially disjoint rectangles may have overlapping boundaries, but no overlap of their interiors.)

A set $\mathcal{S} \subseteq \mathbb{R}^n$ is called a **simple set** if $\mathcal{S}$ is the union of a finite collection of rectangles. It turns out that every simple set $\mathcal{S}$ can in fact be written as the union of a finite collection of pairwise essentially disjoint rectangles. Furthermore, given any two such ways of writing $\mathcal{S}$, the total volumes of the rectangles are the same, and so we may define

$$\mathrm{vol}(\mathcal{S}) = \sum_{i=1}^{m} \mathrm{vol}(\mathcal{R}_i),$$

whenever $\mathcal{S} = \bigcup_{i=1}^{m} \mathcal{R}_i$ and $\mathcal{R}_1, \dots, \mathcal{R}_m$ are essentially disjoint rectangles in $\mathbb{R}^n$.

We would like to define the **volume** of an arbitrary set $\Omega \subseteq \mathbb{R}^n$ as

$$\mathrm{vol}(\Omega) = \lim_{k \to \infty} \mathrm{vol}(\mathcal{S}_k), \tag{6.11}$$

where $(\mathcal{S}_k)_{k \geq 1}$ is a sequence of simple sets which approximates Ω, just as in calculus, area under a curve is defined as the limit of the total area of an approximating set of rectangles.

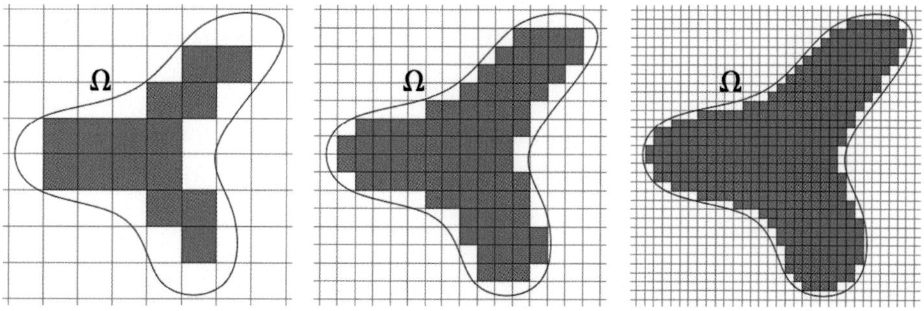

Figure 6.1 Approximation of Ω by simple sets.

There is an important technical caveat: the limit in equation (6.11) may not exist, or may be different depending on which sequence $(\mathcal{S}_k)_{k \geq 1}$ is used. A set $\Omega \subseteq \mathbb{R}^n$ is called **Jordan measurable** if, whenever Ω is approximated by a sequence of simple sets, the volumes of the simple sets actually do approach a limit, which is independent of the sequence. That is, Ω is Jordan measurable iff the limit in equation (6.11) is well-defined. Jordan measurability is a technical condition that we have to think about when proving theorems, but isn't something to worry about a lot in practice. Most sets you come across whose volume you'd like to talk about are Jordan measurable, and from now on we restrict our attention to those sets.

When volume is defined as described above, several familiar properties we generally assume can be proved. In particular, volume is translation-invariant; i.e., the volume of a set does not change based on the position of the set. Volume is also additive: if a set $\mathcal{A}$ can be written as a (finite or countable) union of essentially disjoint sets $\mathcal{S}_k$, then

$$\mathrm{vol}(\mathcal{A}) = \sum_{k} \mathrm{vol}(\mathcal{S}_k).$$

The connection between volume and determinants is the following.

Theorem 6.27 *Let $T \in \mathcal{L}(\mathbb{R}^n)$ and let $\Omega \subseteq \mathbb{R}^n$ be Jordan measurable. Then*

$$\mathrm{vol}\big(T(\Omega)\big) = |\det T| \,\mathrm{vol}(\Omega).$$

As discussed above, a completely rigorous development of volume depends on analytic ideas that are beyond the scope of this book. Therefore we will give only a sketch of the proof of Theorem 6.27, emphasizing the ideas from linear algebra and skipping over some analytic subtleties.

Sketch of proof We begin by supposing that the matrix A of T is an elementary matrix and that $\mathcal{R}$ is a rectangle

$$\mathcal{R} = [a_1, b_1] \times \cdots \times [a_n, b_n].$$

There are three cases to consider, corresponding to the three types of elementary matrices:

- $A = P_{c,i,j}$. For simplicity, we just consider the case of $P_{c,1,2}$ for $c > 0$ and $n = 2$. Then $T(\mathcal{R})$ is a parallelogram $\mathcal{P}$:

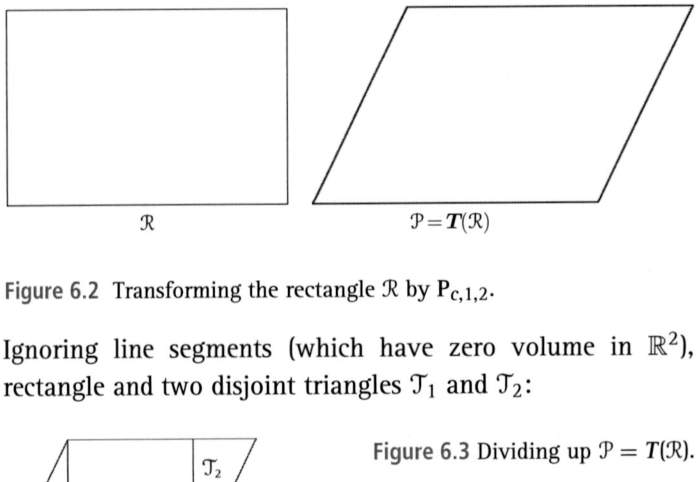

Figure 6.2 Transforming the rectangle $\mathcal{R}$ by $P_{c,1,2}$.

Ignoring line segments (which have zero volume in $\mathbb{R}^2$), $\mathcal{P}$ consists of a rectangle and two disjoint triangles $\mathcal{T}_1$ and $\mathcal{T}_2$:

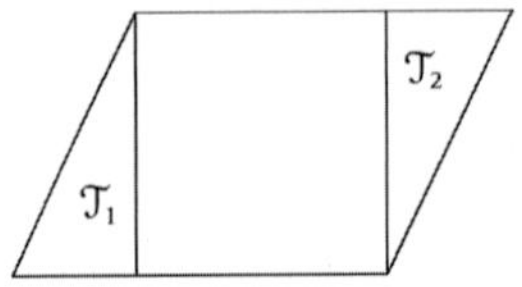

Figure 6.3 Dividing up $\mathcal{P} = T(\mathcal{R})$.

Translating $\mathcal{T}_2$ as shown:

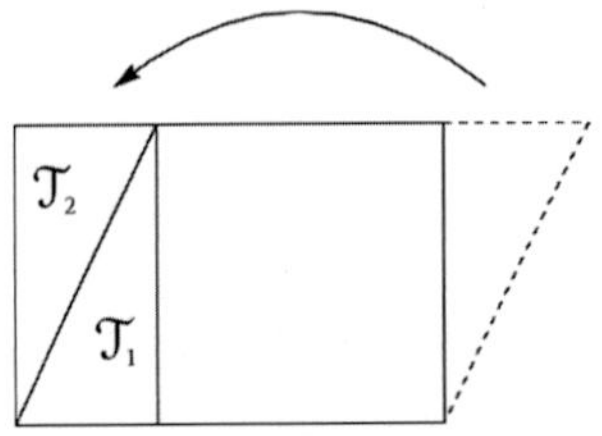

Figure 6.4 Moving $\mathcal{T}_2$ to turn $\mathcal{P}$ back into $\mathcal{R}$.

we see that $T(\mathcal{R})$ has the same area as $\mathcal{R}$. Thus

$$\operatorname{vol}\bigl(T(\mathcal{R})\bigr) = \operatorname{vol}(\mathcal{R})$$

(recall that in this case, $\det T = 1$, so this is what we wanted).
When $n > 2$, the discussion above describes what happens in the cross-sections of $T(\mathcal{R})$ in planes parallel to the x_i–x_j plane. The equality $\operatorname{vol}\bigl(T(\mathcal{R})\bigr) = \operatorname{vol}(\mathcal{R})$ then follows by computing the volume of $T(\mathcal{R})$ by slices.

- $\mathrm{A} = \mathrm{Q}_{c,i}$. In this case

$$T(\mathcal{R}) = [a_1, b_1] \times \cdots \times [ca_i, cb_i] \times \cdots \times [a_n, b_n]$$

if $c > 0$, and

$$T(\mathcal{R}) = [a_1, b_1] \times \cdots \times [cb_i, ca_i] \times \cdots \times [a_n, b_n]$$

if $c < 0$. In either case,

$$\operatorname{vol}\bigl(T(\mathcal{R})\bigr) = |c|\,(b_1 - a_1)\cdots(b_n - a_n) = |c| \operatorname{vol}(\mathcal{R}) = |\det T| \operatorname{vol}(\mathcal{R}).$$

- $\mathrm{A} = \mathrm{R}_{i,j}$.

Quick Exercise #14. Prove that in this case, $\operatorname{vol}\bigl(T(\mathcal{R})\bigr) = \operatorname{vol}(\mathcal{R})$.

In all three cases, by Lemma 6.14, we have

$$\operatorname{vol}\bigl(T(\mathcal{R})\bigr) = |\det T| \operatorname{vol}(\mathcal{R}) \tag{6.12}$$

when $\mathcal{R}$ is a rectangle and the matrix A of T is elementary.

We can immediately extend to simple sets: if A is still elementary and $\mathcal{S}$ is the simple set

$$\mathcal{S} = \bigcup_{i=1}^{m} \mathcal{R}_i,$$

where $\mathcal{R}_1, \ldots, \mathcal{R}_m$ are essentially disjoint rectangles, then

$$T(\mathcal{S}) = T\left(\bigcup_{i=1}^{m} \mathcal{R}_i\right) = \bigcup_{i=1}^{m} T(\mathcal{R}_i),$$

and the sets $T(\mathcal{R}_1), \ldots, T(\mathcal{R}_m)$ are still essentially disjoint. Thus

$$\operatorname{vol}\bigl(T(\mathcal{S})\bigr) = \sum_{i=1}^{m} \operatorname{vol}\bigl(T(\mathcal{R}_i)\bigr) = \sum_{i=1}^{m} |\det T| \operatorname{vol}(\mathcal{R}_i) = |\det T| \operatorname{vol}(\mathcal{S}). \tag{6.13}$$

If $\Omega \subseteq \mathbb{R}^n$ is Jordan measurable and is approximated by a sequence $(\mathcal{S}_k)_{k\ge 1}$ of simple sets, and A is still elementary, then

$$\operatorname{vol}\bigl(T(\Omega)\bigr) = \lim_{k\to\infty} \operatorname{vol}\bigl(T(\mathcal{S}_k)\bigr) = |\det T| \lim_{k\to\infty} \operatorname{vol}(\mathcal{S}_k) = |\det T| \operatorname{vol}(\Omega) \tag{6.14}$$

QA #14: If $i < j$, then $R_{i,j}(\mathcal{R}) = [a_1, b_1] \times \cdots \times [a_j, b_j] \times \cdots \times [a_i, b_i] \times \cdots \times [a_n, b_n]$.

Finally, let $T \in \mathcal{L}(\mathbb{R}^n)$ be an arbitrary linear map. Since $T(\Omega) \subseteq \operatorname{range} T$, if $\operatorname{rank} T < n$ then $T(\Omega)$ lies in a lower-dimensional subspace of $\mathbb{R}^n$ and so $\operatorname{vol}(T(\Omega)) = 0$. Also, if $\operatorname{rank} T < n$, then by Corollary 6.6, $\det T = 0$, and so in this case

$$\operatorname{vol}(T(\Omega)) = |\det T| \operatorname{vol}(\Omega) = 0.$$

If on the other hand $\operatorname{rank} T = n$, then T is invertible by Corollary 3.36, and so there are elementary matrices $\mathrm{E}_1, \ldots, \mathrm{E}_k$ such that

$$\mathrm{A} = \mathrm{E}_1 \cdots \mathrm{E}_k$$

(since the RREF of A is I_n). Let T_{E} denote the linear map with matrix E. Applying equation (6.14) k times (with the linear maps $T_{\mathrm{E}_1}, \ldots, T_{\mathrm{E}_k}$ in place of T), we have

$$\begin{aligned}
\operatorname{vol}(T(\Omega)) &= \operatorname{vol}(T_{\mathrm{E}_1} \circ \cdots \circ T_{\mathrm{E}_k}(\Omega)) \\
&= |\det \mathrm{E}_1| \operatorname{vol}(T_{\mathrm{E}_2} \circ \cdots \circ T_{\mathrm{E}_k}(\Omega)) \\
&= |\det \mathrm{E}_1|\, |\det \mathrm{E}_2| \operatorname{vol}(T_{\mathrm{E}_3} \circ \cdots \circ T_{\mathrm{E}_k}(\Omega)) \\
&\;\;\vdots \\
&= |\det \mathrm{E}_1| \cdots |\det \mathrm{E}_k| \operatorname{vol}(\Omega) \\
&= |\det \mathrm{E}_1 \cdots \mathrm{E}_k| \operatorname{vol}(\Omega).
\end{aligned}$$

Then

$$\operatorname{vol}(T(\Omega)) = |\det(\mathrm{E}_1 \cdots \mathrm{E}_k)| \operatorname{vol}(\Omega) = |\det \mathrm{A}| \operatorname{vol}(\Omega) = |\det T| \operatorname{vol}(\Omega).$$

▲

Cramer's Rule

The next result shows how we can come full circle and use determinants to solve linear systems. The proof illustrates an interesting trick – sometimes the best way to prove something about a vector is to build a whole matrix out of it.

Theorem 6.28 (Cramer's rule) *Suppose that* $\mathrm{A} \in \mathrm{M}_n(\mathbb{F})$, $\det \mathrm{A} \neq 0$, *and* $\mathbf{b} \in \mathbb{F}^n$. *For* $i = 1, \ldots, n$, *define* A_i *to be the* $n \times n$ *matrix*

$$\mathrm{A}_i = \begin{bmatrix} | & & | & | & | & & | \\ \mathbf{a}_1 & \cdots & \mathbf{a}_{i-1} & \mathbf{b} & \mathbf{a}_{i+1} & \cdots & \mathbf{a}_n \\ | & & | & | & | & & | \end{bmatrix}$$

obtained from A *by replacing the ith column with* $\mathbf{b}$.

Then the unique solution of the $n \times n$ *linear system* $\mathrm{A}\mathbf{x} = \mathbf{b}$ *is given by*

$$x_i = \frac{\det \mathrm{A}_i}{\det \mathrm{A}}$$

for each $i = 1, \ldots, n$.

Proof Since $\det A \neq 0$, we know that A is invertible, and so there is indeed a unique solution x of the linear system $Ax = b$. For each $i = 1, \dots, n$, define X_i to be the $n \times n$ matrix

$$X_i = \begin{bmatrix} | & & | & | & | & & | \\ e_1 & \cdots & e_{i-1} & x & e_{i+1} & \cdots & e_n \\ | & & | & | & | & & | \end{bmatrix}$$

$$= \begin{bmatrix} | & & | & | & | & & | \\ e_1 & \cdots & e_{i-1} & \sum_{k=1}^{n} x_k e_k & e_{i+1} & \cdots & e_n \\ | & & | & | & | & & | \end{bmatrix}.$$

Then $\det X_i = x_i$ because the determinant is multilinear and alternating. Furthermore, AX_i is exactly A_i.

Quick Exercise #15. Verify that $AX_i = A_i$.

Therefore $\det A_i = \det AX_i = (\det A)(\det X_i) = (\det A)x_i$. ▲

Example Recall our very first linear system, describing how much bread and beer we could make with given quantities of barley and yeast:

$$x + \frac{1}{4}y = 20$$
$$2x + \frac{1}{4}y = 36.$$

In this system,

$$A = \begin{bmatrix} 1 & \frac{1}{4} \\ 2 & \frac{1}{4} \end{bmatrix}, \qquad A_1 = \begin{bmatrix} 20 & \frac{1}{4} \\ 36 & \frac{1}{4} \end{bmatrix}, \qquad A_2 = \begin{bmatrix} 1 & 20 \\ 2 & 36 \end{bmatrix},$$

and so

$$\det A = -\frac{1}{4}, \qquad \det A_1 = -4, \qquad \det A_2 = -4.$$

It follows from Theorem 6.28 that the unique solution to this system is

$$x = y = \frac{(-4)}{\left(-\frac{1}{4}\right)} = 16.$$ ▲

Cofactors and Inverses

There is a classical formula for the inverse of a matrix in terms of determinants, which can be proved using Theorem 6.28. To state it we first need a definition.

QA #15: The jth column of AX_i is A times the jth column of X_i. If $i \neq j$, this is $Ae_j = a_j$; if $i = j$, this is $Ax = b$.

Definition Let $A \in M_n(\mathbb{F})$. The **cofactor matrix** of A is the matrix $C \in M_n(\mathbb{F})$ with $c_{ij} = (-1)^{i+j} \det A_{ij}$.
The **adjugate*** of A is the matrix $\operatorname{adj} A = C^T$.

Examples

1. The cofactor matrix of $A = \begin{bmatrix} a & b \\ c & d \end{bmatrix}$ is $\begin{bmatrix} d & -c \\ -b & a \end{bmatrix}$, and its adjugate is $\operatorname{adj} A = \begin{bmatrix} d & -b \\ -c & a \end{bmatrix}$.

2. The cofactor and adjugate matrices of $A = \begin{bmatrix} a & b & c \\ d & e & f \\ g & h & i \end{bmatrix}$ are

$$C = \begin{bmatrix} ei - fh & fg - di & dh - eg \\ ch - ib & ai - cg & bg - ah \\ bf - ce & cd - af & ae - bd \end{bmatrix} \qquad \operatorname{adj} A = \begin{bmatrix} ei - fh & ch - ib & bf - ce \\ fg - di & ai - cg & cd - af \\ dh - eg & bg - ah & ae - bd \end{bmatrix}.$$

▲

Theorem 6.29 *If* $A \in M_n(\mathbb{F})$ *is invertible, then*

$$A^{-1} = \frac{1}{\det A} \operatorname{adj} A.$$

Proof The jth column of A^{-1} is $A^{-1}e_j$, which is the unique solution of the linear system $Ax = e_j$. Writing $B = A^{-1}$, by Cramer's rule,

$$b_{ij} = \frac{\det A_i}{\det A},$$

where

$$A_i = \begin{bmatrix} | & & | & | & | & & | \\ a_1 & \cdots & a_{i-1} & e_j & a_{i+1} & \cdots & a_n \\ | & & | & | & | & & | \end{bmatrix}.$$

By Laplace expansion along the ith column,

$$\det A_i = (-1)^{i+j} \det A_{ji},$$

*Not to be confused with the adjoint! The adjugate is sometimes also called the **classical adjoint**, because it used to be known as the adjoint, before that word came to be used for the completely different thing it refers to now.

and so $b_{ij} = \frac{1}{\det A} c_{ji}$, where C is the cofactor matrix of A. In other words,

$$A^{-1} = B = \frac{1}{\det A} C^T = \frac{1}{\det A} \operatorname{adj} A. \qquad \blacktriangle$$

Example By Theorem 6.29 and the last example, if $A = \begin{bmatrix} a & b \\ c & d \end{bmatrix}$ has nonzero determinant, then

$$A^{-1} = \frac{1}{\det A} \begin{bmatrix} d & -c \\ -b & a \end{bmatrix}^T = \frac{1}{ad - bc} \begin{bmatrix} d & -b \\ -c & a \end{bmatrix},$$

giving us the familiar formula, which we first saw in (2.14), for the inverse of a 2×2 matrix. ▲

KEY IDEAS

- When a linear operator T is used to transform $\mathbb{R}^n$, the volumes of sets change by a factor of $|\det T|$.
- Cramer's rule: if A is invertible, the unique solution to $Ax = b$ is given by

$$x_i = \frac{\det(A_i(b))}{\det A},$$

  where $A_i(b)$ is the matrix gotten by replacing the ith column of A with b.
- $A^{-1} = \frac{1}{\det A} \operatorname{adj} A$, where adj A is the adjugate of A, whose (i,j) entry is $(-1)^{i+j} \det(A_{ji})$.

EXERCISES

6.4.1 Given $a, b > 0$, calculate the area of the ellipse

$$\mathcal{E} = \left\{ \begin{bmatrix} x \\ y \end{bmatrix} \in \mathbb{R}^2 : \left(\frac{x}{a}\right)^2 + \left(\frac{y}{b}\right)^2 \leq 1 \right\}.$$

You may use the fact that the unit circle has area π.

6.4.2 Given $a, b, c > 0$, calculate the volume of the ellipsoid

$$\mathcal{E} = \left\{ \begin{bmatrix} x \\ y \\ z \end{bmatrix} \in \mathbb{R}^3 : \left(\frac{x}{a}\right)^2 + \left(\frac{y}{b}\right)^2 + \left(\frac{z}{c}\right)^2 \leq 1 \right\}.$$

You may use the fact that the unit ball has volume $\frac{4}{3}\pi$.

6.4.3 Calculate the area of the ellipse

$$\mathcal{E} = \left\{ \begin{bmatrix} x \\ y \end{bmatrix} \in \mathbb{R}^2 : (x + y)^2 + 4(x - y)^2 \leq 4 \right\}.$$

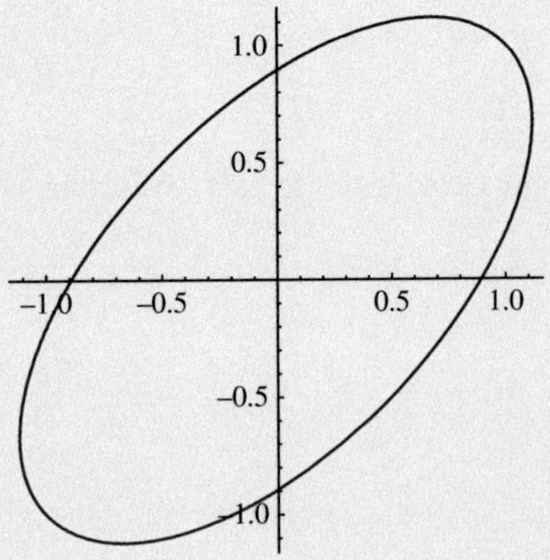

You may use the fact that the area of the unit disc is π.

6.4.4 Use Cramer's rule to solve each of the following linear systems.

(a) $\sqrt{2}x + 2y = -1$
$x - 3\sqrt{2}y = 5$

(b) $ix + (1-i)y = 0$
$(2+i)x + y = 4 + 3i$

(c) $y + z = 2$
$x + z = 1$
$x + y = -1$

(d) $x + z = 1$, over $\mathbb{F}_2$
$x + y + z = 0$
$y + z = 1$

6.4.5 Use Cramer's rule to solve each of the following linear systems.

(a) $x + 2y = 3$
$-x + y = -4$

(b) $x + (2-i)y = 3$
$-ix + y = 1 + i$

(c) $2x - y + z = -1$
$x - 3y + 2z = 2$
$3x + 4y - z = 1$

(d) $2x + z - 3w = 0$
$y + -3z + w = -1$
$-x + 2y - z = 2$
$x + y - 2w = 1$

6.4.6 Use Theorem 6.29 to invert each of the following matrices.

(a) $\begin{bmatrix} 1 & 1 & 0 \\ 1 & 1 & 1 \\ 0 & 1 & 1 \end{bmatrix}$ (b) $\begin{bmatrix} 1 & -2 & 0 & 0 \\ -1 & 0 & 0 & 2 \\ 0 & 0 & 1 & 1 \\ 0 & 1 & 2 & 0 \end{bmatrix}$

(c) $\begin{bmatrix} 1 & 1 & 1 & 1 \\ 1 & 2 & 2 & 2 \\ 1 & 2 & 3 & 3 \\ 1 & 2 & 3 & 4 \end{bmatrix}$ (d) $\begin{bmatrix} 0 & 1 & 0 & 0 & 0 \\ 1 & 0 & 1 & 0 & 0 \\ 0 & 1 & 1 & 1 & 0 \\ 0 & 0 & 1 & 0 & 1 \\ 0 & 0 & 0 & 1 & 0 \end{bmatrix}$

6.4.7 Use Theorem 6.29 to invert each of the following matrices.

(a) $\begin{bmatrix} 2 & -1 & 2 \\ 3 & 0 & 1 \\ 1 & -2 & 4 \end{bmatrix}$ (b) $\begin{bmatrix} 1 & 1 & 0 & 0 \\ 1 & 1 & 1 & 0 \\ 0 & 1 & 1 & 1 \\ 0 & 0 & 1 & 1 \end{bmatrix}$ (c) $\begin{bmatrix} 2 & 1 & 0 & 1 \\ 1 & 1 & -1 & 1 \\ 2 & -1 & 3 & 0 \\ 1 & 0 & 1 & 2 \end{bmatrix}$

(d) $\begin{bmatrix} 0 & -1 & 0 & 2 & 1 \\ 0 & 3 & 0 & 0 & -1 \\ 2 & 0 & -2 & 1 & 0 \\ -1 & 1 & 2 & 0 & 0 \\ 1 & 0 & -2 & 0 & 0 \end{bmatrix}$

6.4.8 Show that if $T \in \mathcal{L}(\mathbb{R}^n)$ is an isometry and $\Omega \subseteq \mathbb{R}^n$ is Jordan measurable, then $\operatorname{vol}(T(\Omega)) = \operatorname{vol}(\Omega)$.

6.4.9 Give a heuristic argument based on the singular value decomposition explaining appearance of $|\det T|$ in Theorem 6.27. (See Exercise 6.2.3.)

6.4.10 Show that if $T \in \mathcal{L}(\mathbb{R}^n)$ and $\Omega \subseteq \mathbb{R}^n$ is Jordan measurable, then

$$\sigma_n^n \operatorname{vol}(\Omega) \le \operatorname{vol}(T(\Omega)) \le \sigma_1^n \operatorname{vol}(\Omega),$$

where $\sigma_1 \ge \cdots \ge \sigma_n \ge 0$ are the singular values of T.

6.4.11 Show that the volume of the n-dimensional parallelepiped spanned by $\mathbf{a}_1, \ldots, \mathbf{a}_n \in \mathbb{R}^n$ is

$$\left| \det \begin{bmatrix} | & & | \\ \mathbf{a}_1 & \cdots & \mathbf{a}_n \\ | & & | \end{bmatrix} \right|.$$

6.4.12 Use Theorem 6.29 to prove that if $A \in M_n(\mathbb{F})$ is upper triangular and invertible, then A^{-1} is also upper triangular. What are the diagonal entries of A^{-1}?

6.4.13 Consider an $n \times n$ linear system $A\mathbf{x} = \mathbf{b}$. Show that if A and $\mathbf{b}$ have only integer entries, and $\det A = \pm 1$, then the unique solution $\mathbf{x}$ also has only integer entries.

6.4.14 Suppose that $A \in M_n(\mathbb{R})$ has only integer entries. Prove that A^{-1} has only integer entries if and only if $\det A = \pm 1$.

6.4.15 In the sketch of the proof of Theorem 6.27, we asserted without proof that a line segment in $\mathbb{R}^2$ has zero volume (i.e., area). Prove this by showing that for every n, a given line segment can be covered by n rectangles, in such a way that the total area of the rectangles goes to 0 as $n \to \infty$.

PERSPECTIVES: Determinants

- The determinant is the unique alternating multilinear function on $M_n(\mathbb{F})$ which takes the value 1 on I_n.
- For any i, $\det \mathrm{A} = \sum_{j=1}^{n} (-1)^{i+j} a_{ij} \det(\mathrm{A}_{ij})$.
- For any j, $\det \mathrm{A} = \sum_{i=1}^{n} (-1)^{i+j} a_{ij} \det(\mathrm{A}_{ij})$.
- $\det \mathrm{A} = \sum_{\sigma \in S_n} (\operatorname{sgn} \sigma) a_{1,\sigma(1)} \cdots a_{n,\sigma(n)} = \sum_{\sigma \in S_n} (\operatorname{sgn} \sigma) a_{\sigma(1),1} \cdots a_{\sigma(n),n}$.
- $\det \mathrm{A} = (-1)^k \dfrac{b_{11} \cdots b_{nn}}{c_1 \cdots c_m}$, where A can be row-reduced to the upper triangular matrix B, and in the process, the c_i are the constants used in row operation R2 and rows are swapped k times.
- If A is upper triangular, $\det \mathrm{A} = a_{11} \cdots a_{nn}$.
- $\det \mathrm{A} = \prod_{j=1}^{n} \lambda_j$, where λ_j are the eigenvalues of A, counted with multiplicity.

We conclude by repeating some earlier perspectives, with new additions involving determinants.

PERSPECTIVES: Eigenvalues

A scalar $\lambda \in \mathbb{F}$ is an eigenvalue of $T \in \mathcal{L}(V)$ if any of the following hold.

- There is a nonzero $v \in V$ with $Tv = \lambda v$.
- The map $T - \lambda I$ is not invertible.
- $\det(T - \lambda I) = 0$.

PERSPECTIVES: Isomorphisms

Let V and W be n-dimensional vector spaces over $\mathbb{F}$. A linear map $T : V \to W$ is an isomorphism if any of the following hold.

- T is bijective.
- T is invertible.
- T is injective, or equivalently $\text{null}(T) = 0$.
- T is surjective, or equivalently $\text{rank}(T) = n$.
- If $(v_1, \ldots, v_n)$ is a basis of T, then $(T(v_1), \ldots, T(v_n))$ is a basis of W.
- If A is the matrix of T with respect to any bases on V and W, then the columns of A form a basis of $\mathbb{F}^n$.
- If A is the matrix of T with respect to any bases on V and W, the RREF of A is the identity I_n.
- $V = W$ and $\det T \neq 0$.

Appendix

A.1 Sets and Functions

Basic Definitions

Definition A set S is a collection of objects, called **elements.** If s is an element of S, we write $s \in S$.

A **subset** T of S is a set such that every element of T is an element of S. If T is a subset of S, we write $T \subseteq S$.

The subset T is a **proper subset** of S if there is at least one element of S which is not in T; in this case, we write $T \subsetneq S$.

The **union** of sets S_1 and S_2 is the collection of every element which is in either S_1 or S_2; the union of S_1 and S_2 is denoted $S_1 \cup S_2$.

The **intersection** of sets S_1 and S_2 is the collection of every element which is in both S_1 and S_2; the intersection of S_1 and S_2 is denoted $S_1 \cap S_2$.

This is a rather unsatisfying definition; if we need to define the word set, defining it as a collection of things seems a bit cheap. Nevertheless, we need to start somewhere.

Some sets which come up so often that they have their own special notation are the following.

- $\mathbb{R}$: the set of real numbers
- $\mathbb{C}$: the set of complex numbers
- $\mathbb{N}$: the set of natural numbers $1, 2, 3, \ldots$
- $\mathbb{Z}$: the set of integers (natural numbers, their negatives, and zero)
- $\mathbb{Q}$: the set of rational numbers; i.e., real numbers that can be expressed as fractions of integers
- $\mathbb{R}^n$: the set of n-dimensional (column) vectors with real entries

When we want to describe sets explicitly, we often use so-called "set-builder notation," which has the form*

*Some people use a colon in the middle instead of a vertical bar.

$$S = \{\text{objects} \mid \text{conditions}\}.$$

That is, inside the curly brackets to the left of the bar is what kind of thing the set elements are, and to the right of the bar are the requirements that need to be satisfied so that the object on the left is in the set. *Every* object of the type on the left which satisfies the conditions on the right is in S. For example, the set

$$S = \{t \in \mathbb{R} \mid t \geq 0\}$$

is the set of all nonnegative real numbers, which you may also have seen denoted $[0, \infty)$.

Set-builder notation doesn't give a unique way of describing a set. For example,

$$\left\{ \begin{bmatrix} 1+z \\ z-6 \\ z \end{bmatrix} \middle| \; z \in \mathbb{R} \right\} = \left\{ \begin{bmatrix} x \\ y \\ z \end{bmatrix} \in \mathbb{R}^3 \; \middle| \; x = 1+z, \; y = z-6 \right\}.$$

As in this example, if there are multiple conditions to the right of the bar separated by a comma (or a semi-colon), it means that they *all* have to be satisfied.

As with the definition of a set, we will give a slightly evasive definition of our other fundamental mathematical object, a function.

Definition A **function** is a rule that assigns an output to each element of a set of possible inputs. To be a function, it must always assign the same output to a given input.

The set of possible inputs is called the **domain** of the function, and the set of all outputs is called the **range** of the function.

Saying that a function **is defined on** a set X means that X is the set of inputs; saying that a function **takes values in** a set Y means that all of the possible outputs are elements of Y, but there may be more elements of Y which do not actually occur as outputs. If a function takes values in Y, then Y is called the **codomain** of the function.

Strictly speaking, if the same rule is thought of as producing outputs in two different sets, then it is defining two different functions (because the codomains are different). For example, the rule which assigns the output x^2 to the input x (for $x \in \mathbb{R}$) could be thought of as having codomain $\mathbb{R}$ or codomain $[0, \infty)$.

It is standard to summarize the name of a function, its domain, and codomain as follows:

If f is a function defined on X and taking values in Y, then we write

$$f : X \to Y.$$

If we want to describe what a function does to an individual element, we may use a formula, as in $f(x) = x^2$, or we may use the notation $x \mapsto x^2$, or $x \overset{f}{\mapsto} x^2$, if we want the name of the function made explicit.

Definition If $f : X \to Y$ is a function and $f(x) = y$, we call y the **image** of x. More generally, if $A \subseteq X$ is any subset of X, we define the **image of** A to be the set

$$f(A) := \{f(x) \mid x \in A\}.$$

Definition Let $f : X \to Y$ be any function. We say that f is **injective** or **one-to-one** if

$$f(x_1) = f(x_2) \qquad \Longrightarrow \qquad x_1 = x_2.$$

We say that f is **surjective** or **onto** if for every $y \in Y$, there is an $x \in X$ so that

$$f(x) = y.$$

We say that f is **bijective** or is a **one-to-one correspondence** if f is both injective and surjective.

Intuitively, an injective function $f : X \to Y$ loses no information: if you can tell apart two things that go into f, you can still tell them apart when they come out. On the other hand, if $f : X \to Y$ is surjective, then it can tell you about every element of Y. In this sense, if there is a bijective function $f : X \to Y$, then, using f, elements of Y contain no more and no less information than elements of X.

Composition and Invertibility

Definition

1. Let X, Y, and Z be sets and let $f : X \to Y$ and $g : Y \to Z$ be functions. The **composition** $g \circ f$ of g and f is defined by

$$g \circ f(x) = g\big(f(x)\big).$$

2. Let X be a set. The **identity function** on X is the function $\iota : X \to X$ such that for each $x \in X$, $\iota(x) = x$.
3. Suppose that $f : X \to Y$ is a function. A function $g : Y \to X$ is called an **inverse function** (or simply **inverse**) of f if $g \circ f$ is the identity function on X and $f \circ g$ is the identity function on Y. That is, for each $x \in X$,

$$g \circ f(x) = x,$$

and for each $y \in Y$,

$$f \circ g(y) = y.$$

In that case we write $g = f^{-1}$. If an inverse to f exists, we say that f is **invertible.**

It's important to watch out for a funny consequence of the way function composition is written: $g \circ f$ means do f first, then g.

Lemma A.1

1. *Composition of functions is* **associative**. *That is, if* $f : X_1 \to X_2$, $g : X_2 \to X_3$, *and* $h : X_3 \to X_4$, *then*

$$h \circ (g \circ f) = (h \circ g) \circ f.$$

2. *If* $f : X \to Y$ *is invertible, then its inverse is unique.*

Proof Exercise. ▲

Proposition A.2 *A function* $f : X \to Y$ *is invertible if and only if it is bijective.*

Proof Suppose first that f is bijective. We need to show that f has an inverse $g : Y \to X$.

Since f is surjective, for each $y \in Y$, there is an $x \in X$ such that $f(x) = y$. Furthermore, since f is injective, there is only one such x. We can therefore unambiguously define a function $g : Y \to X$ by $g(y) = x$. We then have that

$$f \circ g(y) = f(g(y)) = f(x) = y.$$

On the other hand, given any $x \in X$, let $y = f(x)$. Then $g(y) = x$ by the definition of g, so

$$g \circ f(x) = g(f(x)) = g(y) = x.$$

Conversely, suppose that $f : X \to Y$ is invertible, and let $g : Y \to X$ be its inverse. If $x_1, x_2 \in X$ are such that

$$f(x_1) = f(x_2),$$

then

$$x_1 = g(f(x_1)) = g(f(x_2)) = x_2,$$

and so f is injective. Next, for each $y \in Y$,

$$y = f(g(y)),$$

and so f is surjective. ▲

A.2 Complex Numbers

The set $\mathbb{C}$ of complex numbers is

$$\mathbb{C} := \{a + ib \mid a, b \in \mathbb{R}\},$$

where i is defined by the fact that $i^2 = -1$. Addition and multiplication of elements of $\mathbb{C}$ are defined by

$$(a + ib) + (c + id) := a + c + i(b + d)$$

and

$$(a + ib) \cdot (c + id) = ac - bd + i(ad + bc).$$

For geometric purposes, we sometimes identify $\mathbb{C}$ with $\mathbb{R}^2$: if $z = a + ib \in \mathbb{C}$, we can associate z with the ordered pair (a, b) of real numbers. But it is important to remember that $\mathbb{C}$ is not $\mathbb{R}^2$: the multiplication defined above is a crucial difference.

Definition For $z = a + ib \in \mathbb{C}$, the **real part** of z is $\operatorname{Re} z = a$ and the **imaginary part** of z is $\operatorname{Im} z = b$. The **complex conjugate** $\overline{z}$ of z is defined by

$$\overline{z} = a - ib.$$

The **absolute value** or **modulus** $|z|$ is defined by

$$|z| = \sqrt{a^2 + b^2}.$$

Geometrically, the complex conjugate of z is the reflection of z across the x-axis (sometimes called the real axis in this setting). The modulus of z is its length as a vector in $\mathbb{R}^2$.

The following lemma gives some of the basic properties of complex conjugation.

Lemma A.3

1. $\overline{w + z} = \overline{w} + \overline{z}$.
2. $\overline{wz} = \overline{w}\,\overline{z}$.
3. $z \in \mathbb{R}$ *if and only if* $\overline{z} = z$.
4. $z\overline{z} = |z|^2$.
5. $z + \overline{z} = 2 \operatorname{Re} z$.
6. $z - \overline{z} = 2i \operatorname{Im} z$.

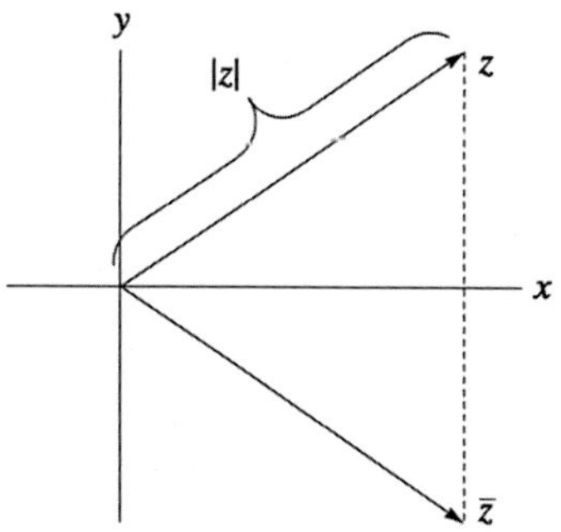

Figure A.1 $\overline{z}$ is the reflection of z across the x-axis, and $|z|$ is the length of z.

Proof Exercise. ▲

The study of complex-valued functions of a complex variable is called complex analysis. We use very little complex analysis in this book, but it is occasionally convenient to have access to the complex exponential function e^z. This function can be defined in terms of a power series as in the real case:

$$e^z = \sum_{j=0}^{\infty} \frac{z^j}{j!}.$$

In particular, if $\operatorname{Im} z = 0$, this is the same as the usual exponential function. More generally, the complex exponential function shares with the usual exponential the property that

$$e^{z+w} = e^z e^w.$$

The following identity relating the complex exponential to the sine and cosine functions is fundamental.

Theorem A.4 (Euler's formula) *For all $t \in \mathbb{R}$,*

$$e^{it} = \cos(t) + i\sin(t).$$

It follows in particular that $e^{\pi i} = -1$ and $e^{2\pi i} = 1$.

Recall that in $\mathbb{R}^2$ there is a system of polar coordinates, in which a vector is identified by its length, together with the angle it makes with the positive x-axis. It follows from Euler's formula that $e^{i\theta}$ is the unit vector which makes angle θ with the positive x-axis, and since $|z|$ is the length of the complex number z, thought of as a vector in $\mathbb{R}^2$, we can write z in polar form as

$$z = |z|\, e^{i\theta},$$

where θ is the angle z makes with the positive real axis; θ is called the **argument** of z.

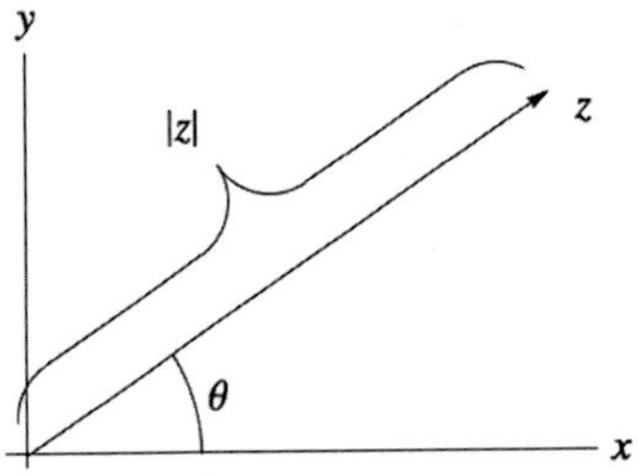

Figure A.2 The modulus and argument of z.

The following formulas follow immediately from Euler's formula, making use of the trigonometric identities $\cos(-x) = \cos(x)$ and $\sin(-x) = -\sin(x)$.

Corollary A.5 *For all $t \in \mathbb{R}$,*

1. $\overline{e^{it}} = e^{-it}$,
2. $\cos(t) = \dfrac{e^{it} + e^{-it}}{2}$,
3. $\sin(t) = \dfrac{e^{it} - e^{-it}}{2i}$.

A.3 Proofs

Logical Connectives

Formal logic is a system created by mathematicians in order to be able to think and speak clearly about truth and falseness. The following five basic concepts, called logical connectives, are the foundations of this system.

Definition The five **basic logical connectives** are:

- not,
- and,
- or,
- if..., then ...,
- if and only if.

The first three are fairly self-explanatory: the statement "not A" is true if the statement "A" is false, and vice versa. For example, the statement "x is not a natural number" (abbreviated $x \notin \mathbb{N}$) is true exactly when the statement "x is a natural number" ($x \in \mathbb{N}$) is false.

The statement "A and B" is true if both A and B are true, but false otherwise. For example, the statement

$$x > 0 \quad \text{and} \quad x^2 < 0$$

is always false because the second part ($x^2 < 0$) is always false; the statement

$$x > 0 \quad \text{and} \quad x < 0$$

is also always false because the two statements cannot both be true for any value of x.

The statement "A or B" is true if at least* one of A and B is true, and false otherwise. So

$$x > 0 \quad \text{or} \quad x < 0$$

is true for every value of x except $x = 0$, whereas

$$x > 0 \quad \text{or} \quad x^2 < 0$$

is true exactly when $x > 0$ is true, because $x^2 < 0$ is always false.

A statement using connective "if..., then..." is called a **conditional** statement. The connective is often abbreviated with the symbol $\Rightarrow$; we also use the word "implies," as in "A implies B," which means the same thing† as "If A, then B," which is also the same as "A $\Rightarrow$ B." A statement of this form is true if whenever A is true, then B must be true as well. For example,

$$x > 2 \qquad \Rightarrow \qquad x > 0$$

is a true conditional statement, because whenever $x > 2$, then it is also true that $x > 0$.

The connective "if and only if" is often abbreviated "iff" or with the symbol $\Leftrightarrow$. It is the connective which describes **equivalence**: two statements are equivalent if they are always either both true or both false. For example, the statements $x > 5$ and $x - 5 > 0$ are equivalent, because for every real number x, either both statements are true or both statements are false. So we might write

$$x > 5 \quad \Leftrightarrow \quad x - 5 > 0.$$

Statements with the connective "if and only if" are sometimes called **biconditional** statements, because they are the same thing as two conditional statements: "A iff B" is the same thing as "A implies B and B implies A." This is neatly captured by the arrow notation.

Quantifiers

Quantifiers are phrases which are applied to objects in mathematical statements in order to distinguish between statements which are always true and statements which are true in at least one instance.

*The mathematical definition of "or" is **inclusive**: "A or B" means "A or B or both."

† It is important to note that in mathematical English, "implies" **does not** mean the same thing as "suggests." "Suggests" still means what it means in regular English, but "A implies B" means that if A is true, B **must** be true. You will sometimes hear mathematicians say things like "This suggests but does not imply that..."

Definition The two **quantifiers** are

- there exists (denoted $\exists$),
- for all (denoted $\forall$).

For example, the statement "There exists an $x \in \mathbb{R}$ with $x^2 = 2$" is the assertion that the number 2 has at least one* square root among the real numbers. This statement is true (but not trivial!).

The statement "For all $x \in \mathbb{R}$, $x^2 \geq 0$" is the assertion that the square of *any* real number is nonnegative, that is, that the statement $x^2 \geq 0$ is always true when x is a real number.

Contrapositives, Counterexamples, and Proof by Contradiction

One of the most important logical equivalences is the contrapositive, defined as follows.

Definition The **contrapositive** of the conditional statement "A $\Rightarrow$ B" is the conditional statement "(not B) $\Rightarrow$ (not A)."

The contrapositive is equivalent to the original statement: "A $\Rightarrow$ B" means that whenever A is true, B must be as well. So if B is false, then A must have been as well.

Whenever we have a statement which claims to *always* be true, it is false if and only if there is a **counterexample**: that is, one instance in which the statement is false. For example, the statement "For all $x \in \mathbb{R}$, $x^2 > 0$." is false because there is a counterexample, namely $x = 0$. Conditional statements are a kind of statement claiming to be always true: "A $\Rightarrow$ B" means that whenever A is true, B is always true as well. So again, a conditional statement is false if and only if there is a counterexample. The statement

$$x^2 > 4 \quad \Rightarrow \quad x > 2$$

is false because there is a counterexample: $x = -3$ (of course, there are infinitely many, but we only need one). On the other hand, the statement

$$(x \in \mathbb{R} \quad \text{and} \quad x^2 < 0) \quad \Rightarrow \quad 3 = 2$$

is true, because there is no counterexample; a counterexample would consist of a value of $x \in \mathbb{R}$ with $x^2 < 0$, together with the falseness of the statement $3 = 2$.

*For a statement beginning "there exists" to be a true statement, what follows "there exists" only needs to be true once. If it's true more often than that, that's fine.

We now turn to proof by contradiction, which used to be known by the charming phrase of "proof by reduction to the absurd." The basic idea is to suppose that the statement you're trying to prove is false, and use that assumption to prove something which is known to be false. If every step in between was valid, the only problem must be with the starting point. We illustrate with a famous example, the irrationality of the square root of two.

Proposition A.6 *If $x \in \mathbb{R}$ with $x^2 = 2$, then $x \notin \mathbb{Q}$.*

Proof Suppose not; that is, suppose that there is an $x \in \mathbb{R}$ with $x^2 = 2$ such that $x \in \mathbb{Q}$. Since $x \in \mathbb{Q}$, we may write $x = \frac{p}{q}$ for some integers p, q. We may assume that p and q have no common prime factors, that is, that we have written x as a fraction in lowest terms. By assumption,

$$x^2 = \frac{p^2}{q^2} = 2.$$

Multiplying through by q^2 gives that

$$p^2 = 2q^2.$$

From this equation we can see that p^2 is divisible by 2. This means that p itself is divisible by 2 as well; if p is odd, then so is p^2. We may therefore write $p = 2k$, so that $p^2 = 4k^2$. Dividing the equation $4k^2 = 2q^2$ through by 2, we see that $q^2 = 2k^2$, so q^2 is also even. But then so is q, and so p and q do have a common factor of 2, even though earlier we assumed that x had been written in lowest terms. We thus have a contradiction, and so in fact x is irrational. ▲

This example illustrates a useful rule of thumb, which is that proof by contradiction is generally most helpful when assuming the negation of what you're trying to prove gives you something concrete to work with. Here, by assuming that x was rational, we got the integers p and q to work with.

Something to take note of in the proof above is the first two words: "suppose not"; this is short for "suppose the statement of the proposition is false," and is typical for a proof by contradiction. It is also typical, and a good idea, to follow the "suppose not" with a statement of exactly what it would mean to say that the statement of the theorem was false. This is called the **negation** of the statement. In this particular example, the statement to be proved is a conditional statement: **If** $x \in \mathbb{R}$ with $x^2 = 2$, **then** $x \notin \mathbb{Q}$. As was discussed above, a conditional statement is false if and only if there is a counterexample; assuming as we did above that there is an $x \in \mathbb{R}$ with $x^2 = 2$ such that $x \in \mathbb{Q}$ was exactly assuming that there was a counterexample.

Mathematical statements can of course become quite complicated; they are built from the five basic logical connectives and the quantifiers, but all of those pieces can be nested and combined in complicated ways. Still, negating a statement comes down to simple rules:

- The negation of "A and B" is "(not A) or (not B)."
- The negation of "A or B" is "(not A) and (not B)."
- The negation of a conditional statement is the existence of a counterexample.
- The negation of "For all x, $Q(x)$ is true" is "For some x, $Q(x)$ is false."
- The negation of "There is an x such that $Q(x)$ is true" is "For all x, $Q(x)$ is false."

Example The negation of the statement

"For all $\epsilon > 0$ there is a $\delta > 0$ such that if $|x - y| < \delta$, then $|f(x) - f(y)| < \epsilon$"

is

"There exists an $\epsilon > 0$ such that for all $\delta > 0$ there exist x and y with $|x - y| < \delta$ and $|f(x) - f(y)| \geq \epsilon$."

Working through to confirm this carefully is an excellent exercise in negation.

Proof by Induction

Induction is a technique used to prove statements about natural numbers, typically something of the form "For all $n \in \mathbb{N}$, ..." One first proves a "base case"; usually, this means that you prove the statement in question when $n = 1$ (although sometimes there are reasons to start at some larger value, in which case you will prove the result for natural numbers with that value or larger). Next comes the "induction"; this is usually the main part of the proof. In this part, one assumes that the result is true up to a certain point, that is, that the result holds for $n \leq N$. This is called the **induction hypothesis**. One then uses the induction hypothesis to prove that the result also holds when $n = N + 1$. In this way, the result is proved for all $n \in \mathbb{N}$: first it was proved for $n = 1$. Then by the inductive step, knowing it for $n = 1$ meant it was true for $n = 2$, which meant it was true for $n = 3$, and then $n = 4$, and so on.

Example We will prove by induction that

$$1 + 2 + \cdots + n = \frac{n(n+1)}{2}.$$

First, the base case: if $n = 1$, the left-hand side is 1 and the right-hand side is $\frac{(1)(2)}{2} = 1$.

Now suppose that the formula is true for $n \le N$. Then

$$\begin{aligned} 1 + 2 + \cdots + N + (N+1) &= \frac{N(N+1)}{2} + (N+1) \\ &= (N+1)\left(\frac{N}{2} + 1\right) \\ &= \frac{(N+1)(N+2)}{2}, \end{aligned}$$

where we used the induction hypothesis to get the first equality and the rest is algebra.

Addendum

Truthfully, we never actually make bread from barley flour.* Here is the recipe for sandwich bread in the Meckes household, adapted from a recipe from King Arthur Flour.

19.1 oz flour
$5\frac{1}{4}$ oz rolled oats
1.5 oz butter
$2\frac{1}{4}$ tsp salt
2 tsp instant (i.e., rapid rise) yeast
$\frac{1}{4}$ c packed brown sugar
15 oz lukewarm milk

Makes a large loaf – we use a long pullman loaf pan.

1. Mix (either by hand or in a stand mixer) all the ingredients except the milk in a large bowl.
2. Stir in the milk and mix until well blended.
3. Knead by hand or with the dough hook of a mixer for around 10 minutes. The dough should be fairly smooth and elastic (although broken up some by the oats) and should no longer be sticky.
4. Let the dough rise in an oiled bowl covered with plastic wrap or a damp towel for around an hour.
5. Form into a loaf and place, seam side down, into the greased loaf pan. Let rise another 45 minutes to an hour – in our pan, it's time to put it in the oven when the top-most part of the dough is just below the edge of the pan. It will continue to rise in the oven.
6. Bake at 350°F for 50 minutes to an hour. Remove from the pan and let cool for at least 45 minutes before slicing.

*We don't make our own beer or cheese either.

Hints and Answers to Selected Exercises

1 Linear Systems and Vector Spaces

1.1 Linear Systems of Equations

1.1.1 (a) You can make 8 loaves of bread, 76 pints of beer, and 21 rounds of cheese.

(b) You'll use 27 pounds of barley.

1.1.3 (a) 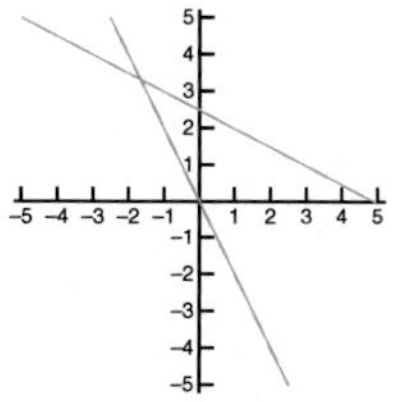 Consistent with a unique solution.

(b) 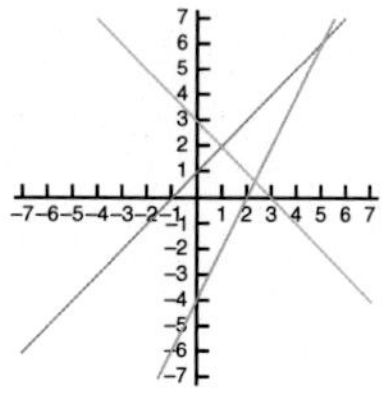 Inconsistent.

(c) 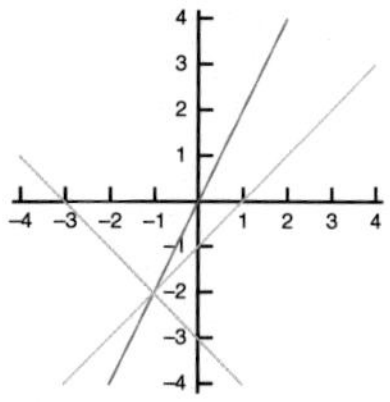 Consistent with a unique solution.

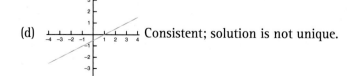

(d) 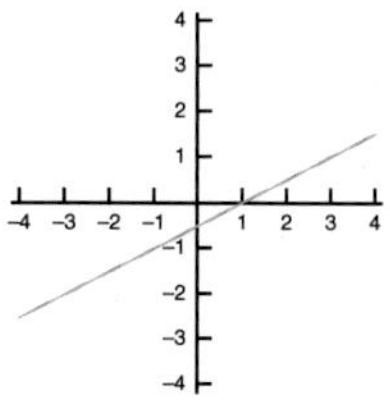 Consistent; solution is not unique.

(e) 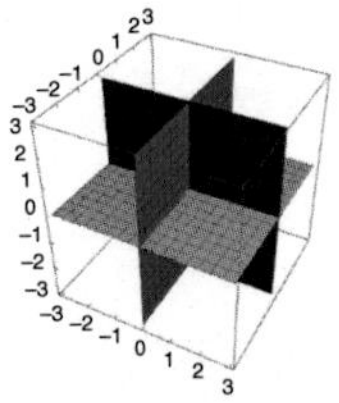 Consistent with a unique solution.

(f) 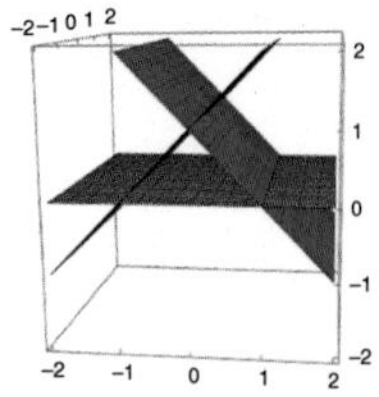 Inconsistent.

1.1.5 (a) The x–y plane. (b) The x-axis. (c) The origin. (d) The origin. (e) No point satisfies all four equations.

1.1.7 (a) $a - b + c = 1$ (b) $f(x) = \frac{3}{2}x^2 + \frac{1}{2}x$.

$$c = 0$$

$$a + b + c = 2.$$

1.1.9 Remember the Rat Poison Principle – just plug it in.

1.1.11 For each i,

$$\begin{aligned} &a_{i1}(tc_1 + (1-t)d_1) + \cdots + a_{in}(tc_n + (1-t)d_n) \\ &= t(a_{i1}c_1 + \cdots + a_{in}c_n) + (1-t)(a_{i1}d_1 + \cdots + a_{in}d_n) \\ &= tb_i + (1-t)b_i = b_i. \end{aligned}$$

1.2 Gaussian Elimination

1.2.1 (a) R1 (b) R3 (c) R1 (d) R1 (e) R2 (f) R2 (g) R1 (h) R1

1.2.3 (a) $\begin{bmatrix} 1 & 0 & -\frac{1}{2} \\ 0 & 1 & -\frac{7}{6} \end{bmatrix}$ (b) $\left[\begin{array}{cccc|c} 1 & 0 & -1 & 0 & 3 \\ 0 & 1 & 2 & 0 & -1 \\ 0 & 0 & 0 & 1 & 2 \end{array}\right]$ (c) $\begin{bmatrix} 1 & 0 \\ 0 & 1 \\ 0 & 0 \end{bmatrix}$ (d) $\begin{bmatrix} 1 & 0 & 0 \\ 0 & 1 & 0 \\ 0 & 0 & 1 \end{bmatrix}$

(e) $\begin{bmatrix} 1 & 0 & 0 & 0 \\ 0 & 1 & 0 & 0 \\ 0 & 0 & 1 & 0 \\ 0 & 0 & 0 & 1 \end{bmatrix}$ (f) $\left[\begin{array}{cccc|c} 1 & 0 & 0 & 0 & 0 \\ 0 & 1 & 0 & -1 & 0 \\ 0 & 0 & 1 & 2 & 0 \\ 0 & 0 & 0 & 0 & 1 \end{array}\right]$

1.2.7 (a) $x = -\frac{7}{2} + z$, $y = \frac{15}{4} - 2z$, $z \in \mathbb{R}$. (b) $x = 1 + \frac{1}{7}z - 2w$, $y = \frac{5}{7}z - w$, $z, w \in \mathbb{R}$. (c) Solutions are $x = 2 - z$, $y = 1 - z$, $z \in \mathbb{R}$. (d) $x = 1$, $y = 0$, $z = 1$. (e) There are no solutions.

1.2.9 $a = -1$, $b = 0$, $c = 4$, $d = 2$

1.2.11 No.

1.2.13 (a) $x + y + z = 1$ (b) Impossible by Corollary 1.4. (c) $x + y = 0$

$$x + y + z = 2$$

(d) $x = 1$

$$x = 2$$

(e)

$$\begin{aligned} x + y &= 1 \\ 2x + 2y &= 2 \\ 3x + 3y &= 3 \end{aligned}$$

1.2.15 $b_1 + b_2 - 2b_3 - b_4 = 0$

1.3 Vectors and the Geometry of Linear Systems

1.3.1 (a)

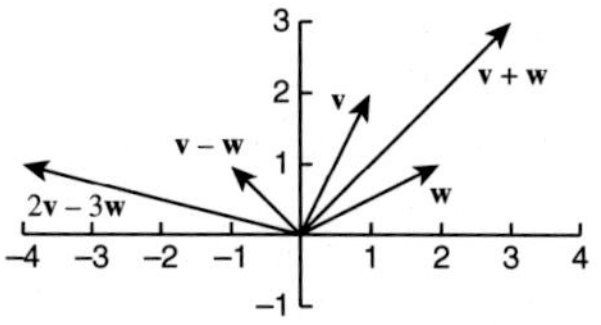

(b)

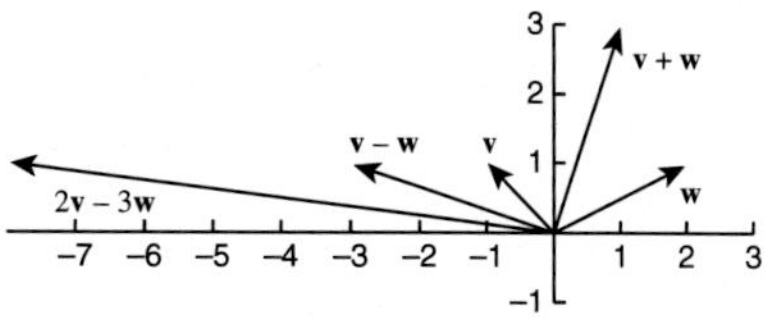

(c)

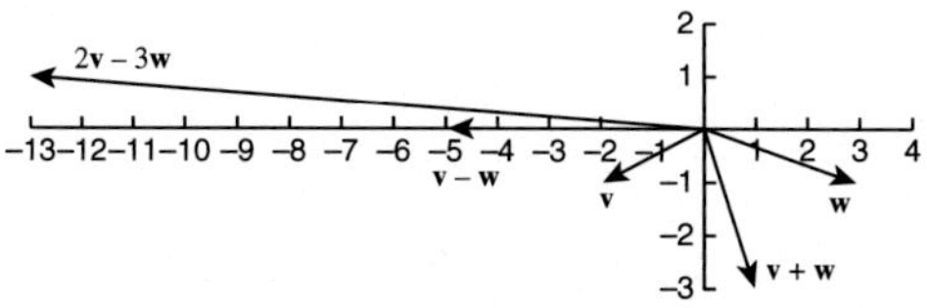

(d)

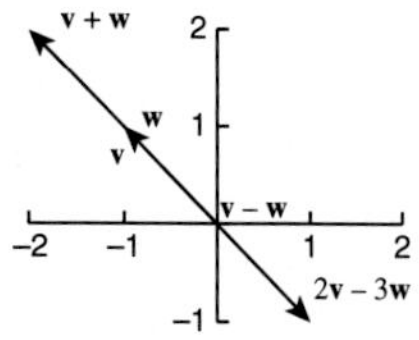

(e)

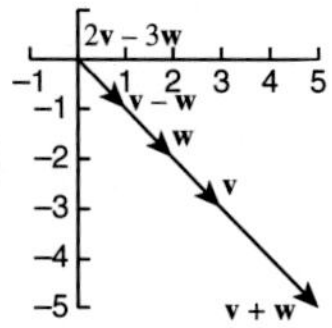

1.3.3 (a) $\begin{bmatrix} x \\ y \\ z \end{bmatrix} = \begin{bmatrix} 2 \\ 0 \\ -1 \end{bmatrix} + c \begin{bmatrix} 1 \\ 1 \\ -1 \end{bmatrix}$ for some $c \in \mathbb{R}$.

(b) $\begin{bmatrix} x \\ y \\ z \end{bmatrix} = \begin{bmatrix} -1 \\ 0 \\ 2 \end{bmatrix} + c \begin{bmatrix} 1 \\ 1 \\ -1 \end{bmatrix}$ for some $c \in \mathbb{R}$.

(c) $\begin{bmatrix} x \\ y \\ z \end{bmatrix} = \left(\frac{1}{5}\right) \begin{bmatrix} 0 \\ -1 \\ 6 \end{bmatrix} + c \begin{bmatrix} -5 \\ 2 \\ 3 \end{bmatrix}$ for some $c \in \mathbb{R}$.

(d) $\begin{bmatrix} x_1 \\ x_2 \\ x_3 \\ x_4 \end{bmatrix} = \left(\frac{1}{3}\right) \begin{bmatrix} -1 \\ 0 \\ 0 \\ 4 \end{bmatrix} + a \begin{bmatrix} 1 \\ -2 \\ 1 \\ 0 \end{bmatrix} + b \begin{bmatrix} 2 \\ -3 \\ 0 \\ 1 \end{bmatrix}$ for some $a, b \in \mathbb{R}$.

(e) $\begin{bmatrix} x \\ y \end{bmatrix} = c \begin{bmatrix} 2 \\ 1 \end{bmatrix}$ for some $c \in \mathbb{R}$.

(f) $\begin{bmatrix} x \\ y \end{bmatrix} = \left(-\frac{1}{2}\right) \begin{bmatrix} 0 \\ 1 \end{bmatrix} + c \begin{bmatrix} 2 \\ 1 \end{bmatrix}$ for some $c \in \mathbb{R}$.

1.3.5 (a) $b_1 + b_2 - 2b_3 - b_4 = 0$ (b) Check each vector using part (a).

1.3.7 For any $z \in \mathbb{R}$,

$$\begin{bmatrix} -1 \\ 1 \end{bmatrix} = \left(z + \frac{7}{2}\right)\begin{bmatrix} 1 \\ 2 \end{bmatrix} + \left(-2z - \frac{3}{2}\right)\begin{bmatrix} 3 \\ 4 \end{bmatrix} + z\begin{bmatrix} 5 \\ 6 \end{bmatrix}.$$

1.3.9 For any $a, b \in \mathbb{R}$,

$$\begin{bmatrix} 1 \\ 2 \\ 3 \\ 4 \end{bmatrix} = (a-2)\begin{bmatrix} 1 \\ 1 \\ 0 \\ 0 \end{bmatrix} + (b-1)\begin{bmatrix} 1 \\ 0 \\ 1 \\ 0 \end{bmatrix} + (4-a-b)\begin{bmatrix} 1 \\ 0 \\ 0 \\ 1 \end{bmatrix}$$
$$+ (4-a-b)\begin{bmatrix} 0 \\ 1 \\ 1 \\ 0 \end{bmatrix} + b\begin{bmatrix} 0 \\ 1 \\ 0 \\ 1 \end{bmatrix} + a\begin{bmatrix} 0 \\ 0 \\ 1 \\ 1 \end{bmatrix}.$$

1.3.11 Write $(x, y, z) = (x, y, 0) + (0, 0, z)$. Apply the Pythagorean Theorem to get that $\|(x, y, 0)\| = \sqrt{x^2 + y^2}$, and then again to get $\|(x, y, z)\| = \sqrt{\|(x, y, 0)\|^2 + \|(0, 0, z)\|^2} = \sqrt{x^2 + y^2 + z^2}$.

1.3.13 The vectors $\mathbf{v} = (v_1, v_2)$ and $\mathbf{w} = (w_1, w_2)$ span $\mathbb{R}^2$ if and only if the linear system

$$\left[\begin{array}{cc|c} v_1 & w_1 & x \\ v_2 & w_2 & y \end{array}\right]$$

has a solution for all x and y. First assume that both v_1 and v_2 are nonzero. Row-reduce the system to see what is required for a solution to always exist, and do a little algebra to see that the condition you get is equivalent to $\mathbf{v}$ and $\mathbf{w}$ being noncollinear. Consider the cases of $v_1 = 0$ and $v_2 = 0$ separately.

1.3.15 If $k = 0$, there is a single solution: one point. If $k = 1$, the set of solutions is a line. If $k = 2$, the set of solutions is a plane. If $k = 3$, then every point is a solution.

1.4 Fields

1.4.1 $0 = 0 + 0\sqrt{5}$. If $a, b, c, d \in \mathbb{Q}$, then $(a + b\sqrt{5})(c + d\sqrt{5}) = (ac + 5bd) + (ad + bc)\sqrt{5} \in \mathbb{F}$. If $a, b \in \mathbb{Q}$, then $\frac{1}{a+b\sqrt{5}} = \frac{a-b\sqrt{5}}{a^2-5b^2} \in \mathbb{F}$. Note that the denominator has to be nonzero because a and b are rational. The remaining parts are similar.

1.4.3 (a) Doesn't contain 0. (b) Doesn't have all additive inverses. (c) Not closed under multiplication (multiply by i, for example). (d) Doesn't have all multiplicative inverses.

1.4.5 (a) $\frac{1}{4}\begin{bmatrix} -1-i \\ 5+3i \end{bmatrix}$ (b) $\begin{bmatrix} 1 \\ i \\ 0 \end{bmatrix} + c\begin{bmatrix} -i \\ -1 \\ 1 \end{bmatrix}$ (c) $\frac{1}{13}\begin{bmatrix} 27 \\ -6i \end{bmatrix}$

1.4.7 Use Gaussian elimination in each case.

1.4.9 (a) If it has a solution over $\mathbb{R}$, Gaussian elimination will find it. But if the coefficients are all natural numbers, then Gaussian elimination can't produce a solution which is not rational.
(b) If $\mathbf{x}$ is the unique solution in $\mathbb{Q}$, then the system has no free variables, whether we consider it over $\mathbb{Q}$ or $\mathbb{R}$, so $\mathbf{x}$ is the only solution.
(c) $2x = 1$

1.4.11 (a) $\begin{bmatrix} 1 & 0 & 0 \\ 0 & 1 & 0 \\ 0 & 0 & 1 \end{bmatrix}$ (b) $\begin{bmatrix} 1 & 0 & 1 \\ 0 & 1 & 1 \\ 0 & 0 & 0 \end{bmatrix}$

1.4.13 If the solution to the system when thought of as a system over $\mathbb{F}$ is unique, what does that tell you about the RREF of the coefficient matrix?

1.4.15 Mod 4, $2 \cdot 2 = 0$, but $2 \neq 0$. This cannot happen in a field (see part 9 of Theorem 1.5).

1.4.17 For $a \in \mathbb{F}$, suppose that $a + b = a + c = 0$. Add c to both sides.

1.4.19 $a + (-1)a = (1-1)a = 0a = 0$, so $(-1)a$ is an additive inverse of a.

1.5 Vector Spaces

1.5.1 (a) Yes. (b) No (not closed under scalar multiplication). (c) No (not closed under addition). (d) No (doesn't contain 0).

1.5.3 The zero matrix has trace zero, and if $\text{tr}(\text{A}) = \text{tr}(\text{B}) = 0$, then $\text{tr}(c\text{A} + \text{B}) = \sum_i (ca_{ii} + b_{ii}) = c\,\text{tr}(\text{A}) + \text{tr}(\text{B}) = 0$.

1.5.5 (a) The properties are all trivial to check.
(b) If you multiply a "vector" (element of $\mathbb{Q}$) by a "scalar" (element of $\mathbb{R}$), you don't necessarily get another "vector" (element of $\mathbb{Q}$): for example, take $v = 1$ and $c = \sqrt{2}$.

1.5.7 All the properties of a vector space involving scalars from $\mathbb{F}$ are just special cases of those properties (assumed to hold) for scalars from $\mathbb{K}$.

1.5.9 Pointwise addition means $(f+g)(a) = f(a) + g(a)$ (with the second $+$ in $\mathbb{F}$), so $f + g$ is a function from A to $\mathbb{F}$. Commutativity and associativity of vector addition follows from commutativity and associativity of addition in $\mathbb{F}$. The zero vector is the function $z : A \to \mathbb{F}$ with $z(a) = 0$ for each a. The rest of the properties are checked similarly.

1.5.11 The subspaces U_1 and U_2 both contain 0, so we can take $u_1 = u_2 = 0$ to see $0 = 0 + 0 \in U_1 + U_2$. If $u_1 + u_2 \in U_1 + U_2$ and $c \in \mathbb{F}$, then $c(u_1 + u_2) = cu_1 + cu_2 \in U_1 + U_2$, since $cu_1 \in U_1$ and $cu_2 \in U_2$. Showing $U_1 + U_2$ is closed under addition is similar.

1.5.13 Addition and scalar multiplication: $\begin{bmatrix} v_1 \\ \vdots \\ v_n \end{bmatrix} + \begin{bmatrix} w_1 \\ \vdots \\ w_n \end{bmatrix} = \begin{bmatrix} v_1 w_1 \\ \vdots \\ v_n w_n \end{bmatrix}$, which has strictly positive entries, and $\lambda \begin{bmatrix} v_1 \\ \vdots \\ v_n \end{bmatrix} = \begin{bmatrix} v_1^\lambda \\ \vdots \\ v_n^\lambda \end{bmatrix}$, which has strictly positive entries. Commutativity and associativity of vector addition follow from commutativity and associativity of multiplication. If the "zero vector" is $\begin{bmatrix} 1 \\ \vdots \\ 1 \end{bmatrix}$, then $\begin{bmatrix} 1 \\ \vdots \\ 1 \end{bmatrix} + \begin{bmatrix} v_1 \\ \vdots \\ v_n \end{bmatrix} = \begin{bmatrix} 1 \cdot v_1 \\ \vdots \\ 1 \cdot v_n \end{bmatrix} = \begin{bmatrix} v_1 \\ \vdots \\ v_n \end{bmatrix}$. The additive inverse of $\begin{bmatrix} v_1 \\ \vdots \\ v_n \end{bmatrix}$ is $\begin{bmatrix} \frac{1}{v_1} \\ \vdots \\ \frac{1}{v_n} \end{bmatrix}$. Multiplication by 1 and associativity of multiplication: $1 \begin{bmatrix} v_1 \\ \vdots \\ v_n \end{bmatrix} = \begin{bmatrix} v_1^1 \\ \vdots \\ v_n^1 \end{bmatrix} = \begin{bmatrix} v_1 \\ \vdots \\ v_n \end{bmatrix}$, and $a \left(b \begin{bmatrix} v_1 \\ \vdots \\ v_n \end{bmatrix} \right) = a \begin{bmatrix} v_1^b \\ \vdots \\ v_n^b \end{bmatrix} = \begin{bmatrix} v_1^{ab} \\ \vdots \\ v_n^{ab} \end{bmatrix} = (ab) \begin{bmatrix} v_1 \\ \vdots \\ v_n \end{bmatrix}$.

Distributive laws: $a \left(\begin{bmatrix} v_1 \\ \vdots \\ v_n \end{bmatrix} + \begin{bmatrix} w_1 \\ \vdots \\ w_n \end{bmatrix} \right) = a \begin{bmatrix} v_1 w_1 \\ \vdots \\ v_n w_n \end{bmatrix} = \begin{bmatrix} (v_1 w_1)^a \\ \vdots \\ (v_n w_n)^a \end{bmatrix} = \begin{bmatrix} v_1^a \\ \vdots \\ v_n^a \end{bmatrix} + \begin{bmatrix} w_1^a \\ \vdots \\ w_n^a \end{bmatrix} = a \begin{bmatrix} v_1 \\ \vdots \\ v_n \end{bmatrix} + a \begin{bmatrix} w_1 \\ \vdots \\ w_n \end{bmatrix}$, and $(a+b) \begin{bmatrix} v_1 \\ \vdots \\ v_n \end{bmatrix} = \begin{bmatrix} v_1^{a+b} \\ \vdots \\ v_n^{a+b} \end{bmatrix} = \begin{bmatrix} v_1^a v_1^b \\ \vdots \\ v_n^a v_n^b \end{bmatrix} = \begin{bmatrix} v_1^a \\ \vdots \\ v_n^a \end{bmatrix} + \begin{bmatrix} v_1^b \\ \vdots \\ v_n^b \end{bmatrix} = a \begin{bmatrix} v_1 \\ \vdots \\ v_n \end{bmatrix} + b \begin{bmatrix} v_1 \\ \vdots \\ v_n \end{bmatrix}$.

1.5.17 For $v \in V$, suppose $w + v = u + v = 0$. Add w to both sides.

1.5.19 $0v = (0+0)v = 0v + 0v$. Now subtract (i.e., add the additive inverse of) $0v$ from both sides.

2 Linear Maps and Matrices

2.1 Linear Maps

2.1.1 (a)

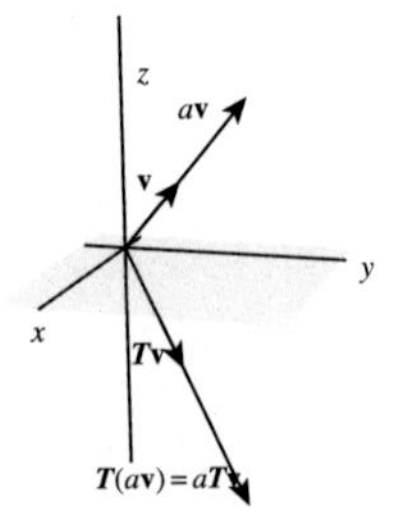

(b)

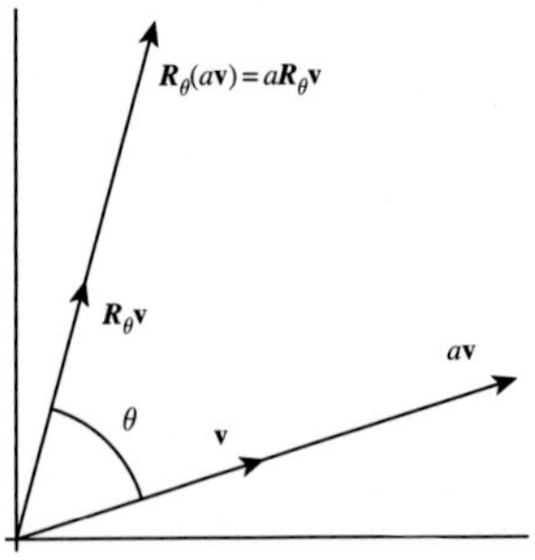

2.1.3 (a) $\begin{bmatrix} 9 \\ 37 \end{bmatrix}$ (b) Not possible (c) $\begin{bmatrix} 2 + \frac{2\pi}{3} + 2\sqrt{2} \\ \frac{130}{7} \end{bmatrix}$ (d) 70 (e) $\begin{bmatrix} 8.03 \\ 7.94 \\ -1.11 \end{bmatrix}$

(f) $\begin{bmatrix} -7 + 3i \\ -5 + 11i \end{bmatrix}$

2.1.5 For continuous functions f and g, and $a \in \mathbb{R}$, $[T(af + g)](x) = (af + g)(x)\cos(x) = [af(x) + g(x)]\cos(x) = af(x)\cos(x) + g(x)\cos(x) = a[Tf](x) + [Tg](x)$.

2.1.7 -2

2.1.9 (a) Multiples of $\begin{bmatrix} 1 \\ 1 \end{bmatrix}$ have eigenvalue 2; multiples of $\begin{bmatrix} 1 \\ -1 \end{bmatrix}$ have eigenvalue 0.

(b) Multiples of $\begin{bmatrix} -1 \\ 1 \end{bmatrix}$ have eigenvalue -2; multiples of $\begin{bmatrix} \frac{3}{5} + \frac{4}{5}i \\ 1 \end{bmatrix}$ have eigenvalue 2.

(c) Multiples of $\begin{bmatrix} 1 \\ 0 \end{bmatrix}$ have eigenvalue 2 (this is the only eigenvalue).

(d) Multiples of $\begin{bmatrix} 0 \\ 1 \\ 0 \end{bmatrix}$ have eigenvalue -2; multiples of $\begin{bmatrix} 1 \\ 0 \\ -1 \end{bmatrix}$ have eigenvalue -1; and multiples of $\begin{bmatrix} 1 \\ 0 \\ 1 \end{bmatrix}$ have eigenvalue 1.

2.1.11 Vectors on the line L have eigenvalue 1; vectors on the line perpendicular to L have eigenvalue 0.

2.1.13 $\mathbf{b} = \mathbf{0}$

2.1.15 First use the fact that each standard basis vector is an eigenvector to get that A is diagonal. Then use the fact that $(1, \ldots, 1)$ is an eigenvector to show that all the diagonal entries are the same.

2.1.17 (b) If λ is an eigenvalue, there there is a nonzero sequence $(a_1, a_2, \ldots)$ such that $\lambda a_1 = 0$ and $\lambda a_i = a_{i-1}$ for every $i \geq 2$.

2.2 More on Linear Maps

2.2.1 (a) Yes. (b) No. (c) No.

2.2.3 The set of all solutions of the system is $\left\{\begin{bmatrix} 0 \\ 0 \\ -\frac{1}{2}-\frac{1}{2}i \\ -\frac{2}{3}i \end{bmatrix} + a\begin{bmatrix} 1-i \\ 1-i \\ 1 \\ 0 \end{bmatrix} + b\begin{bmatrix} -1+i \\ -1-i \\ 0 \\ \frac{2}{3} \end{bmatrix} : a, b \in \mathbb{R}\right\}$. Check that the map T defined by $T((z_1, z_2)) = \begin{bmatrix} 0 \\ 0 \\ -\frac{1}{2}-\frac{1}{2}i \\ -\frac{2}{3}i \end{bmatrix} + z_1\begin{bmatrix} 1-i \\ 1-i \\ 1 \\ 0 \end{bmatrix} + z_2\begin{bmatrix} -1+i \\ -1-i \\ 0 \\ \frac{2}{3} \end{bmatrix}$ is the required bijective function.

2.2.5 $\begin{bmatrix} 0 & 1 & 0 \\ 0 & 0 & 1 \\ 0 & 0 & 0 \end{bmatrix}$

2.2.7 $\begin{bmatrix} -\frac{1}{\sqrt{2}} & -\frac{1}{\sqrt{2}} \\ -\sqrt{2} & \sqrt{2} \end{bmatrix}$

2.2.9 Write $Tv = \lambda v$; apply T^{-1} to both sides and divide both sides by λ.

2.2.11 The line segments making up the sides get mapped to line segments by Exercise 2.2.10, so the image of the unit square is a quadrilateral (with one corner at the origin). Call the right-hand side of the square R and the left-hand side L; then $L = \left\{t\begin{bmatrix} 0 \\ 1 \end{bmatrix} : t \in [0,1]\right\}$ and $R = \left\{\begin{bmatrix} 1 \\ 0 \end{bmatrix} + t\begin{bmatrix} 0 \\ 1 \end{bmatrix} : t \in [0,1]\right\}$. This means $T(R) = \left\{T\left(\begin{bmatrix} 1 \\ 0 \end{bmatrix}\right) + tT\begin{bmatrix} 0 \\ 1 \end{bmatrix} : t \in [0,1]\right\} = T\left(\begin{bmatrix} 1 \\ 0 \end{bmatrix}\right) + T(L)$, and both $T(R)$ and $T(L)$ have direction $T\begin{bmatrix} 0 \\ 1 \end{bmatrix}$. The same argument shows that the images of the top and bottom are parallel.

2.2.13 (a) Evaluate Av for each of the given vectors to check; the eigenvalues are -1 and 2.

(b)

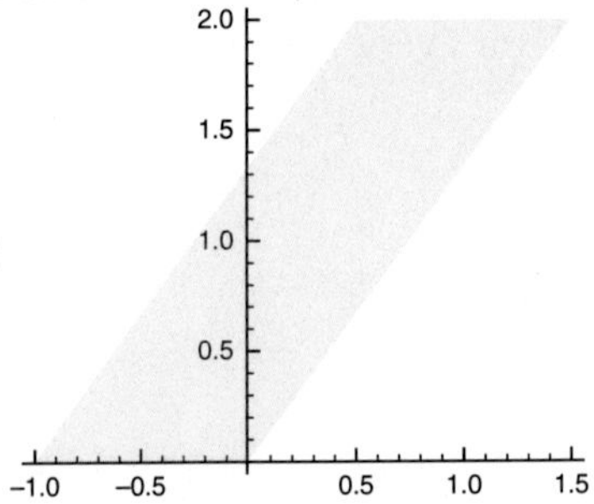

2.2.15 Consider $f(x) = e^{\lambda x}$.

2.2.17 (a) Prove the contrapositive: if T is not injective, there are $u_1 \neq u_2$ such that $Tu_1 = Tu_2$. Now apply S to see that ST can't be injective.
(b) If ST is surjective, then every $w \in W$ can be written $w = STu = S(Tu)$ for some u. This shows that $w = Sv$ for $v = Tu$.

2.2.19 $S\left[T(cu_1 + u_2)\right] = S(cTu_1 + Tu_2) = cSTu_1 + STu_2$.

2.3 Matrix Multiplication

2.3.1 (a) $\begin{bmatrix} 8 & 5 \\ 20 & 13 \end{bmatrix}$ (b) $\begin{bmatrix} -3+6i & 2-2i \\ -2-11i & 8-5i \end{bmatrix}$ (c) Not possible (d) -12
(e) Not possible (multiplying the first two works, but then you're stuck)

2.3.5 $\frac{1}{\sqrt{2}}\begin{bmatrix} -1 & -2 \\ -1 & 2 \end{bmatrix}$

2.3.7 $\operatorname{diag}(a_1b_1, \dots, a_nb_n)$

2.3.9 $[a]$ is invertible if and only if $a \neq 0$. If $a \neq 0$, $[a]^{-1} = [a^{-1}]$.

2.3.11 Write $\mathbf{b}$ as $\mathbf{ABb}$; this says $\mathbf{x} = \mathbf{Bb}$ is a solution to $\mathbf{Ax} = \mathbf{b}$.

2.3.15 $(\mathbf{A}^{-1})^{-1}$ is defined to be any matrix $\mathbf{B}$ such that $\mathbf{BA}^{-1} = \mathbf{A}^{-1}\mathbf{B} = \mathbf{I}$. The matrix $\mathbf{B} = \mathbf{A}$ does the trick.

2.4 Row Operations and the LU Decomposition

2.4.1 (a) $\mathbf{Q}_{2,1}\mathbf{Q}_{3,2}\mathbf{P}_{-\frac{1}{2},1,2}\mathbf{I}_2$ (b) $\mathbf{P}_{3,2,1}\mathbf{P}_{-1,1,2}\mathbf{Q}_{-2,2}\mathbf{I}_2$ (c) $\mathbf{P}_{-2,2,1}\begin{bmatrix} 1 & -2 & -1 \\ 0 & 0 & 0 \end{bmatrix}$

(d) $\mathbf{P}_{2,2,1}\mathbf{P}_{-1,3,1}\mathbf{P}_{1,3,2}\begin{bmatrix} 1 & -2 & 0 & 2 \\ 0 & 0 & 1 & -1 \\ 0 & 0 & 0 & 0 \end{bmatrix}$

2.4.3 (a) $\begin{bmatrix} -2 & 1 \\ \frac{3}{2} & -\frac{1}{2} \end{bmatrix}$ (b) $\begin{bmatrix} 0+0i & 2/5 - 1/5i \\ -1/3+0i & 1/15+2/15i \end{bmatrix}$ (c) $\begin{bmatrix} 1 & 1 & -1 \\ 1 & -1 & 1 \\ -1 & 1 & 1 \end{bmatrix}$

(d) Singular (e) Singular (f) $\begin{bmatrix} 0 & 1 & 1 & 0 \\ -1 & 2 & 1 & -2 \\ 1 & -1 & 0 & 1 \\ -2 & 3 & 2 & -2 \end{bmatrix}$

2.4.5 $\begin{bmatrix} 1 & 0 & 3 \\ 0 & 1 & 1 \\ -1 & 0 & 2 \end{bmatrix}^{-1} = \frac{1}{5}\begin{bmatrix} 2 & 0 & -3 \\ -1 & 5 & -1 \\ 1 & 0 & 1 \end{bmatrix}$

(a) $(x, y, z) = \frac{1}{5}(-7, 6, 4)$ (b) $(x, y, z) = \frac{1}{10}(-31, 43, -13)$

(c) $(x, y, z) = \frac{1}{5}(2\sqrt{2} - 249, \pi - \sqrt{2} - 83, \sqrt{2} + 83)$

(d) $(x, y, z) = \frac{1}{5}(2a - 3c, 5b - a - c, a + c)$

2.4.7 (a) $\begin{bmatrix} 1 & 0 \\ -1 & 1 \end{bmatrix}\begin{bmatrix} 2 & 3 \\ 0 & 4 \end{bmatrix}$ (b) $\begin{bmatrix} 1 & 0 \\ 2 & 1 \end{bmatrix}\begin{bmatrix} 2 & -1 \\ 0 & -1 \end{bmatrix}$

(c) $\begin{bmatrix} 1 & 0 & 0 \\ -1 & 1 & 0 \\ 0 & -2 & 1 \end{bmatrix}\begin{bmatrix} 2 & 0 & -1 \\ 0 & -1 & 2 \\ 0 & 0 & 1 \end{bmatrix}$ (d) $\begin{bmatrix} 1 & 0 \\ 1 & 1 \end{bmatrix}\begin{bmatrix} -1 & 1 & 2 \\ 0 & 2 & -1 \end{bmatrix}$

2.4.9 (a) $(x, y) = (-\frac{5}{4}, \frac{3}{4})$ (b) $(x, y) = (1, 2)$ (c) $(x, y, z) = (0, 3, 2)$
(d) $(x, y, z) = \frac{1}{5}(0, -9, 17) + c(5, 1, 2)$ for any $c \in \mathbb{R}$

2.4.11 (a) $\mathbf{L} = \mathbf{I}_2$; $\mathbf{U} = \begin{bmatrix} 2 & 3 \\ 0 & 1 \end{bmatrix}$; $\mathbf{P} = \begin{bmatrix} 0 & 1 \\ 1 & 0 \end{bmatrix}$.

(b) $\mathbf{L} = \begin{bmatrix} 1 & 0 & 0 \\ 0 & 1 & 0 \\ -1 & 0 & 1 \end{bmatrix}$; $\mathbf{U} = \begin{bmatrix} 2 & -1 & 2 \\ 0 & 2 & -1 \\ 0 & 0 & 1 \end{bmatrix}$; $\mathbf{P} = \begin{bmatrix} 1 & 0 & 0 \\ 0 & 0 & 1 \\ 0 & 1 & 0 \end{bmatrix}$.

2.4.13 $(x, y, z) = \left(\frac{3}{4}, \frac{5}{2}, 3\right)$

2.4.15 These are easiest to see by considering the effect of multiplying an arbitrary matrix $\mathbf{A}$ by these products. $\mathbf{R}_{i,j}\mathbf{P}_{c,k,\ell}$ performs **R1** with rows k and ℓ, and then switches rows i and j, so if $\{i, j\} \cap \{k, \ell\} = \emptyset$, it doesn't matter which order you do these operations in: the matrices commute. If $i = k$ but $j \neq \ell$, then $\mathbf{R}_{i,j}\mathbf{P}_{c,k,\ell}$ adds c times row ℓ to row k, then switches row k with row j. This is the same as first switching row k with row j, then adding c times row ℓ to row j: $\mathbf{R}_{k,j}\mathbf{P}_{c,k,\ell} = \mathbf{P}_{c,j,\ell}\mathbf{R}_{k,j}$. The other two are similar.
Alternatively, you can just write everything out in components and confirm the formulas.

2.4.17 The $\mathbf{L}$ from the LU decomposition of $\mathbf{A}$ already works. The $\mathbf{U}$ is upper triangular but may not have 1s on the diagonal. Use row operation **R2** on $\mathbf{U}$ to get a new $\mathbf{U}$ of the desired form; encoding the row operations as matrices will give you $\mathbf{D}$.

2.4.21 Write $\mathbf{Q}_{c,i} = [q_{jk}]_{j,k=1}^{m}$ and $\mathbf{A} = [a_{jk}]_{\substack{1 \le j \le m \\ 1 \le k \le n}}$. Then $[\mathbf{Q}_{c,i}\mathbf{A}]_{jk} = \sum_{\ell=1}^{m} q_{j\ell}a_{\ell k}$. Since $\mathbf{Q}_{c,i}$ is zero except on the diagonal, $q_{j\ell} = 1$ if $j = \ell$ but $j \neq i$, and $q_{ii} = c$, the only surviving term from the sum is $q_{jj}a_{jk}$, which is just a_{jk}, except when $j = i$, when it is ca_{ik}. This exactly means that the ith row of $\mathbf{A}$ has been multiplied by c.

2.5 Range, Kernel, and Eigenspaces

2.5.1 (a) $\left\langle\begin{bmatrix}1\\-2\\1\end{bmatrix}\right\rangle$ (b) $\left\langle\begin{bmatrix}-6\\3\\5\end{bmatrix}\right\rangle$ (c) $\left\langle\begin{bmatrix}2\\-3\\1\\0\end{bmatrix},\begin{bmatrix}1\\-2\\0\\1\end{bmatrix}\right\rangle$ (d) $\langle\mathbf{0}\rangle$ (e) $\left\langle\begin{bmatrix}1\\1\\1\end{bmatrix}\right\rangle$

2.5.3 (a) Yes. (b) Yes. (c) No. (d) No. (e) Yes.

2.5.5 (a) Yes; $\mathrm{Eig}_3(\mathrm{A}) = \left\langle\begin{bmatrix}2\\1\end{bmatrix}\right\rangle$. (b) No. (c) Yes; $\mathrm{Eig}_{-2}(\mathrm{A}) = \left\langle\begin{bmatrix}0\\1\\1\end{bmatrix}\right\rangle$.

(d) Yes; $\mathrm{Eig}_1(\mathrm{A}) = \left\langle\begin{bmatrix}1\\3\\0\end{bmatrix},\begin{bmatrix}-2\\0\\3\end{bmatrix}\right\rangle$.

(e) Yes; $\mathrm{Eig}_{-3}(\mathrm{A}) = \left\langle\begin{bmatrix}1\\0\\-1\\0\end{bmatrix},\begin{bmatrix}0\\1\\0\\-1\end{bmatrix}\right\rangle$.

2.5.7 (a) $\left\{\begin{bmatrix}0\\-\frac{1}{5}\\-\frac{13}{5}\end{bmatrix}+c\begin{bmatrix}-5\\3\\4\end{bmatrix} : c\in\mathbb{R}\right\}$ (b) $\left\{\begin{bmatrix}-2\\-1\\1\end{bmatrix}+c\begin{bmatrix}-3\\-2\\1\end{bmatrix} : c\in\mathbb{R}\right\}$

(c) $\left\{\begin{bmatrix}0\\\frac{1}{3}\\0\\-\frac{2}{3}\end{bmatrix}+c\begin{bmatrix}-2\\-1\\1\\0\end{bmatrix}+d\begin{bmatrix}-3\\-1\\0\\1\end{bmatrix} : c,d\in\mathbb{R}\right\}$

(d) $\left\{\begin{bmatrix}0\\\frac{3}{2}\\-\frac{1}{2}\end{bmatrix}+c\begin{bmatrix}2\\-1\\1\end{bmatrix} : c\in\mathbb{R}\right\}$

(e) $\left\{\begin{bmatrix}\frac{3}{2}\\\frac{1}{2}\\0\\\frac{1}{2}\end{bmatrix}+c\begin{bmatrix}1\\-1\\2\\1\end{bmatrix} : c\in\mathbb{R}\right\}$

2.5.9 (a) Just plug it in and see that it works.
(b) $Tf = f'' + f$ is a linear map on $C^2(\mathbb{R})$; the set of solutions to equation (2.22) is the set of $f \in C^2(\mathbb{R})$ with $Tf = g$, where $g(t) = 4e^{-t}$.
(c) $f(t) = 2e^{-t} + k_1 \sin(t) - \cos(t)$ for some $k_1 \in \mathbb{R}$.
(d) Such a solution must have the form $f(t) = 2e^{-t} + k_1 \sin(t) + k_2 \cos(t)$ with $k_1 = -2e^{-\frac{\pi}{2}} + b$ and $k_2 = -2 + a$.

2.5.11 (a) If $w \in \mathrm{range}(S + T)$, there is a $v \in V$ such that $w = S(v) + T(v) \in \mathrm{range}\, S + \mathrm{range}\, T$. (b) $C(\mathrm{A}) = \mathrm{range}(\mathrm{A})$ as an operator on $\mathbb{F}^n$, so this is immediate from the previous part.

2.5.13 The matrix $\mathbf{AB} = \mathbf{0}$ if and only if $\mathbf{A}(\mathbf{Bv}) = \mathbf{0}$ for every $\mathbf{v}$, if and only if $\mathbf{Bv} \in \ker \mathbf{A}$ for every $\mathbf{v}$. Explain why this is enough.

2.5.15 Let v be an eigenvector with eigenvalue λ, and apply T k times.

2.5.17 No. For example, the set of eigenvectors together with 0 of $\begin{bmatrix} 2 & 0 \\ 0 & 1 \end{bmatrix}$ is the x- and y-axes (and nothing else), which is not a subspace of $\mathbb{R}^2$.

2.5.19 $C(\mathbf{AB})$ is the set of linear combinations of the columns of $\mathbf{AB}$. Since each column of $\mathbf{AB}$ is itself a linear combination of columns of $\mathbf{A}$, every element of $C(\mathbf{AB})$ is a linear combination of columns of $\mathbf{A}$.

2.5.21 $T(0) = 0$, so $0 \in \mathrm{Eig}_\lambda(T)$. If $v_1, v_2 \in \mathrm{Eig}_\lambda(T)$ and $c \in \mathbb{F}$, then $T(cv_1 + v_2) = cTv_1 + Tv_2 = c\lambda v_1 + \lambda v_2$, so $cv_1 + v_2 \in \mathrm{Eig}_\lambda(T)$.

2.6 Error-correcting Linear Codes

2.6.1 (a) $(1, 0, 1, 1, 1)$ (b) $(1, 1, 1, 0, 0, 0, 1, 1, 1, 1, 1, 1)$ (c) $(1, 0, 1, 1, 0, 0, 1)$

2.6.3 (a) $\mathbf{x} = \begin{bmatrix} 1 \\ 0 \end{bmatrix}$ (b) $\mathbf{x} = \begin{bmatrix} 1 \\ 0 \end{bmatrix}, \begin{bmatrix} 0 \\ 1 \end{bmatrix}$ or $\begin{bmatrix} 1 \\ 1 \end{bmatrix}$ (c) $\mathbf{x} = \begin{bmatrix} 1 \\ 0 \\ 0 \end{bmatrix}$

(d) $\mathbf{x} = \begin{bmatrix} 1 \\ 1 \\ 0 \end{bmatrix}, \begin{bmatrix} 1 \\ 0 \\ 1 \end{bmatrix}, \begin{bmatrix} 0 \\ 1 \\ 1 \end{bmatrix}$, or $\begin{bmatrix} 1 \\ 1 \\ 1 \end{bmatrix}$

2.6.5 (a) $(1, 0, 1, 1)$ (b) $(0, 0, 1, 1)$ (c) $(1, 1, 0, 0)$ (d) $(0, 1, 0, 1)$ (e) $(1, 0, 1, 0)$

2.6.7 If errors occur in bits i_1 and i_2, then $\mathbf{z} = \mathbf{y} + \mathbf{e}_{i_1} + \mathbf{e}_{i_2}$ and $\mathbf{Bz} = \mathbf{By} + \mathbf{b}_{i_1} + \mathbf{b}_{i_2}$. Since $\mathbf{b}_{i_1} + \mathbf{b}_{i_2} \neq \mathbf{0}$ (explain why not!), we can tell errors occured but not how many or where. If three or more errors occur, we can't even tell for sure that there were any (because it *is* possible that $\mathbf{b}_{i_1} + \mathbf{b}_{i_2} + \mathbf{b}_{i_3} = \mathbf{0}$).

2.6.9 There are 2^4 vectors in $\mathbb{F}_2^4$. Each can lead to $8 = 2^3$ received messages (no errors or an error in one of the seven transmitted bits). Since these vectors are all distinct (why?), there are 2^7 possible received messages, which is all of $\mathbb{F}_2^7$.

2.6.11 (a) $\tilde{\mathbf{A}} = \begin{bmatrix} 1 & 0 & 0 & 0 \\ 0 & 1 & 0 & 0 \\ 0 & 0 & 1 & 0 \\ 0 & 0 & 0 & 1 \\ 1 & 1 & 1 & 0 \\ 1 & 1 & 0 & 1 \\ 1 & 0 & 1 & 1 \\ 1 & 1 & 1 & 1 \end{bmatrix}$ (b) $\tilde{\mathbf{B}} = \begin{bmatrix} 1 & 1 & 1 & 0 & 1 & 0 & 0 & 0 \\ 1 & 1 & 0 & 1 & 0 & 1 & 0 & 0 \\ 1 & 0 & 1 & 1 & 0 & 0 & 1 & 0 \\ 1 & 1 & 1 & 1 & 0 & 0 & 0 & 1 \end{bmatrix}$

3 Linear Independence, Bases, and Coordinates

3.1 Linear (In)dependence

3.1.1 (a) Linearly independent (b) Linearly dependent (c) Linearly independent (d) Linearly dependent (e) Linearly dependent

3.1.3 We wanted to see the last equation in (3.1) as a linear combination of the first three, so we wanted a, b, c so that $2x + \frac{1}{4}y + z = a\left(x + \frac{1}{4}y\right) + b\left(2x + \frac{1}{4}y\right) + c(x + 8z)$. By equating coefficients of x, y, and z we get $a + 2b + c = 2$, $\frac{1}{4}a + \frac{1}{4}b = \frac{1}{4}$, and $8c = 1$.

3.1.5 Take the linear combination of $(0, v_1, \ldots, v_n)$ given by $1(0) + 0(v_1) + \cdots + 0(v_0)$.

3.1.7 (a) The collinear condition is a linear dependence of v and w. If v and w are linearly independent, then $c_1v + c_2w = 0$ for some c_1, c_2 which are not both zero. Divide by whichever is nonzero and solve for that vector in terms of the other.
(b) $(1, 0), (1, 1), (0, 1)$

3.1.9 Suppose there are scalars so that $0 = c_1Tv_1 + \cdots + c_nTv_n = T[c_1v_1 + \cdots + c_nv_n]$. By injectivity, this means $c_1v_1 + \cdots + c_nv_n = 0$, and since the v_i are linearly independent, this means all the c_i are 0.

3.1.11 If the matrix is upper triangular with nonzero entries on the diagonal, it is already in REF, and each diagonal entry is a pivot.

3.1.13 Use Algorithm 3.4.

3.1.15 Consider $D : C^\infty(\mathbb{R}) \to \mathbb{C}^\infty(\mathbb{R})$ defined by $[Df](x) = f'(x)$.

3.1.17 Scalar multiples of eigenvectors are eigenvectors (with the same eigenvalue).

3.1.19 T is injective iff $\ker T = \{0\}$. Now apply Proposition 3.1.

3.2 Bases

3.2.1 (a) Yes. (b) No. (c) Yes. (d) Yes.

3.2.3 (a) $\left\{\begin{bmatrix}1\\0\\2\end{bmatrix}, \begin{bmatrix}0\\-3\\1\end{bmatrix}\right\}$ (b) $\left\{\begin{bmatrix}4\\-2\\2\end{bmatrix}, \begin{bmatrix}0\\3\\-2\end{bmatrix}\right\}$ (c) $\left\{\begin{bmatrix}2\\0\\-1\\3\end{bmatrix}, \begin{bmatrix}1\\1\\0\\1\end{bmatrix}, \begin{bmatrix}2\\1\\-2\\0\end{bmatrix}\right\}$

(d) $\left\{\begin{bmatrix}1\\-2\\3\\-4\end{bmatrix}, \begin{bmatrix}4\\-3\\2\\-1\end{bmatrix}, \begin{bmatrix}1\\0\\1\\0\end{bmatrix}\right\}$

3.2.5 (a) $\left\{\begin{bmatrix}1\\-2\\1\\0\end{bmatrix}, \begin{bmatrix}2\\-3\\0\\1\end{bmatrix}\right\}$ (b) $\left\{\begin{bmatrix}-2\\5\\1\\0\end{bmatrix}, \begin{bmatrix}3\\-3\\0\\1\end{bmatrix}\right\}$ (c) $\left\{\begin{bmatrix}2\\1\\0\\0\\0\end{bmatrix}, \begin{bmatrix}1\\0\\1\\2\\-1\end{bmatrix}\right\}$

(d) $\left\{\begin{bmatrix}1\\1\\1\\3\\0\\0\end{bmatrix}, \begin{bmatrix}1\\13\\-2\\0\\-6\\0\end{bmatrix}, \begin{bmatrix}7\\1\\4\\0\\0\\-6\end{bmatrix}\right\}$

3.2.7 (a) $\frac{3}{2}\begin{bmatrix}1\\1\end{bmatrix} - \frac{1}{2}\begin{bmatrix}1\\-1\end{bmatrix}$ (b) $\frac{3}{2}\begin{bmatrix}2\\3\end{bmatrix} - \frac{1}{2}\begin{bmatrix}4\\-5\end{bmatrix}$ (c) $\frac{3}{5}\begin{bmatrix}-1\\2\end{bmatrix} + \frac{4}{5}\begin{bmatrix}2\\1\end{bmatrix}$

(d) $-\frac{3}{4}\begin{bmatrix}2\\-1\end{bmatrix} + \frac{5}{8}\begin{bmatrix}4\\2\end{bmatrix}$

3.2.9 Every element of $\mathbb{C}$ can be uniquely written as $a + ib$ for some $a, b \in \mathbb{R}$.

3.2.11 $\begin{bmatrix}-4\\5\end{bmatrix}$

3.2.13 For $1 \le j \le k$, let $(e_i^j)_{i\ge1}$ be the sequence with $1 = e_j^j = e_{j+k}^j = e_{j+2k}^j = \cdots$ and $e_i^j = 0$ otherwise. Then V is spanned by $(e_i^1)_{i\ge1}, \ldots, (e_i^k)_{i\ge1}$.

3.2.15 Suppose that there are $c_1, \ldots, c_n$ so that $c_1(v_1 + v_2) + c_2(v_2 + v_3) + \cdots + c_{n-1}(v_{n-1} + v_n) + c_n v_n = 0$. Rearrange this to get $c_1 v_1 + (c_1 + c_2)v_2 + \cdots + (c_{n-2} + c_{n-1})v_{n-1} + (c_{n-1} + c_n)v_n = 0$. By linear independence, all the coefficients are 0. In particular, $c_1 = 0$, which then means $c_2 = 0$ and so on. So the new list is linearly independent. To show that it spans, use the same idea: write a general vector as a linear combination with unknown coefficients, rewrite so that you have a linear combination of the original list, use the fact that the original list spans V, and then solve a linear system to recover the unknown coefficients.

3.2.17 $\mathcal{B}$ is linearly independent by assumption and spans $\langle \mathcal{B} \rangle$ by definition.

3.2.19 The right-hand side is $\mathbf{V}\mathbf{x}$, where $\mathbf{V}$ is the matrix with columns $\mathbf{v}_1, \ldots, \mathbf{v}_n$ and $\mathbf{x} = \begin{bmatrix}x_1\\ \vdots \\ x_n\end{bmatrix}$. The matrix $\mathbf{V}$ is invertible because the $\mathbf{v}_i$ are a basis of $\mathbb{F}^n$. If we take $\mathbf{x} = \mathbf{V}^{-1}\mathbf{b}$, then the right-hand side becomes $\mathbf{V}\mathbf{x} = \mathbf{V}\mathbf{V}^{-1}\mathbf{b} = \mathbf{b}$.

3.2.21 This is just the special case of Theorem 3.14, with $W = \mathbb{F}$. It wouldn't hurt to specialize the proof of the theorem to this case to make sure you understand it.

3.2.23 The vectors 1 and 2 span $\mathbb{R}$ as a vector space over itself. There is no linear map $T : \mathbb{R} \to \mathbb{R}$ such that $T(1) = T(2) = 1$.

3.2.25 Proposition 3.9 says $(\mathbf{v}_1, \ldots, \mathbf{v}_n)$ is a basis for $\mathbb{F}^m$ if and only if the matrix $\mathbf{V}$ with columns $\mathbf{v}_i$ has RREF given by the identity. Given $\mathbf{w} \in \mathbb{F}^n$, finding a representation of $\mathbf{w}$ as a linear combination of the $\mathbf{v}_i$ is equivalent to solving the linear system $\mathbf{Vc} = \mathbf{w}$ for the coefficient vector $\mathbf{c}$. This system has a unique solution for every $\mathbf{w}$ if and only if the RREF of $\mathbf{V}$ is $\mathbf{I}_n$.

3.3 Dimension

3.3.1 (a) 2 (b) 3 (c) 2 (d) 2 (e) 3 (f) 2 (g) 2

3.3.3 Exercise 3.1.14 shows that, for any n, $(\sin(x), \sin(2x), \ldots, \sin(nx))$ is a linearly independent list in $C(\mathbb{R})$. Use Theorem 3.18.

3.3.5 Use the Linear Dependence Lemma.

3.3.7 Let $(v_1, \ldots, v_k)$ be a basis for U_1 and $(w_1, \ldots, w_\ell)$ a basis for U_2. Then $(v_1, \ldots, v_k, w_1, \ldots, w_\ell)$ spans $U_1 + U_2$ (why?).

3.3.9 Let $(v_1, \ldots, v_k)$ be a basis for V, and show that $(Tv_1, \ldots, Tv_k)$ spans range T.

3.3.11 Over $\mathbb{C}$, $\mathbb{C}^2$ has the basis $\left(\begin{bmatrix}1\\0\end{bmatrix}, \begin{bmatrix}0\\1\end{bmatrix}\right)$ (this is a special case of $\mathbb{F}^n$ as a vector space over $\mathbb{F}$ for any n and any field $\mathbb{F}$). Over $\mathbb{R}$, show that $\left(\begin{bmatrix}1\\0\end{bmatrix}, \begin{bmatrix}i\\0\end{bmatrix}, \begin{bmatrix}0\\1\end{bmatrix}, \begin{bmatrix}0\\i\end{bmatrix}\right)$ is a basis of $\mathbb{C}^2$.

3.3.13 Eigenvectors corresponding to distinct eigenvalues are linearly independent; if there are n distinct eigenvalues, there is a list of n linearly independent eigenvectors. Since $\dim(\mathbb{F}^n) = n$, this list is a basis.

3.3.15 The one-dimensional subspaces are illustrated in the picture. Each $v = (p_1, \ldots, p_n) \in V$ defines a one-dimensional subspace $\langle v \rangle = \left\{\frac{(p_1^\lambda, \ldots, p_n^\lambda)}{p_1^\lambda + \cdots + p_n^\lambda} : \lambda \in \mathbb{R}\right\}$. Each of these subspaces is a curve inside the simplex. They all contain the point $\left(\frac{1}{n}, \ldots, \frac{1}{n}\right)$ (corresponding to $\lambda = 0$). If $p_i > p_j$ for all $i \neq j$, then as $\lambda \to \infty$, the curve approaches $\mathbf{e}_i$, which is one of the vertices of the simplex. Similarly, if $p_i < p_j$ for all $i \neq j$, then as $\lambda \to -\infty$, the curve approaches $\mathbf{e}_i$. What happens if there is no single largest or smallest coordinate?

3.3.17 Use Lemma 3.22: the dimension of the intersection of two planes in $\mathbb{R}^3$ is at least 1, so it contains a line.

3.3.19 (a) A linearly independent list of length 1 can be anything but zero. Given a linearly independent list of length $k - 1$, its span is all linear combinations of those $k - 1$ vectors: you have $k - 1$ coefficients to choose from the field of q elements, so there are q^{k-1} vectors in the span.

(b) A matrix in $\mathrm{M}_n(\mathbb{F})$ is invertible if and only if its columns form a basis of $\mathbb{F}^n$.

3.3.23 Since $\mathcal{B} = (v_1, \ldots, v_n)$ is linearly independent, Theorem 3.27 says it can be extended to a basis. But since it already has n elements, you can't actually add anything or the dimension of V would be more than n. So it must already be a basis (the extending is trivial).

3.4 Rank and Nullity

3.4.1 (a) rank = 2, null = 1 (b) rank = 3, null = 1 (c) rank = 2, null = 3 (d) rank = 3, null = 0 (e) rank = 2, null = 1

3.4.3 $\operatorname{rank} D = n$, $\operatorname{null} D = 1$

3.4.5 $C(\mathbf{AB}) \subseteq C(\mathbf{A})$, so $\operatorname{rank} \mathbf{AB} = \dim C(\mathbf{AB}) \leq \dim C(\mathbf{A}) = \operatorname{rank} \mathbf{A}$. Also, $C(\mathbf{AB})$ is the range of the restriction of $\mathbf{A}$ to range($\mathbf{B}$), and so $\dim(C(\mathbf{AB})) \leq \dim(\operatorname{range}(\mathbf{B})) = \operatorname{rank} \mathbf{B}$.

3.4.7 If $\mathbf{x} = (x_1, \ldots, x_5) \in S$ and $\mathbf{v}_0 = (y_1, \ldots, y_5)$, then $3x_1 + 2x_4 + x_5 = 3y_1 + 2y_4 + y_5 = 10$ and $x_2 - x_3 - 5x_5 = y_2 - y_3 - 5y_5 = 7$, and so

$$3(x_1 - y_1) + 2(x_4 - y_4) + (x_5 - y_5) = 0 \text{ and } (x_2 - y_2) - (x_3 - y_3) - 5(x_5 - y_5) = 0.$$

Conversely, if $\mathbf{x}$ satisfies $3x_1 + 2x_4 + x_5 = 0$ and $x_2 - x_3 - 5x_5 = 0$, then $\mathbf{x} = (\mathbf{x} + \mathbf{v}_0) - \mathbf{v}_0$; check that $\mathbf{x} + \mathbf{v}_0 \in S$.

3.4.9 (a) Since $\operatorname{range}(S + T) \subseteq \operatorname{range} S + \operatorname{range} T$,

$$\operatorname{rank}(S + T) = \dim(\operatorname{range}(S + T)) \leq \dim(\operatorname{range} S + \operatorname{range} T).$$

By Exercise 3.3.7, $\dim(\operatorname{range} S + \operatorname{range} T) \leq \dim(\operatorname{range} S) + \dim(\operatorname{range} T) = \operatorname{rank} S + \operatorname{rank} T$.

3.4.11 Since $\mathbf{Ax} = \mathbf{b}$ does not have a unique solution, $\operatorname{null} \mathbf{A} \geq 1$, so $\operatorname{rank} \mathbf{A} \leq 4$.

3.4.13 If $\operatorname{rank} T = \dim V$, then by the Rank–Nullity Theorem, $\operatorname{null} T = 0$ and so T is injective. If $\operatorname{rank} T = \dim(\operatorname{range} T) = \dim W$, then $\operatorname{range} T = \dim W$ and so T is surjective.

3.4.15 If T has m distinct nonzero eigenvalues $\lambda_1, \ldots, \lambda_m$, then the corresponding eigenvectors $v_1, \ldots, v_m$ are linearly independent. range T contains $\lambda_1 v_1, \ldots, \lambda_m v_m$, none of which are zero, and so $\operatorname{rank} T \geq m$.

3.4.17 Row-reducing $\mathbf{A}$ is the same whether we do it in $\mathbb{F}$ or in $\mathbb{K}$. The rank of $\mathbf{A}$ is the number of pivots and the nullity is the number of free variables, so these numbers are the same regardless of whether we view $\mathbf{A}$ as an element of $\mathrm{M}_{m,n}(\mathbb{F})$ or $\mathrm{M}_{m,n}(\mathbb{K})$.

3.4.19 (a) Let $\mathbf{a}_j$ be the jth column of $\mathbf{A}$; since $\mathbf{a}_j \in C(\mathbf{A})$, there are coefficients b_{ij} so that $\mathbf{a}_j = b_{1j}\mathbf{c}_1 + \cdots + b_{rj}\mathbf{c}_r$, which is the same as saying $\mathbf{a}_j = \mathbf{C}\mathbf{b}_j$, where $\mathbf{b}_j = \begin{bmatrix} b_{1j} & \cdots & b_{rj} \end{bmatrix}^{\mathrm{T}}$. This means $\mathbf{A} = \mathbf{CB}$.

(b) By Corollary 2.32, $\operatorname{rank} \mathbf{A}^{\mathrm{T}} = \operatorname{rank}(\mathbf{B}^{\mathrm{T}}\mathbf{C}^{\mathrm{T}}) \leq \operatorname{rank}(\mathbf{B}^{\mathrm{T}})$. By Theorem 3.32, $\operatorname{rank}(\mathbf{B}^{\mathrm{T}}) \leq r$ since $\mathbf{B}^{\mathrm{T}} \in \mathrm{M}_{n,r}(\mathbb{F})$.

(c) Applying the previous part twice, $\operatorname{rank} \mathbf{A}^{\mathrm{T}} \leq \operatorname{rank} \mathbf{A}$ and $\operatorname{rank} \mathbf{A} = \operatorname{rank}(\mathbf{A}^{\mathrm{T}})^{\mathrm{T}} \leq \operatorname{rank} \mathbf{A}^{\mathrm{T}}$, and so $\operatorname{rank} \mathbf{A}^{\mathrm{T}} = \operatorname{rank} \mathbf{A}$.

3.4.21 Let $\mathbf{A} \in \mathrm{M}_{m,n}(\mathbb{F})$. If $n > m$ then $\ker \mathbf{A} \neq \{0\}$, and if $m > n$ then $C(\mathbf{A}) \subsetneq \mathbb{F}^m$.

3.5 Coordinates

3.5.1 (a) $\begin{bmatrix} 2 \\ 3 \end{bmatrix}$ (b) $\begin{bmatrix} 1 \\ 2 \end{bmatrix}$ (c) $\begin{bmatrix} 7 \\ 12 \end{bmatrix}$ (d) $\begin{bmatrix} 3 \\ 2 \end{bmatrix}$ (e) $\begin{bmatrix} -11 \\ -16 \end{bmatrix}$

3.5.3 (a) $\frac{1}{2}\begin{bmatrix} 1 \\ 1 \\ 1 \end{bmatrix}$ (b) $\frac{1}{2}\begin{bmatrix} -1 \\ 1 \\ -1 \end{bmatrix}$ (c) $\begin{bmatrix} -3 \\ -1 \\ -3 \end{bmatrix}$ (d) $\begin{bmatrix} 1 \\ 0 \\ 3 \end{bmatrix}$ (e) $\frac{1}{2}\begin{bmatrix} 9 \\ 3 \\ 1 \end{bmatrix}$

3.5.5 (a) $\begin{bmatrix} 1 & 1 \\ 3 & 0 \\ 0 & -4 \end{bmatrix}$ (b) $\begin{bmatrix} 1 & -1 \\ 0 & 0 \\ -1 & 2 \end{bmatrix}$ (c) $\begin{bmatrix} 1 & 0 \\ 0 & 0 \\ 0 & 1 \end{bmatrix}$ (d) $\begin{bmatrix} 1 & 1 & 0 \\ 0 & 1 & -1 \end{bmatrix}$

(e) $\begin{bmatrix} 1 & 0 & 0 \\ 2 & -1 & 1 \end{bmatrix}$ (f) $\begin{bmatrix} 1 & 0 & 1 \\ -1 & 1 & -2 \end{bmatrix}$

3.5.7 (a) (i) $\frac{1}{2}\begin{bmatrix} 3 \\ -1 \end{bmatrix}$ (ii) $\frac{1}{2}\begin{bmatrix} 1 \\ -5 \end{bmatrix}$ (iii) $\begin{bmatrix} 0 \\ 1 \\ 2 \end{bmatrix}$ (iv) $\begin{bmatrix} 0 \\ 3 \\ -2 \end{bmatrix}$ (v) $\begin{bmatrix} 0 & 1 & 1 \\ 1 & 0 & 1 \end{bmatrix}$

(vi) $\begin{bmatrix} 0 & \frac{1}{2} & \frac{1}{2} \\ 0 & -\frac{1}{2} & \frac{1}{2} \end{bmatrix}$ (vii) $\begin{bmatrix} \frac{1}{2} & \frac{1}{2} & 1 \\ -\frac{1}{2} & \frac{1}{2} & 0 \end{bmatrix}$

(b) (i) $\begin{bmatrix} 3 \\ 2 \end{bmatrix}$ (ii) $\frac{1}{2}\begin{bmatrix} -1 \\ 3 \end{bmatrix}$

3.5.9 (a) $\left(\begin{bmatrix} 2 \\ 1 \\ 0 \end{bmatrix}, \begin{bmatrix} -3 \\ 0 \\ 1 \end{bmatrix}\right)$ (b) (i) $\begin{bmatrix} -1 \\ -1 \end{bmatrix}$ (ii) Not in P (iii) $\begin{bmatrix} -2 \\ -3 \end{bmatrix}$

3.5.11 $[T] = \begin{bmatrix} 0 & 0 & & 0 \\ 1 & 0 & & 0 \\ 0 & 1 & & 0 \\ \vdots & & \ddots & \\ 0 & 0 & & 1 \end{bmatrix} \in \mathrm{M}_{n+1,n}(\mathbb{R})$; $[DT] = \begin{bmatrix} 1 & 0 & 0 & & 0 \\ 0 & 2 & 0 & & 0 \\ 0 & 0 & 3 & & 0 \\ \vdots & & & \ddots & \\ 0 & 0 & & & n \end{bmatrix} \in \mathrm{M}_{n,n}(\mathbb{R})$;

$$[TD] = \begin{bmatrix} 0 & 0 & 0 & & 0 \\ 0 & 1 & 0 & & 0 \\ 0 & 0 & 2 & & 0 \\ \vdots & & & \ddots & \\ 0 & 0 & & & n \end{bmatrix} \in \mathrm{M}_{n+1}(\mathbb{R}).$$

3.5.13 $\mathrm{diag}(\lambda_1^{-1}, \ldots, \lambda_n^{-1})$

3.5.15 $[P]_{\mathcal{E}} = \begin{bmatrix} 1 & 0 & 0 \\ 0 & 1 & 0 \\ 0 & 0 & 0 \end{bmatrix}$

3.5.17 $[\mathbf{u}]_{\mathcal{B}} = \begin{bmatrix} c_1 \\ \vdots \\ c_n \end{bmatrix}$ if and only if $\mathbf{u} = c_1\mathbf{v}_1 + \cdots + c_n\mathbf{v}_n = \mathrm{A}\begin{bmatrix} c_1 \\ \vdots \\ c_n \end{bmatrix}$.

3.5.19 If S and T are both diagonalized by $\mathcal{B} = (v_1, \ldots, v_n)$, then for each $i \in \{1, \ldots, n\}$ there are λ_i and μ_i so that $Sv_i = \lambda_i v_i$ and $Tv_i = \mu_i v_i$. Then $STv_i = \lambda_i \mu_i v_i$, and so ST is diagonalized by $\mathcal{B}$.

3.6 Change of Basis

3.6.1 (a) $[I]_{\mathcal{B},\mathcal{C}} = \frac{1}{2}\begin{bmatrix} -13 & -19 \\ -19 & -13 \end{bmatrix}$, $[I]_{\mathcal{C},\mathcal{B}} = \begin{bmatrix} 13 & -19 \\ -9 & 13 \end{bmatrix}$

(b) $[I]_{\mathcal{C},\mathcal{B}} = \frac{1}{2}\begin{bmatrix} 5 & 4 \\ -1 & 0 \end{bmatrix}$, $[I]_{\mathcal{B},\mathcal{C}} = \frac{1}{2}\begin{bmatrix} 0 & -4 \\ 1 & 5 \end{bmatrix}$

(c) $[I]_{\mathcal{B},\mathcal{C}} = \begin{bmatrix} 1 & 2 & 0 \\ -2 & -3 & -1 \\ -4 & -6 & -1 \end{bmatrix}$, $[I]_{\mathcal{C},\mathcal{B}} = \begin{bmatrix} -3 & 2 & -2 \\ 2 & -1 & 1 \\ 0 & -2 & 1 \end{bmatrix}$

(d) $[I]_{\mathcal{B},\mathcal{C}} = \begin{bmatrix} 2 & 0 & 0 \\ -1 & 2 & -1 \\ 0 & -1 & 1 \end{bmatrix}$, $[I]_{\mathcal{C},\mathcal{B}} = \frac{1}{2}\begin{bmatrix} 1 & 0 & 0 \\ 1 & 2 & 2 \\ 1 & 2 & 4 \end{bmatrix}$

3.6.3 (a) $[I]_{\mathcal{B},\mathcal{E}} = \begin{bmatrix} 2 & -1 \\ -3 & 2 \end{bmatrix}$, $[I]_{\mathcal{E},\mathcal{B}} = \begin{bmatrix} 2 & 1 \\ 3 & 2 \end{bmatrix}$

(b) (i) $\begin{bmatrix} 2 \\ 3 \end{bmatrix}$ (ii) $\begin{bmatrix} 1 \\ 2 \end{bmatrix}$ (iii) $\begin{bmatrix} 7 \\ 12 \end{bmatrix}$ (iv) $\begin{bmatrix} 3 \\ 2 \end{bmatrix}$ (v) $\begin{bmatrix} -11 \\ -16 \end{bmatrix}$

3.6.5 (a) $[I]_{\mathcal{B},\mathcal{E}} = \begin{bmatrix} 1 & 1 & 0 \\ -1 & 0 & 1 \\ 0 & 1 & -1 \end{bmatrix}$, $[I]_{\mathcal{E},\mathcal{B}} = \frac{1}{2}\begin{bmatrix} 1 & -1 & -1 \\ 1 & 1 & 1 \\ 1 & 1 & -1 \end{bmatrix}$

(b) (i) $\frac{1}{2}\begin{bmatrix} 1 \\ 1 \\ 1 \end{bmatrix}$ (ii) $\frac{1}{2}\begin{bmatrix} -1 \\ 1 \\ -1 \end{bmatrix}$ (iii) $\begin{bmatrix} -3 \\ -1 \\ -3 \end{bmatrix}$ (iv) $\begin{bmatrix} 1 \\ 0 \\ 3 \end{bmatrix}$ (v) $\frac{1}{2}\begin{bmatrix} 9 \\ 3 \\ 1 \end{bmatrix}$

3.6.7 (a) $[I]_{\mathcal{B},\mathcal{E}} = \begin{bmatrix} 2 & 1 \\ 1 & 1 \end{bmatrix}$, $[I]_{\mathcal{E},\mathcal{B}} = \begin{bmatrix} 1 & -1 \\ -1 & 2 \end{bmatrix}$, $[I]_{\mathcal{C},\mathcal{E}} = \begin{bmatrix} 1 & 0 & 1 \\ 3 & 2 & 0 \\ 0 & 3 & -4 \end{bmatrix}$,

$[I]_{\mathcal{E},\mathcal{C}} = \begin{bmatrix} -8 & 3 & -2 \\ 12 & -4 & 3 \\ 9 & -3 & 2 \end{bmatrix}$ (b) (i) $\begin{bmatrix} 1 & 1 \\ 3 & 0 \\ 0 & -4 \end{bmatrix}$ (ii) $\begin{bmatrix} 1 & -1 \\ 0 & 0 \\ -1 & 2 \end{bmatrix}$ (iii) $\begin{bmatrix} 1 & 0 \\ 0 & 0 \\ 0 & 1 \end{bmatrix}$

(iv) $\begin{bmatrix} 1 & 1 & 0 \\ 0 & 1 & -1 \end{bmatrix}$ (v) $\begin{bmatrix} 1 & 0 & 0 \\ 2 & -1 & 1 \end{bmatrix}$ (vi) $\begin{bmatrix} 1 & 0 & 1 \\ -1 & 1 & -2 \end{bmatrix}$

3.6.9 (a) $\begin{bmatrix} 0 & 1 \\ 1 & 0 \end{bmatrix}$ (b) $[I]_{\mathcal{B},\mathcal{C}} = \begin{bmatrix} 1 & 1 \\ -2 & -1 \end{bmatrix}$, $[I]_{\mathcal{C},\mathcal{B}}\begin{bmatrix} -1 & -1 \\ 2 & 1 \end{bmatrix}$ (c) $\begin{bmatrix} 1 & 0 \\ -3 & -1 \end{bmatrix}$

3.6.11 (a) $\mathcal{B} = \left(\begin{bmatrix}3\\1\end{bmatrix}, \begin{bmatrix}1\\-3\end{bmatrix}\right)$ (b) $\begin{bmatrix}3 & 1\\1 & -3\end{bmatrix}$ (c) $\frac{1}{10}\begin{bmatrix}3 & 1\\1 & -3\end{bmatrix}$ (d) $\frac{1}{10}\begin{bmatrix}9 & 3\\3 & 1\end{bmatrix}$

3.6.13 $[I]_{\mathcal{B},\mathcal{B}'} = \begin{bmatrix}1 & 0 & -\frac{1}{2}\\0 & 1 & 0\\0 & 0 & \frac{3}{2}\end{bmatrix}$, $[I]_{\mathcal{B}',\mathcal{B}} = \begin{bmatrix}1 & 0 & -\frac{1}{2}\\0 & 1 & 0\\0 & 0 & \frac{3}{2}\end{bmatrix}$

3.6.15 The traces are different.

3.6.17 A has two distinct eigenvalues (1 and 2).

3.6.19 A has no real eigenvalues (any eigenvalue λ has to satisfy $\lambda^2 = -1$).

3.6.21 $\operatorname{tr} R = 1$

3.6.23 If T has n distinct eigenvalues $\lambda_1, \ldots, \lambda_n$, then V has a basis $\mathcal{B} = (v_1, \ldots, v_n)$ of eigenvectors: $Tv_j = \lambda_j v_j$. Then $[T]_{\mathcal{B}} = \mathbf{diag}(\lambda_1, \ldots, \lambda_n)$.

3.6.25 $A = \begin{bmatrix}1 & 1\\0 & 2\end{bmatrix} \sim \begin{bmatrix}1 & 0\\0 & 2\end{bmatrix} = D$ since A has distinct eigenvalues 1 and 2. But $\operatorname{sum} A = 4 \neq 3 = \operatorname{sum} D$.

3.6.27 (a) Take $S = I_n$. (b) We have S and T invertible such that $B = SAS^{-1}$ and $C = TBT^{-1}$, so $C = TSAS^{-1}T^{-1} = (TS)A(TS)^{-1}$ and thus $C \sim A$.

3.6.29 A is diagonalizable if and only if it is similar to a diagonal matrix D, which by Theorem 3.54 means that there is a basis $\mathcal{B}$ of $\mathbb{F}^n$ such that $D = [A]_{\mathcal{B}}$. By Proposition 3.45, this means the elements of $\mathcal{B}$ are eigenvectors of A and the diagonal entries are the corresponding eigenvalues. So $A = SDS^{-1}$ with $S = [I]_{\mathcal{B},\mathcal{E}}$, whose columns are exactly the elements of $\mathcal{B}$ expressed in the standard basis.

3.7 Triangularization

3.7.1 (a) $\left(1, \left(\begin{bmatrix}1\\0\end{bmatrix}\right)\right), \left(3, \left(\begin{bmatrix}1\\1\end{bmatrix}\right)\right)$ (b) $\left(1, \left(\begin{bmatrix}1\\0\\0\end{bmatrix}\right)\right), \left(4, \left(\begin{bmatrix}2\\3\\0\end{bmatrix}\right)\right), \left(6, \left(\begin{bmatrix}16\\25\\10\end{bmatrix}\right)\right)$

(c) $\left(-2, \left(\begin{bmatrix}1\\0\\0\end{bmatrix}\right)\right), \left(3, \left(\begin{bmatrix}1\\5\\0\end{bmatrix}\right)\right)$ (d) $\left(1, \left(\begin{bmatrix}-1\\1\\0\\0\end{bmatrix}\right)\right), \left(2, \left(\begin{bmatrix}1\\0\\0\\0\end{bmatrix}\right)\right)$

3.7.3 (a) Diagonalizable (b) Diagonalizable (c) Not diagonalizable (d) Diagonalizable

3.7.5 Show that a lower triangular matrix is upper triangular in the basis $(e_n, \ldots, e_1)$.

3.7.7 Use the fact that the eigenvalues of an upper triangular matrix are exactly the diagonal entries. If A and B are upper triangular, what are the diagonal entries of $A + B$ and AB?

3.7.9 The eigenvalues of an upper triangular matrix are exactly the diagonal entries, so A has n distinct eigenvalues, and hence a basis of eigenvectors.

3.7.11 Since $\mathbb{F}$ is algebraically closed, T can be represented in some basis by an upper triangular matrix. Prove that if A is upper triangular, then $p(\mathrm{A})$ is upper triangular with diagonal entries $p(a_{11}), \dots, p(a_{nn})$. Then apply Theorem 3.64.

3.7.15 Consider the polynomial $p(x) = x^2 - 2$.

4 Inner Products

4.1 Inner Products

4.1.3 (a) $-\frac{3}{2}$ (b) 13

4.1.5 (a) $\langle \mathrm{Re\,A}, \mathrm{Im\,A}\rangle_F = \frac{i}{4}\left[\|\mathrm{A}\|_F^2 - \|\mathrm{A}^*\|_F^2 - 2i\,\mathrm{Im}\,\langle \mathrm{A}, \mathrm{A}^*\rangle_F\right]$
$= \frac{i}{4}\left[\mathrm{tr}(\mathrm{AA}^*) - \mathrm{tr}(\mathrm{A}^*\mathrm{A}) - 2i\,\mathrm{Im}\,\mathrm{tr}(\mathrm{A}^2)\right] = 0$ (b) Apply Theorem 4.4.

4.1.7 (a) $\sqrt{\pi}$ (b) $\sqrt{\pi}$ (c) 0
$\|af + bg\| = \sqrt{a^2\|f\|^2 + b^2\|g\|^2 - 2ab\,\langle f, g\rangle} = \sqrt{\pi(a^2 + b^2)}$

4.1.9 $\langle v_1, v_2\rangle_T$ is trivially a scalar; $\langle v_1 + v_2, v_3\rangle_T = \langle T(v_1 + v_2), Tv_3\rangle = \langle v_1, v_3\rangle_T + \langle v_2, v_3\rangle_T$ by linearity of T; homogeneity is similar. Symmetry and nonnegativity are easy. Definiteness requires the injectivity of T: if $\langle v, v\rangle_T = 0$, then $Tv = 0$, which means $v = 0$ since T is injective.

4.1.11 (a) Let $T(x_1, x_2) := (x_1, 2x_2)$. (b) $(1, 1)$ and $(1, -1)$ (c) $(2, 1)$ and $(2, -1)$

4.1.13 If $\mathrm{A} = \begin{bmatrix} 1 & 1 \\ 0 & i \end{bmatrix}$ and $\mathrm{D} = \begin{bmatrix} 1 & 0 \\ 0 & i \end{bmatrix}$, then $\mathrm{A} \sim \mathrm{D}$ (why?), but $\|\mathrm{A}\|_F = \sqrt{3}$ and $\|\mathrm{D}\|_F = \sqrt{2}$.

4.1.15 Let $\mathrm{x} = (a_1, \dots, a_n) \in \mathbb{R}^n$ and $\mathrm{y} = (b_1, \dots, b_n) \in \mathbb{R}^n$. Then the result is immediate from the Cauchy–Schwarz inequality with the standard inner product on $\mathbb{R}^n$.

4.1.17 Let $\mathrm{x} = (a_1, \sqrt{2}a_2, \dots, \sqrt{n}a_n)$ and $\mathrm{y} = \left(b_1, \frac{b_2}{\sqrt{2}}, \dots, \frac{b_n}{\sqrt{n}}\right)$ and apply the Cauchy–Schwarz inequality.

4.1.19 The properties mostly follow easily from the properties of the original inner product over $\mathbb{C}$. For definiteness, observe that $\langle v, v\rangle$ is necessarily real, so if $\langle v, v\rangle_{\mathbb{R}} = 0$, then $\langle v, v\rangle = 0$.

4.1.21 If $w = 0$, then any $v \in V$ is already orthogonal to w and so taking $u = v$ and any choice of $a \in \mathbb{F}$ works.

4.2 Orthonormal Bases

4.2.1 In each case, the list is the appropriate length, so it's enough to check that each vector has unit length and each pair is orthogonal.

4.2.3 (a) $\begin{bmatrix} -\frac{29}{\sqrt{13}} \\ -\frac{24}{\sqrt{13}} \end{bmatrix}$ (b) $\begin{bmatrix} \frac{2}{\sqrt{30}} \\ \frac{4}{\sqrt{5}} \\ \frac{8}{\sqrt{6}} \end{bmatrix}$ (c) $\begin{bmatrix} \frac{2}{\sqrt{3}} \\ \frac{1+e^{4\pi i/3}}{\sqrt{3}} \\ \frac{1+e^{2\pi i/3}}{\sqrt{3}} \end{bmatrix}$ (d) $\begin{bmatrix} \frac{5}{\sqrt{3}} \\ -\frac{4}{\sqrt{15}} \\ \frac{1}{\sqrt{35}} \\ -\frac{9}{\sqrt{7}} \end{bmatrix}$

4.2.5 (a) $\begin{bmatrix} -1 & 0 \\ 0 & 1 \end{bmatrix}$ (b) $\begin{bmatrix} \frac{1}{6} & 0 & \frac{\sqrt{5}}{6} \\ 0 & 1 & 0 \\ \frac{\sqrt{5}}{6} & 0 & \frac{5}{6} \end{bmatrix}$ (c) $\begin{bmatrix} 1 & 0 & 0 \\ 0 & e^{2\pi i/3} & 0 \\ 0 & 0 & e^{4\pi i/3} \end{bmatrix}$

(d) $\begin{bmatrix} \frac{2}{3} & \frac{2}{3\sqrt{5}} & \frac{\sqrt{7}}{\sqrt{15}} & 0 \\ \frac{2}{3\sqrt{5}} & \frac{2}{15} & -\frac{8}{5\sqrt{21}} & \frac{9}{\sqrt{105}} \\ \frac{\sqrt{7}}{\sqrt{15}} & -\frac{8}{5\sqrt{21}} & -\frac{18}{35} & -\frac{6}{7\sqrt{5}} \\ 0 & \frac{9}{\sqrt{105}} & -\frac{6}{7\sqrt{5}} & -\frac{2}{7} \end{bmatrix}$

4.2.7 (a) U is a plane (hence two-dimensional), so it's enough to check that the vectors are orthogonal and each have unit length.

(b) $\frac{1}{\sqrt{2}}\begin{bmatrix} -1 \\ 3\sqrt{3} \end{bmatrix}$ (c) $\frac{1}{2}\begin{bmatrix} -1 & \sqrt{3} \\ -\sqrt{3} & -1 \end{bmatrix}$

4.2.9 $\begin{bmatrix} 0 & 2\sqrt{3} & 0 \\ 0 & 0 & 2\sqrt{15} \\ 0 & 0 & 0 \end{bmatrix}$

4.2.11 (a) $\left(\frac{1}{\sqrt{2}}\begin{bmatrix} 1 \\ 0 \\ -1 \end{bmatrix}, \frac{1}{\sqrt{3}}\begin{bmatrix} 1 \\ 1 \\ 1 \end{bmatrix}, \frac{1}{\sqrt{6}}\begin{bmatrix} 1 \\ -2 \\ 1 \end{bmatrix}\right)$

(b) $\left(\frac{1}{\sqrt{5}}\begin{bmatrix} 0 \\ -1 \\ 2 \end{bmatrix}, \frac{1}{\sqrt{30}}\begin{bmatrix} 5 \\ -2 \\ -1 \end{bmatrix}, \frac{1}{\sqrt{6}}\begin{bmatrix} 1 \\ 2 \\ 1 \end{bmatrix}\right)$

(c) $\left(\frac{1}{\sqrt{3}}\begin{bmatrix} 0 \\ 1 \\ 1 \\ 1 \end{bmatrix}, \frac{1}{\sqrt{15}}\begin{bmatrix} 3 \\ -2 \\ 1 \\ 1 \end{bmatrix}, \frac{1}{\sqrt{35}}\begin{bmatrix} 3 \\ 3 \\ -4 \\ 1 \end{bmatrix}, \frac{1}{\sqrt{7}}\begin{bmatrix} 1 \\ 1 \\ 1 \\ -2 \end{bmatrix}\right)$

4.2.13 (a) Linearly independent (b) Linearly independent (c) Linearly dependent

4.2.15 (a) $\left(1, 2\sqrt{3}x-\sqrt{3}, 6\sqrt{5}x^2-6\sqrt{5}x+\sqrt{5}, 20\sqrt{7}x^3-30\sqrt{7}x^2+12\sqrt{7}x-\sqrt{7}\right)$

(b) $\left(\frac{1}{\sqrt{2}}, \sqrt{\frac{3}{2}}x, \frac{3\sqrt{5}}{2\sqrt{2}}x^2 - \frac{\sqrt{5}}{2\sqrt{2}}, \frac{5\sqrt{7}}{2\sqrt{2}}x^3 - \frac{3\sqrt{7}}{2\sqrt{2}}x\right)$

(c) $\left(\sqrt{\frac{3}{2}}, \sqrt{\frac{5}{2}}x, \frac{5\sqrt{7}}{2\sqrt{2}}\left(x^2-\frac{3}{5}\right), \frac{21}{2\sqrt{2}}x^3 - \frac{15}{2\sqrt{2}}\right)$

4.2.17 $\left(\frac{1}{\sqrt{2}}\begin{bmatrix} 1 \\ 0 \\ 0 \end{bmatrix}, \frac{1}{\sqrt{6}}\begin{bmatrix} -1 \\ 2 \\ 0 \end{bmatrix}, \frac{1}{\sqrt{12}}\begin{bmatrix} 1 \\ 1 \\ -3 \end{bmatrix}\right)$

4.2.19 $(\sin(nx))_{n\geq 0}$ is an infinite orthogonal (hence linearly independent) list.

4.2.21 Take the inner product of $\tilde{e}_j = v_j - \langle v_j, e_1\rangle e_1 e_1 - \cdots - \langle v_j, e_{j-1}\rangle e_{j-1}$ with v_j to get $\langle v_j, \tilde{e}_j\rangle = \|v_j\|^2 - |\langle v_j, e_1\rangle|^2 - \cdots - |\langle v_j, e_{j-1}\rangle|^2$. Because $(e_1, \ldots, e_n)$ is an orthonormal basis, $\|v_j\|^2 = |\langle v_j, e_1\rangle|^2 + \cdots + |\langle v_j, e_n\rangle|^2$, and so $\langle v_j, \tilde{e}_j\rangle = |\langle v_j, e_j\rangle|^2 + \cdots + |\langle v_j, e_n\rangle|^2 \geq 0$. It cannot be that $\langle v_j, \tilde{e}_j\rangle = 0$ (why not?), and so $\langle v_j, \tilde{e}_j\rangle > 0$. Now divide by $\|\tilde{e}_j\|$.

4.3 Orthogonal Projections and Optimization

4.3.1 (a) $\frac{1}{35}\begin{bmatrix} 26 & -11 & 7 & 8 \\ -11 & 6 & -7 & 2 \\ 7 & -7 & 14 & -14 \\ 8 & 2 & -14 & 24 \end{bmatrix}$ (b) $\frac{1}{23}\begin{bmatrix} 18 & -2-i & 6-7i \\ -2+i & 22 & 1-4i \\ 6+7i & 1+4i & 6 \end{bmatrix}$

(c) $\frac{1}{266}\begin{bmatrix} 66 & 74 & 82 & 26 & 18 \\ 74 & 87 & 100 & 9 & -4 \\ 82 & 100 & 118 & -8 & -26 \\ 26 & 9 & -8 & 111 & 128 \\ 18 & -4 & -26 & 128 & 150 \end{bmatrix}$ (d) $\frac{1}{n}\begin{bmatrix} 1 & 1 & \cdots & 1 \\ 1 & 1 & \cdots & 1 \\ \vdots & \vdots & \ddots & \vdots \\ 1 & 1 & \cdots & 1 \end{bmatrix}$

4.3.3 (a) $\frac{1}{14}\begin{bmatrix} 13 & -2 & -3 \\ -2 & 10 & -6 \\ -3 & -6 & 5 \end{bmatrix}$ (b) $\frac{1}{35}\begin{bmatrix} 26 & 3 & 15 \\ 3 & 34 & -5 \\ 15 & -5 & 10 \end{bmatrix}$

(c) $\frac{1}{7}\begin{bmatrix} 3 & -2+2i & 2i \\ -2-2i & 5 & -1+i \\ -2i & -1-i & 6 \end{bmatrix}$ (d) $\frac{1}{195}\begin{bmatrix} 121 & -32 & -19 & -87 \\ -32 & 139 & -82 & -6 \\ -19 & -82 & 61 & 33 \\ -87 & -6 & 33 & 69 \end{bmatrix}$

4.3.5 (a) $\frac{1}{35}\begin{bmatrix} 57 \\ -12 \\ -21 \\ 66 \end{bmatrix}$ (b) $\frac{1}{23}\begin{bmatrix} 61-9i \\ -5+21i \\ 20+22i \end{bmatrix}$ (c) $\frac{1}{35}\begin{bmatrix} 44 \\ 32 \\ 20 \end{bmatrix}$ (d) $\frac{1}{195}\begin{bmatrix} -467 \\ 109 \\ 83 \\ 339 \end{bmatrix}$

4.3.7 $y = \frac{8}{5}x + 4$

4.3.9 The line is spanned by $\begin{bmatrix} 1 \\ m \end{bmatrix}$; the matrix of orthogonal projection onto the line is $\frac{1}{1+m^2}\begin{bmatrix} 1 & m \\ m & m^2 \end{bmatrix}$.

4.3.11 Take the orthonormal basis $(1, 2\sqrt{3}x - \sqrt{3})$ of the linear polynomials with respect to the given inner product. The best approximation $p(x)$ for e^x is $p(x) = \langle e^x, 1\rangle\, 1 + \langle e^x, 2\sqrt{3}x - \sqrt{3}\rangle\,(2\sqrt{3}x - \sqrt{3}) = 6(e-3)x + 4e - 10$.

4.3.13 $\mathcal{B} = \left(\frac{1}{\sqrt{2\pi}}, \frac{1}{\sqrt{\pi}}\sin(x), \frac{1}{\sqrt{\pi}}\sin(2x), \frac{1}{\sqrt{\pi}}\cos(x), \frac{1}{\sqrt{\pi}}\cos(2x)\right)$ is an orthonormal basis, and the orthogonal projection of x onto its span is $g(x) = \pi - 2\sin(x) - \sin(2x)$.

4.3.15 Let $v \in U_2^\perp$. Then $\langle v, u\rangle = 0$ for every $u \in U_2$, so in particular $\langle v, u\rangle = 0$ for every $u \in U_1$.

4.3.17 Expand $\langle v, w\rangle = \langle v_1 + \cdots + v_n, w_1 + \cdots w_n\rangle$ by linearity; since $\langle v_i, w_j\rangle = 0$ if $i \neq j$, only the terms of the form $\langle v_j, w_j\rangle$ survive. The statement about $\|v\|^2$ is the special case $w = v$.

4.3.19 For each i, let $(e_1^i, \ldots, e_{n_i}^i)$ be an orthonormal basis of U_i (so $n_i = \dim U_i$). Show that $(e_1^1, \ldots, e_{n_1}^1, \ldots, e_1^m, \ldots, e_{n_m}^m)$ is an orthonormal basis of $V = U_1 \oplus \cdots \oplus U_m$.

4.3.21 Use part 5 of Theorem 4.16.

4.3.23 From part 2 of Theorem 4.16, $P_U(cv + w) = \sum_{j=1}^{m} \langle cv + w, e_j \rangle e_j = c\sum_{j=1}^{m} \langle v, e_j \rangle e_j + \sum_{j=1}^{m} \langle w, e_j \rangle e_j = cP_U(v) + P_U(w)$.

4.4 Normed Spaces

4.4.1 In both parts, the parallelogram identity fails with $v = \mathbf{e}_1$ and $w = \mathbf{e}_2$.

4.4.3 Given $v \in V$, $\|Tv\| = \|v\| \left\| T\left(\frac{v}{\|v\|}\right) \right\| \le \|v\| \, \|T\|_{op}$. On the other hand, if C is such that $\|Tv\| \le C \|v\|$ for all $v \in V$, then in particular, $\|Tv\| \le C$ for all unit vectors v, and so $\|T\|_{op} \le C$.

4.4.5 If $\mathbf{e}$ is the vector of errors, then the error in $\mathbf{x}$ is $\mathrm{A}^{-1}\mathbf{e}$, and $\left\|\mathrm{A}^{-1}\mathbf{e}\right\| \le \left\|\mathrm{A}^{-1}\right\|_{op} \|\mathbf{e}\| \le \left\|\mathrm{A}^{-1}\right\|_{op} \epsilon\sqrt{m}$.

4.4.7 $\left\|\mathbf{a}_j\right\| = \left\|\mathrm{A}\mathbf{e}_j\right\|$ and $\mathbf{e}_j$ is a unit vector; use the definition of $\|\mathrm{A}\|_{op}$.

4.4.9 If $u \in U$ is a unit vector, then $\|TSu\| \le \|T\|_{op} \|Su\| \le \|T\|_{op} \|S\|_{op}$.

4.4.11 (a) Let $u \in \mathbb{C}^n$ be a unit vector; since A is invertible, there is a $\mathbf{v} \in \mathbb{C}^n$ with $\mathbf{u} = \mathrm{A}\mathbf{v}$. Then $\|\mathbf{u}\| = \|\mathrm{A}\mathbf{v}\| \le \|\mathrm{A}\|_{op} \|\mathbf{v}\| = \|\mathrm{A}\|_{op} \left\|\mathrm{A}^{-1}\mathbf{u}\right\| \le \|\mathrm{A}\|_{op} \left\|\mathrm{A}^{-1}\right\|_{op}$.

(b) Since $\mathrm{A}\mathbf{x} = \mathbf{b}$, $\|\mathbf{b}\| \le \|\mathrm{A}\|_{op} \|\mathbf{x}\|$, and so $\frac{1}{\|\mathbf{x}\|} \le \frac{\|\mathrm{A}\|_{op}}{\|\mathbf{b}\|}$. Combine this with the fact that $\left\|\mathrm{A}^{-1}\mathbf{h}\right\| \le \left\|\mathrm{A}^{-1}\right\|_{op} \|\mathbf{h}\|$.

4.4.15 Let $\mathbf{x} = (x_1, \ldots, x_n)$. Then $\|\mathbf{x}\|_\infty = \sqrt{\max_{1 \le j \le n} |x_j|^2} \le \sqrt{|x_1|^2 + \cdots + |x_n|^2} = \|\mathbf{x}\|_2$. On the other hand, $\sqrt{|x_1|^2 + \cdots + |x_n|^2} \le \sqrt{n|x_j|^2}$, where $\|\mathbf{x}\|_\infty = |x_j|$.

4.4.17 Let $\mathbf{a}_j$ denote the jth column of A and $\mathbf{x} = (x_1, \ldots, x_n)$, and write $\mathrm{A}\mathbf{x} = \sum_{j=1}^{n} x_j \mathbf{a}_j$. Apply the triangle inequality followed by the Cauchy–Schwarz inequality to show that $\|\mathrm{A}\mathbf{x}\| \le \|\mathrm{A}\|_F \|\mathbf{x}\|$ for every $\mathbf{x} \in \mathbb{R}^n$. For the second inequality, use Exercise 4.4.7.

4.4.19 (a)

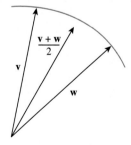

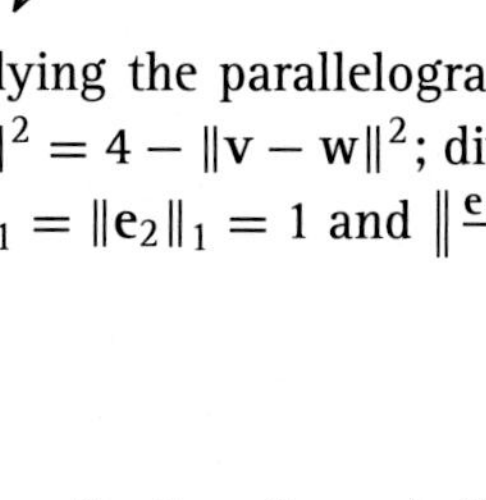

(b) Applying the parallelogram identity to unit vectors $\mathbf{v}, \mathbf{w} \in \mathbb{R}^n$ gives $\|\mathbf{v} + \mathbf{w}\|^2 = 4 - \|\mathbf{v} - \mathbf{w}\|^2$; divide through by 4 and take square roots.

(c) $\|\mathbf{e}_1\|_1 = \|\mathbf{e}_2\|_1 = 1$ and $\left\|\frac{\mathbf{e}_1 + \mathbf{e}_2}{2}\right\|_1 = 1$.

4.5 Isometries

4.5.1 (a) R_U is reflection through U.

(b) Take an orthonormal basis $\mathcal{B} = (e_1, \ldots, e_n)$ of V such that $(e_1, \ldots, e_m)$ is an orthonormal basis of U. Then R_U sends $\mathcal{B}$ to

$(e_1, \ldots, e_m, -e_{m+1}, \ldots, -e_n)$, which is also orthonormal. Apply Theorem 4.28.

4.5.5 Either argue geometrically by finding the reflections of the standard basis vectors across the line making angle $\frac{\theta}{2}$ with the x-axis, or first find the matrix of the projection onto that line and then use Exercise 4.5.1.

4.5.7 (a) $\mathbf{Q} = \frac{1}{\sqrt{10}}\begin{bmatrix} 1 & 3 \\ 3 & -1 \end{bmatrix}, \mathbf{R} = \sqrt{\frac{2}{5}}\begin{bmatrix} 5 & 7 \\ 0 & 1 \end{bmatrix}$

(b) $\mathbf{Q} = \begin{bmatrix} 0 & 1/\sqrt{2} & 1/\sqrt{2} \\ \sqrt{2/3} & -1/\sqrt{6} & 1/\sqrt{6} \\ 1\sqrt{3} & 1\sqrt{3} & -1\sqrt{3} \end{bmatrix}, \mathbf{R} = \begin{bmatrix} \sqrt{2} & 1/\sqrt{2} & 1/\sqrt{2} \\ 0 & \sqrt{3/2} & 1/\sqrt{6} \\ 0 & 0 & 2/\sqrt{3} \end{bmatrix}$

(c) $\mathbf{Q} = \begin{bmatrix} 1/\sqrt{2} & 0 & 1/\sqrt{2} \\ 0 & 1 & 0 \\ -1/\sqrt{2} & 0 & -1\sqrt{2} \end{bmatrix}, \mathbf{R} = \begin{bmatrix} \sqrt{2} & -\sqrt{2} & \sqrt{2} \\ 0 & 2 & -1 \\ 0 & 0 & \sqrt{2} \end{bmatrix}$

(d) $\mathbf{Q} = \frac{1}{2}\begin{bmatrix} 1 & 1 & 1 & 1 \\ 1 & 1 & -1 & -1 \\ 1 & -1 & 1 & -1 \\ 1 & -1 & -1 & 1 \end{bmatrix}, \mathbf{R} = \begin{bmatrix} 2 & 0 & 0 & 0 \\ 0 & 2 & 0 & 0 \\ 0 & 0 & 2 & 0 \\ 0 & 0 & 0 & 2 \end{bmatrix}$

4.5.9 (a) $\begin{bmatrix} 2 \\ 1 \end{bmatrix}$ (b) $\begin{bmatrix} 1 \\ -1 \\ 0 \end{bmatrix}$

4.5.11 The proof of Theorem 4.32 gives $\mathbf{A} = \begin{bmatrix} 1 & 0 \\ 0 & -1 \end{bmatrix}\begin{bmatrix} 2 & 3 \\ 0 & 4 \end{bmatrix}$. Since A is already upper triangular, $\mathbf{A} = \mathbf{I}\mathbf{A}$ is a QR-decomposition.

4.5.13 All that is needed is to show that T is surjective; since $\dim V = \dim W$, this is equivalent to showing T is injective. If $Tv = 0$, then $\|Tv\| = \|v\| = 0$, and so $v = 0$.

4.5.15 Let $\mathbf{u} \in \mathbb{C}^n$ with $\|\mathbf{u}\| = 1$. Then $\|\mathbf{UAVu}\| \le \|\mathbf{U}\|_{op} \|\mathbf{A}\|_{op} \|\mathbf{V}\|_{op} \|\mathbf{u}\| = \|\mathbf{A}\|_{op}$. Since $\mathbf{U}^{-1}$ and $\mathbf{V}^{-1}$ are unitary, the same argument shows $\|\mathbf{Au}\|_{op} = \left\|\mathbf{U}^{-1}\mathbf{UAVV}^{-1}\mathbf{u}\right\|_{op} \le \|\mathbf{UAV}\|_{op}$. For the Frobenius norm, $\|\mathbf{UAV}\|_F = \mathrm{tr}(\mathbf{UAVV}^*\mathbf{A}^*\mathbf{U}^*) = \mathrm{tr}(\mathbf{UAA}^*\mathbf{U}^*) = \mathrm{tr}(\mathbf{U}^*\mathbf{UAA}^*) = \mathrm{tr}(\mathbf{AA}^*) = \|\mathbf{A}\|_F$.

4.5.17 $\left\|\begin{bmatrix} 1 & 1 \\ 1 & -1 \end{bmatrix}\begin{bmatrix} x \\ y \end{bmatrix}\right\|_\infty = \max\{|x+y|, |x-y|\}$. Treat the cases of x and y having the same sign or different signs separately.

4.5.19 By Exercise 4.5.15, $\|\mathbf{QR}\|_{op} = \|\mathbf{R}\|_{op}$ and $\left\|(\mathbf{QR})^{-1}\right\|_{op} = \left\|\mathbf{R}^{-1}\mathbf{Q}^{-1}\right\|_{op} = \left\|\mathbf{R}^{-1}\right\|_{op}$.

4.5.21 $\left\|\mathbf{a}_j\right\| = \left\|\mathbf{A}\mathbf{e}_j\right\| = \left\|\mathbf{QR}\mathbf{e}_j\right\| = \left\|\mathbf{R}\mathbf{e}_j\right\| = \left\|\mathbf{r}_j\right\|$, where $\mathbf{r}_j$ is the jth column of R, and $\left\|\mathbf{r}_j\right\| = \sqrt{\sum_{i=1}^{j} |r_{ij}|^2} \ge |r_{jj}|$.

5 Singular Value Decomposition and the Spectral Theorem

5.1 Singular Value Decomposition of Linear Maps

5.1.1 T has rank 2. Take $\left(\frac{1}{2}\begin{bmatrix}1\\1\\1\\1\end{bmatrix}, \frac{1}{2}\begin{bmatrix}1\\1\\-1\\-1\end{bmatrix}, \frac{1}{2}\begin{bmatrix}1\\-1\\1\\-1\end{bmatrix}, \frac{1}{2}\begin{bmatrix}1\\-1\\-1\\1\end{bmatrix}\right)$ as right singular vectors (check that they are orthonormal) to see that $\sigma_1 = \sigma_2 = \sqrt{2}$ and the left singular vectors are $\left(\frac{1}{\sqrt{2}}\begin{bmatrix}1\\1\end{bmatrix}, \frac{1}{\sqrt{2}}\begin{bmatrix}1\\-1\end{bmatrix}\right)$ (again, orthonormal).

5.1.3 T has rank 1. The orthonormal basis $\left(\frac{1}{2\sqrt{2}}\begin{bmatrix}1+\sqrt{3}\\1-\sqrt{3}\end{bmatrix}, \frac{1}{2\sqrt{2}}\begin{bmatrix}1-\sqrt{3}\\-1-\sqrt{3}\end{bmatrix}\right)$ (the counterclockwise rotation of $\left(\frac{1}{\sqrt{2}}\begin{bmatrix}1\\1\end{bmatrix}, \frac{1}{\sqrt{2}}\begin{bmatrix}1\\-1\end{bmatrix}\right)$ by $\frac{\pi}{3}$) is mapped to $\left(\frac{1}{\sqrt{2}}\begin{bmatrix}1\\1\end{bmatrix}, 0\right)$.

5.1.5 The basis $\left(\frac{1}{\sqrt{2\pi}}, \frac{1}{\sqrt{\pi}}\sin(x), \frac{1}{\sqrt{\pi}}\sin(2x), \ldots, \frac{1}{\sqrt{\pi}}\sin(nx), \frac{1}{\sqrt{\pi}}\cos(x), \frac{1}{\sqrt{\pi}}\cos(2x), \ldots, \frac{1}{\sqrt{\pi}}\cos(nx)\right)$ is orthonormal and D maps it to $\left(0, \frac{1}{\sqrt{\pi}}\cos(x), \frac{2}{\sqrt{\pi}}\cos(2x), \ldots, \frac{n}{\sqrt{\pi}}\cos(nx), -\frac{1}{\sqrt{\pi}}\sin(x), -\frac{2}{\sqrt{\pi}}\sin(2x), \ldots, -\frac{n}{\sqrt{\pi}}\sin(nx)\right)$. The singular values are thus $\sigma_1 = \sigma_2 = n$, $\sigma_3 = \sigma_4 = n-1, \ldots, \sigma_{2n-1} = \sigma_{2n} = 1, \sigma_{2n+1} = 0$. The original basis gives the right singular vectors (rearranged to get the singular values in decreasing order) and the orthonormal basis with negative cosines gives the left singular vectors.

5.1.7 If v is a unit eigenvector for T with eigenvalue λ and $(e_1, \ldots, e_n)$ are the right singular vectors of T, then $|\lambda| = \|Tv\| = \sqrt{\sum_{j=1}^n \langle v, e_j\rangle^2 \sigma_j^2} \le \sqrt{\sigma_1^2 \sum_{j=1}^n \langle v, e_j\rangle^2} = \sigma_1$. The lower bound is similar.

5.1.9 Let $(e_1, \ldots, e_n)$ and $(f_1, \ldots, f_n)$ be orthonormal bases and $\sigma_1 \ge \cdots \ge \sigma_n > 0$ such that $Te_j = \sigma_j f_j$ for each j. Then $T^{-1}f_j = \sigma_j^{-1}e_j$ for each j, so $(f_n, \ldots, f_1)$ are the right singular vectors of T^{-1}, $(e_n, \ldots, e_1)$ are the left singular vectors, and the singular values are $\sigma_n^{-1} \ge \cdots \ge \sigma_1^{-1} > 0$.

5.1.11 Use Theorem 4.28.

5.1.13 k is the number of distinct nonzero singular values; V_k is the span of the right singular values that get scaled by σ_1, V_{k-1} is the span of those right singular values scaled by the second-largest singular value, and so on. $V_0 = \ker T$.

5.1.15 Show that $(f_1, \ldots, f_k)$ is an orthonormal basis of range T.

5.2 Singular Value Decomposition of Matrices

5.2.1 (a) 3, 1 (b) 6, 2 (c) $2\sqrt{2}, 2$ (d) $2\sqrt{2}, 2, 0$ (e) 2, 1, 1

5.2.3 The eigenvalues of A_z are just 1 and 2; $\|A_z\|_F^2 = \sigma_1^2 + \sigma_2^2 = 5 + z^2$, so the singular values must depend on z.

5.2.5 If $A = U\Sigma V^*$, then $A^* = V\Sigma^T U^*$, and for $1 \le j \le \min\{m, n\}$, $[\Sigma^T]_{jj} = \sigma_{jj}$.

5.2.7 If $A = U\Sigma V^*$, then $AA^* = U\Sigma\Sigma^* U^*$ is similar to $\Sigma\Sigma^*$.

5.2.9

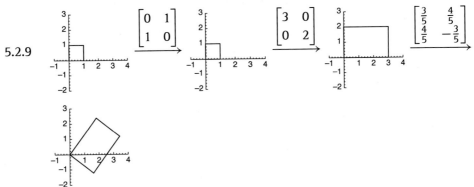

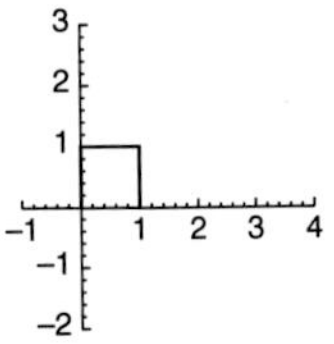

$\xrightarrow{\begin{bmatrix} 0 & 1 \\ 1 & 0 \end{bmatrix}}$

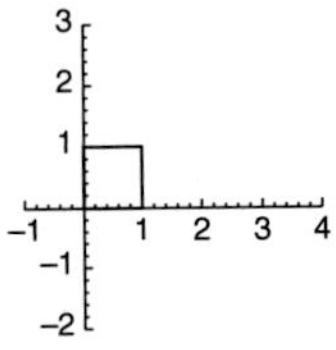

$\xrightarrow{\begin{bmatrix} 3 & 0 \\ 0 & 2 \end{bmatrix}}$

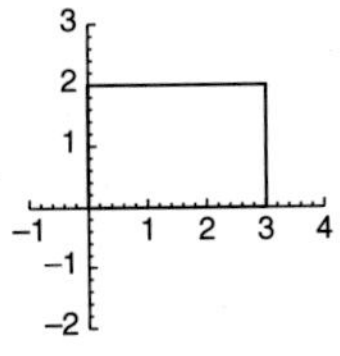

$\xrightarrow{\begin{bmatrix} \frac{3}{5} & \frac{4}{5} \\ \frac{4}{5} & -\frac{3}{5} \end{bmatrix}}$

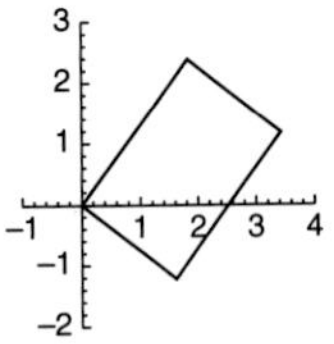

5.2.11 $\kappa(A) = \|A\|_{op}\left\|A^{-1}\right\|_{op}$; $\|A\|_{op} = \sigma_1$ and you should have found in Exercise 5.1.9 that $\left\|A^{-1}\right\|_{op} = \sigma_n^{-1}$.

5.2.13 Since A is invertible, it has singular values $\sigma_1 \ge \cdots \ge \sigma_n > 0$. A rank k matrix B is singular if and only if $k < n$; by Theorem 5.8, the smallest $\|A - B\|_{op}$ is when $k = n - 1$ and then $\|A - B\|_{op} = \sigma_n$ is achieved by $B = \sum_{j=1}^{n-1} \sigma_j \mathbf{u}_j \mathbf{v}_j^*$. Using Theorem 5.9 instead gives the Frobenius norm version.

5.2.15 (a) If A is invertible, then $\Sigma^\dagger = \mathbf{diag}(\sigma_1^{-1}, \dots, \sigma_n^{-1})$ and $V\Sigma^\dagger U^* A = V\Sigma^\dagger U^* U\Sigma V^* = I_n$. (b) $\Sigma\Sigma^\dagger$ is the $m \times m$ matrix with 1 in the (j, j) entry for $1 \le j \le r$ and zeroes otherwise; $\Sigma^\dagger\Sigma$ is the $n \times n$ matrix with 1 in the (j, j) entry for $1 \le j \le r$ and zeroes otherwise.
(c) $AA^\dagger = U\Sigma\Sigma^\dagger U^*$ and it follows from the previous part that $\left(\Sigma\Sigma^\dagger\right)^* = \Sigma\Sigma^\dagger$.
(d) Observe that $\Sigma\Sigma^\dagger\Sigma = \Sigma$.

5.2.17 (a) Find the orthogonal projection of $\begin{bmatrix} 1 \\ 1 \end{bmatrix}$ onto $\left\langle \begin{bmatrix} 2 \\ 1 \end{bmatrix} \right\rangle$.
(b) Write a general point on the line as $\left(t, \frac{t}{2}\right)$, break into cases $t < 1$, $1 \le t < 2$ and $2 \le t$ and compute $\left\|\left(t - 1, \frac{t}{2} - 1\right)\right\|_1$ and minimize over t.
(c) Use the same strategy as the previous part.

5.2.19 $\|A\|_F^2 = \left\|\sum_{j=1}^r \sigma_j \mathbf{u}_j \mathbf{v}_j^*\right\|_F^2 = \operatorname{tr}\left(\sum_{j,\ell=1}^r \sigma_j \sigma_\ell \mathbf{u}_j \mathbf{v}_j^* \mathbf{v}_\ell \mathbf{u}_\ell^*\right)$. Since the columns of V are orthonormal, $\mathbf{v}_j^* \mathbf{v}_\ell = \delta_{j\ell}$. Now use that the columns of U are unit vectors.

5.2.21 $A - B = \sum_{j=k+1}^r \sigma_j \mathbf{u}_j \mathbf{v}_j^*$. If $\mathbf{w} = \sum_{j=1}^n \langle \mathbf{w}, \mathbf{v}_j \rangle \mathbf{v}_j$, then $\|(A - B)\mathbf{w}\| = \left\|\sum_{j=k+1}^r \sigma_j \langle \mathbf{w}, \mathbf{v}_j \rangle \mathbf{u}_j\right\| = \sqrt{\sum_{j=k+1}^r \sigma_j^2 \langle \mathbf{w}, \mathbf{v}_j \rangle^2} \le \sigma_{k+1} \sqrt{\sum_{j=k+1}^r \langle \mathbf{w}, \mathbf{v}_j \rangle^2} \le \sigma_{k+1} \|\mathbf{w}\|$.

5.3 Adjoint Maps

5.3.1 (a) $U = \frac{1}{\sqrt{2}}\begin{bmatrix} 1 & -1 \\ 1 & 1 \end{bmatrix}$, $\Sigma = \text{diag}(3,1)$, $V = \frac{1}{\sqrt{2}}\begin{bmatrix} 1 & -1 \\ 1 & 1 \end{bmatrix}$

(b) $U = \frac{1}{\sqrt{2}}\begin{bmatrix} i & 1 \\ i & -1 \end{bmatrix}$, $\Sigma = \text{diag}(6,2)$, $V = \frac{1}{\sqrt{2}}\begin{bmatrix} 1 & -1 \\ 1 & 1 \end{bmatrix}$

(c) $U = \frac{1}{\sqrt{2}}\begin{bmatrix} -1 & 1 \\ 1 & 1 \end{bmatrix}$, $\Sigma = \begin{bmatrix} 2\sqrt{2} & 0 & 0 \\ 0 & 2 & 0 \end{bmatrix}$, $V = \frac{1}{\sqrt{2}}\begin{bmatrix} 0 & 1 & 1 \\ 0 & -1 & 1 \\ -\sqrt{2} & 0 & 0 \end{bmatrix}$

(d) $U = \frac{1}{\sqrt{2}}\begin{bmatrix} 1 & -1 & 0 \\ -1 & -1 & 0 \\ 0 & 0 & \sqrt{2} \end{bmatrix}$, $\Sigma = \text{diag}(2\sqrt{2}, 2, 0)$,

$V = \frac{1}{\sqrt{2}}\begin{bmatrix} 0 & -1 & 1 \\ 0 & 1 & 1 \\ \sqrt{2} & 0 & 0 \end{bmatrix}$

(e) $U = \begin{bmatrix} 0 & -1 & 0 \\ 1 & 0 & 0 \\ 0 & 0 & 1 \end{bmatrix}$, $\Sigma = \text{diag}(2,1,1)$, $V = \begin{bmatrix} 0 & 0 & 1 \\ 1 & 0 & 0 \\ 0 & 1 & 0 \end{bmatrix}$

5.3.3 $\frac{1}{25}\begin{bmatrix} 1 & 2 \\ -2 & -4 \end{bmatrix}$

5.3.5 Since $R^2 = I$ and R is an isometry, $\langle v, Rw \rangle = \langle R^2 v, Rw \rangle = \langle Rv, w \rangle$. Alternatively, write $R = 2P_L - I$ and use that P_L is self-adjoint.

5.3.7 (a) Integration by parts.
(b) Imitate the proof of Quick Exercise 10 to show that part (a) implies that all eigenvalues of D are purely imaginary.
(c) Integration by parts (twice).

5.3.9 If $u \in U$ and $v \in V$, $\langle Ju, v \rangle = \langle u, v \rangle = \langle u, P_U v + P_{U^\perp} v \rangle = \langle u, P_U v \rangle$.

5.3.13 If $T^*T = 0$, then for all v, $\langle T^*Tv, v \rangle = \|Tv\|^2 = 0$ so $T = 0$. The other direction is trivial.

5.3.15 $\|T^*T\|_{op} \leq \|T^*\|_{op} \|T\|_{op} = \|T\|_{op}^2$ by Exercise 5.3.14. If e_1, f_1 are the first right and left singular vectors, respectively, show that $T^* f_1 = \sigma_1 e_1$ so that $T^*Te_1 = \sigma_1^2 e_1$.

5.3.17 Pick orthonormal bases and relate this to the Frobenius inner product on $M_{m,n}(\mathbb{F})$.

5.3.19 If v_1, v_2 are eigenvectors with respective eigenvalues λ_1, λ_2, then $\lambda_1 \langle v_1, v_2 \rangle = \langle \lambda_1 v_1, v_2 \rangle = \langle Tv_1, v_2 \rangle = -\langle v_1, Tv_2 \rangle = -\langle v_1, \lambda_2 v_2 \rangle = -\overline{\lambda_2} \langle v_1, v_2 \rangle$. Either $\langle v_1, v_2 \rangle = 0$ or $\lambda_1 = -\overline{\lambda_2}$. But by Exercise 5.3.18, λ_1 and λ_2 are purely imaginary, so $-\overline{\lambda_2} = \lambda_2 \neq \lambda_1$.

5.3.21 For part 4, $\langle u, T^*R^*w \rangle = \langle Tu, R^*w \rangle = \langle RTu, w \rangle$. For part 5, first observe that if T is invertible (so that range T is all of V), then $\ker T^* =$

(range $T)^{\perp} = \{0\}$, and so T^* is invertible as well. Let $u_1 \in U$ and $Tu_1 = v_1$, and let $v_2 \in V$ and $T^*v_2 = u_2$. Rewriting $\langle Tu_1, v_2\rangle = \langle u_1, T^*v_2\rangle$ in terms of v_1 and u_2 gives $\langle v_1, (T^*)^{-1}u_2\rangle = \langle T^{-1}v_1, u_2\rangle$.

5.4 The Spectral Theorems

5.4.1 (a) $\lambda_1 = 5$, $\lambda_2 = 0$, $\mathbf{U} = \frac{1}{\sqrt{5}}\begin{bmatrix} 1 & 2 \\ 2 & -1 \end{bmatrix}$

(b) $\lambda_1 = 1+i$, $\lambda_2 = 1-i$, $\mathbf{U} = \frac{1}{\sqrt{2}}\begin{bmatrix} 1 & 1 \\ -i & i \end{bmatrix}$

(c) $\lambda_1 = -1$, $\lambda_2 = 1$, $\lambda_3 = 3$, $\mathbf{U} = \begin{bmatrix} 0 & \frac{1}{\sqrt{2}} & \frac{1}{\sqrt{2}} \\ 1 & 0 & 0 \\ 0 & -\frac{1}{\sqrt{2}} & \frac{1}{\sqrt{2}} \end{bmatrix}$

(d) $\lambda_1 = 2$, $\lambda_2 = 1$, $\lambda_3 = -1$, $\mathbf{U} = \begin{bmatrix} \frac{1}{\sqrt{3}} & \frac{1}{\sqrt{2}} & \frac{1}{\sqrt{6}} \\ \frac{1}{\sqrt{3}} & -\frac{1}{\sqrt{2}} & \frac{1}{\sqrt{6}} \\ \frac{1}{\sqrt{3}} & 0 & -\frac{2}{\sqrt{6}} \end{bmatrix}$

5.4.3 (a) $\mathbf{U} = \begin{bmatrix} 0 & 1 \\ 1 & 0 \end{bmatrix}$, $\mathbf{T} = \begin{bmatrix} 3 & 2 \\ 0 & 1 \end{bmatrix}$ (b) $\mathbf{U} = \frac{1}{\sqrt{2}}\begin{bmatrix} 1 & 1 \\ -1 & 1 \end{bmatrix}$, $\mathbf{T} = \begin{bmatrix} 2 & 1 \\ 0 & -1 \end{bmatrix}$

5.4.5 Use the Spectral Theorem.

5.4.7 If A and B commute, then $e^{\mathbf{A}+\mathbf{B}} = \sum_{k=0}^{\infty} \frac{1}{k!} \sum_{j=0}^{k} \binom{k}{j} \mathbf{A}^k \mathbf{B}^{j-k} = \sum_{j=0}^{\infty} \sum_{k=j}^{\infty} \frac{1}{j!(k-j)!} \mathbf{A}^j \mathbf{B}^{k-j} = \sum_{j=0}^{\infty} \sum_{k=j}^{\infty} \frac{1}{j!(k-j)!} \mathbf{A}^j \mathbf{B}^{k-j} = \sum_{j=0}^{\infty} \sum_{\ell=0}^{\infty} \frac{1}{j!\ell!} \mathbf{A}^j \mathbf{B}^{\ell}$.

5.4.9 Let $\mathbf{u}_1, \ldots, \mathbf{u}_n$ be an orthonormal basis of $\mathbb{C}^n$ with $\mathbf{A}\mathbf{u}_j = \lambda_j \mathbf{u}_j$. Then $\langle \mathbf{A}\mathbf{x}, \mathbf{x}\rangle = \sum_{j=1}^{n} \lambda_j \left|\langle \mathbf{x}, \mathbf{u}_j\rangle\right|^2$.

5.4.11 Write $\mathbf{A} = \mathbf{B}^2$, with B positive definite; note that B is Hermitian, so that you can actually write $\mathbf{A} = \mathbf{B}^*\mathbf{B}$. Find the QR decomposition of B.

5.4.13 Use part 2 of Theorem 5.20.

5.4.15 (a) If $i \leq j$, $[\mathbf{AB}]_{ij} = \sum_{k=1}^{i-1} a_{n-(i-k-1)} b_{j-k+1} + \sum_{k=i}^{j} a_{k-i+1} b_{j-k+1} + \sum_{k=j+1}^{n} a_{k-i+1} b_{n-(k-j-1)}$. Whenever a pair $a_r b_s$ appears, then either $r+s = n+j-i+2$ (the terms in the first and third sums) or $r+s = j-i+2$ (the terms in the middle sum). Moreover, every a_r and every b_s does appear. So $a_r b_s$ appears in the sum if and only if $a_s b_r$ does.

(b) If A is a circulant matrix, so is $\mathbf{A}^*$.

5.4.17 (a) If there is an orthonormal basis of eigenvectors of A, then $\mathbf{A} = \mathbf{UDU}^*$ for a unitary U and diagonal D. Then $\mathbf{A}^* = \mathbf{UD}^*\mathbf{U}^* = \mathbf{UDU}^*$ since the entries of D are assumed to be real.

5.4.19 (a) If T is self-adjoint, then by the Spectral Theorem, there is an orthonormal basis of V of eigenvectors of T. Conversely, if V is the orthogonal direct sum of the eigenspaces of T, then one can choose an orthonormal basis of eigenvectors; the matrix of T in this basis is diagonal, hence symmetric. Apply Theorem 5.13. (b) and (c) are similar.

5.4.21 Since T is Hermitian, there is an orthonormal basis of eigenvectors of T. Let $\lambda_1, \ldots, \lambda_m$ denote the distinct eigenvalues of T and let U_j be the span of those eigenvectors in the orthonormal basis with eigenvalue λ_j. Then V is the orthogonal direct sum of the U_j, and so for $v \in V$, $v = \sum_{j=1}^{m} P_{U_j} v$. Apply T and use the fact that $P_{U_j} v$ is an eigenvector.

5.4.23 (a) Write $v = c_j e_j + \cdots + c_n v_n$, evaluate $\langle Tv, v \rangle$ and use the fact that $\lambda_j \geq \lambda_k$ for all $k \geq j$.
(b) Same idea as (a).
(c) Use Lemma 3.22.

5.4.25 If A is upper triangular, the diagonal entries are the eigenvalues: $\sqrt{\sum_{j=1}^{n} |\lambda_j|^2} = \sqrt{\sum_{j=1}^{n} |a_{jj}|^2} \leq \sqrt{\sum_{i,j=1}^{n} |a_{ij}|^2} = \|\mathrm{A}\|_F$. In general, $\mathrm{A} = \mathrm{UTU}^*$ has the same eigenvalues and same Frobenius norm (see Exercise 4.5.15) as T.

5.4.27 (a) First observe that $[\mathrm{A}^*\mathrm{A}]_{11} = |a_{11}|^2$ and $[\mathrm{AA}^*]_{11} = \sum_{k=1}^{n} |a_{1k}|^2$, so if A is normal, then $a_{1k} = 0$ for $k \geq 2$. Continue in this way.
(b) Write $\mathrm{A} \in \mathrm{M}_n(\mathbb{C})$ as UTU^* for T upper triangular and U unitary. Show that if A is normal, then so is T; part (a) then implies that T is actually diagonal.

6 Determinants

6.1 Determinants

6.1.1 (a) -2 (b) $6 + 6i$ (c) -31 (d) 1

6.1.3 (a) -14 (b) -49

6.1.5 (a) Let $\mathbf{C}$ have first column $\mathbf{a}_1$ and second column $\mathbf{b}_2$ and let $\mathbf{D}$ have first column $\mathbf{b}_1$ and second column $\mathbf{a}_2$. Then $D(\mathrm{A} + \mathrm{B}) = D(\mathrm{A}) + D(\mathrm{B}) + D(\mathrm{C}) + D(\mathrm{D})$.

6.1.7 0

6.1.9 Because T has n distinct eigenvalues, there is a basis of V of eigenvectors, so that the matrix of T is $\mathbf{diag}(\lambda_1, \ldots, \lambda_n)$. By multilinearity, $\det \mathbf{diag}(\lambda_1, \ldots, \lambda_n) = \lambda_1 \cdots \lambda_n \det(\mathrm{I}_n) = \lambda_1 \cdots \lambda_n$.

6.1.11 Since A is Hermitian, $\mathrm{A} = \mathrm{UDU}^*$ for some U unitary and D diagonal with real entries. By Corollary 6.7, $\det(\mathrm{A}) = \det(\mathrm{D}) \in \mathbb{R}$.

6.1.13 If there is some pair $i \neq j$ with $a_{ki} = a_{kj}$ for each k, then the inner sum in $f(\mathrm{A})$ is zero. $f(\mathrm{I}_2) = 2$, and $f\left(\begin{bmatrix} 2 & 0 \\ 0 & 1 \end{bmatrix}\right) = 3$, so f is not multilinear.

6.1.15 Add a linear combination of the remaining columns to column i, expand by multilinearity, and observe that all the new terms have a repeated column in the argument and so give no contribution.

6.1.17 (a) The set of functions on $\mathrm{M}_n(\mathbb{F})$ is a vector space, so it's enough to check that the sum of alternating multilinear functions is alternating and multilinear, and that a scalar multiple of an alternating multilinear function is alternating and multilinear. (b) Corollary 6.4 shows that W is spanned by the determinant.

6.2 Computing Determinants

6.2.1 (a) 36 (b) 0 (c) -6 (d) -6 (e) -2 (f) 0

6.2.3 (a) $1 = \det(I_n) = \det(UU^*) = \det(U)\overline{\det(U)} = |\det(U)|^2$
(b) $|\det(A)| = |\det(U\Sigma V^*)| = \det(\Sigma) = \sigma_1 \cdots \sigma_n$

6.2.5 $\det(LDU) = \det(L)\det(D)\det(U)$; since D is diagonal, $\det(D) = d_{11} \cdots d_{nn}$, and L and U are triangular with ones on the diagonals, so $\det(L) = \det(U) = 1$.

6.2.9 Use the Schur decomposition, and recall that if $[T]_{\mathcal{B}}$ is upper triangular, the eigenvalues of T are exactly the diagonal entries of $[T]_{\mathcal{B}}$.

6.2.11 If $A = UDU^*$ has distinct eigenvalues $\lambda_1, \ldots, \lambda_r$, then $\det A = \lambda_1^{k_1} \cdots \lambda_r^{k_r}$, where k_j is the number of times λ_j appears on the diagonal of D.

6.2.13 In the sum over permutations expansion, the only permutation that gives a contribution which is linear in t is the identity, and the coefficient of t in $\prod_{j=1}^{n}(1 + ta_{jj})$ is $\operatorname{tr}(A)$.

6.2.15 The fact that any permutation matrix satisfies these criteria follows from the definition and the fact that a permutation is a bijection. For the other direction, argue that given a matrix with these properties, you can uniquely reconstruct a permutation from it.

6.2.17 Observe that if τ is a transposition and A_τ is the corresponding permutation matrix, then $\det A_\tau = -1$. Rewrite the equality in terms of permutation matrices and take the determinant of both sides.

6.3 Characteristic Polynomials

6.3.1 (a) $p(\lambda) = \lambda^2 - \lambda - 2$, $\left(2, \begin{bmatrix} 1 \\ 1 \end{bmatrix}\right)$, $\left(-1, \begin{bmatrix} 1 \\ -2 \end{bmatrix}\right)$

(b) $p(\lambda) = \lambda^2 - 2\cos(\theta)\lambda + 1$, $\left(\cos(\theta) - i\sin(\theta), \begin{bmatrix} 1 \\ i \end{bmatrix}\right)$,
$\left(\cos(\theta) + i\sin(\theta), \begin{bmatrix} 1 \\ -i \end{bmatrix}\right)$

(c) $p(\lambda) = -\lambda^3 + 6\lambda^2 - 11\lambda + 6$, $\left(1, \begin{bmatrix} -1 \\ 1 \\ 1 \end{bmatrix}\right)$, $\left(2, \begin{bmatrix} -2 \\ 3 \\ 3 \end{bmatrix}\right)$, $\left(3, \begin{bmatrix} -3 \\ 7 \\ 9 \end{bmatrix}\right)$

(d) $p(\lambda) = -\lambda^3 + 12\lambda + 16$, $\left(-2, \begin{bmatrix} 0 \\ 1 \\ 0 \end{bmatrix}, \begin{bmatrix} 1 \\ 0 \\ -1 \end{bmatrix}\right)$, $\left(4, \begin{bmatrix} 1 \\ 0 \\ 1 \end{bmatrix}\right)$

6.3.3 (b) The 1-eigenspace is one-dimensional for A and two-dimensional for I_2.

6.3.5 $-\frac{3}{2} \pm \frac{\sqrt{3}}{2}i$

6.3.7 (a) $p(\lambda) = \lambda^2 - 5\lambda - 2$

(b) By the Cayley–Hamilton Theorem, $A^2 = 5A + 2I_2$, so $A^3 = 5A^2 + 2A = 5(5A + 2I_2) + 2A = 27A + 10I_2 = \begin{bmatrix} 37 & 54 \\ 81 & 118 \end{bmatrix}$.

6.3.9 $p_A(0) = 0$ if and only if 0 is a root of p_A, i.e., an eigenvalue, which is true if and only if A has a non-trivial null space.

6.3.11 First prove the statement when A is diagonal, then use the Spectral Theorem.

6.3.13 Write $A = U\Sigma V^*$ and express the characteristic polynomial of AA^* in terms of Σ.

6.3.15 Use the Cayley–Hamilton Theorem and induction to show that for any $k \in \mathbb{N}$, $A^k \in \langle I_n, A, A^2, \ldots, A^{n-1} \rangle$.

6.3.19 Show that $t_A(x) = \operatorname{tr}(A) - nx$ so that $t_A(A) = \operatorname{tr}(A)I_n - nA$. So $t_A(A) = 0$ if and only if $A = \frac{\operatorname{tr}(A)}{n} I_n$.

6.4 Applications of Determinants

6.4.1 $ab\pi$

6.4.3 Show that if $\mathcal{C}$ is the unit circle centered at the origin, then $\mathcal{E} = T(\mathcal{C})$, where $T \in \mathcal{L}(\mathbb{R}^2)$ has matrix $\begin{bmatrix} 1 & 1/2 \\ 1 & -1/2 \end{bmatrix}$.

6.4.5 (a) $\left(\frac{11}{3}, -\frac{1}{3}\right)$ (b) $\left(-\frac{1}{4} - \frac{1}{4}i, \frac{5}{4} + \frac{3}{4}i\right)$ (c) $(-3, 5, 10)$ (d) $(2, 3, 2, 2)$

6.4.7 (a) $\begin{bmatrix} \frac{2}{3} & 0 & -\frac{1}{3} \\ -\frac{11}{3} & 2 & \frac{4}{3} \\ -2 & 1 & 1 \end{bmatrix}$ (b) $\begin{bmatrix} 1 & 0 & -1 & 1 \\ 0 & 0 & 1 & -1 \\ -1 & 1 & 0 & 0 \\ 1 & -1 & 0 & 1 \end{bmatrix}$ (c) $\begin{bmatrix} -\frac{3}{2} & \frac{5}{2} & 1 & -\frac{1}{2} \\ \frac{9}{2} & -\frac{11}{2} & -2 & \frac{1}{2} \\ \frac{5}{2} & -\frac{7}{2} & -1 & \frac{1}{2} \\ -\frac{1}{2} & \frac{1}{2} & 0 & \frac{1}{2} \end{bmatrix}$

(d) $\begin{bmatrix} -\frac{1}{2} & -\frac{1}{2} & 1 & 1 & 0 \\ 0 & 0 & 0 & 1 & 1 \\ -\frac{1}{4} & -\frac{1}{4} & \frac{1}{2} & \frac{1}{2} & -\frac{1}{2} \\ \frac{1}{2} & \frac{1}{2} & 0 & -1 & -1 \\ 0 & -1 & 0 & 3 & 3 \end{bmatrix}$

6.4.9 If the matrix of T were just $\Sigma = \operatorname{\mathbf{diag}}(\sigma_1, \ldots, \sigma_n)$, then for a cube C with each face in some $\langle e_j \rangle^\perp$, it is clear that $\operatorname{vol}(T(C)) = \sigma_1 \cdots \sigma_n \operatorname{vol}(C)$; this is then true for Jordan measurable sets by approximation. By SVD, T is related to Σ by isometries, which don't change volume, so we should have $\operatorname{vol}(T(\Omega)) = \sigma_1 \cdots \sigma_n \operatorname{vol}(\Omega)$ in general. Recall that $|\det T| = \sigma_1 \cdots \sigma_n$.

6.4.11 The n-dimensional parallelepiped spanned by $a_1, \ldots, a_n$ is exactly $\begin{bmatrix} | & & | \\ a_1 & \cdots & a_n \\ | & & | \end{bmatrix} C$, where C is the parallelepiped spanned by $e_1, \ldots, e_n$.

6.4.13 Use Cramer's rule.

6.4.15 Let ℓ denote the length of the line segment L. Subdivide L into n pieces of length $\frac{\ell}{n}$; cover each piece with a rectangle whose opposite corners lie on the line. The area of any such rectangle is bounded by $\left(\frac{\ell}{n}\right)^2$ and there are n rectangles, so the total area is bounded by $\frac{\ell^2}{n}$.

Index

$\Leftarrow$, 385
$\Leftrightarrow$, 385
$\Rightarrow$, 385
$\cap$, 378
$\cup$, 378
$\exists$, 386
$\forall$, 386
$\in$, 5, 378
$\subseteq$, 378
$\subsetneq$, 378
V, 51
$\mathbb{C}$, 40, 46
$\mathbb{F}$, 39
$\mathbb{F}^\infty$, 56
$\mathbb{F}^n$, 52, 59
$\mathbb{F}_2$, 41, 46
$\mathbb{Q}$, 41, 46
$\mathbb{R}$, 5, 46
$\mathbb{R}^n$, 25, 35
$\mathbb{Z}$, 41
$C[a, b]$, 50
$D[a, b]$, 57
$\mathcal{L}(V, W)$, 64
$\mathcal{L}(V)$, 64
$M_{m,n}(\mathbb{F})$, 44
$M_{m,n}(\mathbb{R})$, 10
$\mathcal{P}_n(\mathbb{F})$, 57
c_0, 56
$\operatorname{Im} z$, 382
$\operatorname{Re} z$, 382
$|z|$, 226, 382
$\bar{z}$, 226, 382
$f(A)$, 380
$f : X \to Y$, 379
$f \circ g$, 381
f^{-1}, 381
$x \mapsto f(x)$, 380
$\langle v, w \rangle$, 226
$\|v\|$, 267
  in an inner product space, 229
$\langle v_1, v_2, \ldots, v_k \rangle$, 53
$\operatorname{diag}(d_1, \ldots, d_n)$, 68
A^*, 97
A^T, 96
A^{-1}, 97
$\|A\|_F$, 234
$\|A\|_{op}$, 272
T^*, 311
$\operatorname{tr} A$, 208
$\operatorname{tr} T$, 209
a_{ij}, 10
$p(T)$, $p(A)$, 217
$C(A)$, 116
$\dim V$, 164
$\operatorname{Eig}_\lambda(T)$, $\operatorname{Eig}_\lambda(A)$, 120
$\ker T$, $\ker A$, 118
$\operatorname{null} T$, $\operatorname{null} A$, 175
$\operatorname{rank} T$, $\operatorname{rank} A$, 173
n-dimensional, 164
$P_{c,i,j}$, 102
$Q_{c,i}$, 102
$R_{i,j}$, 102
$[T]_{\mathcal{B}_V,\mathcal{B}_W}$, 187
$[T]_{\mathcal{B}_V,\mathcal{B}_W}$, 195
$[v]_{\mathcal{B}}$, 185, 195
L^1 norm, 268
ℓ^1 norm, 267
ℓ^∞ norm, 267
S_n, 351
ι, 351
$\operatorname{sgn}(\sigma)$, 352
$\det(A)$, 336
A_{ij}, 339
$p_A(x)$, 358

Printed in the United States
by Baker & Taylor Publisher Services